Bridge Management

Bridge Management

Second Edition

M. J. Ryall

ELSEVIER

AMSTERDAM • BOSTON • HEIDELBERG • LONDON • NEW YORK • OXFORD
PARIS • SAN DIEGO • SAN FRANCISCO • SINGAPORE • SYDNEY • TOKYO
Butterworth-Heinemann is an imprint of Elsevier

Butterworth-Heinemann is an imprint of Elsevier
The Boulevard, Langford Lane, Kidlington, Oxford OX5 1GB, UK
Radarweg 29, PO Box 211, 1000 AE Amsterdam, The Netherlands

First edition 2001
Second edition 2010

British Library Cataloguing in Publication Data
A catalogue record for this book is available from the British Library

Library of Congress Cataloging-in-Publication Data
A catalog record for this book is availabe from the Library of Congress

ISBN-13: 978-0-7506-8511-5

For information on all Butterworth-Heinemann publications visit our website at books.elsevier.com

Printed and bound in Italy

10 11 12 13 14 10 9 8 7 6 5 4 3 2 1

Contents

Preface to the Second Edition

Since the first edition of this book, the scrutiny of our bridges has become a race against time. Why? because we have realized that bridges really 'do not last forever'. Many bridge owners all over the world have realized that their stock of bridges was in a pretty bad state and needed attention. Not only that, but they did not have the wherewithal in both expertise and funds to deal with the situation. Most (if not all) Governments and Local authorities were running on very tight budgets, and they did not take kindly to the request for *extra* money. This hiatus resulted in the spawning of Bridge Management Conferences and Courses throughout the world in order to focus on *all* of the perceived problems.

The knowledge gained has not only brought about a greater awareness of potential problems that can arise, but – which is greater by far – has instilled in the bridge fraternity the **sine qua non** that inspections must be carried out on a regular and thorough basis before any works commence. As a newly elected British Prime Minister once said on tour, his number one priority is *'education, education and education'*. Ours must be *'inspection, inspection and inspection.'* So that we can be sure that we do not miss any latent or incipient failure mechanism.

So, why a second edition? Quite simply because things have changed drastically in the last 10 years, resulting in a plethora of articles found in locations as listed in the Preface to the First Edition. Instead of having to trawl through masses of information on numerous web sites, my main aims have been to amend, add and revise as concisely but as comprehensively as I could to save both time and money. I hope and trust these aims have been met.

M. J. Ryall
2009

Preface to the First Edition

The management of bridges is a fascinating and challenging subject. It may well be that in the early days of bridge construction when masonry was the prime construction material and the arch was the predominant structural form, very little thought, if any, was given for ensuring the longevity of a bridge. Certainly with masonry the general view was that it was virtually imperishable and – barring an act of God – would last forever. Sadly, we have learnt by bitter experience that no bridge will last forever and that proper management contains elements of both a preventive and curative nature to ensure that a bridge will perform properly over its 120-year nominal life and beyond.

But what are the management tasks and where does one find information on them? When I first came to write this book the tasks were broadly categorised as inspection, assessment and repair, and the information was to be found in conference proceedings; seminar papers; symposia; Highways Agency advice and design notes; British Standards; articles in professional magazines; text books; research papers; newspapers and in the minds of individuals working within the bridge fraternity worldwide. There was clearly the need to provide some sort of comprehensive reference work that would in essence bring together all of the tasks encountered in the management process. A definitive work was out of the question – that would require many tomes and take a decade to write – but I thought it was possible to produce, if you like, a *guide* book which could be of use both to beginners and seasoned professionals alike.

Bridge Management is the result. It has proven to be a challenging yet rewarding task and resulted in many a sleepless night and the looking up of hundreds of references. There is no doubt that many readers will have wished for either more information on a particular topic or another topic itself, but I have had to draw the line somewhere, and, as it is, the publishers have allowed me to extend the number of words and chapters to what I now think is a useful and manageable volume.

Most of what is contained in the book is due to the direct endeavours of leading academics and practitioners in the field (both in government and industry). I have drawn extensively upon published material relating to the more mundane tasks of inspection and assessment to the more sexy tasks of repair using Fibre Reinforced Plastics (FRPs) and reliability analysis. Where the reader wishes for more information not included in the text, I have provided a list of references at the end of every chapter, which, I hope, will provide a profitable trail along which to travel to find what is being sought.

I have received extensive help along the way from idea to manuscript from individuals, companies, libraries and bridge authorities who have provided information and photographs, and I am deeply grateful for all of them. Finally, in the tradition of all authors, I am indebted to my wife Beverley and my children for their patience and long-suffering whilst I was working on the book.

M. J. Ryall
2001

Acknowledgements

My deepest and most heartfelt thanks for my wife Beverley, whose love I did not deserve, and who showed, uncommon kindness, patience and forebearance whilst I was writing this book.

A big THANK YOU must go to my three lovely daughters Marianne, Elizabeth and Caroline who continually praised me up when I was feeling down.

Author Biographies

- Comprehensive specialist information for the bridge engineer
- Covers not only testing and assessment, but also financial and management issues
- Highly illustrated with full colour diagrams and photographs

As the emphasis in construction moves from building new bridges to maintenance and rehabilitation of existing stock, bridge management is becoming an increasingly important subject. This is the definitive, single-volume reference for professionals and postgraduates on bridge management. The book focuses on inspection, assessment, testing, evaluation, repair, as well as financial aspects such as whole life costing.

Highly illustrated and in full colour, this revised and updated second edition has been expanded to include new chapters on deterioration modelling, asset management and the impact of heavy goods vehicles. As this edition includes examples of practice and techniques drawn from around the world, this will be an invaluable source for the modern bridge engineer.

CHAPTER 1

Bridge Management Systems

1.1 INTRODUCTION

CONTENTS

Bridges cannot last forever. Whatever form of construction is used and whatever materials are adopted, sooner or later the effects of degradation begin to appear. There are many contributory factors that affect the nature and degree of degradation such as the structural form, construction materials, quality of construction, design and detailing, atmospheric environment, scour, fire, fatigue, earthquakes, floods, weather, and nature and intensity of the imposed traffic loading.

The early timber bridges built up to and including the nineteenth century have all but disappeared, but many of the masonry, cast iron, wrought iron, steel, composite steel-and-concrete, reinforced concrete, and prestressed concrete ones remain. The final choice for a given bridge, was in most cases, made on the basis of economy, that is, the cheapest first-cost, and very little thought was given to whole life costing to take into account 'in service' expenditure deriving from maintenance requirements. This appears to have been the case certainly in the last generation (especially, the heady days of the 1960s and 1970s) and may well have been the view taken before that time.

Winter salting of roads in many countries of the world has become universal and commonplace, and the problems that ensue from sodium chloride solutions leaching into concrete decks and piers have only been fully appreciated in the last 20 years or so. Steel bridges have rusted and suffered severe corrosion at badly detailed joints, which allowed rain water and salt water to collect.

Yet, despite the known enemies we have often viewed bridges as *permanent* artifacts not realising that they are subject to deterioration and need a special kind of care and attention. Maxwell (1990) admits that 'The problems of deterioration now with us suggest a failure in bridge management', and

McIntyre (1997) said that we have inherited a legacy resulting from many years of neglect: failure to fund and implement steady state maintenance on our oldest bridge stock.

In his book *Engineers of Dreams*, Henry Petroski (1995) writes:

> *'...bridges are affected by their environment no less than people are, and the wear and tear of traffic, pollution, abuse, neglect, and just plain old age take their toll. It is implicit, and often made quite explicit, in the design of every product of engineering that there are limits to its health and strength, and therefore limits to what it can be subjected to. A recognition of those limits and regular check-ups and inspections of the artifact are required, as is a certain amount of preventive maintenance and repair. To neglect this common sense is to find ourselves in the position in which we now are in America, with roughly one out of every five of our bridges is said to be structurally deficient'.*

This may seem a negative start to the subject of bridge management, but it gives us some idea of the consequences we face if we insist on only addressing the issues of function and structural economy and neglect the weightier issues of durability and low maintenance. It is like bringing a baby into the world and abandoning it. The management of a bridge does not stop the moment it goes into service.

The Romans appear to have been the first to recognise this. In addition to insisting that a bridge be certified by a council of 'experts' before commission, they applied a punitive clause in the contract to the effect that the builder was held responsible, at his own cost, for its stability for 40 years, at which time he was able to retrieve a deposit required of him at the start of the contract (Shirley-Smith, 1964). Bridges like the Alcantara bridge over the River Tagus in Spain (more than 2000 years old) and the Pont du Gard near Nimes in southern France (see Figure 1.1) are examples of well-constructed and maintained Roman bridges.

In Europe, in the Middle Ages, the construction of bridges seems to have been taken up by the Church, and at one time the Pope himself was referred to as Pontifex Maximus, the 'supreme bridge builder'. The order was always to *'build and maintain'* and the *monies to pay for the upkeep of bridges* built at that time came from endowments, tolls levied on travellers and vehicles and, if a chapel was built on the bridge, income from offerings and the sale of indulgences. Shirley-Smith (1964) states that one of the richest bridges in England at that time was the multispan stone arch bridge at Bideford (still standing), which had accrued such a surplus of funds by the 'caretakers' that they were able to spend money on education and charity! In France, emperor

FIGURE 1.1
Pont du Gard.

Charlemange created a *toll system for the maintenance of bridges* and personally supervised the work.

The first-named bridgemaster in England was Peter of Colechurch who was chaplain of St Mary Colechurch, and it was he who supervised the construction (1176–1209) of the London Bridge and *provided for its future maintenance*. It was the first bridge across the River Thames to be built from stone, and it lasted for more than 600 years until 1831 when it was replaced. Its longevity is due not only to the durability of the materials and the soundness of the design, but also to the good management and maintenance funds available from many sources including church endowments, national and private donations, and a wool tax imposed by Henry II.

Bridges are important links in any national road or rail network, and the capital outlay required to build them is high. If their carrying capacity is impaired or if they collapse, the resulting cost as a result of road closure and rehabilitation could double. As Petroski (1995) says, 'barring accidents, bridges, like health, are most appreciated when they begin to deteriorate and fail. Thus politicians seemed more interested in bridges when they found that so many of them were structurally deficient'. This was a major problem when funds are requested to repair them because the bridge engineer has to explain why the deficiencies have occurred in the first place!

1.2 BRIDGE MANAGEMENT

Bridge management is the means by which a bridge stock is cared for from conception to the end of its useful life. Unfortunately, many politicians and bridge authorities throughout the world while acknowledging the need for regular inspection and maintenance during the service life of their bridges, failed to appreciate the need for forward planning at the conception and design stages to ensure that sound principles were applied, which would maximise their long-term durability and very often preferred 'neglect', euphemistically called 'deferred maintenance', which is only postponement of the inevitable' (Petroski, 1995). Consequently, the present generation of bridge engineers has inherited a legacy of badly deteriorating bridges that now have to be repaired, strengthened, replaced, propped, or have a weight restriction posted on them. We should be very grateful to men such as the New York Senator Moynihan who, since the late 1980s have been ceaselessly campaigning to save the Hell Gate Bridge, built in 1918, and which had not been given a coat of paint since the 1930s (Petroski, 1995). A sum of $55 million has since been appropriated from the government to repair and repaint it. Many of the world's long steel bridges tend to impose tolls to *pay for maintenance*, and with luck and some foresight the annual revenue can exceed the basic maintenance costs and be invested in new equipment, upgrading, or, indeed, education and charity.

A countrywide random survey of 200 concrete highway bridges in the United Kingdom for the Department of Transport in 1989 (Wallbank, 1989) suggested that many of these structures have deteriorated at a faster rate than originally expected, and that the design life of 120 years will not be realised.

Badly conceived structural form, awkward details, the proliferation of joints in multispan bridges, poor construction, failure to understand the potential damage that can be caused in reinforced and prestressed concrete bridges from chloride contamination during winter salting, wind-blown salt water, and water itself, and, finally, the unexpected rise in the volume and density of traffic have resulted in many seriously weakened bridges and increased the level of fatal accident risk.

The present situation has resulted in a flurry of research around the world and many exciting developments are taking place that will ensure, at least, that the next generation of bridges will not deteriorate at an alarming rate and will be strong enough to support the projected levels of road traffic using our roads in the next 30–40 years.

In practice, *bridge management* is necessary to *coordinate and implement* the tasks associated with the care of our bridges, such as

- Collection of inventory data
- Regular inspection
- Assessment of condition and strength
- Repair, strengthening, or replacement
- Prioritising allocation of funds
- Safety

The mechanism by which the coordination and implementation is achieved is the *Bridge Management System (BMS)*, with the specific aims of assisting bridge managers and managing agencies:

- To have a clear picture of all the bridges being managed and to *prioritise* them in terms of importance relative to the overall road and rail traffic infrastructure
- To understand the maintenance needs of a particular bridge and by considering a number of intervention strategies to *optimise the cost benefit ratio*
- To initiate and control the chosen maintenance action
- To assess the value of the bridges on a periodic basis by the inclusion of performance indicators.

1.3 BASIC MANAGEMENT

The six critical stages in the life of a bridge are shown in Figure 1.2.

The main management activities begin once the bridge is commissioned. This entails an initial logging of the basic bridge data in the form of a *database* followed by the regular collection and recording of information through *inspections* to appraise the *condition* of the bridge and the *rate of deterioration*. This and other information are then used to *assess* its load-carrying capacity and also to help with the *prioritising of financial resources* for *maintenance* purposes.

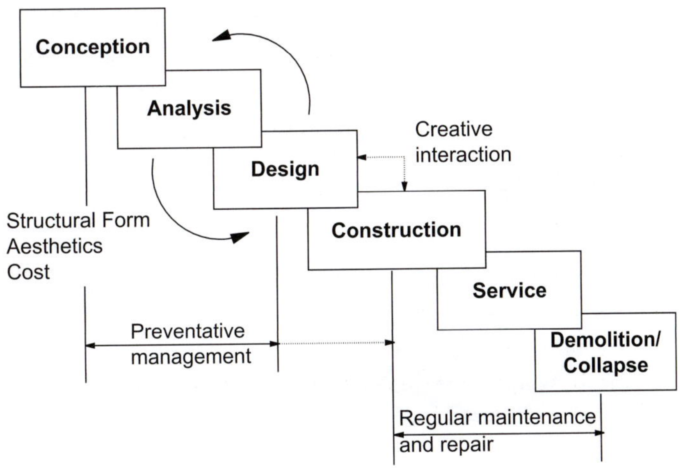

FIGURE 1.2

The six stages in the life of a bridge.

1.3.1 Card index system

A bridge manager needs a way of keeping track of the bridges in his charge and will require a minimum level of recorded information to carry out his work efficiently and effectively. The most rudimentary system is by means of *file cards* (rather like a card index) that record the location and type of bridge, together with any handwritten notes about inspections and the type and cost of repairs (see Figure 1.3).

The file cards together form a *register* of bridges in the stock. This may seem very primitive but it works very well in some developing countries or small counties where there is a relatively small number of bridges; indeed in the county of Surrey in the United Kingdom, a card indexing system was in use prior to 1980 and served well for the amount of work and budget levels at the time (Palmer & Cogswell, 1990).

The record or file cards form the core of any BMS and take the form of a record containing the data appertaining to a particular bridge. The master cards should be retained by the national bridge or transportation authority and a copy kept by the local authority. *The resulting file is the substance of the bridge register.*

A more sophisticated paper system might record a more comprehensive amount of data such as the following:

- Location plan
- Bridge number
- Bridge name

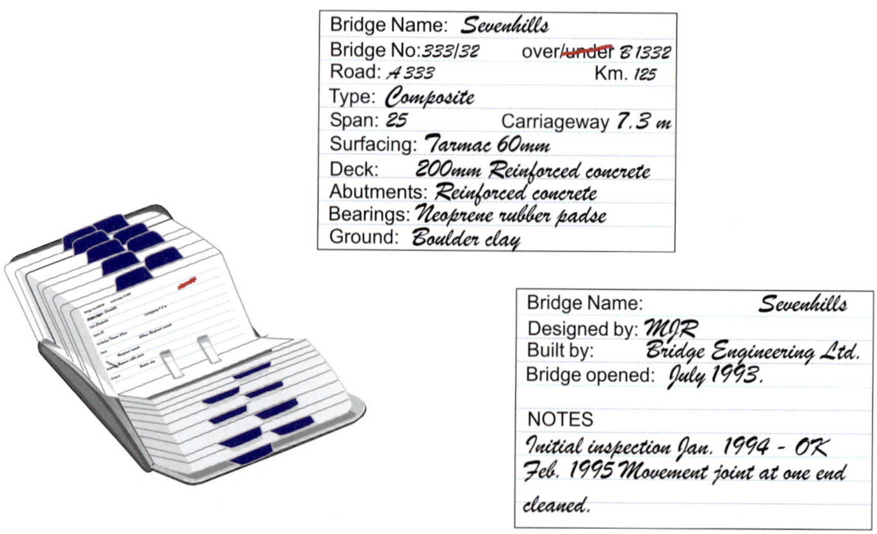

FIGURE 1.3
Card Index System.

- Picture or photograph
- General arrangement drawing
- Whether an over- or underbridge
- Obstacle being crossed
- Type of bridge
- Number and dimensions of spans
- Whether simply supported or continuous
- Carriageway width
- Services
- Live load capacity
- Abnormal load capacity

Sketches can be made or photographs can be taken to augment the index cards–one from the side (or preferably each side) and one on top looking along the length of the bridge–so that someone coming fresh to the bridge has an instant visual image, which aids the interpretation of what is on the record cards.

A second card (or the other side of the card) can be used to record inspection personnel and inspection dates, together with observations of cracking, rusting, bulging, buckling, rotting, etc. and events that *change* the bridge in some way due to:

- Repair or strengthening works
- Posting of a weight limit
- Widening
- Demolition and rebuilding
- Demolition of a redundant bridge
- Temporary propping

Any of these items may mean that original drawings will have to be amended. A typical record card might look like the one shown in Figure 1.4a and b. The bridge depicted is more than 144 years old, and its first recorded inspection was in 1964 when a rough sketch with principal dimensions was drawn up. Thereafter, irregular visits were made to the bridge over a period of 34 years, and any defects noted by the various inspectors were recorded in separate 'Bridge Inspection Reports' and put in a general bridge file (N I Roads Service, 2000).

1.4 MODERN BMSs

With the advent of powerful personal computers in the mid-1980s came the possibility to progress to *electronically based* BMSs capable of processing a lot more data than a paper system and in a lot less time. At the heart of

Bridge Number: 60534 Bridge Name: Tattinweer

Design, As Built Drawings and Calculation References

No original drawings available
No as-built drawings available
MEXE calculations in general bridge file
Original bridge information sheet (1964) in general bridge file

Inspections on file

1964 Original inspection: H Lucy
1967 General inspection: R Armstrong
1970 General inspection: R Armstrong
1970 Principal inspection: R Armstrong
1972 General inspection: R Armstrong
1975 General inspection: A Boyd
1977 General inspection: A Boyd
1983 Principal inspection: E Duke
1987 General inspection: E Black
1988 General inspection: M McClure
1989 Principal inspection: R Baird
1992 General inspection: EC & McM.
1993 Principal inspection: E McMorrow
1994 Masonry bridge assessment: I McClang & R Parkinson
1994 General inspection: C & E
1997 General inspection: J Muldoon
1998 General inspection: J Muldoon & D Campbell

FIGURE 1.4a *Front side of a bridge record card.*

Bridge Number: 60534 Bridge Name: Tattinweer

Divisional number	534	Section	Fermanagh
Route No:	B80	OS Sheet No.	34
Grid Ref:	363,419	Orientation	W-E
Over:	River	Navigable	NO

Overall Dimensions

No. of spans:	1	Span(s):	10.8m
Cover:	0.67m	Skew angle:	0
Deck width:	6.5m	Verge:	None
Carriageway:	Asphalt	Structure:	Masonry Arch
Abutments:	Masonry	Pier(s):	None
Foundation:	Bed rock	Wingwalls:	Masonry
Bearings:	None	Parapets:	Masonry
Crash barrier:	None	Services:	Water, BT
Designed by:	Unknown	Built by:	Tho' Kidd
Built:	1,856	Wt. Restriction	None

General Notes

OS Bench Mark on u/s parapet
Water main attached to u/s spandrel

FIGURE 1.4b *Back side of a bridge record card.*

a BMS is the *database* built up using information obtained from the regular inspection and maintenance activities and containing a register of the bridges. A BMS is thus the means by which the bridge manager can keep himself fully informed of the 'health' of the bridges under his control, and make informed decisions about future maintenance activities.

1.4.1 Input information for a BMS

For any BMS to work effectively, it has to have as much pertinent input information about the bridge as possible. The amount of information will depend upon the size and complexity of the system, but basically all systems will have modules dealing with *inventory, inspection, maintenance, cost, and bridge condition.* Embracing, analysing, and processing all of this information will be the *management* control. This is shown diagrammatically in Figure 1.5.

A BMS is more than a collection of facts, it is a system that looks at all the information concerning all the bridges and is able to make comparisons between each to rank its importance within the overall infrastructure with regard to safety and budgetary constraints. *Basically it should be able to tell the bridge manager where he should be spending his funds in the most efficient way.* Not an easy task! Unless the number of bridges is very small, a modern *computer-based* system becomes essential, which can operate on a number of basic components to provide the end information required.

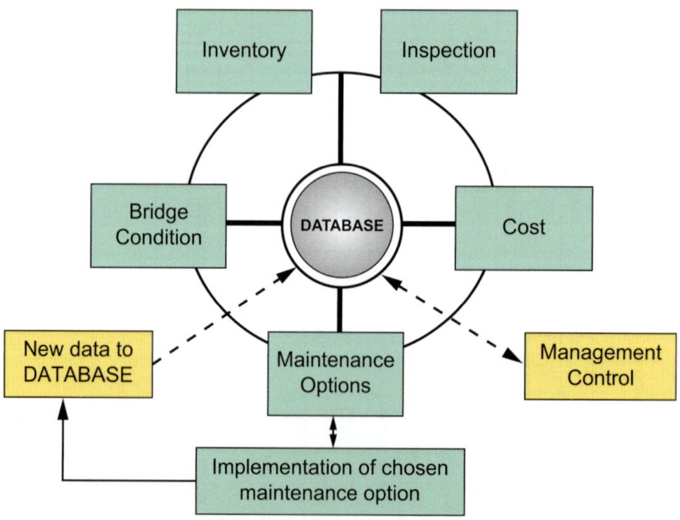

FIGURE 1.5
A basic BMS.

There are a number of basic components comprising a BMS to make it a fully integrated system capable of analysing the *database* and then interacting with other components together with incoming information. The output should ideally be in the form of a limited schedule listing the ailing bridges in priority of need (which requires some form of condition rating) followed by a prediction of the costs of various maintenance strategies.

1.4.2 Inventory component

As we have already seen, this component stores information about the bridge in terms of its name, location, construction, etc. and provides the starting point for the system. It requires reviewing drawings, maintenance records, and a 'walkover survey' to familiarise the user with the bridge (also enabling a check to be made on site of the existing drawings from the drawing register). With luck, there will be two sets of drawings, the *design* set and the *as-built* set with later alterations recorded. If unlucky, there may be no drawings at all. In either case, a site visit is necessary; in the former case to compare the drawing information with what is in situ and to correct it if necessary, and in the latter case to obtain a graphical record of the main dimensions and structural members.

The availability of *calculations* is of great benefit as it enables the bridge engineer to assess whether the bridge has been overdesigned, in which case there may well be enough reserve strength to carry increased traffic loads. If the design was optimised there may be no spare capacity. If the design was inadequate, then....!

Record cards need to be established (or checked if they already exist), and finally a computer-based database established to provide information for the management system. This information only changes as and when more bridges are added to the network and supplies routine enquirers with basic information in pro-forma format. Alternatively, it should be possible to obtain individual reports comprising selected information, for example, about its structural form, when it was last inspected, what maintenance works have been carried out to date or in the current financial year, etc.

1.4.3 Inspection component

This component stores the information from the inspection pro-formas and reports, which includes information about the general condition of the bridge, the specified treatment, the priority given to past remedial works, and the cost. It helps the inspectors who are about to embark on an inspection, and those responsible for manpower and budgetary control.

Some systems include the storage of *inspection drawings* that show in detail the make up of the bridge and identify all of those members needing inspection. These remain valid until updated after subsequent inspections and intervention for repair or strengthening.

Regular inspection of a bridge is vital if its safety and serviceability are to be maintained, because many things can go wrong once a bridge has been designed, built, and been in service for a number of years (see Chapter 2 for more details). Therefore, an inspection seeks to provide the information to be recorded as clearly as possible so that a quick assessment can be made by a qualified engineer as to the action necessary. The inspector, himself, may not be qualified but can nevertheless tender a report on which the engineer can act.

The action taken will range from 'do nothing' to 'demolish and rebuild', and in between these two limits there will be a wide range of propping, repair, and strengthening options.

Prior to 1977 in the United Kingdom there was no authoritative guidance for the inspection of bridges. In that same year, BE4 (1977) was published following an OECD report of 1976 recommending the four types of inspection which, broadly speaking, are followed by most countries of the world: *superficial, general, principal, and special*. The details of each are given in Chapter 2.

1.4.4 Maintenance (options) component

To keep abreast of the condition of a bridge, maintenance records are essential. They will inform the bridge owner of the nature of the maintenance carried out and exactly what is being spent on any given bridge. The system usually adopted is the one-file-per-bridge system that enables the history of any one bridge to be followed.

If there is constant severe deterioration and mounting maintenance costs, then the owner may decide to demolish and rebuild. This was the situation in the case of the ill-fated Taff Fawr bridge in Merthyr Tydfil, Wales, United Kingdom, which carries the A465 Heads of the Valleys trunk road in South Wales (NCE 10/3/83; 29/3/84; 9/5/85; 3/10/85). The bridge was constructed in 1964, and consisted of a three-cell cantilever structure in post-tensioned concrete over a main central span of 65 m and two 39 m side spans. Severe corrosion problems came to light during a minor repairs contract let in 1983. Spalling from the arched soffit was discovered together with wide cracks in the unusual precast flanges and the webs of the generally in situ structure. Further investigation revealed that road salt percolating through the deck waterproofing had caused severe corrosion of the prestressing tendons eventually resulting in cracking. The cost of repairing the bridge was prohibitive and so it was finally demolished in 1985/1986 and replaced in 1987, with a new 15-m wide single cell post-tensioned concrete box structure at a cost of £2 M.

For future planning and monitoring purposes it is necessary to record *the value of estimates, details of the work done in the form of sketches and photographs, starting and finishing dates, and the final costs.* It will also mean copying the changes to the drawings.

If an inspection highlights a *defect*, then a process is put in motion that selects the relevant data from the inspection report as to the *condition rating of the element* under inspection, the type of defect, its cause, the proposed remedy, and possible cost. If the original inspector was unable to decide on the best course of remedial action, then the description of the problem together with site sketches and/or digital photographs is examined with a view to specifying the type, extent, and cost of the works required. The money then has to be found to do the work!

1.4.5 Financial (cost) component

All eyes are on this component, especially in times of financial restraint! It processes all of the cost information from past and present projects and should be able to produce regular and reliable financial reports. This is where a certain amount of *financial engineering and lobbying* are necessary to extract monies from the bridge owners. It is an exhausting and often a soul-destroying task, but if sound data from past and extant projects relating to new-build, operation, and repair/maintenance costs can be proven to be reliable, in the sense that each cost as a percentage of the whole was consistently the same, then more confidence can be placed in the value of the (recurrent/capital cost) ratio, which, in turn, elicits the cooperation and trust of the owner and contractor. To do this each and every element has to be considered, and a very interesting and informative research programme carried out by Bouabaz et al. (2000) showed a marked consistency over many contracts incorporating hundreds of bills, where, for example, a *large* percentage of the contract value was contained in the cost of only a *small* percentage of a particular element.

1.4.6 Bridge condition component

The bridge condition component is based on historical data combined with the latest inspection information, which is then assigned a priority rating that identifies its importance at network and/or the project levels. There are many ways of arriving at a rating level but it is generally agreed that it should be based on the nature and extent of a *defect* in a structural member (or members) and the resulting effect on the bridge should the defect be left in its present state or, indeed, allowed to deteriorate further up to the next planned inspection. Because the condition is of a visual nature only, the information must be analysed by suitably qualified and experienced engineers, who must agree on the governing time scale, nature, and importance of the works.

All agencies and countries will have their own views on how to arrive at a *number* relating to a *term*, which expresses the action to be taken, i.e., **[0] = do nothing** to **[100] = demolish and rebuild.**

1.4.7 Database

The *database* is basically a store containing all of the historical and extant information about the bridges in a particular network. It forms the nucleus or heart of the BMS and contains details of a technical nature as well as administrative and financial information, and even anecdotal information similar to that described in section 1.3.1. It draws from all of the circumferential components and the management control shown in Figure 1.5 when in *analysis* mode, and stores new information from each contract when in *storage* mode.

1.5 THE SYSTEM

All of the components that have been considered so far provide a fund of information and material that have to be analysed and interpreted for rational decisions to be made regarding, for example, the allocation of available funds. This is achieved by means of a *system* where all of the components are linked together and put under strict control to form a living organism.

Ideally any system should be user friendly, and should provide information about the current condition, maintenance history, cumulative costs, and priorities. The way such a system is put together varies throughout the world to meet the needs of a particular country or authority and working within the bounds of available expertise and technology.

Basically, it would have to analyse data from inspections, past maintenance, and repair costs, and then by means of an in-built algorithm (which would consider its structural form, the importance of the bridge, the road class it is serving, its age, the volume of traffic it carries, and potential weaknesses identified from previous inspections), *it should advise where the current budget would be most effectively spent,* and possibly provide information for long-term strategic *planning of maintenance works.* Generally, money is spent on bridges that generate the highest level of benefits, something which is not easily quantifiable and can be overtaken by events, such as a bridge collapse, which, of course, would claim top priority.

There are, in general, two operating levels to be considered: *network* and *project.* The former defines the level where bridges are examined as *a system at their optimum condition*, that is, the bridges are in a condition that is safe and pose little risk to the users. It is not economically viable to 'maintain all bridges in exemplary condition'. This requires a very sensitive juggling act so that bridge standards are maintained and public confidence is sustained. In addition, politicians have to be satisfied that the optimum condition level is reasonable and not conservative. The latter provides bridge engineers with data from *individual bridges or groups of bridges*, which they can use to formulate a maintenance action plan for both the short and long term. The engineer can also interact with the system based on his professional experience and try to establish the priorities.

A more sophisticated BMS might also be able to allow a 'what if' scenario to be played out so that a bridge manager can see what the effect would be if he, for example, put back the date of work on a particular bridge, or decided to override the inspection schedule. It might also consider things like deterioration models, and route management for exceptional loads, and be able to plan a repair/strengthening programme for several years to come.

It cannot be emphasised enough, that all of the activities described above have to be under the strict *management control* of the engineer. All of the information from the active components together with costs and budgetary constraints are analysed and prioritised with a view to achieving the most economical and elegant solution possible, and to take account of other work already in progress or those planned for the future.

1.5.1 Personnel

Suitably qualified personnel to carry out the bridge tasks in hand should be chosen with care to ensure the efficient use of human resources. It normally falls on the *senior engineers* to be responsible for management and programming duties, such as maintenance management, assessment programmes, strengthening and rehabilitation programmes, and to undertake principal and special inspections. *Junior engineers and technicians* would normally be responsible for assisting a senior engineer, gathering information, carrying out general inspections, assessments, and site supervision of remedial and/or strengthening works.

1.6 SOME EXAMPLES OF BMS

1.6.1 Pennsylvania, USA

The Pennsylvania BMS was put into service on 1 March 1987 (McClure & Hoffman, 1990) after satisfying its general objective which was:

> *'to develop a management tool which will enable a systematic determination of present needs for maintenance, rehabilitation and replacement of bridges in Pennsylvania, and to predict future needs using various scenarios, along with a prioritisation for maintenance, rehabilitation and replacement, which will provide guidance in the effective use of designated funds'.*

The overall system is shown in Figure 1.6.

This system was developed from a structural inventory records system (SIRS), which was already in place in the Pennsylvania Department of Transportation (PenDOT). This was modified to be compatible with other systems in the department, and also linked to the cost data inventory file that stores information about labour requirements, materials, and equipment and produces 'unit costs' for maintenance activities, as well as the cost in $/m^2 for rehabilitation, replacement, or widening.

Inspection has been formalised to enable bridge inspectors to identify potential work needed from a menu of 76 maintenance activities. The inspectors also fill in a form giving estimates for the *quantities and priorities* for each listed maintenance requirement. The total cost is simply the product of the estimated quantities and unit costs.

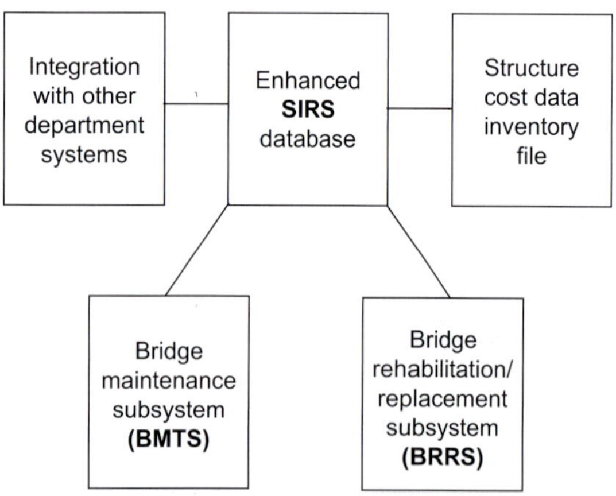

FIGURE 1.6 *The Pennsylvania BMS.*

The BMTS subsystem prioritises the work to 'apportion limited available funding and manpower' because there is a constant backlog of work that the department is never able to match in terms of manpower or finance. A prioritisation method has therefore been developed which considers *(1) the importance of the maintenance activity on the safety of the bridge; (2) the urgency of the repair; (3) the importance of the bridge on the road network; and (4) the current load-carrying capacity of the bridge*.

A *deficiency point assignment* has been developed whereby each of the four components above has relative weights assigned to them and to the elements of their make up. For convenience,

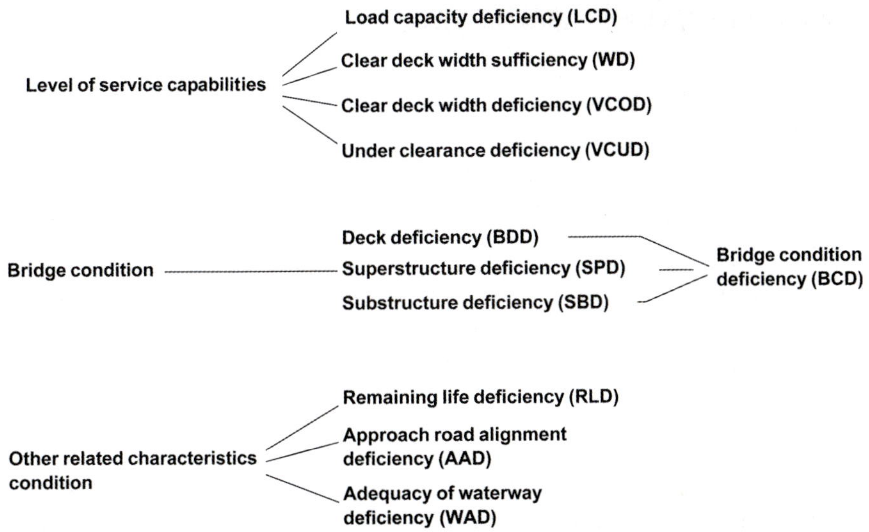

FIGURE 1.7
Factors making up the TDR.

the deficiency point assignment has been arbitrarily given a maximum of 100. The higher the total on a bridge, the higher the priority. The manager is then able to view the relative maintenance needs of one bridge against another in his jurisdiction, and also relate this to the worst possible case of 100 deficiency points.

The BRRS produces a prioritisation list for rehabilitation and replacement, and assigns points to three specified categories (level of service, bridge condition, and other characteristics) to yield a total deficiency rating (TDR) ranging from 0 to 100. The interrelationship between all these components is shown in Figure 1.7, and the TDR given by McClure and Hoffman (1990) is:

$$\mathrm{TDR} = \phi[\mathrm{LCD} + \mathrm{WD} + \mathrm{VCOD} + \mathrm{VCUD} + \mathrm{BCD} + \mathrm{RLD} + \mathrm{AAD} + \mathrm{WAD}]$$

where $\mathrm{BCD} = \mathrm{SPD} + \mathrm{SBD} + \mathrm{BDD}$ and φ is one of four factors for each of the categories of road in the United States and ranges from 0.75 for local roads to 1.00 for interstate roads.

Finally, a table is drawn up listing the deficiencies in order together with the maximum deficiency points that they carry, from which four listing conditions can be calculated taking into account costings and other (unspecified) factors.

This particular BMS (as of 1990) is a sound model of common sense and includes all of the ingredients that make up a useful tool for the management of bridges. It is straightforward and utilises practical 'bridgespeak' rather than unnecessary 'academicspeak'.

1.6.2 Japan

Along with many other countries, Japan's bridge engineers have been frustrated in their attempts to maintain their bridge stock of around 650,000 (20,000 with spans greater than 2 m) because of insufficient funds and reduction in personnel. Meanwhile, traffic volumes increase, and vehicles are getting bigger and heavier, and more and more bridges are being built along stretches of the new national road structure (Yokoyama, Sato, Ogihara, & Toriumi, 1996).

Their present fully operational MS is based on three pertinent BMS modules: inspection, maintenance, and evaluation. In one respect, however, it flies in the face of the nature of the information collected at the inspection stage in that the present performance is based on a visual inspection. Each bridge is treated individually (i.e. at project level) and cannot be used at network level. Each element of the bridge is thus assigned a deficiency rating (a numerical example of this procedure is given in section 2.10.1), and all the values are fed into the BMS that then analyses the data and prepares a repair schedule for each bridge.

Recently, Bakht and Mutsoshi (2005) have undertaken collaborative research to establish the merits and demerits of two BMSs, one in the United States (PONTIS) and one in Japan (J-BMS), each using a different philosophy when evaluating the present performance of bridge elements. The former works at the network level and so the repair schedule is determined on a priority basis; the latter works at the project level on a bridge-by-bridge basis. Quite why this comparison was chosen and what the expected outcome will be are not made clear. No doubt the system that gives the lowest cost will be the favourite.

1.6.3 United Kingdom

In the United Kingdom, High-Point Rendel developed a BMS called HiSMIS (Highways Structures Management Information System), which, since its inception in 1990, was reputed to be the most widely used BMS in the United Kingdom apart from NATS (DoT, 1993), the Department of Transport's in-house system (Blakelock, 1993). The layout of the system is shown in Figure. 1.8.

The System Administration module allowed the user administrator to adjust and maintain the system for its particular use. The actual database was made up of five modules providing all relevant INPUT information for

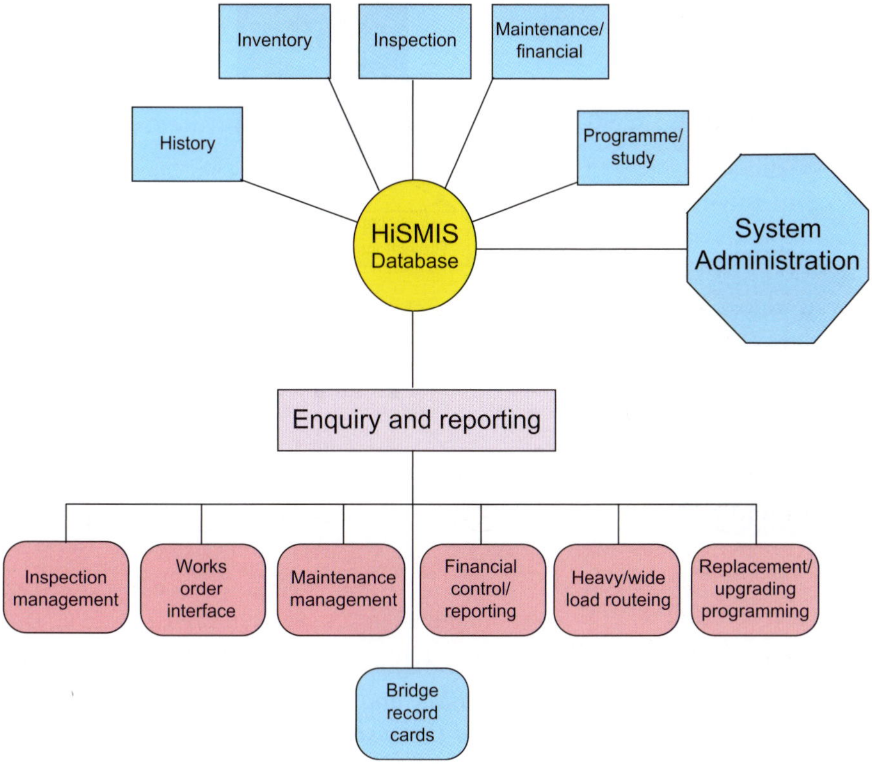

FIGURE 1.8
HiSMIS system.

the system. OUTPUT is via the enquiry and reporting suite made up of six modules.

It is user friendly, has adequate data storage volume, and can deal with structures other than bridges, namely tunnels, culverts and causeways, floodways and fords, gantries, lighting systems, and retaining walls making it one of the more comprehensive systems. The system was not prescriptive, indeed RPT built in some limited facilities to allow users to customise the system for their own use. Blakelock (1993) pointed out that 'the real power of any BMS lies in the ability to process information and produce output'. The enquiry and reporting stage was therefore all important and the output, of course, must be what the enquirer wanted. Experience over a number of years enabled HiSMIS to be continually reviewed and developed by a HiSMIS User Group to meet users' requirements. The group reported regularly to RPT who considered and implemented recommendations it deemed appropriate.

One of the major difficulties that High-Point Rendel came across in the development of HiSMIS was the enormity of the data collection task, and

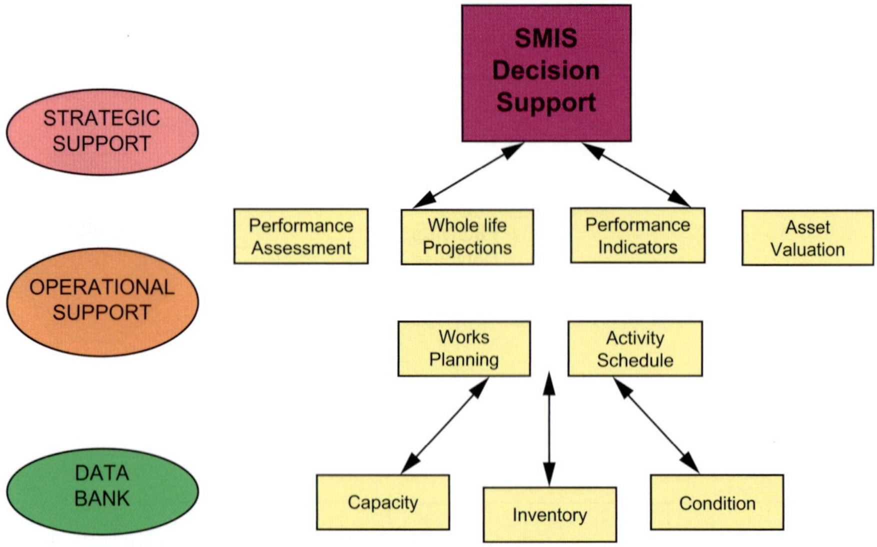

FIGURE 1.9
SMIS system.

Blakelock (1993) reports that *such a major job can cost in man-time terms ten times the cost of buying the BMS and the appropriate hardware and that over half of the users have had the system for less than a year and are still collecting input data!* This is not just the *quantity* of the data but includes the *nature* of the data. The more sophisticated the system, the more detailed the data required to operate the system. Once the data are on board the process of updating is, of course, less time consuming.

Since then, because of outdated software and hardware the system has been replaced by Structures Management Information System (SMIS), which more closely supports the new management processes. Although not an 'expert' system, it is now a well-proven management tool, providing information upon which sound engineering decisions can be made. The system is networked to 16 areas, each being managed by a firm of consulting engineers. Figure 1.9 shows the overall configuration of the system. Continued updating will, it is hoped, ensure that the present system is meeting current requirements as a strategic decision-making tool (Allison, 2005).

1.6.4 Surrey County Council, UK

Surrey County Council (SCC) has a track record of being at the forefront of BMS technology. This is due in part to the fact that, in terms of bridges, they are located in one of the most densely populated parts of the United

Kingdom. In 1980, the Council took active steps to move from a simple record card system to a 'no-nonsense' system, which is based partly on paper and partly electronically. It is a modified form of COSMOS (Computerised System for the Management of Structures), which, in itself, is Bridge Inventory and Management System (Wootton, 2000) and was originally developed by Babtie and first used in Berkshire. The modified system is shown in Figure 1.10.

Although SCC no longer uses it in its present form, a bare outline of its operation is given here as it could still prove useful to bridge agents struggling to maintain a network of bridges but are badly strapped for cash. Inspections are carried out using Paper BE11's and recorded in the BMS and the BE11's filed. Maintenance needs are assessed using a maintenance priority factor (MPF) algorithm described in Chapter 9. COSMOS prioritises on the basis of the road link, location, structural element, and the importance of the link and the computer prints out the MPFs. The MPFs are then transferred to a site instruction (SI) module where they are processed 'and a complete set of site instructions produced' (Wootton, 2000). What is actually in the SI module is not described. After the bridge inspector has read it and added any

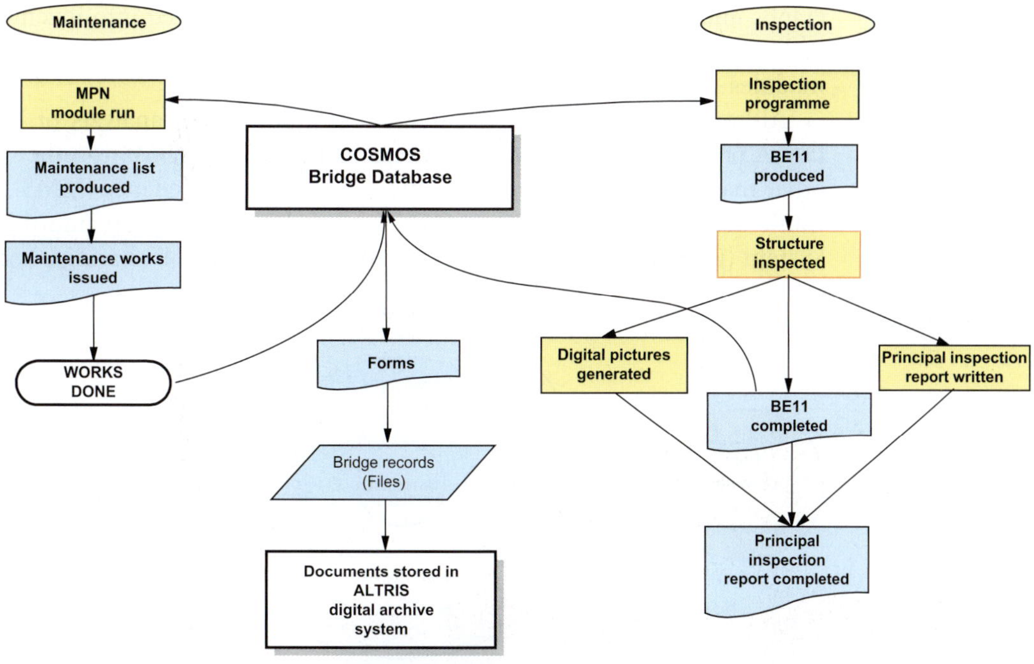

FIGURE 1.10 *Modified COSMOS.*

comments of his own and, presumably after authorisation is given, the works are passed on to the maintenance contractor.

SCC is very forward looking and now uses digital cameras to record before and after images of the major schemes, which are then stored on a CD-Rom and later incorporated into high-quality electronic reports. This saves a lot of storage space and is a convenient way of archiving data and also a suitable format for forwarding to clients.

1.6.5 South Africa

The Centre for Scientific and Industrial Research (CSIR) in Pretoria, South Africa, in partnership with Stewart Scott International (SSI) has recently developed a new BMS for the National Roads Agency (NRA) (Nordengen et al., 2000). With a bridge maintenance allocation of only 2.5% of the road rehabilitation budget, it is vital to have a proper system in place to use the money to the best advantage. Existing BMSs developed for the cities of Cape Town and Spoornet were used as the basis for the new computerised system.

The system is operated in a *paperless* environment *after* data have been collected on site in paper format, and later transferred to the BMS. This has meant the purchase of digital cameras to take standard inventory photographs and identified defects, which are then incorporated into inventories and inspection reports in electronic format. The system comprises five modules arrayed as shown in Figure 1.11.

Although the BMS is a stand-alone system for the analysis of bridge DATA only, it is hoped in the future to incorporate the BMS together with other management systems for highways (HMS), maintenance (MMS), and traffic observation (TOMS) into an Integrated Transportation Information System (ITIS) to manage NRA assets and information, all sharing a central

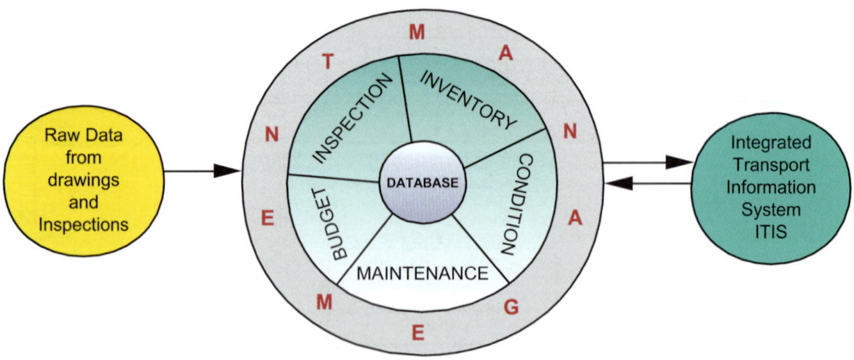

FIGURE 1.11
South African BMS.

Oracle database. Geomedia software at the heart of ITIS will allow Internet and Intranet users the access to data commensurate with their status, for example, 'general public – low level access' and 'consultants under contract – high level access' and so on. Information will be able to flow freely between the BMS and the ITIS database. Path B transfers raw data to ITIS database and path C allows data to be retrieved for analysis in the BMS.

The *condition module* is included to prioritise bridges on the basis of what are defined as a *functional* index and a *priority* index; the former relates to the strategic importance of the bridge and the latter is based on the *condition rating* calculated from the 'Degree (D), Extent (E), and Relevancy (R) of each of the identified defects on each of the 21 predefined inspection items' known as the DER rating system (Nordengen et al., 2000). Weighting factors are given to distinguish between items of importance such as main structural members, and relatively unimportant items such as kerbs, surfacing, and drains.

The South African system rightly places emphasis on the value of personnel who should be properly trained and continually attend refresher courses, and also on the need for inspectors to pay attention to detail. The quality of the inspection and condition rating input is of paramount importance in determining the structural integrity and safety of the bridge.

1.6.6 United States of America

On 15 December 1967, the Silver Bridge (spanning the Ohio River at Point Pleasant, West Virginia, USA) collapsed without warning. In less than 1 min, 46 people lost their lives. According to Czepiel (2005), 'at that time, there was no systematic maintenance programme in place to monitor the condition of the bridge population'. This tragedy acted as a 'wake-up' call, and President Johnson established the National Bridge Inspection Standards (NBIS), which are guidelines for the inspection of all bridges in the United States. Building on this and data from the already existing National Bridge Inventory (NBI), the Federal Highways Association (FHWA) has overseen a national programme for all states to have a fully operational BMS in place by 1 October 1998.

The primary BMS in use in the United States is called PONTIS (Latin for *bridge*), which was developed under the auspices of the Federal Highway Administration (FHWA) to be able to predict needs, future conditions, and carry out strategic economic analysis (Kulkarni, 1998) and which incorporates data from the already existing National Bridge Inventory (NBI). It became commercially available in 1992 (Thompson, 1993), and is now used in a majority of states (Figure 1.12).

FIGURE 1.12
Overall structure of PONTIS.

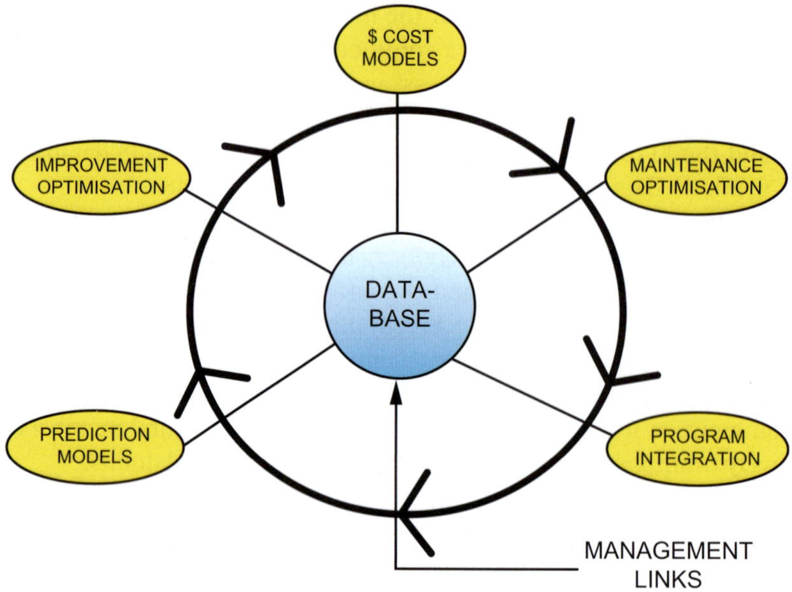

A very sophisticated computer-based platform is used to support PONTIS in a Windows environment, and it is possible to customise the system for each user. Its greatest value is the simple way in which it translates the value of sound engineering decisions into tangible benefits and meaningful maintenance costs.

The analysis of inspection data is carried out at the *project* level on an *element-by-element* basis. Inspectors have a total of 120 basic (commonly recognised CoRe) bridge elements (bearing, stringer, pier column, etc.) to choose from to which they assign (from a maximum of five) a *condition state or rating*. Then by means of a Markovian deterioration model (Thompson, Small, Johnson, & Marshall, 1998) the probability of change between consecutive *condition states* is determined, and thus the degree or rate of deterioration can be assessed. *Each condition state has related maintenance actions (up to a maximum of 3) assigned to it, and a specified unit cost for each action.* An example of this process is given by Thompson et al. (1998).

At the network level, *preservation optimisation* is undertaken to 'find the optimal, long-term policy that minimises expected life-cycle costs while keeping the element out of risk of failure' (Thompson et al., 1998). An improvement model seeks to quantify the benefits of functional improvement in terms of user cost savings as the *sum of accident costs, vehicle operating costs, and travel time costs.* The results of benefit calculations are stored in the database and when the time comes to recommend a maintenance action PONTIS consults them and calculates the net benefit (optimal action costs minus do-nothing costs) and recommends the action that gives the highest net benefit. This process is repeated for each element in the bridge, summed, and compared with each bridge in the inventory. The final choice of whether to proceed with the project depends (as usual) on budgetary constraints, which means that a compromise has to be made or some objective means used to differentiate between the competing bridge works, which is the domain of the engineer and not the computer, although it is claimed that PONTIS can make the choice by means of an 'incremental benefit/cost method' (Thompson et al., 1998). It is a pity that underfunding forces these decisions to be made, but on the other hand, it does ensure that funds are put to the best possible use, and only reflects what we all have to do with our own personal budgets on a month-by-month cycle.

The state of California has enhanced the basic system with a deterioration simulation to calculate the 'health index' (HI) of a bridge on a scale of 0 to 100. A bridge with an HI of 0 is worthless and a bridge with an HI of 100 is in full service. By means of this information in a PONTIS summary report, the Californian Department of Transportation (Caltrans) was able to secure generous funding over a 5-year period from 1998 to 1999 increasing by 25% year on year (Thompson et al., 1998). PONTIS is also being used to calculate the expected annual maintenance costs for alternative bridge designs for new crossings, which are then combined with the initial capital construction costs to obtain the full life cycle costs (LCC) of each bridge and thus to enable a final selection to be made on *economic* grounds.

1.7 ENGINEERING JUDGEMENT

With all BMSs there is always the temptation to believe that the input and output information is unquestionably correct. This is not, or ever will be, true. Sound engineering judgement is required to be able to step back, and through the process of experience and the understanding of structures be able to look at the whole as well as the detail and decide what is *sensible, reasonable, and practical*. If something about the input or output offends the user's engineering sensibilities, then there is a strong chance that the information is wrong.

Final decisions must be left to the engineer because the BMS does not exist to manage our bridges, it is there to assist the engineer by manipulating large amounts of data and to produce reports to aid the engineer in decision making.

1.8 FUNDING

One of the most contentious areas that virtually all bridge managers face is a paucity of funds to carry out their work effectively and efficiently. The money available in most countries of the world depends upon their wealth. The finance ministers of wealthy countries can afford to give generously to national bridge agencies, who, in turn, can give generously to local bridge agencies. Alas, even wealth will not assure us of sufficient funds; it is very difficult to convince the relevant powers that it makes good sense to take care of our bridges, which are of inestimable benefit to the nation. Lynn (1997) has been quite outspoken on this subject in relation to *local authority-owned bridges only* in the United Kingdom. His arguments stem from the Highways Agency's aim in 1989 to complete an 'assessment and strengthening programme roads'. Due to inadequacy of funding for inspection, assessment, strengthening, and continued maintenance not only has the objective not been realised but it has resulted in a backlog of bridges requiring strengthening. The pressure on local authority budgets has been intense and because of the lack of adequate capital and revenue reserves, the bridge management programme has had to suffer.

Lynn's warning (1997) of the 'consequences of our failure as a nation to deliver the bridge programme so that bridges are in a suitable condition to accommodate 40 t lorries from 1st January 1999' has come true. Namely, the Freight Transport Association (FTA) in the United Kingdom (McIntyre, 1997) shares the same frustrations as Lynn (1997). The main thrust of their complaint to the government has been that 'funding for road and bridge maintenance on both trunk and local roads is determined not by need objectively assessed, but by short-term and short-sighted policies intended to minimise what is apparently regarded by central government (and perhaps by local government when the political chips are down) as optional expenditure'.

The fact that the bridge strengthening programme was irretrievably behind schedule and that as a result many bridges were not capable of supporting the 40 t lorries seems to have been completely lost on the government.

It was adamant that 'well maintained roads and bridges minimise congestion, delay accidents and environmental damage. They ensure reliability, predictability and (literally in respect of bridges) continuity in level of service' all vital 'to the UK's competitiveness and environmental well being' (Figure 1.13).

The 'UK government' said McIntyre (1997) 'managed to persuade the rest of Europe that here in the UK our bridges could not cope with (40 t) vehicles'

FIGURE 1.13
Forth Railway Bridge.

and that 'while bridge engineers and the construction industries in many countries had no problem in accommodating 40 t lorries, somehow our engineers and builders were less able'.

In evidence to the Transport Committee Inquiry in 1996, the Highways Agency identified the *minimum* level of funding, which would ensure that all trunk road bridges would be available to 40 t lorries by 1999, but only by resorting to temporary propping, one-way, and single vehicle operation.

Although not technically competent in this area, the FTA alluded to the fact that the Department of the Environment, Transport and the Regions (DETR), through the Highways Agency, is looking at *relaxation of the National Assessment code for the less heavily trafficked bridges*. This reflects a substantial body of opinion among bridge engineers that the existing code is unduly conservative. McIntyre (1997) concludes that 'lorries are being diverted from many bridges by weight limits well below their actual safe load bearing capacity. If this is happening it is a nonsense, and whatever changes are needed to the code and its implementation must be progressed as a matter of urgency'.

1.9 MOBILITY

No matter how sophisticated a BMS may be in prioritising maintenance works and allocating funds, if bridges have to be closed off for works to proceed, then measures must be taken to ensure that the traffic keeps moving. The Crie de coeur of all drivers when they see bridge construction works up ahead is, 'Please dear God let the traffic keep moving…'.

At the network level, a great deal of ingenuity and effort is required to find ways of re-routing the flow of traffic. In the state of Oregon, in America, traffic management is carried out at a dedicated site called The Mobility Center (OMCN, 2005). It is a place where 'planners of bridge works can meet to co-ordinate projects and to avoid conflicts'. This may involve *bridge, road, and traffic management teams working in harmony* to ensure that there is no conflict of interest. Using a computer, these state personnel are able to track road and bridge works in the same way that airplanes are tracked by air-traffic control. Regular 30 reports are posted on a dedicated Website, which aims to incorporate information for at least 10 years ahead. It has proven to be very popular with truck companies that are able to plan ahead with confidence both their route and their estimated time of arrival (ETA), which we all know is *money.*

1.10 PLANNING

Planning plays a very important part in the overall management process, and it is a good idea to lay out in detail what it is hoped will be accomplished in the coming year. A typical 1-year plan (Kindly donated by West Berkshire Country Council) is reproduced in Appendix C.

1.11 CONCLUSIONS

To protect the investment that we make on our bridges, it is essential that they are properly managed and that adequate funding is made available for continued maintenance. Ideally, this should be from conception to completion, but is more likely to be from when the bridge was commissioned, and worst of all it may only be from the time when the bridge had sustained such serious damage or degradation that intervention was inevitable.

BMSs are a means of manipulating information about a bridge so that its long-term health is assured and that maintenance programmes can be formulated in line with budgetary constraints and funding limitations. They are, however, not infallible, and engineering judgement is still necessary to ensure that action based on computer output is sensible and rational.

There are many well-established BMSs in use throughout the world, which can be used without modification, or used as a base onto which other modules can be connected to provide a custom-made BMS. According to Cole (2000), this is becoming the norm for small agencies that have their own particular requirements.

REFERENCES

Allison, B. H. (2005). The structures management information system. Bridge Management Five (pp. 83–91). In: *Fifth international conference on bridge management*. London: Thomas Telford.

Bakht, B., & Mutsoshi H. (2005). Development of bridge management system (BMS) in Japan and USA.

Blakelock, R. (1993). Experience in the development of a computerised bridge management system and of its use in a number of authorities in the UK and overseas (pp. 880–888). In: *Second international conference on bridge management*. London: Thomas Telford.

Bouabaz, M. et al. (2000). Modelling and predicting bridge repair and maintenance costs (pp 187–197). In: *First international conference on bridge management*. London: Elsevier Applied Science.

Cole, G. (2000). Managing substandard bridges (pp. 710–718). In: *Am. Fourth Int. conference in bridge management*. London: Thomas Telford.

Czepiel, E. (2005). Technical report funded by the Infrastructure Technology Institute at Northwestern University.

Department of Transport. (1993). *Trunk road maintenance manual*. London.

Haneef, N., & Chaplin, K. (1998). Bid assessment and prioritisation system (BAPS). In: *The management of highway structures conference*. London: ICE, HA.

Kulkarni, S. R. (1998). Bridge management in the state of Michigan. In: *The management of highway structures conference*. London: ICE, HA.

Lynn, D. (1997). The spending gap widens. In: *Proceedings of the conference on British roads: national asset or national disgrace*. Sponsored by Surveyor Magazine and the Automobile Association.

Maxwell, J. W. S. (1990). Highway bridge management (pp. 113–120). In: *First international conference on bridge management*. London: Elsevier.

McClure, R. M., & Hoffman, G. L. (1990). The Pennsylvania bridge management system (pp. 75–87). In: *First international conference on bridge management*. London: Elsevier.

McIntyre, D. (1997). Weak bridges: the impact on freight movement. In: *Proceedings conference on British roads: national asset or national disgrace*. Sponsored by Surveyor Magazine and the Automobile Association.

OECD. (1976). *Bridge inspection*. A report prepared by an OECD Road Research Group.

OMCN. (2005). *Oregon Mobility Center News* (p. 3).

Palmer, J., & Cogswell, G. (1990). Management of the bridge stock of a UK county for the 1990s (pp. 39–50). In: *First international conference on bridge management*. London: Elsevier.

Petroski, H. (1995). *Engineers of dreams*. New York: Allfred A. Knopf.

Shirley-Smith, H. (1964). *The world's great bridges*. London: Phoenix House.

Thompson, P. D. (1993). PONTIS: the maturing of bridge management systems in the USA. In: *Second international conference on bridge management*.

Thompson, P. D., Small, E. P., Johnson, M., & Marshall, A. R. (1998). The Pontis bridge management system. *Structural Engineering International, 8*(4), 303–308.

Wallbank, E. J. (1989). *The performance of concrete bridges: A survey of 200 highway bridges*. (pp. 1–96). Report prepared for the DoT. London: HMSO.

Wootton, N. (2000). Management of bridge maintenance: a local authority perspective. In: *Fourth international conference on bridge management*. London: Thomas Telford.

Yokoyama, K., Sato, H., Ogihara, K., & Toriumi, R. (1996) Development of a Bridge Management System in Japan (pp. 580–594). In: *Third international conference on bridge management*. London: Spon.

FURTHER READING

(1990). *Proceedings of the first international conference on bridge management*. London: Elsevier.

(1993). *Proceedings of the second international conference on bridge management*. London: Thomas Telford.

(1996). *Proceedings of the third international conference on bridge management*. London: Spon.

(1998). *Structural Engineering International, 8*(4).

(2000). *Proceedings of the fourth international conference on bridge management*. London: Thomas Telford.

(2005). *Proceedings of the fifth international conference on bridge management*. London: Thomas Telford.

Glamorgan. *Proceedings of the institute of civil engineers. Part 1, 84*, 634–669.

Hearn, G., & Shim, H. (1997). *Integration of bridge management systems and nondestructive evaluations*. Recent Advances in Bridge Engineering. Proc. US-Canada-Europe Workshop on

Kahkonen, A., & Marshall, A. R. (1990). Optimization of bridge maintenance appropriations with the help of a management system - development of a bridge management system in Finland (pp. 101–111). In: *First international conference on bridge management*. London: Elsevier.

Lindbladh, L. (1990). Bridge management within the Swedish National Road Administration (pp. 51–61). In: *First international conference on bridge management*. London: Elsevier.

Lowe, D. (2005). *The transport manager's & operator's handbook 2005*. (Chapter 12).

Lutzenberger, S., & Baumgartner, W. (2000). Evaluation of measured bridge responses due to an instrumented truck and free traffic (pp. 71–79). In: *Fourth international conference on bridge management*. London: Thomas Telford.

Marghitu, D. B. (2000). *Mechanical engineers handbook*. Academic Press.

Ryall, M. J., et al. (2000). *Manual of bridge engineering*. (p. 55). London: Thomas Telford.

Transport Research Laboratory (TRL). (1998). Overseas Road Note 7. *A guide to bridge inspection and data systems for district engineers* (Vol. 1). Crowthorne: Author.

APPENDIX A

Some BMSs around the world.

Denmark	DANBRO (DANish Bridges and Roads)
	COWiconsult, 45, Tekikerbyen, DK-2830 Virum, Denmark
Finland	FinnRABMS (Finnish National Roads Administration Bridge Management System)
	Finland Roads and Waterways Administration, Opastinsilta 12, PO Box 33, SF-00521 Helsinki, Finland
Holland	DISC
	Ministry of Transport and Public Works, Rijkswaterstaat, Bridges Department, Voorburg, The Netherlands
Italy	SAMOA (Surveillance, Auscultation and Maintenance Of Structures)
	Autostrada, SpA, Italy
South Africa	BMS.NRA (National Roads Authority)
	Transportek, CSIR, Pretoria, South Africa
	SIHA
Sweden	BMS (Swedish National Road Administration, Head Office, Borlange, Sweden)
United Kingdom	STEG (Structures REGister)
	HiSMIS (Highway Structures Management Information System)
	High-Point Rendel, Southwark Street, London, United Kingdom
	SMIS (Structures Management Information System)
	Highways Agency
	BRIDGEMAN (BRIDGE MANagement system)
	Laboratory, Bridges Department, Crowthorne
	Buckinghamshire, United Kingdom
	COSMOS (Computerised System for the Management Of Structures)
United States of America	PONTIS (Preservation, Optimisation and NeTwork Information System)
	Federal Highways Administration, McLean, Virginia, United States of America
	BRIDGIT (BRIDGe Information Technology)
	Federal Highways Administration, McLean, Virginia, United States of America
	PENBMS (PENnslyvania Bridge Management System)
	Pennsylvania Department of Transportation and Safety Building 1009, Harrisburg, Pennsylvania 17120, United States of America

APPENDIX B

Highways agency – new methodology

The Highways Agency in the United Kingdom is developing a new Bridge Management Procedure that will involve seven distinct stages:

- Inspection
- Assessment
- Maintenance bid
- Prioritisation
- Allocation
- Execution
- Performance indicators and outputs

To aid inspection, a new Bridge Inspection Manual is being produced in six volumes plus an annex to cover common types of bridge structure as in Table B1.1.

Relevant Departmental Assessment Standards will also be available with the aim of rationalising the assessment procedures, and allowing bridge engineers to consider various maintenance options on the basis of whole life assessment as in Table B1.1.

The maintenance bids will be analysed using BAPS, a computerised Bid Assessment and Prioritisation System, which requires bridge engineers to submit a number of bid options to undertake the maintenance work over a period of 30 years (Haneef & Chaplin, 1998) and to determine the Optimum Maintenance Strategy, a strategic analysis will be made at the network level (see 1.5.1). Finally, strategic outputs are derived in the form of reports and graphs to develop performance indicators related to structural

Table B1.1 UK Inspection and Assessment Guides

Bridge Inspection Manual	Departmental Assessment Standards
Vol. 1 General procedures	BD21 Assessment of highway structures
Vol. 2 Concrete bridges	BD44 Assessment of concrete bridges
Vol. 3. Steel, steel/concrete composite bridges	BD48 Assessment and strengthening of bridge supports
Vol. 4 Masonry arch bridges	BD56 Assessment of steel bridges
Vol. 5 Retaining structures	BD61 Assessment of composite bridges
Vol. 6 Miscellaneous structures	BD77 Assessment of steel bridges for fatigue (draft)
Subdivision of structures	BD73 Assessment for scour (draft)
	BA81 Whole life assessment of highway bridges and structures

maintenance to see whether maintenance is being carried out efficiently and effectively.

The whole bridge management methodology is designed to fit within a well-defined programme of inspection and assessment, and represents a welcome rationalisation of the management procedure.

APPENDIX C

The management of bridges and other highway structures in West Berkshire

This outlines West Berkshire Council's procedures to ensure highway structures within the district are safely and cost effectively maintained.

Contents:

1 Introduction

There are in excess of one thousand highway structures in West Berkshire District. These include bridges, retaining walls, culverts, sign gantries, and pedestrian subways.

Not all of these structures are owned by the Council. Some are owned by other organisations such as Network Rail and British Waterways; however, as the high authority, the Council has a duty to ensure they are all maintained in a safe and functional condition.

The Council achieves this by using the following procedures:

- Undertaking routine inspections to ensure public safety and identify maintenance needs
- Undertaking minor maintenance works identified by routine inspections
- Programming and executing maintenance work to minimise inconvenience to the travelling public
- Undertaking reactive and emergency repairs resulting from traffic accidents and vehicle damage
- Undertaking major maintenance and replacement works where structures have reached the end of their service life
- Imposing protection measures such as weight restrictions and traffic signals on weak or impaired bridges
- Retaining structures' data and information on a comprehensive computerised database
- Checking the suitability and construction of all new highway structures through the technical approval process
- Ensuring bridges owned by other organisations are maintained in a sage condition
- Managing the sage movement of oversized or overweight vehicles known as 'abnormal loads' through the district
- Predicting future aspirations for highway structures through consultation and feasibility studies.

2 Inspections

Highway structures are routinely inspected to ensure they remain safe for public use. The inspection also provides the data required to support good management practice in accordance with the Road Liaison Group's document: Management of Highway Structures, A Code of Practice.

Maintenance work to the Council-owned structures is identified by inspection and placed in the works programme according to priority.

Structures owned and maintained by other organisations such as Network Rail and British Waterways are also inspected and any outstanding maintenance works notified to the organisation concerned.

All highway structures are routinely inspected in the following cycle:

Principal inspection	Carried out at intervals not greater than 6 years. A close examination within touching distance of all accessible parts of the structure. Specialised access equipment may be required.
General inspection	Carried out not more than 2 years following the previous general or principal inspection. A visual examination of all parts of the structure where special access equipment is not required.
Superficial inspection	Carried out not more than 2 years following the previous superficial inspection on all private structures. A visual examination of all parts of private structures where special access equipment is not required.
Routine surveillance	General surveillance by highways inspectors as part of regular Highway Safety Inspections. A visual examination identifying obvious defects requiring urgent attention, e.g., damage to parapets, superstructure, joints.
Inspections for assessment	Carried out after a particular defect is identified in a previous inspection that may affect the structures' load-carrying capacity.
Special inspection	Carried out after a particular defect is identified in a previous inspection or event, e.g., flooding, scour, vehicle impact damage.
Acceptance inspections	Carried out due to a transfer of responsibility.

3 Data storage and the BMS

The Council maintains a computerised database that contains information about bridges and other highway structures including information about their condition at the time of their last inspection. The database will also compare the condition of any one structure to the condition of the other structures thereby enabling maintenance works to be prioritised and target the structures in the poorest condition.

The Council is striving to achieve the standards of Asset Management and Performance Measurement of Highway Structures recommended in the recently published Code of Practice and referred to in the Council's LTP2 document.

4 Maintenance

4.1 Programming maintenance works

When programming maintenance works, care is taken to minimise inconvenience to the travelling public by:

- Selecting a repair method or technique that will cause minimal obstruction of the highway
- Providing traffic management appropriate to the type of road and expected traffic flows
- Programming works with regard to other works in the area and the availability of diversion routes.

4.2 Minor maintenance

The bridge inspection regimen is the main source of information identifying the routine maintenance requirements of the structures' stock.

There are complex formulae within the BMS that prioritise the maintenance works identified by the routine inspections and from this an annual programme of maintenance work is drawn up. Typical minor works are repointing and brickwork repairs, the repair and replacement of handrails and parapets, repairs to protective coatings, and the repair or replacement of expansion joints.

Graffiti is removed from structures if it is considered to be offensive or unsightly. Subways most commonly fall victim to graffiti, and to counter this most of them have been painted with antigraffiti paint and are inspected and cleaned on a monthly basis.

4.3 Reactive and emergency repairs

Vehicle collisions can severely damage highway structures and leave them in an unsafe condition. In the worst cases, this results in temporary road closures and diversions until the necessary repairs can be made.

4.4 Major maintenance, strengthening, and replacement

As a structure approaches the end of its design life or service life, it will usually become uneconomical to repair and will require replacement. A structure may also become due for replacement if it represents a restriction to the highway through being, for example, too narrow, under strength or unsuitable for disabled use.

Recently completed bridge replacement and strengthening schemes:

Blackboys Bridge, Newbury
Parkway Bridge, Newbury
Kennet and Avon Canal Bridge, Hungerford High Street

5 Highway structures owned and maintained by third parties

Approximately one third of structures on West Berkshire's highways are owned and maintained by third parties. Network Rail, the Highways Agency, the Environment Agency, British Waterways, local land owners, and businesses all own bridges and structures that affect the highway in some way.

The Council undertakes routine 'superficial' inspections of these structures to ensure they are maintained in a condition that will not endanger the travelling public. The Council will notify the bridge owner if a bridge is found to be in a dangerous condition and take enforcement action if the owner fails to take remedial action.

5.1 The load bearing obligations of Network Rail and British Waterways

Most of the bridges in West Berkshire that have weight restrictions are owned by either Network Rail or British Waterways. Because of an anomaly in highway legislation, these organisations have a legal obligation to maintain their bridges to 24 t to an old assessment standard known as BE4. However, the permissible maximum weight of vehicles using the highway is currently 40 t. The Council as the Highway Authority is required to assess the strength of bridges to a more modern standard than BE4 known as BD21/01. This is more onerous than BE4, hence some Network Rail and British Waterways bridges have weight restrictions on them less than 24 t.

Network Rail's bridges in West Berkshire were recently assessed jointly by Network Rail and West Berkshire Council in accordance with the Bridgeguard Three Agreement. Bridgeguard Three is a protocol agreed nationally between The County Surveyor's Society and Network Rail for the protection of Network Rail's bridges.

Since its creation in 1998, the West Berkshire Council has taken ownership of and strengthened or replaced three strategically important bridges: Blackboys's Bridge (completed December 1999), Basildon Skew Bridge (both bridges were previously owned by Network Rail), and Hungerford Canal Bridge (previously owned by British Waterways). The Council also financed the strengthening of Alfrey's Rail Bridge in Grazeley, in 2006, another Network Rail-owned bridge.

6 Managing structures that cause restrictions to the highway

There are three ways in which bridges can restrict or impair highway usage: weight restrictions, height restrictions, and width restrictions.

6.1 Weight restrictions

There are two types of weight restriction: environmental and structural. Environmental restrictions are applied to lengths of road that are unsuitable to carry heavy vehicles and are not related to bridges. Structural weight restrictions are applied specifically to protect weak bridges.

Table C1 Structural Weight Restrictions on Bridges in West Berkshire

Bridge Name	Parish	Bridge Owner	Type of Restriction	Restriction (in t)
Pinewood North	Burghfield	Network Rail	Temporary	3
Boundary Road	Newbury	Network Rail	Permanent	3
Aldermaston Station	Padworth	Network Rail	Temporary	3
Ufton Swing Bridge	Ufton Nervet	British Waterways	Permanent	5
Tyle Mill Swing Bridge	Sulhamstead	British Waterways	Permanent	5
Theale Swing Bridge	Theale	British Waterways	Permanent	3
Tyle Mill River	Sulhamstead	WBDC	Permanent	7.5
Whitchurch Toll Bridge	Oxfordshire Boundary	Private Ownership	Permanent	7.5

From the above table it can be seen that most of bridges with weight restrictions in West Berkshire are owned by Network Rail and British Waterways for the reasons explained in the subsection *The load bearing obligations of Network Rail and British Waterways.* These bridges are all on minor roads and the weight restrictions have no significant effect on vehicle movements across the district. Tyle Mill River Bridge, owned by West Berkshire Council, adjoins Tyle Mill Swing Bridge and would be of no benefit to strengthen due to the restriction of the swing bridge.

6.2 Height (or headroom) restrictions

Network Rail, however they can be problematic and there are currently no bridges in West Berkshire that warrant their use.

Table C2 Height Restrictions in West Berkshire

Bridge Name	Parish/Town	Bridge Owner	Height Restriction
Marsh Lane	Hungerford	Network Rail	12 feet 6 inches (3.81 m)
Parsonage Lane	Hungerford	Network Rail	11 feet 3 inches (3.42 m)
Croft Road	Hungerford	Network Rail	13 feet 3 inches (4.03 m)
Hungerford High Street	Hungerford	Network Rail	15 feet 6 inches (4.70 m)
Inglewood Gate Arch	Kineterbury	Private	15 feet (4.57 m)

Table C2 Height Restrictions in West Berkshire *(continued)*

Bridge Name	Parish/Town	Bridge Owner	Height Restriction
Marston Road Bridge (disused)	Hermitage	West Berkshire Council	13 feet (3.96 m)
Eling Road (disused)	Hampstead Norreys	West Berkshire Council	12 feet 9 inches (3.88 m)
Compton Bridge (disused)	Compton	West Berkshire Council	13 feet 6 inches (4.1 m)
Pangbourne Station Arch	Pangbourne	Network Rail	3 feet 3 inches (4.03 m)
Whitchurch Railway Arch	Pangbourne	Network Rail	11 feet (3.35 m)
Pound Green Bridge	Wokefield	Network Rail	12 feet 9 inches (3.88 m)

Generally, the cost inhibits raising the headroom under any of the Network Rail-owned structures. There has, in the past, been requests for the demolition of both Compton and Marston Road disused bridges, however, public and Parish Council opinion is currently opposed to this as the bridges are now perceived to be performing important traffic management functions.

6.3 Width restrictions

There are various ways of managing width restrictions to the highway caused by narrow bridges. The most suitable method will depend on various site conditions such as the severity of the restriction, traffic flows, and sight lines. Narrow bridges may restrict either the carriageway width (Aldermaston Wharf Lifting Bridge) or footway widths. Studies have been undertaken to investigate possible improvements at these locations.

7 Technical approval and commuted sums

7.1 Technical approval

West Berkshire Council as the Highway Authority ensures that all structures that are constructed over, under, or next to the public highway are constructed to an acceptable standard by requiring them to achieve 'technical approval'. Some highway structures may remain in private ownership, but in most cases the Council will 'adopt' or take ownership of the new structures. In either instance, technical approval is required.

In normal circumstances, only structures that are a necessary part of the highway and constructed on highway land will be considered for adoption. These may include road bridges, public footbridges, culverts, retaining walls that support the highway, and sign gantries. When the Council agrees to adopt a structure, an additional adoption agreement will be drawn up by the Council.

Structures that remain in private ownership require technical approval only. These may include private road bridges and footbridges over the highway, retaining walls more than 1.5 m high above ground level that support property or land adjacent to the highway, support structures for pipes and conveyers that cross the highway, basements that support the highway, parts of buildings that cross or overhang the highway, and temporary structures intended for public use. Private structures that cross the highway also require a licence, which will be drawn up by the Council.

7.2 Commuted sums

When the Council is required to adopt a new structure from a private enterprise or developer it will usually require them to pay a 'commuted sum' to the Council. The commuted sum is equal to the costs the Council may expect to incur for maintenance, inspection, and renewal of the structure and is usually calculated over a period of 120 years.

8 Abnormal load movements and the assessment of structures

8.1 Abnormal load movements

An abnormal load is a vehicle that is outside the classification of normal permitted traffic by virtue of its gross weight, length, width, or axle configuration according to current road vehicle regulations. In accordance with The Road Vehicle (Authorisation on Special Types) General Order 2003, the movement of abnormal or STGO loads should be notified to the relevant Highway Authority or Authorities.

Jacobs currently acts as West Berkshire Council's appointed agent to receive notifications for the movement of abnormal loads and STGO vehicles on nontrunk roads in West Berkshire. Jacobs are also appointed agents for Reading, Bracknell Forest, Windsor, Maidenhead, and Slough Councils and Rail Property. Notifications should be faxed directly to their Reading office (Fax No. 0118 9881 656).

The movement of abnormal loads needs to be carefully managed so that large and heavy vehicles only use those parts of the road network that can safely accommodate them.

Hauliers of abnormal loads are required to notify the appointed Abnormal Loads Officer at Jacobs. The bridges on the notified route are then assessed in accordance with the current requirements of the Design Manual for Roads and Bridges (BD 86/01) and the Code of Practice for the Management of Highway Structures. If any of the bridges on the route are considered too weak to bear the abnormal load, it will be re-routed. If no suitable alternative route is available, special measures to increase the load-bearing capacity of the bridge may be considered.

8.2 ESDAL

ESDAL stands for 'Electronic Service Delivery for Abnormal Loads Project'. Once fully implemented, ESDAL will be a national one-stop Web-based portal to assist route planning and automated notification of abnormal load movements.

At present, before moving an abnormal load, hauliers must notify a number of authorities, including the Highways Agency, agents, local authorities, private bridge owners, and the police. This is a lengthy process and is entirely manual, based on the use of fax to exchange and agree information. The ESDAL project will simplify this process, significantly reducing the amount of time it takes for everyone involved in the process to plan, notify, and approve the movement of an abnormal load.

8.3 The assessment of structures

Highway structures are assessed to establish their ability to carry the loads, which are imposed upon them. The assessment provides valuable information for managing the safety and serviceability of highway structures. The road bridges in West Berkshire were last assessed under a national programme of assessment undertaken in the mid-1980s.

In accordance with the recommendations of the new Code of Practice, the West Berkshire Council is currently undergoing a structural review. This involves revisiting the existing assessment information and from this developing a prioritised reassessment programme.

9 The management of bridges on or near the district boundaries

West Berkshire has five neighbouring authorities: Wiltshire, Hampshire, Oxfordshire, Reading, and Wokingham. To avoid any confusion, bridges near to or on the boundaries are subject to Maintenance Agreements whereby the maintenance of the bridge is delegated to one or other of the authorities. Agreements are in place with Hampshire and Oxfordshire for bridges over the Rivers Enborne and Thames, respectively. The Council remains in close consultation with all neighbouring authorities regarding bridge restrictions and maintenance works that may affect the highway network beyond West Berkshire's boundaries.

Whitchurch Tool Bridge, which crosses the River Thames at Pangbourne, is currently managed by Oxfordshire County Council on behalf of the bridge owners, 'the company Proprietors of Whitchurch Bridge'. The bridge currently has a 7.5 t weight restriction. Major reconstruction is planned for 2013, which will require the bridge to be closed for approximately 3 months.

10 Future bridge management requirements

In September 2005, the Roads Liaison Group launched a Code of Practice for the Management of Highway Structures. Prior to the publication of the Code of Practice, there was no overarching document available to guide bridge owners on good maintenance practice and this has led to differing practice across the county. The purpose of the code is to standardise practice thereby providing better information regarding the condition of bridges across the county and so determine future funding requirements.

West Berkshire Council's current practice generally aligns very well with the basic recommendations made in the new code of practice. It is the Council's intention to become fully compliant with the code within the recommended time frame of 5 years by the development of asset management principles set out in the new code.

CHAPTER 2

Inspection and Condition Rating

2.1 INTRODUCTION

CONTENTS

One of the most important tasks in the operation of a sound bridge management system (BMS) is inspection of the bridge. It is 'the keystone of our knowledge' of the bridge. Apart from initial recording of the basic bridge data, regular reporting of a bridge's condition provides a way of alerting bridge engineers to *deterioration* of the bridge from whatever cause, be it damage from vehicles, fracture, or material breakdown, and enables a bridge engineer to assess maintenance requirements. A great deal of experience and technical understanding is required to expedite an inspection in a methodical and systematic way. Inspection has, therefore, to be carried out by professional engineers, or at least supervised by a professional engineer. Each bridge is unique, and its form and layout will dictate the focus of the inspection, for example, arch bridges suffer in totally different ways from prestressed concrete box girder bridges and steel is different from timber.

The results of an inspection must be accurately and fully recorded (see also Chapter 1) including nil returns, so that a complete history of the structure is available at any time. Every defect has a *cause*, and, following inspection, this should be identified and rectified to prevent further deterioration. Besides the physical causes due to loading, environment, and accidental impact, defects may appear due to faults in design (poor detailing, inadequate cover, errors in calculation, etc.), materials (poor quality, use of inappropriate admixtures or contaminated water), or workmanship (poor mixing of concrete, compaction, curing, placement of reinforcement, placing of falsework, etc.).

Because resources are not infinite, fully comprehensive inspections of every bridge under an agency's care are not possible. Most authorities, therefore, have a hierarchical system of visual inspection routines with limited testing varying from the superficial to the most detailed, and the lower-order inspections are timed to be more frequent than the higher-order ones. The standard of inspections carried out on similar structures by different people must be consistent, and the results must be of use in assessing the bridge's load-carrying capacity and monitoring its condition.

If available, all of the design information such as drawings, design calculations, and soil investigation reports should be used to help with the inspection.

2.2 WHY INSPECT?

Bridges are vital components in the transportation infrastructure of a nation. They represent a large capital outlay in a road or rail network and are important not only because of their location, but by virtue of the cost implications if their capacity is impaired or if they fail outright. They are of immense economic importance in peacetime to allow the free flow of goods and people and allow communities to flourish and prosper. (Their importance can be judged from their use as primary strategic targets in wartime.) They have, therefore, to be maintained to ensure that they remain in service for as long as possible.

Users expect a bridge to be 'safe', and by that they often mean that it has to *look* safe and *feel* safe. For the bridge engineer, it has to *be* 'safe', in that for its expected life it has to function properly (i.e. be serviceable) under design loads, and should be sufficiently strong so as not to result in injury or loss of life to the users. (This is discussed in greater detail in Chapter 10.)

The two possible policies under which it may be possible to act to ensure adequate safety (OECD, 1976) are:

- Breakdown maintenance
- Systematic inspection

The former policy confines inspection (and maintenance) to the primary load-carrying elements of the bridge. This is not an acceptable policy as it is in effect 'crisis management' (a response when things go wrong) and as such poses unacceptable risks to the users. A typical example of this would be when pieces of concrete fall off a bridge due, for example, to the corrosion of embedded steelwork or when flood waters erode a bridge foundation. When the deterioration is brought to the notice of the managing authority – possibly by a member of the public or a highway inspector – then something is done!

The latter policy is preventative in nature and involves a carefully worked-out system of inspection to enable any structural weaknesses to be seen and logged before safety is compromised. Appropriate measures can then be drawn up to remedy the situation, thereby maintaining the *expected operational life* of the bridge or even prolonging it.

The feedback from regular inspection not only aids maintenance strategy but can often contribute to a greater understanding of structural and material behaviour, and lead to changes of practice in design, construction, and operation, as well as provide meaningful data for research.

Regular inspection, therefore, seeks to provide:

- A consistent record of the state of the structure, which allows the significance of any changes (accidents, overloading, or environmental deterioration) to be analysed and acted upon
- Data upon which the safety and serviceability of the bridge can be assessed
- Information on any potential trouble spots
- Information upon which a consistent maintenance strategy can be established
- Data for monitoring the effect of any changes in traffic loads and the use of new structural forms and materials
- Data for monitoring the behaviour of new strengthening techniques
- Data for research purposes.

2.2.1 Inspection categories

In the United Kingdom at the moment (2000), there are four categories of 'set schedule' inspections listed by Narasimhan and Wallbank (1998), which closely follow the OECD report of 1976:

Superficial inspection

This involves a cursory check for obvious deficiencies that might compromise the integrity of the bridge or lead to accidents or result in potentially high maintenance costs. They are often *undertaken by highway maintenance staff on an ad hoc basis*, usually when a problem has been observed and reported by other staff, the police, or the public. There are no set programmes or reporting requirements.

General inspection

This is a visual examination of representative parts of the bridge to ascertain condition and note items requiring attention. This is normally *undertaken every 2 years*. Inspection is from ground level, using binoculars if necessary, and information is recorded in the field using a checklist that has been jointly prepared for the particular type of bridge by the engineers responsible for both design and maintenance, or in the United Kingdom, for example, on form BE11 (Highways Agency, 1993, 1994), though this allows only one entry for each type of structure element regardless of the number present or the range of defects.

Principal inspection

This involves close examination (within touching distance) of all parts of the bridge against a prepared checklist. There is always an initial principal inspection before handover of a new structure, but thereafter it is *undertaken generally at intervals of 6 years*. Access equipment and traffic management are usually needed to enable all parts of the structure to be inspected. Originally, principal inspections were confined to visual examination, but limited testing (half-cell potential, chloride concentration, cover, and carbonation) has been included in recent years (Department of Transport, 1990). Full written reports with drawings and photographs form part of these inspections.

Special inspection

This is a close inspection of a particular area or defect that is causing concern. It is *undertaken when needed* for a wide variety of reasons, including:

- Following up a defect identified in an earlier inspection
- Investigating a specific problem discovered on similar structures; for example, the sudden failure of an old segmental prestressed concrete bridge resulted in the close inspection of many other such bridges (Woodward & Williams, 1988)
- Checking after flooding
- Subsidence in coal or mineral mining areas
- Exceptional events such as fire or flood
- Seismic activity
- Unexpected settlement
- Bridges subject to critical stress, for example, before, during, and after the passage of an exceptional load
- A bridge due for demolition but expected to last a few more years
- Lively bridges, especially steel ones where fatigue could be a problem

- A low capacity bridge that is temporarily carrying diverted traffic from a main road
- A bridge in poor condition but which is considered 'safe' for the time being.

Some further supplementary inspection or testing may be required to study the extent and cause of the defect, and in this case, a written special report is required detailing the cause and recommended remedial works. In addition to these, there are specialised inspections for joints, underwater structures, and of paint (coating) to steelwork.

Shortcomings have been observed in the implementation of the Bridge Inspection Guide in the United Kingdom (HMSO, 1984), and there is a proposal to replace it by five Bridge Inspection Manuals (Narasimhan & Wallbank, 1998) covering concrete bridges, steel and composite bridges, masonry arch bridges, retaining walls, and miscellaneous structures.

The shortcomings listed in the review were:

- Principal inspections required the whole of the structure to be inspected closely, irrespective of the importance of each component or its likely deterioration.
- Not enough use was made of the growing range of non-destructive testing techniques.
- The limited testing undertaken was not always as effective as it should have been.
- Standards of inspection and reporting vary greatly between maintaining agents.
- Reporting requirements for general inspections did not provide adequate scope for reporting a variety of defects or for dealing with multispan bridges.

The new procedures will aim to provide continuous detailed information about the current condition and rate of deterioration. This will be achieved by tailoring an inspection programme to a particular bridge. It is proposed to rationalise inspections into two groups: those that apply to **all** bridges and those that apply to the **majority** of bridges.

A summary of the proposed changes is given in Table 2.1.

This is still pretty much a 'set schedule' with very little flexibility, and not much scope for the bridge manager to exercise engineering judgement. Does a bridge that is constantly rated as in 'excellent' condition really have to be inspected as frequently as another more dubious one? There is room for

Table 2.1 Current and Proposed Inspection Procedures (after Narasimhan & Wallbank, 1998)

Current Procedures			Proposed Procedures		
Inspection Type	**Interval**	**Remarks**	**Inspection Type**	**Interval**	**Remarks**
Superficial	When needed	Cursory inspection No standard report	Superficial	When needed	Cursory inspection Simple report format
General	2 years	Visual inspection from ground level	General	2 years	Visual inspection from ground level Improved report
Principal	6 years	Close visual inspection All defects recorded		6 years	Visual inspection of other areas
	6 years	Limited testing of specified areas	Benchmark	From 6 to 24 years depending on condition	Close visual inspection All defects recorded
			Particular	From 6 months to 12 years depending on condition	Detailed testing of particular areas to suit structure
Special	When needed	Detailed testing of particular areas to suit structure	Special	When needed	Detailed testing of particular areas to suit structure
Joint	At construction completion	New structures	Special	At construction completion	No change
Initial principal	At the end of maintenance period	New structures	Benchmark	At the end of maintenance period	Close visual inspection
Underwater	6 years	Part of principal inspection	Particular	6 years	No change
Scour	When needed	Special inspection	Special	When needed	Detailed advice to be issued
Paint survey	When needed		Particular or special	When needed	No change

a considerable amount of debate on this issue and one that will smoulder for many years to come. For example, why have principal inspections every 6 years? Why not every 5 or 7?

2.3 WHAT TO LOOK FOR

For the purposes of inspection it is usual to subdivide the bridge into its main constituent parts, namely, the examination of the *superstructure, substructure, and foundations*, and then to further subdivide these into their separate *elements*. In the United States, Frangopol et al. (1996) have proposed that the elements are further subdivided into *segments* bounded by 'physical landmarks' on the bridge or element. The *accessories* (such as parapets, expansion joints) are considered separately.

The primary aim of the inspection is to determine whether any degradation has taken place, and if so the cause and extent of the damage (vehicle impact, river damage, water, earthquakes, corrosion, material reactions, etc.). This will involve both visual examination and recording (*graphic and photographic*), and in some cases testing.

2.3.1 Superstructures

Inspection of the deck of a bridge is usually given the highest priority in that it is the part of the bridge that directly carries the traffic. In the case of older masonry or brick arches with fill between the spandrels, the arch barrel is given priority because it is the main load-carrying element.

The deck can be subdivided into segments (if it is a concrete deck or its individual load-carrying elements) and further subdivided into its constituent materials, if other than concrete. Terminology, for example, cracking, spalling, and delamination, is given in Appendix A together with some extreme examples of the damage described in section 2.3.1.1.

2.3.1.1 Inspection of concrete decks and slabs

Concrete decks (of composite bridges) and slabs should be inspected for the following aspects.

2.3.1.1.1 Cracking

Concrete cracks due to tensile forces from shrinkage, temperature changes, and bending, and these are manifested as a regular pattern of fine cracks and do not necessarily indicate a problem or jeopardise the safety of a bridge. The location, orientation, and size of cracks can all point to the probable cause or causes that may be thermal, shrinkage, freezing and thawing, loading,

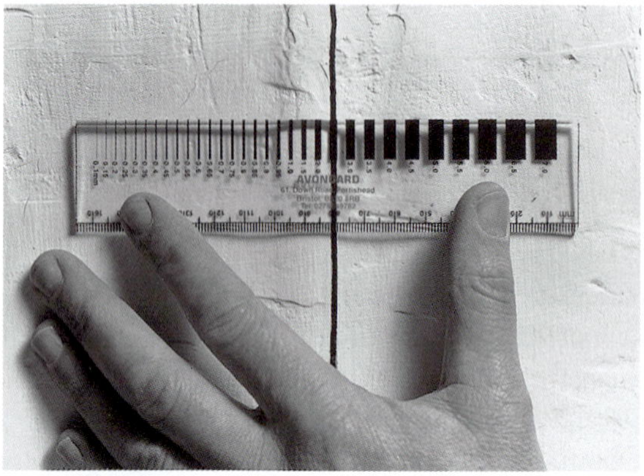

FIGURE 2.1 *Crack width measuring rule (0.1 mm to 7 mm) (courtesy of Avongard).*

FIGURE 2.2 *Cracking to the soffit of a slab.*

corrosion of reinforcement, and sulphate or aggregate reactions. In particular, a basic horizontal and vertical pattern with some branching that surrounds the larger aggregate particles could indicate a chemical attack in the form of calcium sulphoaluminate, which reverts to various forms of calcium aluminate hydrates and in the presence of sulphates in the ground converts to ettringite (Rogers, 1996) – expansive calcium sulphoaluminates – which disrupt the structure of the cement matrix and cause the concrete to crumble. If this is suspected, a *petrographic* examination should be carried out to establish its presence or otherwise.

The characteristic cracking on the surface of concrete due to alkali-silica reaction (ASR) or alkali-aggregate reaction (AAR) may be indistinguishable from other cracks and if suspected then further testing may be necessary if the reaction has been severe enough to cause macrocracking although not found to affect structural performance significantly (Clark, 1996). The width of any unusual cracking to the soffit not associated with the normal functioning of the deck should be measured using an instrument such as that shown in Figure 2.1, which measures crack widths down to 0.1 mm. These are then recorded on an appropriate drawing. It is up to the bridge engineer to interpret the readings and to try and diagnose the cause to ascertain whether it is a superficial crack due to temperature or shrinkage stresses or whether it is structural in which case action needs to be taken.

The ends of the cracks should also be recorded on the structure itself with a small line perpendicular to the cracks and the date. Reflected cracking in the surfacing material may indicate problems on the top side of the deck, and it may be necessary to remove some areas to gauge the extent of the problem. Cracking to the sides may indicate shear failure or loss of prestress. Typical cracking to the underside of a slab is shown in Figure 2.2.

Regular inspection of the cracks will determine whether they are active or dormant.

2.3.1.1.2 Spalling

Under pressure (for example, due to freeze-thaw action) bits of concrete can fall away from the deck leaving a crater, which defines the fracture surface. The cause is often due to corrosion of reinforcement where the volume of the corrosion products is much greater than the virgin steel and the resulting pressure causes local fracture of the concrete (see Figure 2.3). This is recorded by sketching out the affected areas and, if corroding reinforcement is visible, by measuring the cover (the distance from the outer surface of the concrete to the surface of the reinforcement bars).

FIGURE 2.3
Typical staining and spalling due to corroding steel.

2.3.1.1.3 Corrosion of reinforcement

In its early stages this can be detected by surface discolouration and rust stains, and later (when it has advanced) by spalling. The location and extent of any discolouration are recorded and a cover meter survey is carried out (see Figure 2.3).

2.3.1.1.4 Leaching

This phenomenon is due to water passing through the concrete and dissolving constituent material such as calcium hydroxide at crack locations. The result is very unsightly causing staining, efflorescence, encrustation, or at its very worst stalactites. It causes loss of alkalinity of the concrete, which could lead to corrosion of the embedded reinforcement or prestressing cables (see Figure 2.4).

FIGURE 2.4
(a) The effects of leaching on the underside of a concrete deck, and (b) on the underside of a jack-arch bridge.

2.3.1.1.5 Poor quality concrete

Because of bad construction practice, the concrete may be porous, exhibit honeycombing, or have the incorrect cover to the reinforcement. Honeycombing is usually easy to spot, the other two effects are not. Random tests with a cover meter should be made together with a hammer test in areas of suspected porous concrete. All suspect areas need to be identified on a drawing.

2.3.1.1.6 Deflection and vibration

A simple heel drop test can establish whether a deck is 'lively' or not, but more often than not standing centrally on the deck and waiting for a heavy vehicle to cross is more revealing. If the dead load deflection can be seen (without measurement), then it is probably too great and must be surveyed.

2.3.1.1.7 Accidental damage

Bridge bashing by over-height vehicles *under* a bridge can result in scraping and removal of concrete from the soffit (Figure 2.5a) and/or local spalling at the deck edge. Piers can be severely impacted (Figure 2.5c) and vehicles travelling *over* a bridge may cause damage to the surfacing due to skidding, braking, or overturning, and if it is a 'through bridge', out-of-control vehicles can damage the sides of the bridge (Figure 2.5b). Each area of damage should be recorded (see Figure 2.5).

FIGURE 2.5(a)
Example of bridge bashing to the soffit of an overbridge.

FIGURE 2.5(b) *Example of bridge bashing to a through-truss bridge.*

FIGURE 2.5(c) *Example of bridge bashing to a reinforced concrete pier.*

2.3.1.1.8 Deck surface damage

The surfacing material (usually asphalt based) can suffer damage from overheating by the sun causing it to flow and rut (see Figure 2.6). It can also be dislodged by ice beneath the surface where rain water has leaked through and subsequently frozen, or simply be pulled off by traffic due to insufficient adhesion to the deck.

2.3.1.1.9 Salt crystallisation

This appears as a white band just above ground level due to the crystallisation of salt-bearing water that has risen due to capillary action within the concrete.

2.3.1.2 Prestressed and concrete beam and slab bridges

These should be inspected for the same defects as for slabs, but some additional factors need to be considered:

Spalling

The vulnerable areas are at the ends of the beams directly over the bearings, and near the edges on the end face of post-tensioned beams.

FIGURE 2.6
Example of deck surfacing damage adjacent to an expansion joint.

Cracking

As for slabs, these should be carefully mapped over the underside of the slab and the sides and soffits of the beams. Particularly vulnerable areas are at the ends of the beams where diagonal cracks may develop. In large post-tensioned concrete bridges, the critical areas for inspection are along the lines of the tendon, at the end anchorage points, and at intermediate joints. Longitudinal cracks reflected in the top surfacing of the deck develop as a result of *local tendon failure* or insufficient transverse reinforcement. Transverse cracks on the soffits of the beams also indicate the failure of tendons in the midspan regions of a deck of beam. Vertical cracks in the diaphragms or anchorages could indicate either local tendon failure or insufficient transverse reinforcement or prestressing, see Figure 2.7.

FIGURE 2.7
Cracking in a prestressed concrete box girder (cracks highlighted for clarity).

Grout

The presence and condition of the grout in the tendon ducts is of crucial importance in understanding the reason for the cracking. Ideally, the grout should completely fill the duct and be dry. A non-destructive (semi-intrusive) method of examination is explained in sections 4.2.1 and 4.2.14, which enable inspection of all the critical positions along the line of the duct, namely:

- High points
- Low points
- Coupling points
- Anchorage zones
- Joints in segmental construction
- Construction joints *in situ* segmental construction

If it is wet, or shattered or found to contain chlorides, or there are voids, then it is suspect and further intrusive investigation is necessary to augment the visual inspection.

Box girders

The surfaces of the flanges, webs, and diaphragms on the *inside* of box girders need to be inspected and any cracks carefully recorded. The Highways Agency in the United Kingdom has instigated what it calls a 'Special Inspection Programme' for internally post-tensioned concrete bridges, which is described in BA 50/93. The inspection takes place in three phases:

- a *desk study* to discover what is already known about the structure (its design and construction) and its history (previous inspections, etc.)
- a *preliminary site investigation* to decide what action to take ranging from monitoring to lane closure or strengthening
- a *further site investigation* to decide what non-destructive tests or load tests are necessary to discover the integrity of the bridge.

Accidental damage

Beam and slab bridges are particularly vulnerable to collisions by over-height vehicles because individual beams can be almost rendered ineffective if the bottom flange (containing the prestressing) is severely damaged.

No apparent damage

This may seem a strange condition to consider, but a strictly visual survey of a prestressed concrete bridge may not reveal its true condition. For example, the Ynys-y-Gwas bridge in Wales in 1953 (see section 10.5.4) appeared sound when inspected, but collapsed without warning in 1986. It is therefore prudent to take extra special care when considering *older bridges of a segmental type* where a crack-free, spall-free, and stain-free exterior belies the true situation. If drawings are available they should be studied assiduously to discover where the latent weakness in construction lies and what situation (loading or corrosion) would cause failure. The Ynys-y-Gwas bridge was not only vulnerable, it was also *fragile*, and as far as can be ascertained it was not traffic that triggered the collapse but the mechanism of deterioration reducing the cross-sectional area of some of the main tendons to a level where they could not sustain the load in them. Even posting a load restriction would not have prevented collapse.

2.3.1.3 Steel and beams/girders/boxes in composite bridge decks and trusses

2.3.1.3.1 Condition of the protective system

This is the most common form of deterioration in structural steelwork. Both paint systems and galvanising eventually deteriorate and areas of breakdown should not only be recorded, but immediate action should be taken to prevent the problem spreading – see Figure 2.8 showing a lattice truss (a) and a trough deck (b), both showing serious breakdown of the paint protective system.

2.3.1.3.2 Corrosion

This is the 'cancer' of steel and its control is critical to the integrity of the structure. The vulnerable areas are at connections where the fixings (connections, fasteners) can corrode badly, at intersections with concrete or masonry, and in the natural water traps in open steel sections. The location and extent of the damage (which may be rust or delamination) should be recorded and if possible the loss of section assessed. The existence or otherwise of debris and proper drainage points should also be noted. Between pieces of steel joined by rivets or bolts, this can cause complete fracture of the fixings (fasteners) due to the compression exerted between the plates from the expanding corrosion products (rust is seven times greater in volume than original steel) – see Figure 2.9.

Steel box girders should be regularly inspected to ensure that they are free from water condensation or leakage, and if drain holes are provided (which they should be) they should be inspected to ensure that they are not blocked.

FIGURE 2.8
Breakdown of paint protective system.

2.3.1.3.3 Fracture

Although very unusual, steel can fracture at points along the line of individual elements or at joint positions, as a result of brittle fracture, manufacturing flaws, or fatigue. The width, length, and offset dimensions of all cracks should be recorded. This can also occur in bolts or welds at connections. In particular, inspection should be made to see whether cracks due to brittle fracture have propagated after a severe cold spell.

2.3.1.3.4 Deformation

Global deformations in any direction can be measured using surveying equipment, which will reveal the presence of any unusual departure from line and level (both longitudinally and transversely).

FIGURE 2.9
Severe rusting of steel superstructure (a) at a support and (b) main superstructure.

Local deformations in the form of buckling, warping, or kinking are usually indicative of compression failure, but may also be due to vehicle impact. They should be accurately recorded, and later, the relevant members checked to see if there has been any serious loss of strength.

2.3.1.3.5 Loose fixings (fasteners)

Constant vibration can loosen bolts and rivets, a condition not easy to detect unless they are so loose that they rattle when vehicles pass! Luck is required to spot them in truss girder bridges, which have many connections, but it may be possible to check all the bolts at bearings/steel flange and cross bracing positions. Key joints should always be inspected. Even new bridges can be subject to disastrous effects due to the failure of fixings. The 480-km long multispan concrete box girder Jamuna Bridge in Bangladesh is a case in point (Parker, 1998e). Two weeks before it was due to open, the fixings of the whole 4.8 km length of a 750 mm gas pipe hung from beneath the cantilever section ripped away as it was being filled with water in preparation for hydraulic testing! The pipe disappeared beneath the waters of the Jamuna river.

2.3.1.3.6 Accidental damage

These have the same vulnerability as for concrete beam and slab bridges where vehicle impact can severely distort the bottom flanges of the girders (Besem, Woulters, & Warnon, 1990).

2.3.1.4 Masonry arches

The arch barrel is the key component as it is the main load-carrying element, but the spandrel walls and the support for the springings must also be inspected.

2.3.1.4.1 Ice

This can cause problems if the arch is constructed of soft stone or weak bricks. Water percolates through gaps in the joints (or even through the masonry itself) and if it freezes, the water expands into ice, exerting pressure on the masonry, thus causing it to break up by spalling and splitting.

2.3.1.4.2 Vibration

Arch bridges are invariably old, and most were substantially built with high reserves of strength, thus enabling them to carry modern traffic loads. They are not, however, so good at sustaining constant vibration, and this causes the masonry to loosen, joints to open, and cracks to form, thus weakening the arch.

2.3.1.4.3 Weathering

The constant daily variations of temperature, and the seasonal variations of ambient temperature, rain, and wind can seriously weaken the structure by gradual erosion of the masonry. Arches constructed of 'soft' rock, such as sandstone and jointed with lime mortar, can suffer to such a degree that often an outside layer of rock has eroded leaving the hard lime mortar joints still in place! This extreme type of degradation can be arrested by the use of clear epoxy resins or silicon compounds.

2.3.1.4.4 Cracking

This indicates the presence of tensile forces as a result of heavy traffic, vibration, foundation failure, temperature changes, or wetting and drying. Water can percolate into the cracks, and if airborne seeds enter then vegetation springs up, further widening the cracks and weakening the arch. Superficial cracking due to environment changes is not serious, but cracks that are deep and wide (5 mm plus) must be dealt with. All cracks must be recorded showing their location, extent, and width.

Cracking may also be observed at the edges of the arch barrel between rings of masonry (often termed ring separation) and probably means that the bridge has been overloaded. It is a serious defect and the position, length, width, and (if possible) the depth of each crack should be recorded – see Figure 2.10.

2.3.1.4.5 Bulging

Because of the constant passage of traffic, the fill between the spandrel walls is increasingly compacted and exerts horizontal earth pressures on the walls causing them to bulge between tie rods (if they exist). The extent of the bulge (location and distortion) should be measured on a grid and recorded.

2.3.1.4.6 Water

Most masonry arch bridges in existence were not constructed with a waterproof membrane at road level, therefore water can freely percolate through the surfacing into the fill, and if drainage has not been provided (and in most cases it has not) it collects in the space between the abutments and the top

FIGURE 2.10

Example of cracking between the arch barrel and spandrel wall (cracks highlighted for clarity).

side of the arch barrel. This water not only increases the loading on the bridge, but it can saturate the arch masonry (unless it is igneous such as granite) and seriously reduce its load-carrying capacity and even cause local crushing at the springings. Inspection consists of checking the line and level of the surfacing, and if it is rutted or undulating it could signal that moisture beneath has softened the fill and caused the deformation. The moisture content of the bricks on the intrados at the springings can also be recorded, and a record of any leaching of salts from the fill and the bricks in the form of staining, encrustation, and stalactites noted – see Figure 2.11.

2.3.1.4.7 Poor workmanship

This can result in a bridge with no ties between spandrel walls where they were needed, poor pointing of the joints, poor quality and badly coursed masonry, and lack of drainage for water both behind the abutments and in the fill zone.

2.3.1.4.8 Accidental damage

Arches are inherently strong and certainly in the case of arches over roads if they suffer a vehicle impact from below, it is normally the vehicle that sustains most of the damage while the arch is left relatively unscathed.

Vehicle impact on the road over an arch results in damage to the parapet. Generally, traditional masonry parapets are able to contain a 1.5 t vehicle travelling at 60 mph at an angle of incidence of 20° (Galloway, 1996), although some masonry may be displaced. The parapet type and extent of the damage must all be recorded. Photographs are particularly useful in such instances.

FIGURE 2.11
Leakage deposits on the barrel of an arch bridge (courtesy A.S. Glover).

2.3.1.5 Timber bridges

The two primary enemies of timber bridges are *decay and insect attack*. It is possible to tanalise the timber prior to construction but this only protects the outer layers leaving the heartwood vulnerable. It eventually breaks down and the piece has to be renewed. The symptoms to look for are the following.

2.3.1.5.1 Decay

This is a natural process caused by the presence of *fungus*. It causes the timber to soften and become pulpy resulting in a loss of strength (Table 2.2). It can be discovered by the presence of staining, soft spots identified by pushing a metal spike into the wood, and the sight of fungus growing on the wood. Particularly vulnerable are areas such as fixings, splices, parts in contact with ground at the support points, and points on the deck where water and debris can accumulate. Any shakes or splits in the wood can also trap water and encourage decay.

Table 2.2 Diagnosis of Fungi Causing Timber Decay

Fungus	Effect on Timber
Merulius lacrymans	Rotted wood shrinks, splitting into cubical pieces by deep cross-cracking. Generally occurs in damp, not wet, positions.
Poria vaillantii	Rot similar but less widespread than that produced by *Merulius*. All require more moisture than *Merulius*.
Coniophora cerebella	Causes darkening of the wood and longitudinal cracking, little cross-cracking. Usually found in very damp situations (mc > 20).
Paxillus panuoides	Causes a characteristic yellow discolouration in the attacked wood, which finally becomes red–brown.
Lentinus lepideus	Causes an internal cubical rot with a strong aromatic smell.
Trametes serialis	Causes 'dote' in the form of small pockets of brown rot. Usually found in timber from America.

2.3.1.5.2 Insect

Grubs of the anobiidae or lyctidae families of beetles and termites that bore into the wood forming a network of 'tunnels' can seriously weaken it. The visible evidence is the presence of holes on the surface of the wood above ground or water level. A few holes will not pose a problem, a large number of holes will. Below ground or water level, it is more difficult to assess.

There are three standard tests that can be used for diagnosing insect and fungal decay:

- **The Spike Test**
  As the name suggests, this test uses a metal spike that is pushed into the wood at random and suspect locations to check for the presence of soft spots. The spike is square in section so as to distinguish the spike marks from insect holes.

- **The Hammer Test**
  This identifies areas of decay inside the timber. In very badly infected areas, the impact will split the outer crust of wood and become embedded in the pulp beneath; in less severe cases, the impact will be 'thud-like'; and in areas of sound wood, there will be more of a 'ring' on impact.

- **Drilling Test**
  Using a 5-mm diameter drill, a hole is made in a suspect area, and the change in pitch and 'feel' as the drill passes through hard to soft wood indicates decay. All holes should be filled with a preservative and plugged.

Many of the problems due to decay and beetle attack can be avoided by the use of timber classed as *very durable* (Bateman, 1999) such as Opepe, Alformosia, and Balau.

2.3.1.5.3 Loose joints

The members of a timber bridge are often connected at the joints using steel bolts and plates. The bolts may work loose due to vibration from traffic or shrinkage if the timber has not been properly seasoned. The use of *toothed-plate connectors* will counter the first of these effects. The locations of loose bolts should be logged and a note made of the severity or otherwise of their looseness.

2.3.1.5.4 Deck separation

Some timber bridges drive their strength from composite action of the timber deck nailed directly to the top chord of the stiffening trusses. The integrity of the deck/truss connections should be checked by looking for gaps between the deck and the trusses.

2.3.1.5.5 Corrosion

This occurs at the joint positions and is indicated by staining on the surface of the wood. All such areas need to be logged, and, if necessary, some bolts will have to be removed to check the severity of the damage.

2.3.1.5.6 Accidental damage

Timber is not a ductile material, and so vehicle impact normally causes splitting and shattering of the wood and tearing where members are joined at the connections.

2.3.1.6 Suspended bridges

The inspection of steel and concrete in cable stayed and suspension bridges is the same as for that given in sections 2.3.1.1, 2.3.1.2 and 2.3.1.3. There are, however, additional considerations that relate to bridge geometry and type of damage.

- The *cables* are made up of individual wires that can break due to corrosion and if localised deterioration of the cable's protective coating is discovered, it may be prudent to inspect the individual high-tensile wires immediately below the protective soft wire wrapping. This can be accomplished by completely removing the wrapping over a specified length, and if there is no apparent damage, cleaning, and rewrapping.
- The *cable bands* in suspension bridges connect the tops of the suspenders/hangers to the cables. If rust staining is discovered at or near the cable/clamp boundary, then it may be necessary to unclamp a bolt at a time to check its condition. Also, if there is evidence of slippage, this must be recorded as it may mean that there is insufficient axial force in the clamping bolts to provide the necessary friction to resist the horizontal cable forces. Older bridges have chains with vertical tie rods connected to both the chains and the stiffening trusses by cottered pins. These must be inspected for corrosion and integrity.
- A survey of the *vertical deflections of the main span and the horizontal movements of the towers* should also be made under *dead loads only*. These measurements are then compared with results from a non-linear computer analysis corrected for temperature effects and any changes brought about due to strengthening of members, settlement of the foundations, drag of the anchorages, replacement of pins and saddles, and cable slippage. When the computer results are close to the measured ones (incorporated within normally acceptable limits), the dead load stresses can be computed and added to the stresses from critical live load cases. A picture of the 'real' behaviour of the bridge then emerges.
- The embedded steel in the anchorages and the inside of the spun cables of suspension bridges are difficult to inspect, but according to

Birnstiel (1990a, b), in the New York City, the inspectors drive wedges between individual cable wires to expose the inside and enable viewing. They can also be visually examined externally where they enter the sleeves leading into the anchorage chamber. Any signs of fretting should be recorded.

- The bottom connection of hangers to the decks of suspension bridges should be inspected for fraying and pitting of the wires. Recently, all of the 192 hangers in the 35-year-old Forth Road Bridge were replaced following the discovery of such damage, which in some cases amounted to 25% understrength (Haywood, 1999).

2.4 ACCESSORIES

Expansion joints, bearings, parapets, deck waterproofing, and drainage comprise the accessories on a bridge.

2.4.1 Expansion joints

The generic expansion joint types currently in use in the United Kingdom are buried, sphaltic plug, nosing, reinforced elastomeric, elastomeric in metal runners, and cantilever comb or tooth. These are all described in the Manual of Bridge Engineering (Thayre, Jenkins, Broome, & Grant, 2000).

Expansion joints are probably the most vulnerable parts of a bridge, exposed as they are to direct impact loads from traffic, effects of the environment such as severe temperature changes, UV light from the sun, and (in some countries) salt-laden water in the winter months. Add to that degradation from animal faeces and urine on farm access bridges, and vandalism, and it is small wonder that the average service life of such joints is about 10 years (Price, 1984).

Joints should be inspected for the following.

2.4.1.1 Tracking

This is one of the most common faults and is directly linked to the traffic flow over the joint. This is usually worst in the nearside lanes where there is a greater percentage of heavy vehicles. It can cause severe distortion of the joint and a tendency for asphaltic joints, for example, to be plucked out. Areas of tracking need to be accurately recorded – see Figure 2.12.

FIGURE 2.12
Example of tracking and cracking.

2.4.1.2 Cracking

This can appear in all of the types listed above, both in the actual elastomer itself and at the joint surfacing interface. Flexible decks that have a relatively high rotation tend to be worst affected. All cracked sections should be marked on a plan and photographed, if possible.

2.4.1.3 Debonding

This is where the joint material separates from the adjacent surfacing, and is closely associated with cracking. In severe cases where there is substantial cracking, it is possible for the joint to separate from the substrate posing a danger to users. Elastomeric strips can also tend to fill with debris when in the expanded condition, which can cause damage to the strip and corrosion of the metal rails.

2.4.1.4 Excessive or limited movement

If the joints have been incorrectly set at the construction stage, then they may overstretch in cold weather (causing cracking and/or tearing, and constituting a hazard to traffic) or jam tight in hot weather (causing crushing of the surfacing and deck) (see Figure 2.13a). Inspections should be

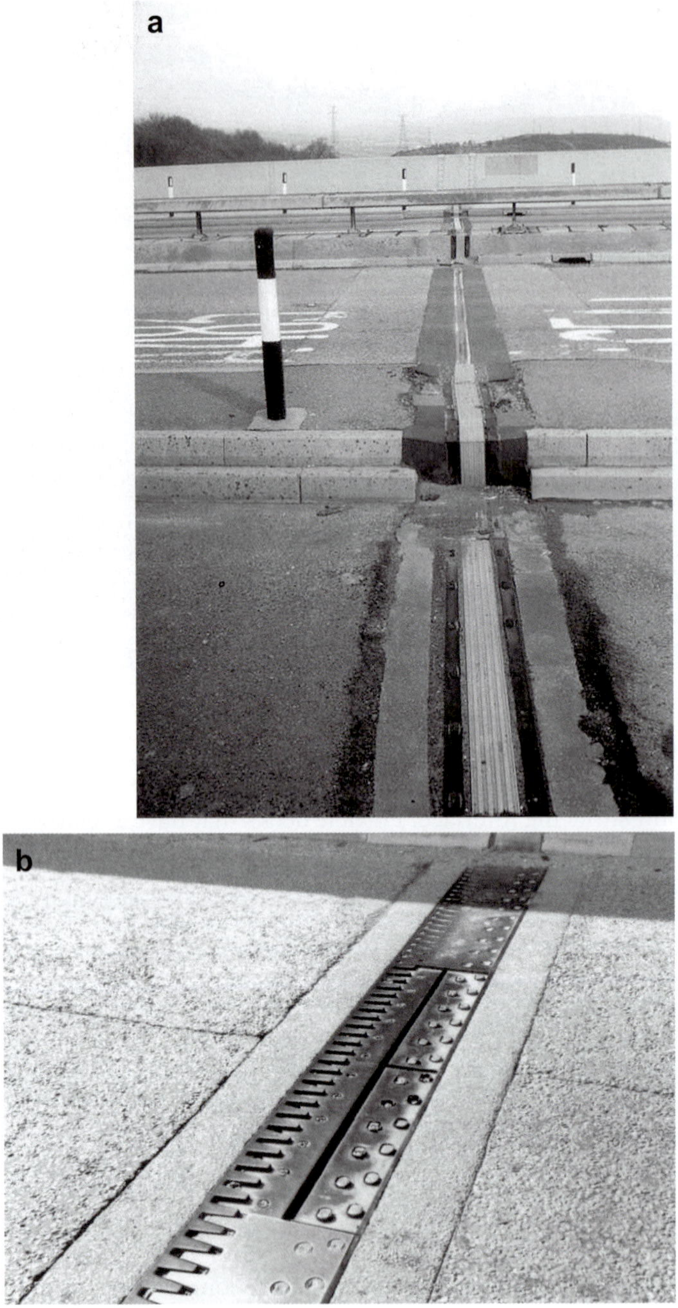

FIGURE 2.13

(a) Break-up of a deck due to a malfunctioning expansion joint; (b) cover plate missing due to poor installation.

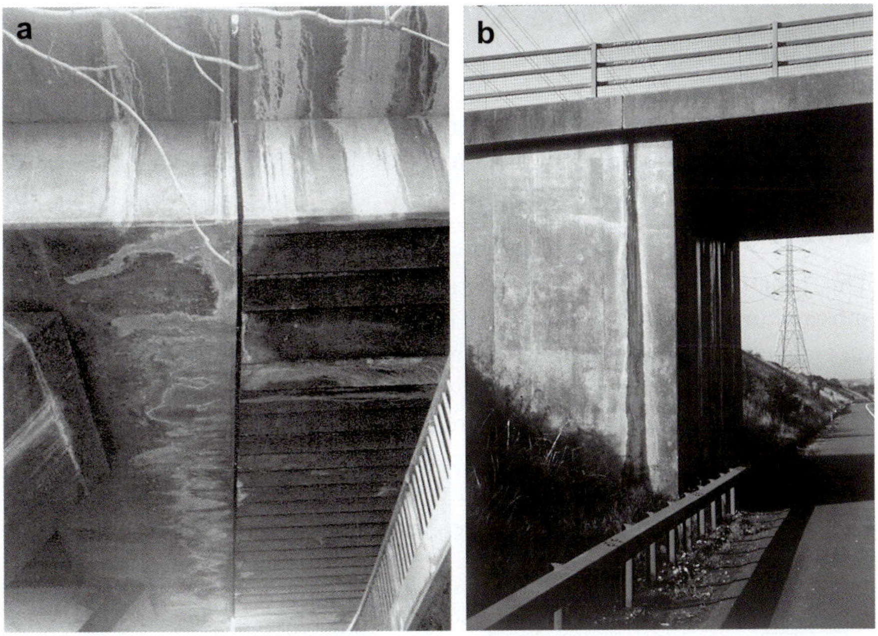

FIGURE 2.14
(a, b) Example of leaking joints.

timed to coincide with such extreme conditions if possible. The joint may also malfunction because the design specification was incorrect – see Figure 2.13b.

2.4.1.5 Leakage

This can occur for a number of reasons, the most common of which are

- Faulty or poor central reserve, kerb, or footway details
- Overstretching of the insert
- Seepage of water under the cast-in rails
- Seepage of water through or under the joint where it was bonded to the deck suggesting either that the secondary waterproof membrane had failed or that it did not exist in the first place.

Leakage tends to be localised, and this is borne out by the many unsightly streaks and stains down abutments and piers (see Figure 2.14). All areas should be photographed and recorded.

2.4.1.6 Misaligned joints

If the joint has been incorrectly set vertically so that the joint is standing proud or below the surfacing or the surfacing each side of the joints has compressed due to impact forces from tyres, then the riding quality is impaired. A level survey should be made of the mismatched areas.

2.4.2 Bearings

The function of bearings is to transmit loads from the deck to the substructure and to accommodate both translational and rotational deck movements. If for any reason these are impaired, then there is the danger of damage to the deck and supporting structure. The most common different types of bearing are shown in Figure 2.15. The generally lighter and more widely used ones are *plane sliding, curved sliding, pot, disk, and elastomeric (both strip and pad) bearings. Other more heavy duty and robust types include pin or knuckle, roller, rocker, leaf, and link*. All of these types are

FIGURE 2.15
Different bearing types (Lee, 1994).

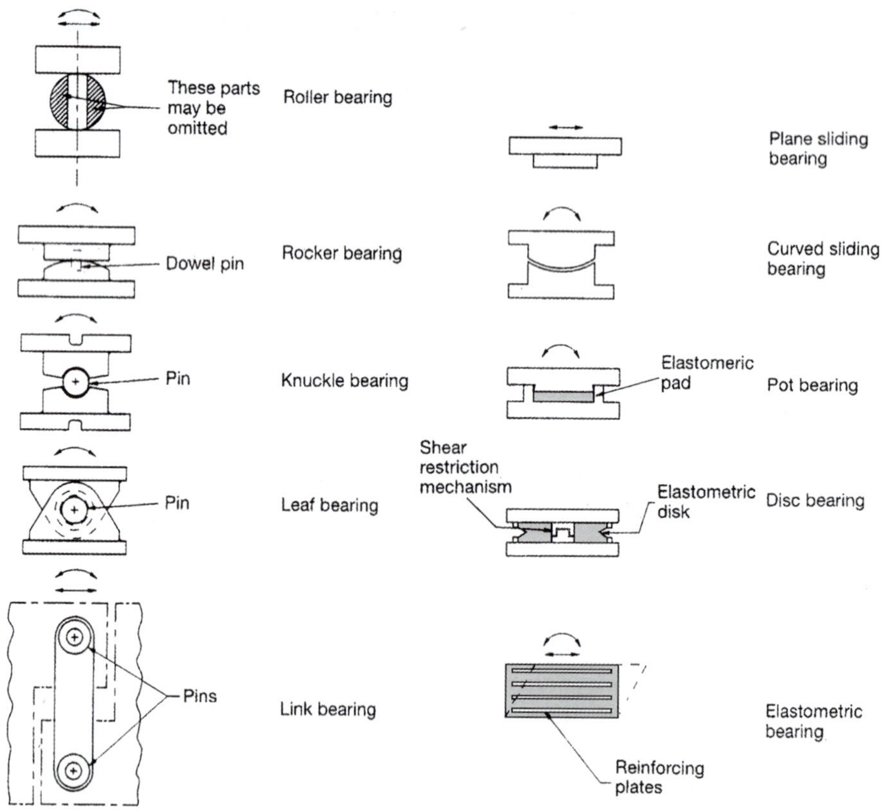

described by Lee (1994). Metal bearings are fixed in the vertical plane, but in the horizontal plane they may be fixed (no translational movement), guided (unidirectional movement), or free (universal movement). Neoprene (elastomeric) bearings (strengthened with steel plates) can accommodate limited rotation and movement in the horizontal and vertical planes.

A thorough inspection should record the following:

Bearing shelves

Debris, bird droppings, and vegetation, can not only damage the bearings but block the drainage channel causing areas of standing water and dampness. They also absorb and hold water providing an ideal environment for corrosion if close to a bearing.

Bearings

Steel sole plates and bearing plates should be checked for corrosion, which may result in high frictional forces in sliding bearings causing additional forces on the substructure. Rollers and rockers can lock due to corrosion, which will prevent movement and result in higher peak moments at internal supports and increased horizontal forces at abutments, resulting in spalling and cracking. Neoprene bearings are subject to tearing, bulging, and cracking, particularly in hot climates, which affect the properties of the elastomers and, in extreme situations, this can cause complete failure of the bearing.

Laminar or spherical bearings utilising stainless steel against polytetrafluoroethylene (PTFE) at movement planes must be kept completely free of dirt if they are to work properly and usually have a protective skirt or other covering. These should be checked to ensure their integrity.

If bearings are discovered to be loose, the cause should be ascertained during the inspection. This may be due to settlement of the substructure, excessive vibration from traffic, loose bolts, or severe corrosion. Some malfunctioning bearings are shown in Figure 2.16.

2.4.3 Waterproofing membrane

The waterproofing membrane may be hand laid sheets or sprayed liquid and is protected by a (20 mm) sand/asphalt carpet over which there is a regulating layer and finally, the surfacing, as in Figure 2.17.

2.4.3.1 Construction

Damage to the membrane can only be detected when leakage has taken place and is identified by damp patches or areas of crystallised salt deposits

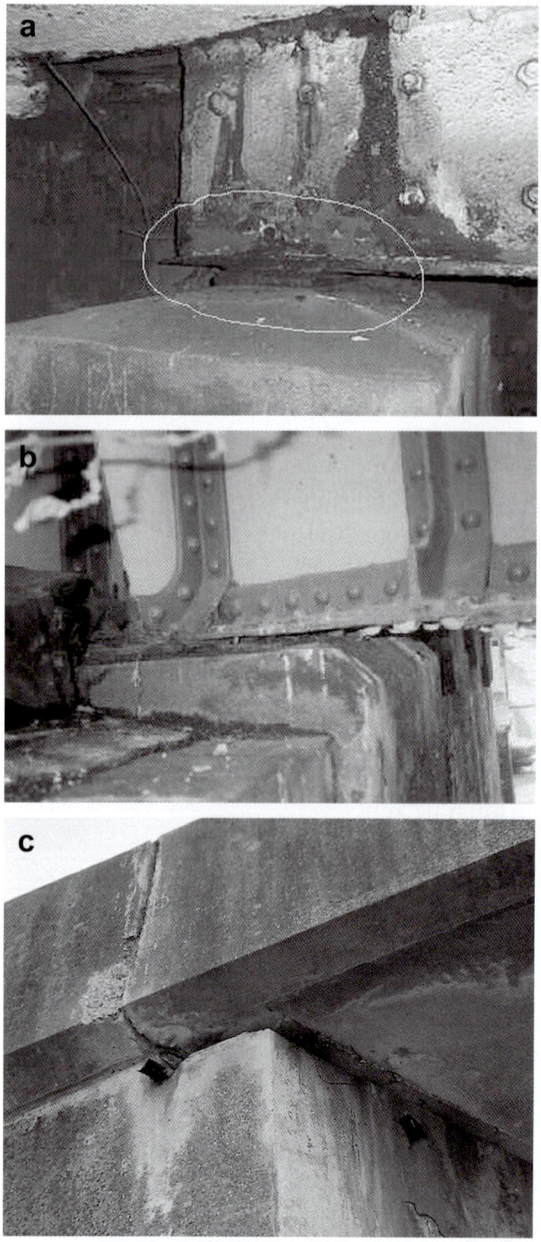

FIGURE 2.16 'Failing' bearings.

on the underside of the deck. The reason for the leak is most likely to be poor construction in that the membrane was not properly adhered to the concrete or that it was pierced by some plant or machinery during construction.

2.4.3.2 Surfacing damage

This is evidenced by undulations in the surfacing at kerbs and expansion joints and is due to traction or braking forces that cause movement of the layers above the waterproof membrane.

2.4.4 Drainage

'Keeping the water out' or away from vulnerable parts of the bridge means that the drainage system should be kept free from blockages and fracture. Inappropriately placed outlets can lead to water blowing back on to the structure causing staining. The inspection should check for the following:

Standing water

This can indicate that falls (slope) on the surfacing are not as designed (possibly due to resurfacing) and the levels should be checked to ensure that water will flow to the gullies. It can also indicate blocked pipes and/or gullies, in which case rodding will be necessary to restore the system.

Steel box and prestressed concrete box structures should be checked to ensure that there is no standing water within the box. If there is, the low-point gullies should be investigated for blockage.

Damp areas or staining

The pipe runs and gully positions should be determined, and if damp areas are in evidence along the route then it is probable that they are leaking either through a damaged joint or a fracture. If scupper-type outlets are positioned over the abutment seating areas and the abutment gullies are blocked, then staining will be in evidence on the face of the abutment where the water has overflowed. Also, if scupper outlets have been constructed with their ends flush with the soffit of the deck slab, water will spread out over the concrete and cause staining.

Inadequate capacity

If the number and size of the drainage runs are not able to cope with heavy rainfalls, then ponding is likely and debris will collect around the gullies.

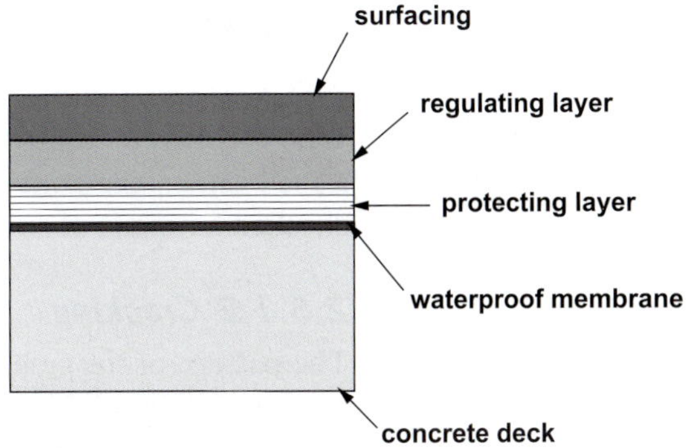

FIGURE 2.17 *Make up of a typical deck topping.*

2.4.5 Parapets

These may be of steel (open post and rail system) or a (closed) precast concrete system. The inspection should record the following:

Corrosion

The parapets are exposed to the aggressive environment of salt water splashed up by passing vehicles in wintertime. This can cause a breakdown of the protective system on steel parapets and subsequent corrosion of all parts including the holding down bolts and base plates. It can also destroy the normally passive environment of reinforcement in concrete parapets causing staining and spalling.

Accidental damage

Vehicle impact damage should be carefully recorded noting (for steel and masonry) the length of parapet damaged and the extent to which it has been laterally displaced. Concrete parapets rarely suffer extensive impact damage, but may have sustained grooving and pitting.

2.5 SUBSTRUCTURES

The substructures include all of the bridge elements below bearing level such as *abutments, wing walls, and piers*. They are generally of concrete and masonry and exhibit defects similar to those found in bridges of the same materials.

2.5.1 Abutments, wing walls, and retaining walls

2.5.1.1 Corrosion

Where concrete abutments are positioned adjacent to the carriageway of a road under, the zone just above ground level (the splash zone) should be checked for signs of corrosion and concrete degradation.

2.5.1.2 Cracking

The pattern of the cracking to the compression faces of concrete abutments and wing walls will give a clue as to its cause (see section 2.3.1.1). Shrinkage cracks are usually vertical or near vertical and propagate from the base (where the concrete is restrained) up towards the top of the stem where they tail off. The diagnosis of ASR cracking is described by Palmer (1992). Horizontal cracking could indicate reverse bending due possibly to compaction of the soil by traffic on the approach to the abutment. Vertical cracking near the top of abutments could indicate that there is insufficient reinforcement to contain the local bursting stresses at bearing positions.

2.5.1.3 Concrete erosion

The first case of *thaumasite sulphate attack* was discovered in the United Kingdom in the below ground section of the columns of the Tredington–Ashchurch overbridge near Tewkesbury, where it was found that in places the concrete had been eroded down to the reinforcement! (Parker, 1998b). Thus, it is potentially a very aggressive form of degradation, because 'the calcium silicate cementing phases are destroyed, leading ultimately to complete disintegration of the concrete' (Sims & Hartshom, 1998). Fortunately, it is still very rare and is found in wet soils containing sulphates in a cold environment (<15°C) and explains why foundations are particularly vulnerable. Concrete using limestone aggregates suffers the worst damage. Such a situation warrants a special inspection, and soil samples and concrete cores will need to be taken to a laboratory to undergo a range of petrographical, chemical, and mineralogical analyses.

Some work was recently published by the Building Research Establishment in the United Kingdom (Crammond & Halliwell, 1999) on laboratory-prepared concretes subject to 1 or 2 years of exposure to $MgSO_4$ (magnesium sulphate) solutions. They found that concretes made with carboniferous, Jurassic, or magnesium limestone were *much more susceptible to the thaumasite form of sulphate attack at 5°C* [cold] *than their siliceous aggregate controls*. The Thaumasite Expert Group (1999) has recently reported on the problem and describes the mechanism of the thaumasite attack in detail, the risk factors that need to be present for the problem to take hold, and, finally, the structural members that are most vulnerable.

2.5.1.3.1 Drainage

Any drain pipes and weep holes should be inspected for blockages. Blocked weep holes can cause ground water to back up behind abutments and wing walls exerting pressures not allowed for in design.

2.5.1.3.2 Masonry

The common defects are listed in section 2.3.1.4.

2.5.1.3.3 Line and level

Retaining walls and headwalls should be surveyed for line and level. Departure from design values could indicate an unexpected build-up of pressure from behind the walls or failing foundations.

2.6 FOUNDATIONS

These are often difficult to inspect because they cannot be seen, and signs of weakness or failure are not readily manifest.

2.6.1 Scour

The effect of *scour* is an obvious example but all too often the problem is discovered after the event has occurred (an example of breakdown maintenance). In Chapter 7 just such a situation is reported by Kittinger and Ashabar (1993) where displacements under one of the piers to the Inn Bridge in Kufstein, Austria were measured as 1.15 m on the downstream side and 0.22 m on the upstream side. Another example is that of the Kingston Bridge, Glasgow (Carruthers & Coutts, 1993) where, following the discovery of 'out of plumb', a flattened main span, a bulging quay wall, concrete crushing at joints, and locked expansion joints, the behaviour of the bridge was 'called into question'. Expensive refurbishment was put in hand to remedy the situation.

Both examples illustrate that regular underwater inspection and regular line and level surveys could well have prevented expensive remedial works. One method employed for the inspection of scour holes in river beds adjacent to piers and abutments is *impulse radar* (Millard & Thomas, 1995). It is virtually identical to the ground penetrating radar (GPR) described in Chapter 4, section 4.2.11, but as yet has not been found to be reliable enough to use confidently.

For scour, in particular, prevention is better than cure (see Chapter 11, section 11.6), and ways of monitoring the extent of scour are what is needed on all bridges where scour of any kind is a real threat be it progressive degradation, general scour, or local scour (Meadowcroft & Whitbread, 1993). Scour is revealed at low tides or periods of low flow, and it is at these times that most information can be gathered.

2.6.2 Other causes

Other causes of movement are mining subsidence, 'soft' spots in the subsoil, seasonal changes in the water table, and, more drastically, earthquakes. Such movements cause cracking and spalling of the concrete, and unusual movements of bearings and expansion joints. Inspection in such cases requires *the use of accurate surveying equipment* to establish the change in geometry of the structure, followed by a close examination of any structural damage and material degradation. Bridges built over water pose a particular problem because of the difficulty of access. At a pier position, for example, it is necessary to inspect above the water line, the splash and tidal zones, and below the water line. The latter requires diving equipment both to inspect and take material samples for laboratory examination.

If settlement is a problem, boreholes can be made to examine the current nature of the subsoil and to compare it with that assumed in the design. Appropriate remedial works can then be put in hand.

2.6.3 Piles

These may be timber, concrete, or steel. Experience has shown that for *all* types, the length of pile actually buried within the soil mass suffers very little degradation (primarily due to the lack of oxygen). *The critical points are at the mud line and water line*.

Steel suffers from corrosion due to the presence of water and oxygen, which induces a form of bacterial attack known as accelerated low water corrosion (ALWC) that can eat away steel at a phenomenal 5 mm per year (Parker, 1998a, c, d). During an inspection 'oxygen and free iron levels can be measured and bacterial activity established'. Concrete piles can suffer normal corrosion damage in the form of cracking and spalling, and from thaumasite sulphate attack (see section 2.5.1). Timber piles are at risk from insect infestation and wet rot.

2.7 MOVABLE BRIDGES

Movable bridges such as swing, vertical lift, and bascule, retractile, and transporter suffer from the same ailments as fixed bridges, but, in addition, deterioration of the machinery needs to be considered during the inspection. Movable bridge terminology in the United States differs from that in the United Kingdom and has changed during the last century. Accepted nomenclature may be found in the work by Hool and Kinne (1943). Because of opening and closing cycles, the machinery is subjected to wear and the superstructures of some types of movable bridges are subjected to reversals of self-weight stresses. Furthermore, during operation, the self-weight causes dynamic stresses in the superstructure as well as the machinery, and these can result in fatigue cracking. Hence, movable bridges have special inspection requirements. In the United States, visual inspections of the structure and machinery controls are conducted biennially as part of the inspection programme mandated by the Federal Government (Birnstiel, 1990a). In the New York State, visible machinery and controls and some hidden components are inspected along with an electrical performance test. Birnstiel (1990a) lists the work items as *bridge operation, performance tests, span drive machinery, stabilising machinery, electrical system, and traffic control devices*. Detailed listing of devices to be inspected is given in the Bridge Inspector's Manual for Movable Bridges (FHWA, 1990) and the AASHTO Movable Bridge Inspection, Evaluation and Maintenance Manual (AASHTO, 1998).

Every 6 years, in-depth machinery inspections are performed in the New York State during which nearly every mechanical and electrical component is inspected and measurements are made to determine wear of mechanical parts and deterioration of major electrical components. During these inspections, special attention is paid to parts that experience has indicated may develop fatigue cracks. The imbalance of bascule and vertical lift bridges is often measured using the dynamic strain gauge technique. In addition to determining imbalance, an examination of the torque–displacement curves often reveals improper operation of limit switches and brakes. The use of such a test in diagnosing a machinery failure has been presented by Birnstiel (1990b).

Further reading of White, Minor, and Derucher (1992) is recommended for more detail.

2.8 INSPECTION EQUIPMENT

2.8.1 Basic equipment

Safety is of paramount importance during a bridge inspection, especially where parts of a bridge are not safely accessible. In many instances, it is necessary to employ the services of specialists adept at climbing or abseiling.

The equipment required will vary from bridge to bridge, but there are some items common to all, such as given below.

2.8.1.1 Safety items

Safety helmet, fluorescent waistcoat, sturdy boots (preferably with steel caps), waterproof clothing, road cones and other traffic warning signs, first aid box, and life jacket.

2.8.1.2 General equipment

Clipboard with waterproof covering, camera, binoculars, measuring tapes, thermometer, ladder, hammer, chisel, timber borer, mirror, wire brush, torch, straight edge, string line, crack width gauge (see Figure 2.1), waterproof pen or crayon, penknife, magnifying glass, trowel, shovel, and square spike.

2.8.1.3 More sophisticated equipment

Digital camera, lap-top computer, data-logger, electronic level and staff, hand-held dictaphone, mobile phone, battery-operated drill, and boat.

2.8.2 Permanent and mobile equipment

The inspection of small bridges can usually be accomplished by a 'walk-around' with access to all parts readily available and visible with the naked eye or through binoculars. The main problem is often the management of traffic during the inspection.

2.8.2.1 Scaffolding/platforms

The most common form of access is by the traditional means of ladders incorporated into scaffolding in the form a simple platform under a bridge or in the form of a tower, for example, where a high pier or abutment is to be inspected. The scaffolding tubes are anchored to the structure at regular intervals and properly braced together in all directions. Rest platforms are provided at intervals in accordance with current regulations. To provide a firm foundation, the uprights are provided with base plates that are founded directly on the ground in the case of granular soils and on timber baulks or concrete pads for weaker material. It is vital that the whole assembly is stable and strong enough to carry all the anticipated loads from wind forces and personnel.

Scaffolding can be erected by semiskilled workmen; it can follow the varying geometry of the superstructure (curves and intersections) and it is demountable and reusable.

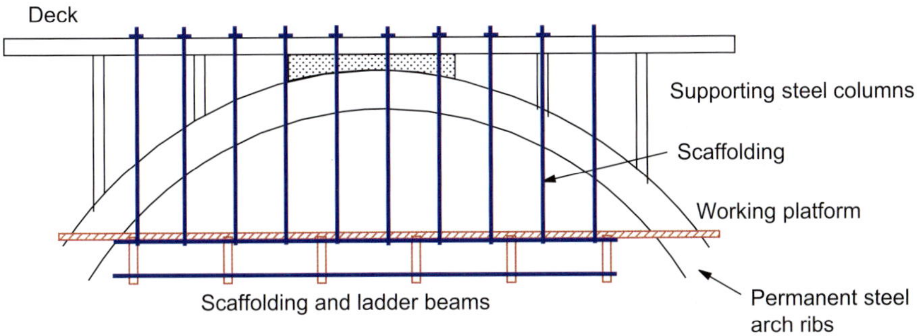

FIGURE 2.18
Access by hanging platform.

Scaffolding tubes can also be used to suspend a platform from a bridge deck to provide access to the underside. It is of particular use over heavily trafficked areas where the platform can be erected piecemeal during the early morning hours until it is complete; the whole assembly can then be made weatherproof by tarpaulins at the sides. Such a platform was provided recently to a steel arch bridge over the A3 near Guildford, UK, both to inspect and repair damaged concrete covering to the arch ribs, which was disintegrating and falling on the carriageway beneath – see Figure 2.18.

High bridges and long viaducts, however, often pose problems that make inspection using scaffolding a hazardous operation. Especially difficult is the inspection of the bearings, and the top of and the underside of a bridge. The solution is to hire specialised inspection equipment and personnel.

2.8.2.2 Truck-mounted mobile gantries

These comprise either the bucket or 'cherry picker' type of equipment or the more sophisticated underslung lattice platform. Both types are also available on a rail-mounted engine for use on railway bridges, and can be used to carry out repairs as well as inspections. Special portal-type inspection units are also available and can be fabricated to customer requirements. Typical examples of each are shown diagrammatically in Figure 2.19. The whole platform arrives folded on a truck or trailer and then is hydraulically unfolded so that the working platform extends below the deck by up to 10 m, and under the bridge by up to 21 m. The bucket type with its three articulated closed steel sections also arrives folded and can be hydraulically opened to positions up to 15 m above and 15 m below the deck, and extended under by up to 12 m (see Figure 2.20). Disruption to traffic is minimal.

2.8.2.3 Permanent gantries

Sometimes it is more economical to provide a permanent gantry operating on runways built onto the side of the bridge. This is certainly the case

FIGURE 2.19
Lattice platform and cherry picker.

for long, heavily trafficked bridges, cable stayed and suspension bridges, and some high arch bridges where the cost of providing such facilities at the construction stage is a small fraction of the total construction costs. Two different types are shown in Figure 2.21.

The Tamar bridge recently had two *aluminium* maintenance/inspection gantries installed (CEI, 1999). They are completely self-contained with their own air-cooled diesel generators for motive power as well as in situ power for tools and the operation of a telescopic personnel lift (see Figure 2.22.).

FIGURE 2.20(a)
Lattice platform for inspecting a prestressed concrete box girder bridge.

The gantries run along runway beams slung under the whole length of the bridge. They can operate independently or they can be linked together to provide a greater working area if required.

So successful have they been that a similar system is to be installed on the Avonmouth Bridge. Prior to installing a permanent inspection gantry on the bridge, the existing rails on the underside of the bridge were being upgraded using a temporary platform that was suspended from the rails and tethered at each end with steel ropes. Sadly on 8 September 1999, four steelworkers were killed when one end of the platform came off the rails in a freak gust of wind. Wind then carried the dangling platform 800 m across the Avon River where it crashed into the scaffolding around an approach span pier. The incident is under investigation, but it appears that only one end was tethered rendering it unsafe (Mylius, 1999). It serves as a timely warning that safety is paramount when planning and undertaking an inspection.

FIGURE 2.20(b) *Cherry picker in use for inspecting a concrete arch bridge.*

2.8.3 Rope access

There are situations where rope access is the only appropriate method of inspection, for example, when inspecting the main cables of a suspension bridge, or inspecting arch foundations cut into the sides of a gorge. Access wires are installed to which the ropes can be clipped, and they can remain in place for subsequent periodic inspections (Fewtrell, A. (2000). Private communication. Director of Up and Under Ltd.).

There are a number of specialist companies that have rigorous training programmes for engineers who wish to learn abseiling techniques and abseilers who wish to learn engineering inspection methods. Most of the work done by such 'rope technicians' is visual in nature, but it is possible to use lightweight equipment for some tests such as cover meters, battery-operated drills for taking core samples, cameras for close up photographs, and laser distance measuring devices. Work is carried out at the end of a rope in a 'comfort chair' to aid posture, using a tool harness around the waist. In

FIGURE 2.21
Examples of permanent gantries: (a) Kohlbrand Bridge near Hamburg; (b) Tay Bridge, Scotland.

addition, the technicians are taught to be safety conscious, and learn how to fix ropes, use ropes in tandem, the rescue techniques, the functions of all equipment, rigging techniques, and how to transfer between lines and angled ascents. Two examples of the use of rope access are shown in Figure 2.23 (Figure 2.23(b) also reproduced in colour plate section).

2.9 PLANNING AN INSPECTION

A desk study of documents relative to the bridge to be inspected is normally carried out in the office. This will include study of the bridge record card (or

FIGURE 2.22
Aluminium gantries slung under the Tamar Bridge.

electronic file), any previous inspection reports, design drawings, 'as-built' drawings, photographs, and discussions with any colleagues who may know the bridge or even have been involved in its original construction.

Prior to an inspection, a site visit is first undertaken by the client's representative and the consulting engineer appointed to carry out the inspection. The purpose of the visit is to get a general 'feel' for the bridge such as its location, type, size, and general condition, and also to decide upon the

FIGURE 2.23
(a) Inspection of a brick arch barrel (courtesy of Up and Under) and (b) inspection of Tower Bridge, London (courtesy of Up and Under) (see colour plate section).

objectives and scope of the inspection, and to draw up a plan of attack. This will ensure that the engineer's time is well spent and that the inspection is thorough. The initial 'walk-over' should establish answers to the questions such as:

- Are there any particular problems with access to any part of the bridge? If so what special equipment is required?
- Will specialist engineers be required, i.e., geotechnical, marine, structural, or materials?
- Will traffic be interrupted? If so what traffic management arrangements are necessary?
- How long is the inspection likely to take?
- Is any non-destructive testing likely to be needed, such as cover meter, concrete coring equipment, carbonation equipment? This can be established at the 'walk-over' stage if staining, efflorescence, or rusting is in evidence.

2.9.1 Documentation

A notebook or dictaphone is indispensable for recording what is seen in written and sketch form – both sound and deteriorating areas – before a standard inspection form is completed.

In the United Kingdom, there is a standard form BE11/94 that can be used for the principal inspections of trunk road and motorway structures (BA63/94) (reproduced in Appendix B). County Councils may use this or devise their own for bridges under their control. Surrey County Council has an inspection form as indicated in Table 2.3. Each structure can be divided into elements, and elements into segments. This will enable the details to be more focused in inspections and assessments. Digital photographic and/or video records can also be taken to be fed into the management programme for processing.

The data from this form are fed into a bridge management programme, which analyses the data along with data from other structures and provides a maintenance strategy in terms of priority and budgetary constraints (see Chapter 1). *It does not, however, give any indication of the strength of the bridge, only its condition*.

2.9.2 Inspection by segments

The division of bridge elements into segments for the purposes of inspection is an idea that was referred to in section 2.3. Each segment can be coded or

Table 2.3 Typical Inspection Form

Inspection Date ________ **Inspection Type** ________ ________

Inspected By ________

Structure Number	Structure Name	C	T	P	M			C	T	P	M
1	Inverts					22	Toughing				
2	Aprons					23	Jack arches				
3	Foundations					24	Superstructure drainage				
4	Cutwaters					25	Waterproofing				
5	Piers or columns					26	Surfacing				
6	Fenders					27	Service ducts/fixings				
7	Wing walls					28	Expansion joints				
8	Retaining walls/sheet piling					29	Parapets/pilasters				
9	Revetments/batter paving					30	Arch springing				
10	Abutments					31	Arch ring				
11	Approach embankments					32	Voissoirs/arch face				
12	Training walls					33	Spandrels				
13	Substructure drainage					34	Tie rods				
14	Bearings					35	Tunnel portals				
15	Crossheads					36	Masts				
16	Main beams					37	Catenary cables				
17	Transverse beams					38	Tunnel linings				
18	Diaphragms or bracings					39	Armco/concrete pipes				
19	Concrete slab					40	Access gantries/walkway				
20	Timber deck					41	Machinery				
21	Metal deck plates					42	Warning signs				

C	Condition Factors	Treatment		Priority
1	Good, <5% defective	R	Repair	H High
2	Fair, 5–20% defective	C	Replace	M Medium
3	Poor, >20% defective	P	Repaint	L Low
4	Very poor major defects Still standing	M	Monitor	
		I	Inspect	
5	Immediate attention			

numbered and marked on an electronic or paper drawing. During an inspection, the rating of each segment is marked on the drawing and the information fed into the BMS. A very detailed record is therefore built up of every part of the bridge and is of particular use where there are different exposure or environmental conditions over a single element. An obvious example is where a bridge spans over a railway line, a road, and a river, and

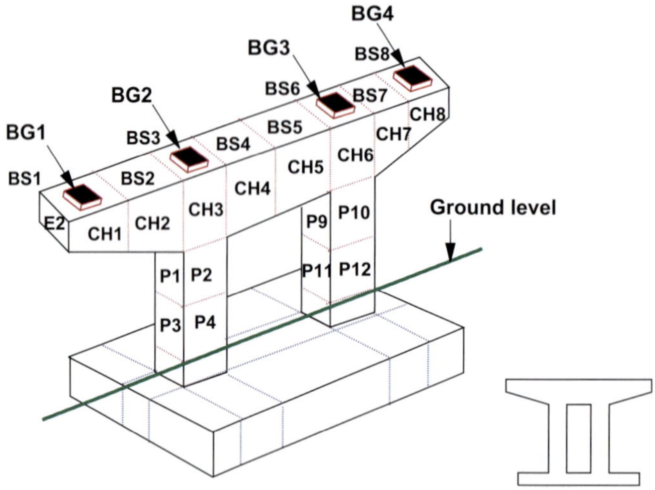

FIGURE 2.24
Example of segmental referencing.

the soffit of the deck is subject to the effects of diesel fumes over the railway, carbon monoxide over the road, and, possibly, high humidity over the river. Also one edge of a deck might be south facing and subject to severe heating and cooling, while the opposite side does not suffer such extremes. Another example is that of an abutment or pier, part of which is above ground level and part below ground level, and part which is directly supporting bearings. Figure 2.24 illustrates the idea. Clearly, it is not possible to examine every part of a foundation below ground level, but if 'as-built' drawings are available, at least every part is referenced should the need arise to inspect.

Hearn (1998) describes a similar method of segmental inspection in relation to a three-span beam-and-slab bridge shown in Figure 2.25.

2.10 CONDITION RATING

One of the words that has particular resonance in the bridge fraternity is the word *condition*. During an inspection, an attempt is made to determine the *condition* of an element based on the subjective opinion of qualified experts. It is usually in the form of a number. No calculations are performed, but the numbers are later used in conjunction with other factors to calculate the *condition rating* (also as a number), which is then further used to decide on the maintenance strategy. Some countries of the world have developed ways within their BMS of providing an assessment of the 'condition' of their bridges in an attempt to prioritise them within the constraints of repair work necessary and a limited budget.

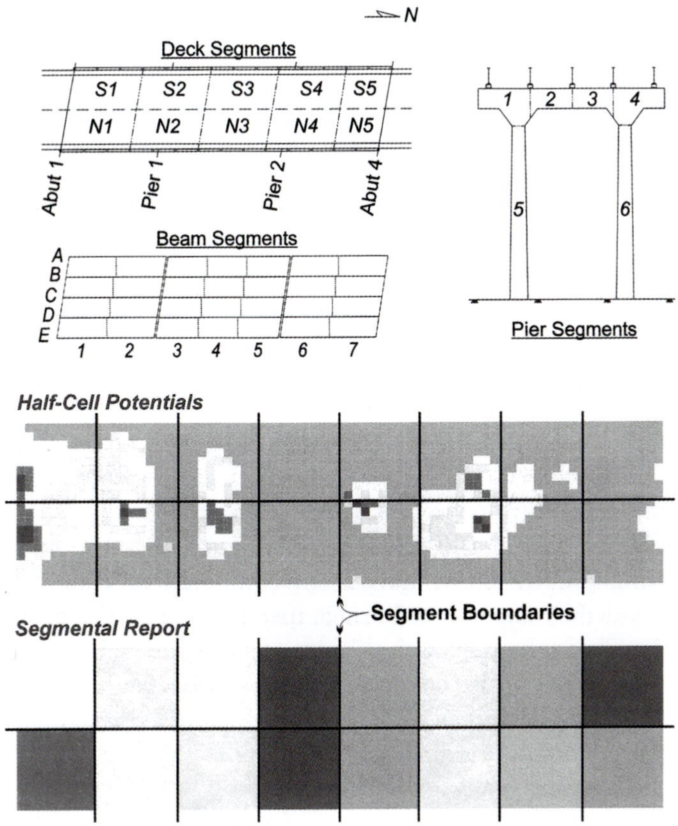

FIGURE 2.25
Segmental inspection of bridge elements (after Hearn, 1998).

2.10.1 Japanese experience

In Japan, for example (Yokoyama, Sato, Ogihara, & Toriumi, 1996) a list of *deficiency ratings* is specified as shown in Table 2.4. Each element of the bridge is evaluated by assigning a deficiency rating for each kind of defect such as cracking, corrosion, and deformation.

Combined with these, a *demerit rating* d_I for each component of damage is calculated from the pre-defined *deficiency rating* within the BMS as shown in Table 2.5. The demerit rating d_I corresponding to deficiency rating I for each component of damage was assigned within the BMS, and the remainder of the demerit ratings are calculated from:

$$d_{II} = d_I \times \alpha_{II} \tag{2.1}$$

$$d_{III} = d_I \times \alpha_{III} \tag{2.2}$$

Table 2.4 List of Deficiency Ratings

Deficiency Rating	Description
I	Serious damage. There is a possibility of danger to the traffic.
II	Damage in a large area. Detailed investigation is required. Following the investigation, the necessity of immediate repair work should be evaluated.
III	Damage. Follow-up investigation is required.
IV	Slight damage. Inspection data are recorded.
OK	No damage.

$$d_{IV} = d_I \times \alpha_{IV} \tag{2.3}$$

$$d_{OK} = 0 \tag{2.4}$$

For example, the d_I for corrosion of a steel girder is given as 70, thus $d_{II} = 0.5 \times 70 = 35$. The demerit rating for each structural element is tabulated, and by a process of reduction a final overall rating is arrived at and subtracted from 100 (the perfect bridge) to give the final bridge condition rating (BCR). These have been compared to condition ratings (CoRe) provided by experienced bridge inspectors, and the process is continually refined year by year with the aim of getting a calibration factor of 1.0.

Once the bridge condition factors have been determined for each bridge under the authority's care, the numbers are fed into a bridge rehabilitation planning module, which screens all of the bridges and then prepares three rehabilitation plans for each bridge:

- Repair only the most damaged part of the bridge.
- Repair all parts where the damage is less than deficiency rating III.
- Replace.

The module then calculates benefits and costs for all plans for all bridges and lists the costs in ascending order and finally calculates an incremental benefit–cost ratio for each plan. If the ratio is less than 1, it is removed. Further refinement of all plans continues until a ranking order is arrived at and continues until 'the total cost of selected plans matches the annual budget'.

2.10.2 The New York experience

In another system developed by the New York City (Yanev, 1997), all components in all spans are inspected at least once every 2 years and rated as

Table 2.5 Deficiency Ratings and Reducing Ratios

Deficiency rating	χ	I	II	III	IV	OK
Reducing ratio	α_χ	1	0.5	0.2	0.05	0

follows: 7, new; 5, functioning well; 3, not functioning as designed; 1, failed; the even numbers 6, 4, and 2 denote intermediate conditions.

The various components of the bridge are then assigned a weighting as shown in Table 2.6, given by

$$\text{Bridge condition rating (BCR)} = \sum(\text{Component rating} \times \text{Weight})/\sum \text{Weightings}$$

After examining 720 of the city's 860 bridges and correlating the BCR with worst deterioration rates, Yanev (1997) discovered the information shown in Table 2.7.

What this shows is that slowing down the deterioration rate can significantly improve the BCR by improving the performance of decks, expansion joints, scupper outlets, and drainage systems.

2.10.3 UK local authority experience

The Surrey County Council has a system in place, which aims to determine a maintenance priority number (MPN) for each bridge under its jurisdiction.

Each structure is broken down into its elements (42 for Surrey's bridges and 33 for Highways Agency bridges).

Each element is awarded a *condition value* (CV) between 1 and 5. Condition 1 is GOOD and condition 5 is CRITICAL. The CV is then converted to a *condition factor* (CF) for use in determining the MPN (Table 2.8).

Table 2.6 Component Weightings

Component	Weighting
Bearings	6
Back wall	5
Abutments	8
Wing walls	5
Piers	8
Primary members	10
Secondary members	5
Deck	8
Kerb	1
Wearing surface	4
Bridge seats	6
Sidewalks	2
Joints	4

Table 2.7 General Observations from Field Data

Primary members	Steel and concrete deteriorate at a nearly constant rate from new (7) to defunct (1) in approximately 30 years
Bridge deck	Decks with separate overlay have a minimum useful life of 40 years without joints and 30 years with joints
Bridge seats, bearings, piers, sidewalks	Ratings drop from 7 to 4 (3 for bearings) in less than 5 years Thereafter, there is a slower rate of decline to 1 after 30 years
Joints	Joints begin to fail after 10 years although experience suggests even worse performance in the field

- The location of each element in the structure is given a location factor (LF) between 5 and 10 depending upon its importance within the structure as shown in Table 2.9
- A road factor (RF) is assigned to each type of road between 9 and 14 depending upon its relative importance. Values of 9 or 10 are assigned to motorway and trunk road bridges. Values of 11 to 14 are assigned to County Roads and Rights of Way, namely
  - County Principal Roads (A roads) 11
  - County Non-principal Roads (B and C roads) 12
  - Unclassified Roads (D roads) 13
  - Rights of Way 14
- The MPN is then calculated from:

$$MPN = CF \times LF \times RF/14$$

Thus, for example, the *highest* MPN (for a non-critical element in good condition on a Rights of Way bridge) is given as:

$$10 \times 10 \times 14/14 = 100$$

Thus, a high number corresponds to a component in good condition. The lowest MPN (for a critical element in poor condition on a motorway bridge) is given as:

$$5 \times 1 \times 9/14 = 3.2$$

Table 2.8 Conversion of CV to CF

Condition value	5	4	3	2	1
Condition factor	1	2	4	7	10

Table 2.9 Location Factors for Bridge Elements

Element	Location Factor	Justification
Main beams, transverse beams, and piers	5	Main structural component. Deterioration may lead to reduced load-carrying capacity Designed for 120 years overall life span
Concrete slab, metal deck plate, and bearing	6	As above
Abutments, wing walls, and approach embankment	7	
Foundation	8	
Waterproofing, surfacing, parapets and handrails, and expansion joints	9	These elements do not contribute to the strength of the bridge deck but provide containment of traffic (safety), durability, movement, and a smooth ride
Drainage system, inverts, aprons, and service ducts	10	

2.10.3.1 Deterioration model

Each element in each structure can therefore have an MPN calculated for it, and an MPN can also be calculated for the structure as a whole or a group of structures or the whole bridge stock.

The rate of change of these numbers or the *rate of deterioration* of the bridge can now be plotted. Consider the following example (Table 2.10):

If these figures are now plotted, the resulting graph represents a *deterioration model* for that element over the given age span and represents what would happen if there was no intervention (Figure 2.26). If maintenance is carried out at 24 years giving an improvement in the value of 1 and a life of 6 years, then the MPN values change as in Table 2.11.

If these revised figures are plotted, then the improvement in life span can be clearly seen. The age at which the deterioration reaches state 4 has been moved from 26 years to 38 years, thus extending the life of the structure by some 12 years. Intervention at the right time is therefore clearly desirable.

Table 2.10 MPN Values of a Bridge Element (No Intervention)

Age of element	10	12	14	16	18	20	22	24	26	28	30	32
Condition value	1	2	2	2	3	3	3	3	4	4	4	4
MPN	47	33	33	33	19	19	19	19	9	9	9	9

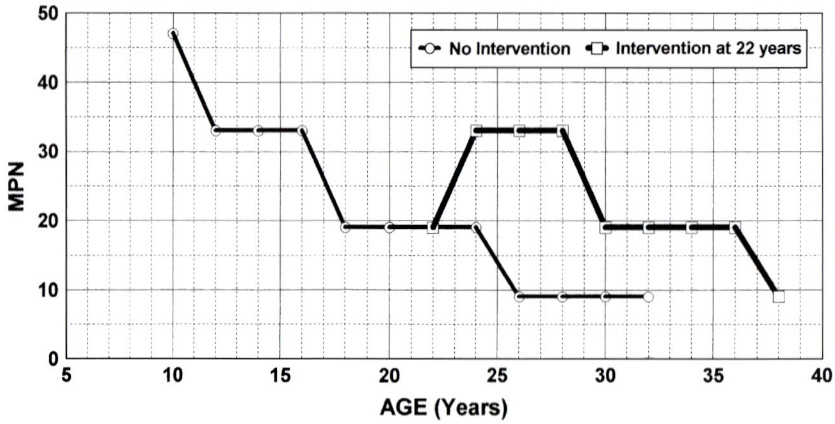

FIGURE 2.26
Deterioration model (no intervention).

2.10.4 Bridge condition index

In the United Kingdom, at national level the concept of the *Bridge Condition Index* (BCI) is being used in an attempt to provide a means of monitoring the change in condition state of a bridge element, or a single bridge within an entire bridge stock (Blakelock, Day, & Chadwick, 1998) and could be used to address the problem of prioritisation of funding. It is numerical and a 100% score is described 'as new'. It is also subjective but during trials carried out to calibrate the BCI, Blakelock et al. (1998) have introduced the idea that the BCI calculated for a particular bridge *should accord with the interpretation of condition that a reasonably experienced engineer would have in inspecting the bridge*. The BCI equation has been developed for the *United Kingdom only* and is applicable to a *large* stock of bridges. It cannot and should not be used for assessing the condition of an isolated bridge or a small stock of bridges, but it should help a bridge manager who, for example, for a number of years has spent say *x*% of his budget annually on bridge maintenance but sees the average BCI of his stock going down, to argue for more funds.

In the current County Inspection Forms (see Figure 2B.1, Appendix B), there are seven fields to record:

- Element type
- Estimated cost

Table 2.11 MPN Values of a Bridge Element (with Intervention)

Age of element	10	12	14	16	18	20	22	24	26	28	30	32	34	36	38
Condition value	1	2	2	2	3	3	3	2	2	2	3	3	3	3	4
MPN	47	33	33	33	19	19	19	33	33	33	19	19	19	19	9

- Extent (A–D)
- Severity (1–4)
- Work
- Priority
- Comment

In the new proposals, estimated cost and comments have been omitted, extent and severity have been combined, and work is omitted. In addition, the description of element type will include factors relating to *primary elements* (foundations, arch barrel, main tension boom, etc.) and *secondary elements* (see Table 2.12).

The algorithm finally arrived at was

$$BCI = 100 - F_1 \times \left[F_2 \frac{S(E_{fp} \times S_f)}{N_p} + F_3 \frac{S(E_{fp} \times S_f)}{N_s} \right] \tag{2.5}$$

except if an element has an E5 extent/severity factor, then $BCI = 0$.
where BCI = condition index; E_{fp} = element factor for primary elements; E_{fs} = element factor for secondary elements; S_f = extent/severity factor; N_p = number of primary elements on the bridge; N_S = number of secondary elements on the bridge; and F_1, F_2, and F_3 are a series of factors. If the whole structure is A1 'as new', then the $BCI = 100$ and clearly to try and maintain a bridge at that level is impossible. A limited amount of deterioration is acceptable, which implies a $BCI < 100$. An age factor has therefore been introduced so that

$$\text{Target } BCI = \text{Current } BCI \times \text{Age factor}/100 \tag{2.6}$$

Table 2.12 Bridge Condition Assessment

Old (Prior to 1984)	Current	Proposed
Element type	Element type	Element type
Estimated cost	Estimated cost	
Condition (good, fair, poor, very poor, immediate attention)	Extent (A–D) Severity (1–4)	Extent/severity (1–5) 1. No significant defect 2. Minor defects causing no damage. 3. Moderate defects that appear to be causing damage 4. Severe defects causing damage. 5. Element nonfunctional
Treatment (repair, replace, repaint, monitor, inspect)	Work	
Priority	Priority	
Comment	Comment	

Also, to take account of the 'non-reporting' of elements, a confidence factor has been introduced defined as

$$\mathrm{CF} = 100 \times \left[\frac{\text{number of elements reported}}{\text{number of elements that exist in the bridge}}\right]$$

For this approach to be successful, Blakelock et al. (1998) stress that *all* elements must be reported on whether visible or not.

2.10.5 Condition states

Another foray into the question of the collection, presentation, and interpretation of condition data has been made by Hearn (1998). The current practice in the United States is to assign CoRe from a visual inspection for use with the Pontis BMS (see section 1.5.6), but Hearn has proposed the use of 'new condition states defined as *stages* of service life'. The premise is that service life is a progression of stages and Hearn (1998) illustrates with two examples reproduced in Table 2.13 and compared directly with the existing condition states.

The CoRe rating relates only to the *amount* of damage not on the *stage* that the element has reached regardless of its age. The proposal is to use a scale of 1–5 representing the stages of service life as shown in Table 2.13, and it is clear that specific maintenance actions relate to the condition state.

For example, a reinforced concrete member has five distinct stages in its life (Hearn, 1998) as defined in row 3 of Table 2.13. A particular element may be *exposed* (no sealer), but tests show no sign of chloride contamination, therefore the best maintenance action for an *old* bridge may be simply to monitor it. For a *young* bridge it may be appropriate to take *preventative action* by applying a sealer. On the other hand, if there is sign of

Table 2.13 Service Life Stages Compared with Existing Condition States

Existing condition states

1	CoRe condition rating	1	2	3	4	5
2	Extent of spalls or area	0%	<2%	<10%	<25%	>25%

Proposed condition states

	Stages of Service Life	Protected	Exposed	Vulnerable	Attacked	Damaged
3	RC deck with sealer	Good sealer	Failing sealer	Cl contamination	Active corrosion	Spalls
4	Painted steel element	Good paint	Failing paint	Staining	Surface corrosion	Section loss
5	CoRe ratings in stages			1		2,3,4,5

contamination then action is taken on the basis of the depth of the chloride ingress. For deep contamination, this would mean breaking out spoiled concrete, chloride extraction, and concrete repair. For shallow penetration then re-application of the sealer is appropriate.

In any given bridge stock, there are bridges of *differing age,* many *different bridge types* with *different elements* in *different environments,* and at *different stages* of their service life. To enable a bridge manager to keep track of the progression of CoRe and to predict future condition, recourse is often made to a Markov Chain, which is a way of accounting for all of the different processes involved in deterioration, and is much more reliable than simple extrapolation of CoRe.

2.10.6 The Markov Chain

The typical form of a Markov Chain or Event Tree is shown in Figure 2.27, where each branch represents a condition state (or stage) and where there are a number of possible paths via probability nodes for any chosen time period. The various progression paths through the tree are called *condition state trajectories*.

The probabilities at each node are Pi,i, which is the probability of remaining in state *i* the following year, and $Pi,i + 1$ *is the probability of changing state the following year.*

The Markov Chain assumes that 'future states of the process depend only on current states, and that the rates of transition from one state to another remain constant throughout the time' (Ng & Moses, 1996). At the nodes (or bifurcation points) there are two maintenance options (although there may be more) and one of them is the 'do-nothing' option, that is, zero cost. If Figure 2.27 represents the tree for a particular element, then starting from NOW, the probability of remaining in condition 2 in a year's time is compared with the probability of deteriorating to condition 3, and so the first step is highlighted. This process is continued up to the fourth year (or any other chosen year). The thickened line represents the highest probable progression of stages of service life from the inspection date (now) to year four. The actual transition probabilities at each node are based on information gained over the

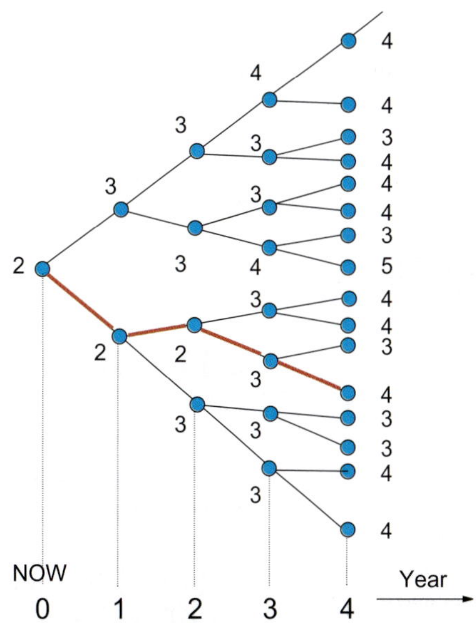

FIGURE 2.27 *Markov Chain.*

years from other similar elements in similar bridges in a similar environment and stored in the BMS database. All the possible pathways are considered and the costs of each are calculated. The cheapest is usually the option that is implemented.

The aim is to carry out the right repairs at the right time by optimising the maintenance programme for a stock of bridges so that over the life of an element or bridge the costs are minimised. This is done at the maintenance planning stage by preventing deterioration from proceeding beyond a specified target value.

For example, for the service stage method proposed by Hearn (1998) it might be decided that the element should not go beyond stage 3 (vulnerable) before intervention. For critical structural members, this might be stage 1 or 2. For a bearing, however, stage 5 might be acceptable, at which point – providing there is no major disruption to traffic – the bearing is replaced. If disruption is expected then a lower condition (or stage) value would be more appropriate. It is all a matter of engineering judgement based on experience. If it is predicted that the target value is going to be exceeded then do-nothing is not an option and maintenance must be carried out.

2.10.7 Reliability-based performance measure

Many BMSs, including Pontis, utilise the concept of the Markov Chain, but Ng and Moses (1996) have proposed an alternative approach 'based on time-dependent reliability theory which amounts to the generalisation of the Markov Chain model'. It is argued that the stochastic model does not differentiate between bridges of different ages, and that condition rating is not adequate as a performance measure as it does not reflect the structural integrity of the bridge as a whole, its improvement needs, or the load rating for the bridge. The details of the method are given in Ng and Moses (1996).

2.11 CONCLUDING REMARKS

The role of the bridge inspector is to record as accurately as possible any defects found in a bridge, to note the probable cause, and to advise on the priority level of any required maintenance work or the need for a special inspection.

Inspection is not an exact science, and many judgements have to be made as to the extent and severity of any degradation or defect. For example, when recording the condition factor on the inspection form in Table 2.3, considerations of the percentage of the bridge that is defective is partly a *subjective* judgement. Close observation is vital and, of course, accuracy increases with experience. *Remember, the inspection report provides the raw information on which important managerial decisions regarding maintenance are made commensurate with available resources*. Any intervention will involve time, money, and possible inconvenience and cost to the bridge users.

REFERENCES

AASHTO. (1998). *Movable bridge inspection, evaluation and maintenance manual.* Washington, DC: American Association of State Highway and Transportation.

Bateman, D. (1999). Protecting our piers. *The Structural Engineer, 77*(13), 16–22.

Besem, P. H., Woulters, M., & Warnon, C. (1990). The repair of a composite concrete-steel bridge (pp. 747–764). In: *First international conference on bridge management*, Guildford, Surrey, UK. Elsevier.

Birnstiel, C. (1990a). Movable bridge machinery inspection and rehabilitation (pp. 295–304). *First international conference on bridge management*, Guildford, Surrey, UK. Elsevier.

Birnstiel, C. (1990b). Operational tests of swing span with multiple independent drives. *Third Biennial Symposium of Heavy Movable Structures/Movable Bridges Affiliate*, St. Petersburg, FL, 12–15 November (Paper 3–12).

Blakelock, R., Day, W., & Chadwick, R. (1998). Bridge condition index. *The management of highway structures*. Highways Agency.

Carruthers, D., & Coutts, D. (1993). Refurbishment aspects of Kingston Bridge, Glasgow (pp. 604–613). In: *Proceedings of the Second International Conference on Bridge Management*, Guildford, Surrey, UK. London: Thomas Telford.

CEI. (1999). Safely down under. *Concrete Engineering International, 3*(3), 24.

Clark, L. A. (1996). Assessment of concrete bridges with ASR (pp. 19–28). In: *Proceedings of the Second International Conference on Bridge Management*, Guildford, Surrey, UK. London: Thomas Telford.

Crammond, N., & Halliwell, M. (1999). *The thaumasite form of sulfate attack in laboratory-prepared concretes*. Building Research Establishment (Report No. 09:50 306).

Department of Transport. (1990). BA35/90. The inspection and repair of concrete highway structures Vol 3: *Design manual for roads and bridges*. London: HMSO.

Galloway, R. M. (1996). Assessment and design of unreinforced masonry parapets (pp. 696–703). *Proceedings of the Third International Conference on Bridge Management*, Guildford, Surrey, UK. E&FN Spon.

Haywood, D. (1999). Contractor slammed for Forth delays. *New Civil Engineer.* 21 January, 8.

Hearn, G. (1998). Condition data and bridge management systems. *Structural Engineering International, 8*(3), 221–225.

Highways Agency. (1993). BA50/93, Post-tensioned concrete bridges – planning, organization and method of inspection. Vol. 3: *Design manual for roads and bridges*. London: SO.

Highways Agency. (1994). BA63/94. Inspection of highway structures. Vol. 3: *Design manual for roads and bridges*. London: SO.

HMSO. (1984). *Bridge inspection guide.* London: Author.

Hool, G. A., & Kinne, W. S. (1943). *Movable and long-span bridges.* New York: McGraw-Hill.

Kittinger, W., & Ashabar, M. (1993). Inn Bridge, Kufstein, Austria: Bridge damage and repair (pp. 787–793). In: *Proceedings of the second international conference on bridge management*, Guildford, Surrey, UK. London: Thomas Telford.

Lee, D. L. (1994). *Bridge bearings and expansion joints* (2nd ed.). London: E&FN Spon.

Meadowcroft, J. C., & Whitbread, J. E. (1993). Assessment and monitoring of bridges for scour (pp. 128–136). *Proceedings of the Second International Conference on Bridge Management*, Guildford, Surrey, UK. London: Thomas Telford.

Millard, S. G., & Thomas, C. (1995). Assessment of bridge pier scour using impulse codes. In: *Proceedings of the institute of civil engineers*. 216–227.

Mylius, A. (1999). Avonmouth disaster probe focuses on work methods. *New Civil Engineer.* 16 September, 4–5.

Narasimhan, S., & Wallbank, J. (1998). Inspection manuals for bridges and associated structures. *The management of highways structures*. London: ICE.

Ng, S. K., & Moses, F. (1996). Prediction of bridge service life using time-dependent reliability analysis (pp. 26–33). *Proceedings of the Third International Conference on Bridge Management*, Guildford, Surrey, UK. E&FN Spon.

OECD. (1976). *Bridge inspection*. A report prepared by an OECD Road Research Group. France: Author.

Palmer, D. (1992). *The diagnosis of alkalia–silica reaction*. British Cement Association.

Parker, D. (1998a). Limpet dam technology can slow bacterial corrosion. *New Civil Engineer* 21 January, 38.

Parker, D. (1998b). Sulphate attack hits M5 bridges. *New Civil Engineer* 2 April, 3.

Parker, D. (1998c). Expert group to gather corrosion data. *New Civil Engineer* 2 April, 5.

Parker, D. (1998d). Biocorrosion – time for the state to step in. *New Civil Engineer* 22 October, 9.

Parker, D. (1998e). Weight probe on collapsed Jamuna gas pipe. *New Civil Engineer* 25 June, 3.

Price, A. R. (1984). *The performance in service of bridge deck expansion joints.* TRL Report LR1104. Crowthorne, UK: TRL.

Rogers, E. (1996). The association of ettringite with the cracking of in-situ concrete (pp. 219–224). *Proceedings of the Third International Conference on Bridge Management*, Guildford, Surrey, UK. E&FN/Spon.

Sims, I., & Hartshom, S. (1998). Recognising thaumasite. *Concrete Engineering International, 2*(8).

The Thaumasite Expert Group. (1999). The structural implications of the thaumasite form of sulfate attack. *The Structural Engineer, 77*(4), 10–11.

Woodward, R. J., & Williams, F. W. (1988). The collapse of the Ynys-y-Gwas Bridge, West Glamorgan. In: *Proceedings of the institute of civil engineers*. Part 1, 84, 634–669.

Yanev, B. (1997). Life-cycle performance of bridge components in New York City. In: U. Meier & R. Betti (Eds.), *Recent advantages in bridge engineering* (pp. 385–392).

Yokoyama, K., Sato, H., Ogihara, K., & Toriumi, R. (1996). Development of a bridge management system in Japan (pp. 580–586). In: *Proceedings of the third international conference on bridge management*, Guildford, Surrey, UK. E&FN Spon.

FURTHER READING

Ramberger, Gunter. (2002). Structutal Bearings and Expansion joints for Bridges. IABSE Structural Engineering Documents, *6*.

APPENDIX 2A

The definitions used in the text relating to concrete elements are shown in Figure 2A.1. Some examples of very severe damage to concrete structures are shown in Figures A.2–A.6.

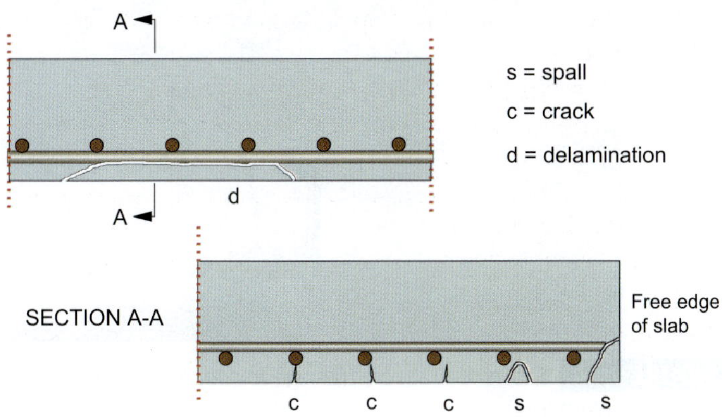

FIGURE 2A.1

Definitions of damage to concrete (courtesy of Dr Florian Burtescu, Romania).

FIGURE 2A.2 Example of severe damage to concrete elements (courtesy of Dr Florian Burtescu, Romania).

FIGURE 2A.3 Example of severe damage to concrete elements (courtesy of Dr Florian Burtescu, Romania).

FIGURE 2A.4 Example of severe damage to concrete elements (courtesy of Dr Florian Burtescu, Romania).

FIGURE 2A.5 *Example of severe damage to concrete elements (courtesy of Dr Florian Burtescu, Romania).*

FIGURE 2A.6 *Example of severe damage to concrete elements (courtesy of Dr Florian Burtescu, Romania).*

APPENDIX 2B

FIGURE 2b.1
Structure inspection form BE11/94 from BA63/94.

Structure Inspection Report

County BE11

Structure No.		Grid Ref:	
Structure Name		Date of Inspection	Day / Month / Year
Type of Inspection	Risk assessment read and checked Signed:	Inspected by	
Overall Assessment	G F P		

	Exist	Estimate	Extent	Severity	Work	Priority	Comment
Foundations							
Inverts or aprons							
Cutwaters							
Fenders							
Piers or columns							
Abutments							
Wing walls							
Retaining walls or revetments							
Approach embankments							
Bearings							
Mainbeams/Tunnel portals/mast							
Transverse beam/catenary cables							
Diaphragms or bracing							
Concrete slab							
Timber deck							
14 Metal deck plates/Tunnel linings							
15 Jack arches							
16 Arch rings							
Voussoir/Arch face							
17 Spandrels							
18 Tie rods							
19 Drainage systems							
20 Waterproofing							
21 Surfacing							
22 Service ducts							
23 Expansion joints							
24 Parapets/handrails							
25 Access gantries or walkways							
26 Machinery							
32 Dry stone walls							
33 Troughing							
Warning signs/street furniture							

Signed (Inspector) ____________ Date: ____________

CHAPTER 3

Assessment and Evaluation

CONTENTS

3.1 INTRODUCTION

The *strength assessment or evaluation* of an existing road bridge becomes necessary for three main reasons:

1. It is required to carry greater traffic loads than designed for, due to a general increase in traffic weights (principally from heavy lorries or trucks), increased traffic densities (with trucks occupying more lanes than the statutory limits) or from the passage of a single abnormal vehicle transporting, for example, a piece of heavy equipment.
2. The structure has seriously deteriorated or suffered substantial damage resulting in a decrease in strength.
3. There has been a change in design codes which may mean a reduction in acceptable safety levels.

The aim of an assessment is to establish the *safe load carrying capacity* of the bridge and most (if not all) national bridge management systems allow for this eventuality at particular intervals of time. This is usually initially confined to the bridge superstructure which is regarded as the weakest element of most bridges, and then extended to bridge supports and foundations.

Assessment is a complex task because it deals with a bridge *in situ* with all its attendant imperfections. Designing a bridge is a relatively easy task when compared with assessment because it is all done on paper. Changes can therefore be made at the whim of the designer to overcome any theoretical or detailing problems encountered. The designer is theoretically able to create a *perfect* bridge in every respect. Assessment, on the other hand, involves a great deal of detective work and ingenuity in assessing the loads and material strengths, and developing the correct model for the purposes of analysis and design. The assessor deals with a bridge which is *imperfect* in every respect, but which in some ways has been more thoroughly investigated than if it were a new design.

The level of safety is determined at the *ultimate limit state* since that is when extreme loads occur. Therefore, for shallow deck construction the purpose of the assessment is to determine the values of the critical stress resultants, i.e. action effects S (such as bending moment and shear force) and then to compare them with the resistance (R) of the section. The bridge is safe provided that:

$$R - S \geq 0 \tag{3.1}$$

The *serviceability* or working condition of the bridge is normally assessed by a visual inspection. Very often this reveals no evidence of distress within the structure and yet it can very often fail its assessment based on strength alone. Conversely the structure may be visibly in a parlous state of repair and yet still retains its load carrying capacity. Lark and Mawson (2000) address this situation head-on and suggest that there is a need for a *plausibility* check to explain any clear difference between the *real* and *assessed* conditions of the bridge. They propose the introduction of what they term 'change management' whereby the present and future conditions of a structure can be linked using information gathered from monitoring devices in conjunction with theoretical/empirical deterioration models developed from past measurements such as the level of chloride penetration in concrete. This is addressed in more detail in Chapter 2, section 2.9.

A bridge may 'pass' or 'fail' its initial assessment. The consequences of failure can mean the imposition of a weight restriction, the need for propping, monitoring, closing, load testing (see Chapter 5), extensive strengthening or, in extreme cases, demolition and rebuilding. In the UK alone rehabilitation work has been estimated at over £2.2 billion (HMSO, 1996). In the USA this figure is more than double at £5 billion (Chase, 1998).

It is incumbent on bridge engineers to make sure that their assessments employ the best procedures to get good value for money and which allow for the application of reliability theory if necessary and the exercise of sound engineering judgement.

The basic requirement is that a good balance is maintained between *safety and economy* to ensure that assessments are not unduly conservative, resulting in unnecessary remedial action, or so lax that they impose an unacceptable risk to the public.

3.2 BASIC CONSIDERATIONS

There are several very important factors which need consideration before embarking upon a strength assessment. They are: *loading, materials strength* and *structural form*. When *designing* a bridge, the structural form is decided according to functional, aesthetic and economic requirements, whilst the loading and material characteristics are specified in Codes of Practice, and the bridge is modelled in a simplified, idealised way. When *assessing* a bridge, it is often found that material strengths and traffic loads vary significantly from the original design values, and the actual behaviour means that more sophisticated modelling techniques are required.

3.2.1 Loading

A schematic representation of all the loads considered in *design* is shown in Figure 3.1. Initially, the usual loads considered in an *assessment* are those due to gravity which may be *permanent* or *transient* (dead loads and traffic).

3.2.1.1 Permanent loads

Permanent loads usually do not normally vary with time and consist of the self weight of the materials constituting the main load carrying elements (steel girders, reinforced concrete, prestressed concrete or timber materials); superimposed dead loads such as fill, surfacing, paving slabs, parapets,

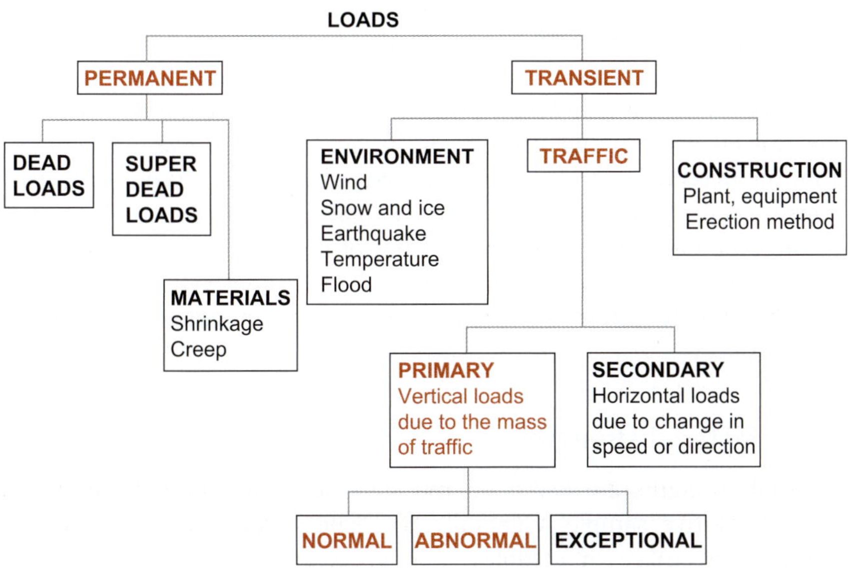

FIGURE 3.1
Loads on bridges.

lighting standards, safety barriers and the like. Unless there is a need to upgrade, such as renewing of the surfacing, then they are likely to remain unchanged for the life of the bridge.

Uncertainties exist in the *material densities* and the *construction dimensions* but they can usually be estimated with a good degree of certainty.

3.2.1.2 Traffic loads

One of the major factors affecting the overall safety of bridge decks, world-wide, is the increased intensity and volume of road traffic. This is particularly so on secondary road systems where many older bridges are located and on primary roads which have attracted a density of commercial vehicles greatly in excess of that envisaged in the original design.

In design, the traffic loads are derived from the consideration of the frequency distribution of vehicle weights and the probability of a mix of those weights under certain loading conditions along and across the bridge deck. They are by nature very difficult to model and predict as they are dependent on many random variables. Load models in design codes are, therefore, based on probabilistic methods using data from a limited number of bridge sites. At the design stage, since there is a high uncertainty of future trends, the traffic models tend to be conservative, though recent experience in the UK has confirmed that they were not conservative enough.

Most normal loading configurations adopted in practice comprise a uniformly distributed lane loading in conjunction with a knife edge load (BD 37-01, 2000; AASHTO, 1992; Eurocode 1, 1993; SIA 462, 1994) which, in varying degrees, take into account longitudinal and lateral bunching, and include amplification factors for dynamic effects.

In assessment it may become necessary to determine the basic static and dynamic effects of traffic at the bridge in question. The *static* effects result from the mass of the traffic using the bridge taking into account the mix of vehicles (number of axles, vehicle weights and dimensions); and the *dynamic* effects which take into account of the number of vehicles on the bridge, the speed, type of suspension and weights of vehicles, and finally the vibration characteristics of the bridge and the roughness of the running surface.

3.2.2 Material strengths

Material strengths are based on the statistical data taken from tests on a representative sample. Generally the 'safe' value is that below which a small percentage of samples is not expected to fall. The degree of manufacturing control also plays an important part in defining the partial safety factor for strength calculations. The specified or characteristic strengths at the design stage are assumed to be constant throughout the bridge and to remain unchanged throughout the life of the bridge.

In a completed bridge this may not be the case and they may have reduced, for example, concrete can be weakened through chloride contamination; masonry/brickwork can degrade due to the ingress of moisture; timber can be weakened by wet rot, and steel can be affected by high temperature. The converse is also true in that enhancement of the characteristic strengths may be possible if, for example, concrete has gained strength with time, or steel beams are of a higher grade than originally specified. The solution is to test a representative sample of each material to get realistic statistical values for use in the assessment.

Material weakness is further compounded by the loss of section of structural elements due to corrosion effects, and this too needs to be taken into account. Concrete is subject to a whole list of ills such as sulphate attack; freeze/thaw damage; alkali–silica reaction; high alumina cement conversion; salt crystallisation and salt scaling (Leeming, 1990). Local pitting corrosion of steel reinforcement is not impossible to detect, but general corrosion can usually be spotted due to the presence of cracking and spalling of concrete.

Corrosion in steel structures can usually be seen and allowed for by cleaning off loose scale products and taking accurate measurements of the steel elements to determine the loss of section.

3.2.3 Structural form

The assessment of a bridge means that it has to be analysed, and the structural form clearly determines the way in which it is modelled. The analysis has to fit the form in a way which can, in the first instance, be *simple* and if the results indicate a 'fail' situation, then progressively *'refined'* analyses can be undertaken to the point where it is deemed that no further refinement can be made. The unspoken assumption is that suitable refinement will result in a more favourable answer, though of course this is not always the case.

Many older bridges are of unusual construction by today's standards and are often difficult to conceptualise for analysis purposes, for example jack-arch bridges; filler beam decks; steel trough and concrete; cast iron beam-and-slab; steel beam and timber composite/non-composite decks, and skew masonry arch bridges (Choo, Coutie & Gong, 1991). In such cases the results of model or real load tests should be sought from available literature for guidance. For example Bakht and Jaeger (1992) have shown that this is especially true in respect of nominally non-composite decks where some composite action has been shown to exist at low load levels, which all but disappears as the load reaches the ultimate limit state levels.

3.3 STRUCTURAL SAFETY

The aim of finding the *structural safety* of a road bridge is to ensure that the probability of loss of human life is at an acceptable level. This is referred to as the *target reliability* which may be based on the safety level of existing bridges or the risk compared with other everyday events. Just how safe bridges are in fact is illustrated by the fatality accident rates shown in Table 3.1 taken from work done by Hambly and Hambly (1994) and Blockley (1992).

Several countries of the world consider that some form of hierarchical procedure is the way forward whereby an assessment is carried out firstly by considering a *simple deterministic* approach and then if the assessment fails, to carry the process forward towards a full *reliability analysis* (Das, 1997; AASHTO, 1989; CSA, 1990). A good example of the general method is given by Bailey (1996) and what follows in the next section is largely attributable to him.

3.3.1 General method

The philosophy is to demonstrate adequate safety as 'easily' as possible, whilst using the most advanced evaluation methods before intervening.

Table 3.1 Risks from Everyday Life

Activity	Deaths per 100 Million Hours of Exposure
Travel by helicopter	500
Travel by aeroplane	120
Walking beside a road	20
Travel by car	15
Construction (average)	5
Building collapse	0.002
Bridge collapse	0.000002

The assessment of a particular bridge is defined by a *Rating Factor* (*RF*) described by Moses and Verma (1990) as:

$$\text{RF} = \text{live load carrying capacity/applied live load} \tag{3.2}$$

$$\text{RF} = (R - G)/Q \tag{3.3}$$

Both Bailey (1996) and Moses and Verma (1990) take into account permanent loads as well as live loads and define the rating factor as:

$$\text{RF} = \text{total load carrying capacity/applied loads} \tag{3.4}$$

$$\text{RF} = R/(Q + G) \tag{3.5}$$

In either case if RF $\geq$ 1.0 then the bridge is safe and no action need be taken. If RF $<$ 1.0 then action is necessary as described in section 3.1. Since the rating factor is proportional to structural safety, greater priority is implied for lower rating factors.

The four methods of assessing structural safety are given as:

- Determine analysis with default models to calculate RF_{det}
- Reliability analysis with default models to calculate RF_{prob} (rarely used)
- Deterministic analysis with updated models to calculate $\text{RF}_{\text{det,upd}}$
- Reliability analysis with updated models to calculate $\text{RF}_{\text{prob,upd}}$

A flow chart showing the hierarchy of increasingly complex actions is shown in Figure 3.2.

3.3.1.1 Method 1

This is the simplest of all the methods and uses codified values of partial safety factors for both loads and material strengths, and dimensions are assumed to be 'as designed' or 'as built'. At the ultimate limit state the following expression has to be satisfied:

$$S_{\text{d}} \leq \frac{R}{\gamma_{\text{R}}} \tag{3.6}$$

$$S(\gamma G \cdot G_{\text{m}}) + \frac{S(\gamma_{\text{Q}} \cdot Q_{\text{r}})}{\alpha_{\text{Q}\cdot\text{def}}} \leq \frac{R}{\gamma_{\text{R}}} \tag{3.7}$$

that is factored dead loads + factored live loads $\leq$ resistance/material safety factor; where S_{d} = design load effect; $S_{()}$ = effect of actions; γ = partial safety factor; G_{m} = average value of permanent actions;

FIGURE 3.2
Bridge assessment flow diagram (after Bailey, 1996).

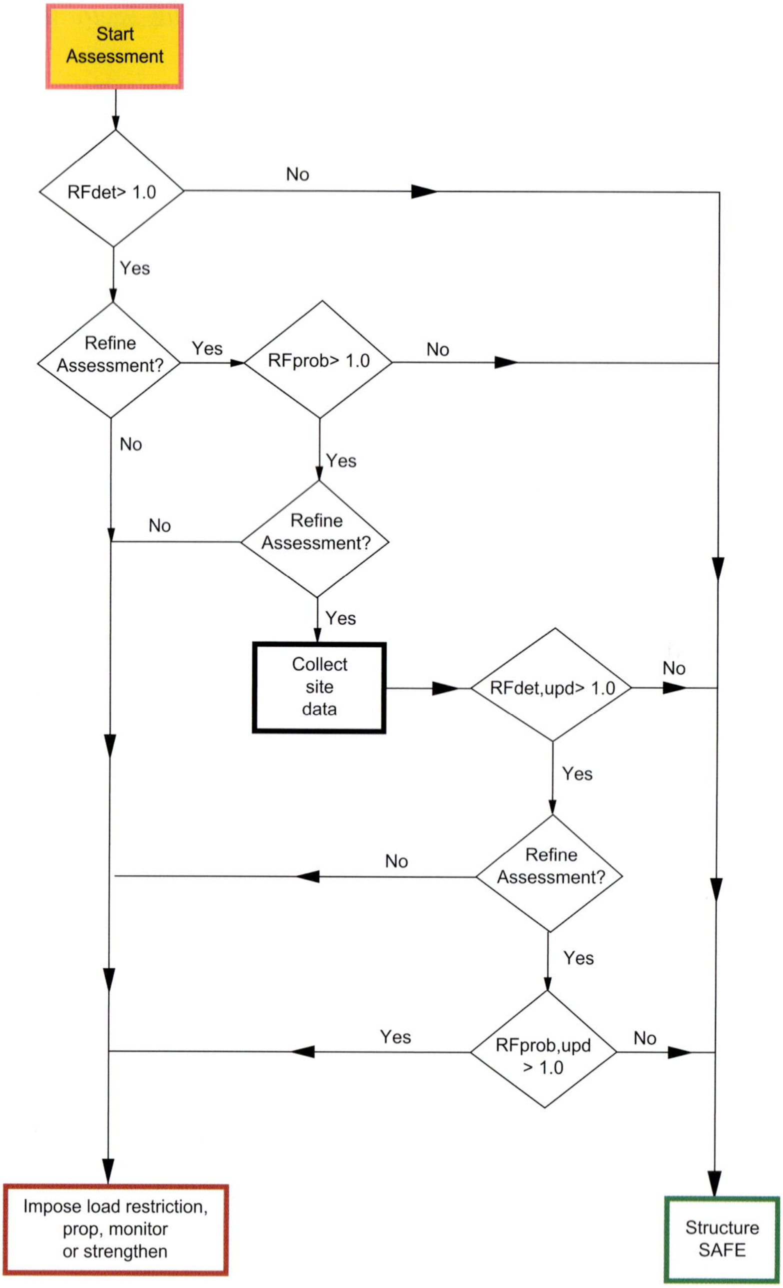

Q_r = representative value of traffic actions; $\alpha_{Q,def}$ = default traffic action effect reduction factor (=1 in the UK) and R = resistance. G, Q and R refer to permanent loads, traffic loads and resistance, respectively.

The default traffic action effect reduction factor ($\alpha_{Q,def}$) can be derived using probabilistic methods (Bailey, 1996) and can be applied if there happens to be a traffic restriction in force at the time of the assessment.

Using Eq. (3.7), the deterministic rating factor (RF_{det}) can be expressed as:

$$RF_{det} = \frac{\dfrac{R}{\gamma R}}{S(\gamma_G \cdot G_m) + \dfrac{S(\gamma_Q \cdot Q_r)}{\alpha_{Q,def}}} \tag{3.8}$$

The appropriate action is then taken depending upon whether $RF_{det} \geq 1.0$ or $RF_{det} < 1.0$. If $RF_{det} < 1.0$ (see Figure 3.2) then it may be possible to:

- collect site data to update load and/or resistance models
- perform a more accurate assessment (Method 2) and
- strengthen the bridge or impose traffic restrictions.

3.3.1.2 Method 2

This method utilises reliability concepts (see section 3.6) and although complex may be worthwhile if relevant data are available, otherwise a more refined deterministic approach as in Method 3 can be used. Method 2 results in the evaluation of a probabilistic rating factor RF_{prob}:

$$RF_{prob} = \frac{\beta_{assess}}{\beta_{design}} \tag{3.9}$$

where β_{assess} is the assessment reliability index and β_{design} is the design reliability index. This probabilistic value can be converted to an equivalent deterministic factor for direct comparison with factors derived from a deterministic assessment (Bailey, 1996).

Probabilistic models of the *permanent* loads are used assuming a normal distribution based on nominal dimensions and densities and modified by the appropriate bias and coefficient of variation. *Traffic actions and structural resistance* are also based on probabilistic models, examples of which are given by Bailey (1996).

The appropriate action is then taken depending upon whether $RF_{prob} \geq 1.0$ or $RF_{prob} < 1.0$. If $RF_{prob} < 1.0$ (see Figure 3.2) then it may be possible to:

- perform a more accurate assessment (Method 3)
- collect site data to update load and/or resistance models
- strengthen the bridge or impose traffic restrictions

3.3.1.3 Method 3

This method is *deterministic* and uses *site-specific data* (see section 3.3.2). Reduction factors are applied to the limit state equation to update the models for the permanent actions and traffic actions such that:

$$\frac{S(\gamma_G \cdot G_m)}{\alpha_G} + \frac{S(\gamma_Q \cdot Q_r)}{\alpha_Q} \leq \frac{R}{\gamma_R} \tag{3.10}$$

Each type of *permanent* action is treated separately. This essentially means that uncertainty is reduced and updated representative values are used which are more accurate than assumed in design.

Reduction factors applied to *static traffic* actions take into account of the difference in both the magnitude and characteristics of the traffic load models assumed in design to those used during assessment. An updated *dynamic* coefficient can also be determined based on the fundamental frequency of the bridge.

Structural resistance data can be collected from the bridge by measurement and testing for use in the assessment. Structural safety is then assessed by calculation of the deterministic rating factor, *viz:*

$$RF_{det,upd} = \frac{\dfrac{R}{\gamma_R}}{\dfrac{S(\gamma_{Gi} \cdot G_{mi})}{\alpha_{\Gamma i}} + \dfrac{S(\gamma_{Gi} \cdot G_{mi})}{\alpha_\theta}} \tag{3.11}$$

The appropriate action is then taken depending upon whether $RF_{det,upd} \geq 1.0$ or $RF_{det,upd} < 1.0$. If $RF_{det,upd} < 1.0$ (see Figure 3.2) then it may be possible to:

- perform a more accurate assessment (Method 4)
- collect site data to update load and/or resistance models and
- strengthen the bridge or impose traffic restrictions.

3.3.1.4 Method 4

This is the most accurate method of bridge assessment which utilises a *reliability analysis using site-specific data* to update probabilistic models of both permanent and traffic actions. A probabilistic assessment and collection of site data will be cheaper than bridge closure or repair/strengthening.

Structural safety is assessed in the same way as for Method 2 except that updated data are used instead of default values, and is given by:

$$\mathrm{RF}_{\mathrm{prob,upd}} = \frac{\beta_{\mathrm{assess,upd}}}{\beta_{\mathrm{design}}} \tag{3.12}$$

As for Method 2, this can be converted to an equivalent deterministic value. The appropriate action is then taken depending upon whether $\mathrm{RF}_{\mathrm{prob,upd}} \geq 1.0$ or $\mathrm{RF}_{\mathrm{prob,upd}} < 1.0$. If $\mathrm{RF}_{\mathrm{prob,upd}} < 1.0$ (see Figure 3.2) then it may be possible to:

- collect more site data to update load and/or resistance models
- strengthen the bridge or impose traffic restrictions

3.3.2 USA and Canada experience

Both Canada (OHBDC, 1991) and the USA (AASHTO, 1994) have utilised reliability analysis methods to determine load and resistance factors for the purposes of bridge assessment based on research for new bridge *designs* adopting the Load and Resistance Factor Design (LRFD) methodology (similar to Limit State design used in the UK and the EU).

The analyses are carried out for *individual structural elements* although it is possible to develop response functions for the complete bridge structure which takes into account of the advantages of redundancy and structural form (Ghosen, Moses & Khedekar, 1994). The application of the methods is clearly described by Nowak (1997) and (Ghosen, 1996) and is based on the calculation of the *reliability index* (β) which is a measure of the probability of failure using the first order reliability method (FORM). This is compared to a *target reliability index* (β_T) which is selected to provide a consistent and uniform safety margin for all bridges. The notional value chosen was $\beta_T = 3.5$.

Nowak (1997) states that 'load factors γ are calculated so that the factored load has a predetermined probability of being exceeded and resistance factors ϕ are calculated so that the structural reliability is close to β_T'.

The output from the reliability analyses for design to AASHTO (1994) resulted in the following strength requirement for girders:

$$1.25D + 1.50D_A + 1.75(L + I) < \phi R_n \tag{3.13}$$

where D = dead load; D_A = dead load due to asphalt wearing surface; L = static live load; I = dynamic load; R_n = resistance or load carrying capacity and ϕ = resistance factor [1.0 for moment and shear in steel girders; 0.9 for moment and shear in reinforced concrete T-beams; 1.0 and 0.9 for moments and shear in prestressed concrete beams respectively]. Bridges *designed* on this basis for all materials and spans have values of β between 3.5 and 4.0 which is close to the target reliability index of 3.5, and the work is deemed to be appropriate for bridge assessment with the variables in the above equation based on a probabilistic approach.

In assessment, if it is found that $\beta_{\text{assess}} < \beta_{\text{design}}$ (i.e. $\beta > 3.5$ or a rating factor < 1.0) then it may be possible to reduce the load and resistance factors based on site-specific data about the actual loads and material strengths.

In the USA at the moment (White, Minor & Derucher, 1992) there are two capacity ratings, namely: the Operating Rating which refers to the absolute maximum load that the bridge deck may carry, and the Inventory Rating which relates to the load that the bridge can be subjected to for an indefinite period of time. Each rating has its own assessment equation.

White et al. (1992) work through a number of very comprehensive examples of load carrying capacity using AASHTO (1989), which are very helpful in understanding the general assessment procedure. They are based on the Rating equation given in AASHTO (1989), *viz*:

$$\phi R_{\text{n}} = \mu_{\text{d}} D + \mu_{\text{L}}(RF)L(1+I) \tag{3.14}$$

or

$$\text{RF} = [\phi R_{\text{n}} - \mu_{\text{d}} D]/[\mu_{\text{L}}(1+I)L] \tag{3.15}$$

where I = 'impact factor' to magnify static loading due to dynamic amplification; L = nominal live load effect; D = nominal dead load effect; RF = rating factor; R_{n} = nominal strength or resistance; μ_{d} = dead load factor; μ_L = live load factor and ϕ = resistance factor (capacity reduction). Values for each of the variable factors are tabulated within the specification. The impact factor takes into account of the condition of the wearing surface (and not the span); the dead load factor is based on knowledge about the actual weights and dimensions; the live load factor depends upon the average daily truck traffic (ADTT), and the resistance factor is a function of the condition of the structure, redundancy, type of inspection, maintenance and type of structure. Further factors can be applied to take into account of the number of lanes on the bridge and the type of analysis used.

3.3.3 UK experience

The UK is undergoing a comprehensive rehabilitation programme for local and trunk road bridges which commenced in 1988 in response to an

EU directive for international transport road vehicles to be allowed into the UK. In particular this referred to 40 tonne vehicles coming onto UK roads from 1 January 1999. Assessments are carried out using limit state principles to the current mandatory assessment code BD21 (1997) supported by a complementary advice note BA16 (1997). The code is generally concerned with older structures using outmoded forms of construction and obsolete materials, and provides specific guidance for assessing the capacity of masonry arch bridges. Guidance is given on inspection; unit weights of materials; live loading; strength of materials and methods of analysis.

At present the assessment live loading is considered at its worst possible level and is applied indiscriminately to motorways and minor roads; to relatively new bridges as well as old ones, and to bridges of different types. This is clearly an unsatisfactory state of affairs and research sponsored by the Highways Agency is now underway with the aim of producing a comprehensive set of assessment documents that will discriminate between bridges of different ages, types and carrying different intensities of traffic, specifically the number of Heavy Goods Vehicles (HGVs) which use the bridge. The new methods will provide default values of load and resistance variables and allow for the incorporation of probabilistic and reliability methods at higher levels of assessment using *bridge specific* data.

Das (1997) advocates the use of a hierarchy of assessment methodologies starting at a very simple level and becoming progressively more sophisticated with the aim of determining more accurately the load carrying capacity of a particular bridge should its initial assessment fail. Five levels are proposed (with a possible *modus operandi* as in Appendix 3C), namely:

Level 1 This employs a *simple* analysis tool, with design based on specified material properties combined with codified partial safety factors. An example may be the analysis of a simple beam analysis using a 'strip' of deck under full assessment loading. This gives information about longitudinal bending only since transverse bending is ignored.

Level 2 This employs a more *refined* linear-elastic analysis technique employing a better structural idealisation such as a grillage or finite element analysis both of which allow for the *transverse distribution of load* due to the presence of transverse bending moments and shear forces.

It may be possible to carry out useful non-destructive tests to establish material properties and also to carry out a load test to establish the transverse load distribution characteristics (see Chapter 5).

Level 3 This level is *bridge specific* in that it uses material properties and loading specific to a particular bridge and uses the results of research into local traffic conditions to derive the assessment live loads. For short span bridges the 1997 version of BD21 takes into account of traffic density and road surface irregularities. For longer bridges other factors derived from traffic surveys are recommended in BD50 (1997). Bridge response data may be derived from load tests if these are considered appropriate.

Level 4 This level aims to use *modified partial safety factors* for both loads and materials (as opposed to the full codified values used in Levels 1–3). The partial safety factors reflect the uncertainty of the given variable and need to be calculated with great care. Der Kiureghan (1989) defines four sources of uncertainty as *natural variability; estimation error; model imperfection and human error* which are defined in Appendix 3A. Shetty, Chubb and Manzocchi (1998) suggest that modifications to partial safety factors can be derived to take into account of the specific characteristics of a particular bridge, which might include, for example:

- Measurements of actual dimensions of the main structural and superimposed dead load elements – this would enable more accurate assessment of dead loads and element strengths.
- Live Load Factor (LLF) = ratio of live load to dead load. Clearly a high LLF would mean that a bridge is less sensitive to increases of live load.
- Age of the structure and its expected remaining service life – a bridge in good condition can be expected to be 'safer' than a similar bridge suffering severe deterioration.
- Reserve strength and redundancy – could take into account of things like membrane action; fixity at supports and improved load distribution.
- Inspection and monitoring regime linked to warning of failure – this increases confidence and enables remedial action to be taken more immediately.
- Consequences of failure – even if there is the loss of one life it is difficult to see how this knowledge can enable a modification of the safety factors.

Level 5 This level incorporates the use of *structural reliability* analysis techniques. It considers the *probability* of failure of an element within the structure (or the overall structure) based on the random uncertainties in load and resistance. The same sources of uncertainty are used as in Level 4 from which a reliability index can be calculated. Shetty et al. (1998) point out that the probability of failure should be 'treated as a "notional" value' and not used as a measure of the frequency of failure which could be expected in service. This is because the probability distributions used in a reliability analysis do not take into account gross errors in design, construction or operation (see Appendix 3A). In assessing the capacity of a bridge Micic (1997) suggests that gross errors can be disregarded since they would have manifested themselves during the previous lifetime of the bridge.

The reliability procedure can be a daunting one, but it is hoped to produce a guidance document so that bridge engineers can work alongside reliability specialists to ensure a consistency of results. In particular this will mean the provision of probability distributions and the target reliability index (β_T). Since the latter has not been specified in the UK, research will have to be carried out to determine it, and could follow the calibration procedure used by Nowak (1997); in particular the selection and study of a wide spectrum of bridges designed according to UK standards. The guidance will include a number of worked examples.

3.4 ANALYSIS METHODS

At whatever level an assessment is carried out, once the loads and strengths have been finally determined it is necessary to carry out an analysis. Normally a global analysis is first carried out followed by a local analysis of the bridge deck using influence surfaces such as those produced by Pucher (1964) or Westergaard (1931). There are essentially six ways of doing this, namely:

- simple strip analysis (*S*)
- grillage analysis (*G*)
- distribution methods (*D*)
- linear-elastic finite element analysis (FE)
- space frame analysis (SF)
- non-linear finite element analysis (NLFE)
- yield-line analysis (YL)
- plastic strip analysis (PS)

All of these methods are described in detail by Hambly and Hambly (1992) but some observations as to their use are included here to highlight the limitations.

3.4.1 Simple strip analysis

This method involves the consideration of a single representative longitudinal element (or strip) of a bridge deck.

The analysis is linear-elastic and can be a *lower bound* one where a static transverse load distribution is assumed (Figure 3.3i), or an *upper bound* one where a perfect transverse load distribution is assumed, that is the whole load is shared equally by every transverse section (Figure 3.3ii). In Figure 3.3(i) the load directly over the width (*s*) of a deck element is considered equal to *w/b*, and in Figure 3.3(ii) the load is first divided over the width *b* and the proportion carried on the 'strip' is *s/b*. This plastic moment is then compared with the flexural capacity of the element. The method is generally considered to be conservative.

In both the UK and the USA simple distribution factors are given for beam spacing and load type to take into account of some transverse distribution.

3.4.2 Grillage analysis

The grillage analysis method is well known among bridge engineers and is widely accepted as an inexpensive and reliable way of determining the elastic stress resultants in a bridge deck (Hambly & Hambly, 1992). It allows for the transverse distribution of load through bending of the deck slab and therefore effectively reduces the load carried by the most heavily loaded beam. The applied loads are 'factored' to the ultimate limit state and an elastic analysis is performed. The results are then examined to find the positions of maximum stress resultant and compared with the capacity of the section at

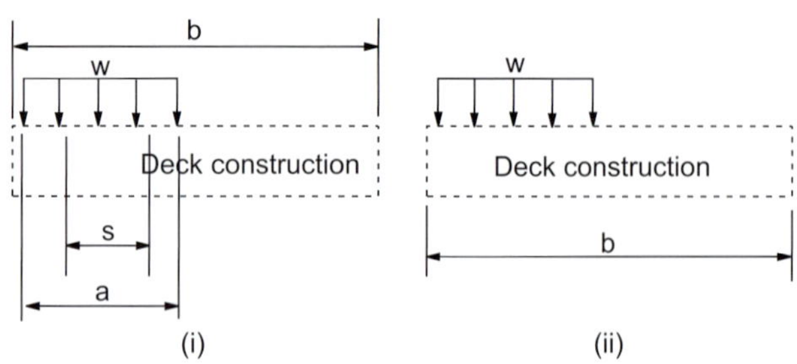

FIGURE 3.3 *Strip analysis – upper and lower bound solutions.*

that location. In many bridges in the USA there is only a nominal amount of transverse reinforcement so a grillage analysis is not appropriate.

For the reinforced concrete elements in the transverse direction, the stiffness properties (according to Jackson, 1996) can be based on either the *uncracked or cracked* sections. Since the amount of reinforcement is usually known it is usual to assume an uncracked section. This can result in a less favourable distribution of load between the beams and may lead to overstressing of the longitudinal beams. Conversely, using an uncracked section will improve the distribution properties of the deck but may lead to overstressing of the transverse reinforcement. One way around this *impasse* is to carry out an uncracked analysis and try to define those areas where the moment which would cause cracking (M_{cr}) is exceeded, and then to use the information to revise the grillage so that there is a mix of members with cracked and uncracked properties.

In reinforced concrete slabs (right or skew) the reinforcement usually consists of two layers (top and bottom) arranged in an orthogonal or skew pattern, and so if possible the grillage beams should be arranged to reflect that pattern. The output from the grillage analysis consists of a moment triad of two bending moments and a twisting moment (e.g. M_x, M_y, M_{xy}) and these can be used to carry out a Mohr's circle analysis to determine the resultant bending moment in each reinforcement direction. If a torsionless grillage is used, and the grillage is a mirror of the reinforcement, then the bending moment output is that directly resisted by the reinforcement.

Whilst allowing for redistribution of load in the elastic range, no allowance is made for redistribution at the ultimate limit state and therefore it is considered to be conservative.

3.4.3 Distribution analysis

Distribution methods offer a quick way of determining the maximum load (or moment) carried by any one girder in *right, simply supported bridge decks*. The method has been codified in both the USA, AASHTO (1992) and Canada, OHBDC (1991) and to a very limited degree in the UK (BA 16, 1997). In the UK, the method has been proposed, researched and disseminated by Ryall (1992) as the *D-Type* method (after Bakht & Jaeger, 1985).

The codified methods enable a useful *first estimate* of the load carrying capacity of a bridge, and can be used as the platform to progress to more sophisticated levels if necessary. The D-Type method is, itself, a sophisticated method, but it is limited to the elastic range and is therefore conservative.

3.4.4 Finite element analysis

An *elastic finite element* analysis (Hambly & Hambly, 1992) is often used to model more accurately *plate-type decks such as slab, or beam-and-slab* which are of an *irregular shape in plan, or box girder type bridges*.

Finite elements of triangular or rectangular shape are normally used to model slabs, which may be isotropic or orthotropic in nature. Initially an isotropic model may suffice, but sound judgement must be excercised. If, for example, the areas of main reinforcement are considerably different to the areas of secondary reinforcement then it may be more 'accurate' to consider an orthotropic model or a grillage analysis. Simple line elements are used to model beams.

Supports should be given a finite size to avoid the anomaly of infinite moments over 'point' supports. If the mesh is made finer in the region of supports, then the mesh size should be similar to the plan size of the support. This helps to minimise errors and avoid unrealistically high moments. In box girders where diaphragms are placed at support positions this is not a problem.

In beam-and-slab bridges the beam type elements model the longitudinal beams and diaphragms, whilst the plate type elements model the deck slab. The model may be 2D with downstand beams or a fully 3D one. The latter, however, only gives stresses and not moments and shears as output, and so these have to be converted to moments by integrating over the whole section. Alternatively, the stresses could be compared with allowable stresses related to the moment capacity at the ultimate limit state. Both provide not only global results but also local results under wheel loads, and so separate analyses for local effects are not necessary.

It is always advisable to carry out a simple hand calculation to check the validity of the results.

3.4.5 Space frame analysis

A 3D space frame analysis may sometimes be an appropriate way of representing a structure such as a composite bridge with deep plate girders, or a prestressed concrete box girder as its closeness to the actual bridge makes the output easier to interpret.

Figure 3.4 shows a typical box girder modelled in this way, and indicates the proportions of the original used to determine the member properties. The output is in the form of direct force and moments which can be converted to stresses in the prototype. The shear stresses are calculated from the vertical component of the diagonals in the webs. Displacements are output directly.

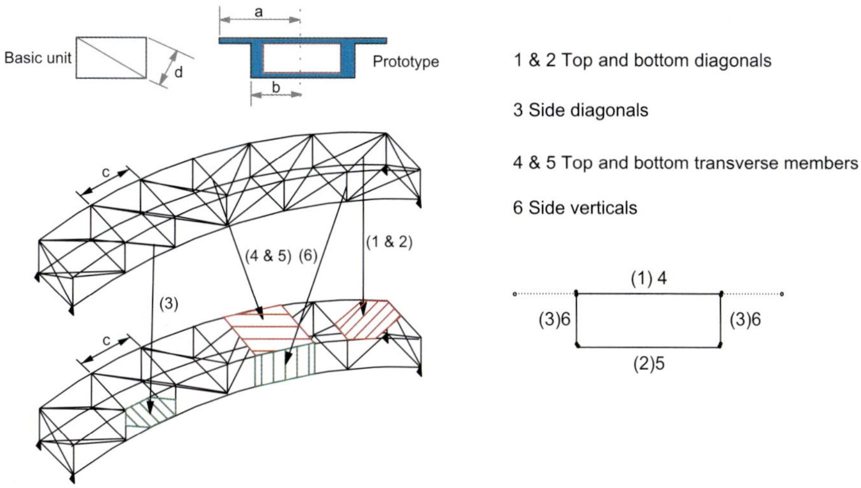

FIGURE 3.4

Typical space frame representation of a box girder bridge.

Hambly and Hambly (1992) describes two other less conventional ways of modelling in 3D, namely the use of a McHenry lattice or a cruciform frame.

3.4.6 Non-linear finite element analysis

NLFE programmes are able to model the non-linear characteristics of a bridge deck under a gradually applied load such as the change of stiffness as concrete cracks; the non-linear nature of the stress–strain curve for steel reinforcement, the non-linear load–displacement relationship that results when deflections become large, in-plane forces etc. In this sense they are a very realistic form of analysis in that they seek to model the bridge behaviour incrementally at every point in the load history right up to collapse. NLFE programmes are, however, very expensive to buy at the moment and require a high level of expertise to use them effectively. They are also sensitive to the choice of material properties and generate much output.

Henriques, Calheiros and Figueiras (1996) describe the use of an NLFE programme to model a two-span twin box concrete girder bridge, but the results are nowhere verified by test data – as for example measurement of deflection under load – and are therefore unreliable. Jackson (1996), on the other hand, used an NLFE programme to predict the behaviour of an in-fill joist type bridge which was actually tested for destruction and he clearly explains the limitations of the analysis. He concludes, that in that particular case, the predicted reserve of strength was justified in that it was conservative compared to the actual value, and therefore it is probable that NLFE programmes have a role to play in assessment. Jackson and Cope (1990) carried out an assessment of two half-scale beam-and-slab decks using various methods including an NLFE analysis. The decks were then tested to failure and the failure loads compared with those obtained from the various methods. The conventional methods based on elastic theory and the yield-line approach all gave conservative results. They found, however, that not only did the NLFE analysis accurately predict the failure *load*, but also the failure *mode*. They conclude that *prediction methods which predict the correct failure load but the wrong mode must be considered highly suspect*.

Use of NLFE programmes is likely to remain in the domain of research for some time to come and in the meantime they should be used with caution and if possible backed up by physical tests.

3.4.7 Yield-line analysis

Yield-line analysis is a familiar technique for determining the bending strength of concrete bridge slabs (Johansen, 1962; Clark, 1983) but is generally not used in practice, probably because it is an upper bound method and also because many possible mechanisms have to be investigated to find the critical one. Many engineers find it both tedious and unnecessarily complex. However, providing that the slab is of a simple geometry and possesses sufficient ductility for a mechanism to form, it could be used when conventional elastic techniques indicate inadequate bending strength. A separate check is required for shear using the conventional elastic approach.

Middleton (1997) argues passionately for the use of yield-line analysis for both slab and beam-and-slab bridge decks and has produced a collapse analysis programme called COBRAS (Concrete BRidge ASsessment). It uses up-to-date graphics and 3D modelling techniques which, it is claimed, has the 'ability to analyse rigorously realistic configurations of loading, bridge geometry, support fixity and failure mechanisms, without the need to derive mathematical expressions describing the interrelationships between these parameters'. A typical bridge assessment can be completed in 'a couple of minutes'.

The programme has been calibrated against published analytical solutions; results from an NLFE analysis, and from test results on 13 model bridge slabs. In all but two cases, the yield-line method gave conservative estimates of strength. Assessments have also been carried out on 21 short-span bridges for various bridge authorities and consultants and in every case the plastic rating from COBRAS exceeded the rating obtained using elastic methods. Fifteen were found to have a 40-tonne capacity whilst having failed their original assessment, which was in one case only 3 tonne! It should be noted that all but three of the bridges were simply supported and the failure mechanism under the HA (normal) assessment load was, in most cases, a full-width transverse yield line at mid-span – the analysis, of which, could have been done by hand, also in 'a couple of minutes'.

The method clearly has potential for short-span slab bridges, but whether there is a need for a sophisticated programme such as COBRAS is open to debate. For simply supported short-span *slab* bridges it should be possible to choose a set of probable failure mechanisms to enable the work equations to be formulated to find the strength of the bridge deck. Some typical examples of failure modes given by Thoft-Christensen, Jensen, Middleton and Blackmore (1997) are reproduced in Figure 3.5.

A *collapse analysis* is then performed by assuming a unit virtual displacement, and the work done by the applied live loads (W_L) is compared with the work done in the rotation of the yield lines (W_Y) to obtain the load factor against collapse, such that:

$$\text{Load factor (LF)} = W_Y/W_L \tag{3.16}$$

The failure mode which gives the lowest calculated load factor is then selected as that governing collapse. The critical mode can then be used in a reliability analysis (section 3.6) to determine a probability of failure. This may be compared with an accepted target value to decide whether the bridge is safe or not.

Ibell and Middleton (1998) have addressed the problem of *plastic shear capacity* to be used in conjunction with yield-line analysis. Many laboratory tests were carried out to investigate failure mechanisms for various quantities of flexural and tensile reinforcements, and to study the enhancement of the shear strengths of internal beams due to the restraint offered by the surrounding structure. This has led to the development of a 3D collapse

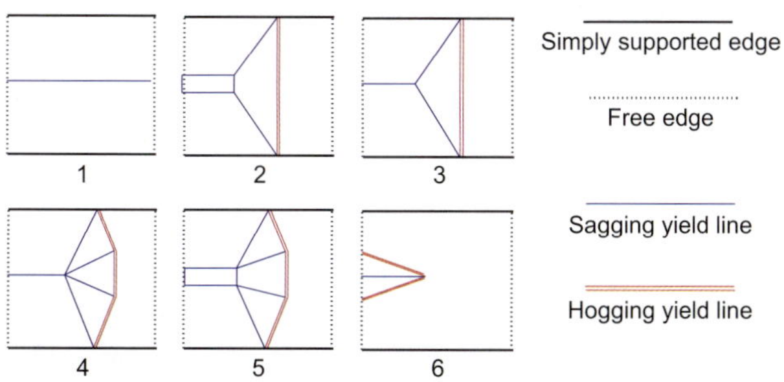

1. Full width failure
2. Partial box failure
3. Partial wedge failure
4. Partial wedge failure 3 wedge fan
5. Partial box failure 3 wedge fan
6. Partial edge wedge failure

FIGURE 3.5 *Typical failure modes for simply supported slab bridges.*

analysis method which has now been incorporated into a computer program suite called BaSiS (Beams-and-Slabs in Shear). The program is recommended for use as an assessment tool in reinforced concrete beam-and-slab bridges but it does rely on the presence of ductility in the form of stirrups (or links) and the need to assume an 'effectiveness factor' *versus* for use in the analysis.

3.4.8 Plastic strip method

This method was pioneered by Hillerborg (1962) and its use relating to bridge deck analysis is described by Clark (1983). It is a simpler alternative to a yield-line analysis and it provides safe lower bound solutions. Difficulties arise, however, in deciding upon the distribution of the loads to produce 'strips' because whereas with traditional buildings many alternatives are possible owing to the four-sided supports, bridges only have two supports.

3.5 RELIABILITY CONCEPTS

Reliability concepts are not generally understood by bridge engineers because they are not explicitly used in the everyday practice of bridge assessment or design. An understanding of reliability methods is, however, vital if the structural safety of a bridge is to be assessed taking into account of all the uncertainties such as traffic loading, material strengths and structural form. This section provides an overview and a fuller account is given in Chapter 12.

3.5.1 Basic concepts

The most universally understood way of ensuring the safety of a bridge is by the use of partial safety factors at the limit state so that:

$$\text{Resistance} \geq \text{Load effects} \tag{3.17}$$

or

$$R \geq S \tag{3.18}$$

where representative values used for R and S are random. Deterministic values of both R and S are, therefore, impossible because in reality they are the products of many uncertainties which are accounted for by the introduction of partial factors which allow for uncertainty and 'provide safety', so that:

$$\frac{R}{\gamma_m} > \gamma_L S \tag{3.19}$$

This enables the calculation of an overall safety factor $\gamma = \gamma_m \cdot \gamma_L$ which ranges from 1.5 to 2.0.

Reliability is another way of taking into account of the same uncertainties in a probabilistic way. If *R* and S *are* assumed to be independent variables which can be expressed as probability distributions similar to that of Figure 3.6, then the *probability of failure is equal to the probability that S > R* given by:

$$P_f = p[S > R] = \sum P[S = x] \cdot P[R < x] \tag{3.20}$$

This describes the probability of failure as equal to the summation (over the range of all possible values of *S*) of the probability that $S = x$ *and* the probability that $R < x$. This is expressed as an integration from $-\infty$ to $+\infty$,

$$P_f = \int_{-\infty}^{\infty} f_s(x) \cdot F_R(x) dx \tag{3.21}$$

where $f_s(x) = P[x < S < x + dx]$ and the hatched area in Figure 3.6 is $F_R(x) = P[R < x]$. The failure zone represented by this equation is shown in Figure 3.7 (See Figure 12.2 in Chapter 12).

The design point $[R^*, S^*]$ is defined as the most probable failure point on the *limit state surface* which is nearest to the origin.

The reliability (or safety) index β is another measure of the safety of a bridge or an element and for a linear limit state function and independent normal variables the reliability index β is then given by:

$$\beta = \frac{\mu_R - \mu_S}{(\sigma_S^2 + \sigma_R^2)^{1/2}} \tag{3.22}$$

where μ are mean values of *R* and *S*, and σ are the variances of *R* and *S*. This value of β is then compared to a target reliability $\beta_T = 3.5$ (still the

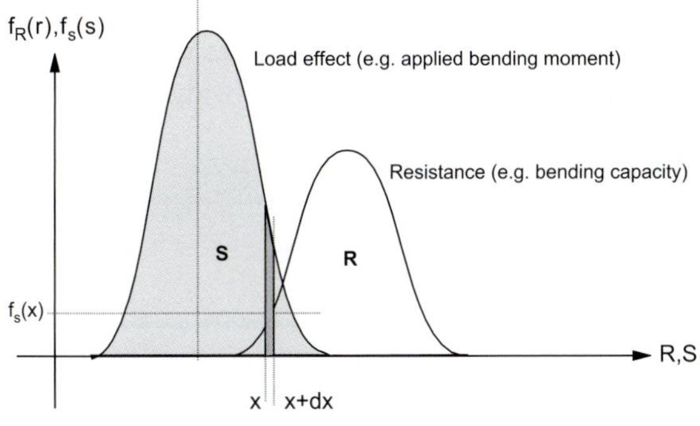

FIGURE 3.6 *Probability distribution of* R *and* S.

FIGURE 3.7
Failure zone with respect to joint probability distribution of R and S.

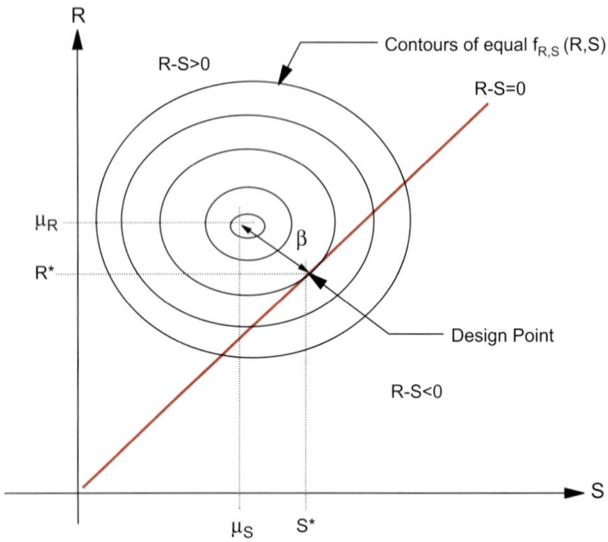

subject of a great deal of discussion – see section 3.3.2). If $\beta < 3.5$ then the bridge can be deemed as 'safe' and it has 'passed' its assessment. Vrouvenvelder and Waarts (1993) relate the target reliability index to a given period of time for which the probability applies. Typical values quoted by them are shown in Table 3.2. Then if $R_d \geq S_d$ (one load and one resistance) the reliability is expressed by:

$$P(S_d > S) \geq (-\alpha_s\, \beta_T) \tag{3.23}$$

where ϕ is standard normal distribution; α_s is influence coefficient for load and β_T is target reliability index. In Codes of Practice any combination of characteristic value and partial safety factor can be used which leads to S_d, and α_s is found from a complete reliability analysis. In the Swiss Code SIA 160 (1989) α_s is given as 0.7.

The actual value of the safety index can be established as mentioned before (section 3.3.2) by an analysis of existing bridges which is then used for

Table 3.2 Typical Value of Target Reliability (After Vrouvenvelder and Waarts, 1993)

Limit State	Reference Period	Accepted Probability of Failure	Target Reliability Index β_T
ULS	100 years	10^{-4}	3.6
FLS	100 years	10^{-4}	3.6
SLS	1 year	0.5	0

calibration as a target value. Further reading of this very important topic is recommended, and two basic texts are given at the end of this chapter.

3.6 ARCHES

3.6.1 Introduction

Because of their unique shape (see Figures 3.8 and 3.9) and behavioural characteristics, arch bridges are normally treated as a special class of bridge for the purposes of assessment.

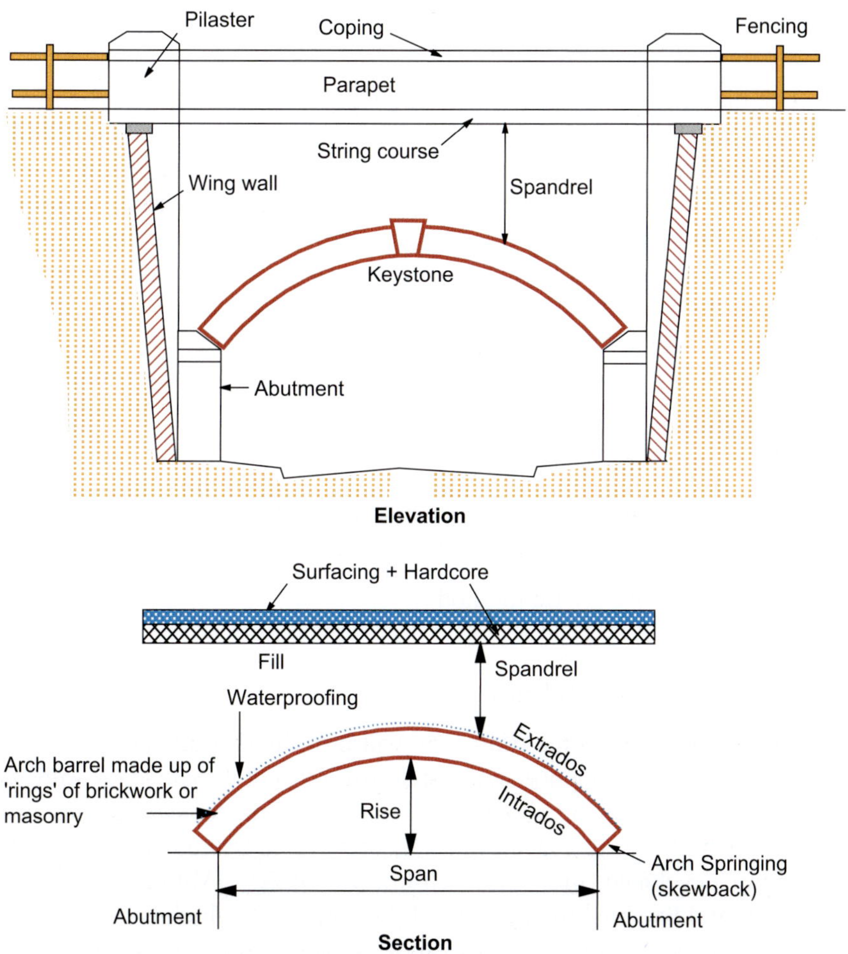

FIGURE 3.8
General arrangement of a masonry arch bridge.

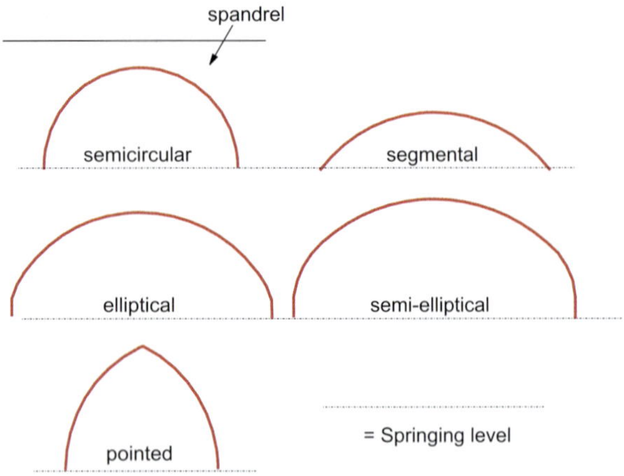

FIGURE 3.9 *Different arch profiles.*

The historical development and application of arch analysis theories are ably presented by both Page (1993) and Sinopoli, Corradi and Foce (1995). In the UK the development of current methods of strength assessment of arches dates back to 1936 (Das, 1993) when Professor Pippard under the auspices of the then Building Research Station undertook research which involved laboratory testing to failure of many arch bridges to devise theoretical and empirical methods for use by practising bridge engineers. The methods developed at that time were directed at finding the strength of the arch barrel which is the key load bearing element in an arch bridge. They were:

1. The elastic method, and
2. The mechanism method

both, of which, have continued – with some modifications – to dominate to the present day. One other method which has developed is:

3. The cracking-elastic method

3.6.2 The elastic method

This method is based on a unit (1 m) width of the (assumed parabolic) arch barrel with pinned supports and a span/rise ratio of 4. Safe loads were originally based on limiting allowable compressive and tensile stresses in the masonry equivalent to a factor of safety of between 3 and 4 at collapse. This has now developed into the well-known MEXE (Military Engineering Experimental Establishment) method (MEXE, 1963) which is fully described in BA 21 (1997). It involves three steps as shown in Figure 3.10.

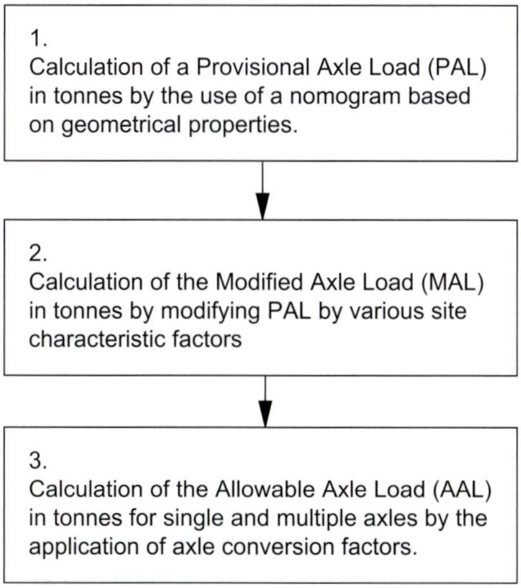

FIGURE 3.10 *Steps in the MEXE method.*

The final AAL is that which can safely be sustained at the crown position. Hughes and Blackler (1997) point out that there are a number of shortcomings with this method, the principal one being the uncertainty surrounding the various modifying factors and the assumptions made in deriving the PAL. A comprehensive analysis of all of the shortcomings is given by Woolfenden (1993). Nevertheless, this method has been shown to give reliable results, and in the UK it is generally the first method used in assessing the capacity of an arch. More sceptical engineers may, however, agree with Harvey (1995) 'MEXE is very close to black magic ...'! The arch idealisation for MEXE and an example of its use are given in Appendix 3B.

In the latest version of BA 16 (1997), details of a computerised version of the MEXE method is described for use with any 2D frame or 3D FE analysis programme. It is usually referred to as pinned-elastic and allows for the live load to be applied over a finite area, and also for dispersion of load through the fill.

3.6.3 The mechanism method

The tests referred to earlier by Professor Pippard showed quite clearly that arches collapse due to the formation of hinges in the arch barrel (Das, 1993). This was investigated further by Professor Heyman who developed what is now known as the mechanism method (Heyman, 1982).

An arch transmits load *via* a line of thrust, the natural path for a load to travel to the foundations, and is the locus of the resultant compression stresses at any section due to applied loads. Provided that the line of thrust remains within the middle-third of the (prismatic) section then the whole of the section is in compression. If it falls outside the middle-third then part of the section is subject to tensile stresses causing part of the masonry to crack. Further application of load will displace the line of thrust even further from the centroid until it reaches the extrados and a 'hinge' is formed. The hinges

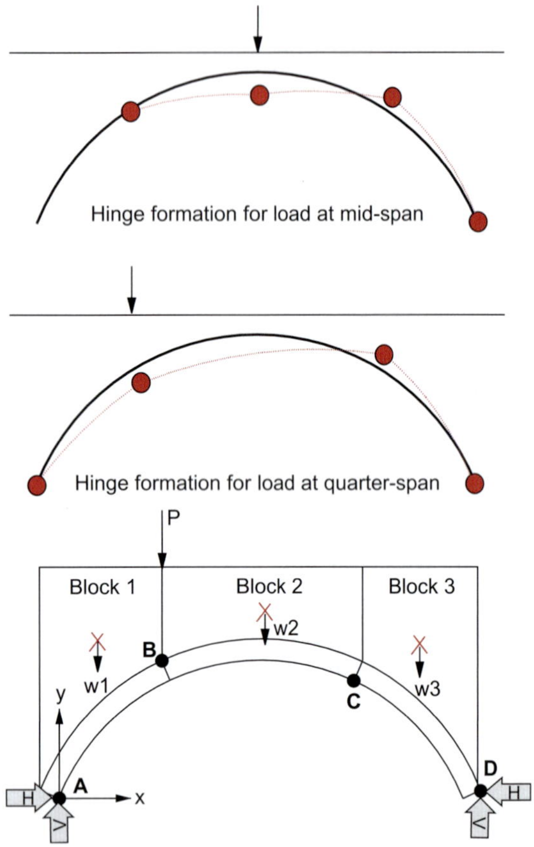

FIGURE 3.11 *Four-hinge mechanisms in masonry arches.*

form and rotate at positions dictated primarily by the position of the applied (axle) load as shown in Figure 3.11. This can be analysed to determine the critical value of P at the quarter point or the mid-point.

$$\sum M \text{ about B: } Hy_B + w_2(x_2 - x_B) + w_3(x_3 - x_B) = V(X_c - X_B) \tag{3.24}$$

$$\sum M \text{ about C: } Hy_c + w_3(x_3 - x_c) = V_D(x_D - x_c) \tag{3.25}$$

Hence H and V_D

$$\sum M \text{ about A: } w_1\, x_1 + Px_B + w_2\, x_2 + w_3\, x_3 = V_D\, x_D \tag{3.26}$$

Hence P.

This failure mechanism has been observed by many researchers such as Harvey (1988); Melbourne (1990); Gilbert and Melbourne (1995); Melbourne and Gilbert (1995) and Sinopoli et al. (1995). The method assumes that the arch has no tensile strength and that each arch ring is rigid (i.e. it possesses infinite stiffness and compressive strength). The method is particularly amenable to progamming using a computer, and one of the first programmes involving the use of finite elements was developed by Chrisfield (1985). Since then other software packages have come on to the market in the UK such as ARCHIE, MAFEA and ARCH all of which have been assessed by the Department of Transport against collapse loads on actual bridges (BA 16, 1997). A typical 2D model for use in such an analysis is shown in Figure 3.12. The programmes utilise an iterative approach and plot the line of thrust under an incrementally increasing load until a four-hinge mechanism has developed causing collapse. Dead loads are applied as a separate load case as vertical loads at the nodes.

Woolfenden (1993) describes the use of 1D, 2D and 3D finite element models which are able to take into account of limited compressive stress in

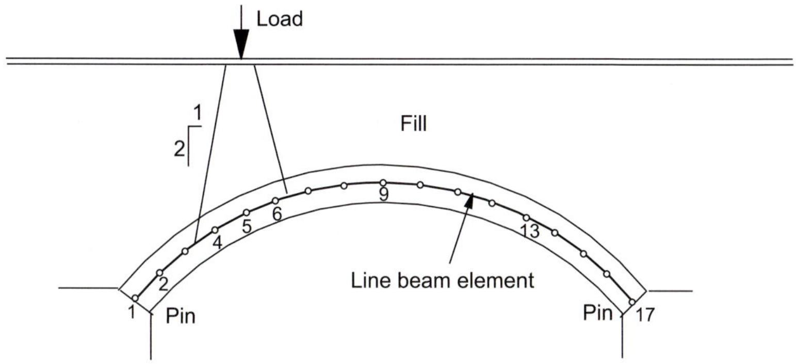

FIGURE 3.12
Typical line representation of arch barrel.

the arch barrel; defects such as cracks, missing bricks and separated arch rings, and also allow for rotated or displaced springings. In addition the 3D analysis can model the effects of skew (Choo & Gong, 1995), stiffening of the fill (Kumar, 1995; Hodgson & Fashole-Luke, 1997), diagonal cracking and spandrel stiffening.

One way of evaluating the accuracy of an analysis method is to calculate the degree of correlation between full-scale tests to failure and FE models. This was done at the TRL for a fully bonded arch, a ring separated arch, and a ring separated and strengthened arch by Sumon (2000). A 2D FE model for each was produced containing both brick and mortar elements. During each test deformation, displacements, stresses and strains and failure load were measured and later the results were compared with those from the FE analysis. Among other things it was found that the mortar/sand joints were very sensitive to the applied loads; the accurate specification of the material properties was crucial; the failure points in the models closely mirrored those of the prototypes. Finally, it was found that the displacements induced in the ring separated arches were very similar between analyses and tests but there was wide difference (7–10 times) in the measured and calculated strains – a totally unexpected result for which no explanation was given. Overall, the exercise was judged a success and further research is underway using 3D modelling.

3.6.4 Cracking-elastic method

This method employs the use of 1D or 2D finite elements arranged in a similar way to that shown in Figure 3.9 (Choo et al., 1991; Gong & Choo, 1994) to examine the stress and deflection at points along the arch barrel to the applied loads, and the resistance of the soil which is modelled by simple spring elements. Stresses enable areas of tensile yielding (and hence cracking) of the masonry to be identified and the section properties modified by thinning of the arch barrel, and since the depths of cracks vary, this leads to tapering of the line elements. The deformations enable modification of the arch geometry. This process is continued up to the point where failure occurs. This method requires very sophisticated programming to continually update the model but it can yield good results.

Based on the geometrical and material properties of a particular bridge, Hughes and Blackler (1997) examined the overall performance of four assessment methods based on their own postulated criteria. Their conclusions act as a good guide when trying to choose a suitable method for assessment and are reproduced in Table 3.3.

By this reckoning it would appear that the cracking-elastic method is the clear winner, but it can be seen to be expensive (at least at the moment) and is no more 'accurate' in assessing the load carrying capacity than the other methods. The reader is recommended to read the paper by Hughes and Blackler (1997) to get the full benefit of the discussion of their results. Their final recommendation is to begin with MEXE, and if there is a lack of capacity, to then follow up with either the mechanism or elastic-cracking methods.

3.6.5 Centrifuge modelling

A very interesting series of tests on models of brick arches in a centrifuge were undertaken by Hughes, Davies and Taunton (1998). The reason for these tests was that since the structural behaviour of arches is dominated by

Table 3.3 Overall Assessment Methods Performance (After Hughes and Blackler, 1997)

No.	Criteria	MEXE	Pinned-elastic	Mechanism	Cracking-elastic
1	Correctly predicting load-carrying capacity	A	A	A	A
2	Allow assessment at reasonable cost	G	A	A	P
3	Correctly model the failure modes	P	P	A	G
4	Properly consider those parameters which influence load capacity	P	P	A	G
5	Minimise the use of additional global factors	P	A	G	G
6	Provide additional appreciation of behaviour	P	P	A	G
7	Allow progressively more detailed investigation of borderline bridges	P	P	A	G
8	Model remedial works	P	A	G	G
9	Allow the modelling of service loads	P	A	P	G
10	Be suitable for improvement with additional knowledge	P	P	A	G

Note: *P, poor; A, average and G, good.*

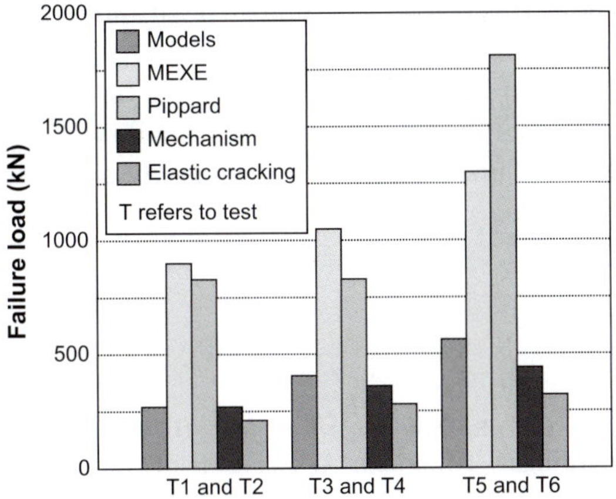

FIGURE 3.13 *Variation of model and predicted assessment loads with masonry and soil type (after Hughes* et al.*, 1998).*

gravity, then conventional modelling of arches does not give a true measure of the stresses in the backfill material and the arch barrel. The use of a centrifuge increases the apparent density of the constructional materials. Hughes et al. (1997) used 1/6 dimensional scale models, and so the centrifuge had to provide a force equivalent to 6 g, thus making the materials seem six times heavier. They compared the model failure loads with those calculated from the four common methods of assessment listed in Table 3.3. The results are shown in Figure 3.13.

From Figure 3.13 it is clear that the mechanism and elastic-cracking methods provide the most accurate assessments of strength, whilst MEXE and Pippard overestimate by two to three times. Perhaps it is now time to abandon the latter two and use the mechanism method in general, and the more sophisticated elastic-cracking method if time and cost allow.

Sicilia, Hughes and Pande (2000) are involved in the continuation of the research of Hughes et al. (1998) and have recently carried out a centrifuge test which they claim is 'the first 3D model equivalent of a masonry arch of significant dimensions tested in a laboratory'.

3.7 CONCLUDING REMARKS

It should be clear from the foregoing sections that the assessment process is a complex one involving many decisions. The objective is to arrive at a satisfactory conclusion regarding the load-carrying capacity of a particular bridge. In most cases this can be accomplished without recourse to reliability methods, and the five phase approach indicated in Appendix 3C is presented as a guide to decision making. At Phase (II), (III) or (IV) a decision can be taken regarding the usefulness and economic viability of carrying out a load test (see Chapter 5) to enable a more complete understanding of the structural action to be made.

REFERENCES

AASHTO. (1989). Guide Specification for Strength Evaluation of Existing Steel and Concrete Bridges, American Association of State Highway and Transportation Officials, Washington, DC.

AASHTO. (1992). Standard Specification for Highway Bridges, American Association of State Highway and Transportation Officials, Washington, DC.

AASHTO. (1994). LFRD Bridge Design Specifications, American Association of State Highway and Transportation Officials, Washington, DC.

BA 16. (1997). *Advice note – The assessment and strengthening of highway structures, Design Manual for Roads and Bridges*. London: HMSO.

Bakht, B., & Jaeger, L. G. (1985). *Bridge Analysis Simplified*. New York: McGraw Hill.

Bahkt, B., & Jaeger, L. G. (1992). Ultimate load test of Slab-and-Girder bridge. *Journal of Structural Engineering, 118*(6), 1608–1624.

Bailey, S. F. (1996). Basic principles and load models for the structural safety evaluation of existing road bridges, Thesis submitted to the Ecole Polytechnique Federal de Lausanne for the award of Doctor in Scientific Techniques.

BD 21. (1997). *Specification – The assessment and strengthening of highway structures, Design Manual for Roads and Bridges*. London: HMSO.

BD 50. (1997). *Implementation of the 15 year bridge rehabilitation programme – Stage 3 long span bridges, Design Manual for Roads and Bridges*. London: HMSO.

Blockley, D. (ed.). (1992). *Engineering Safety*. London: McGraw-Hill.

Chase, S. B. (1998). *The Bridge Maintenance Program of the United States Federal Highways Administration, The Management of Highway Structures Symposium*. London: ICE.

Choo, B. S. and Gong, N. G. (1995). Effect of skew on the strength of masonry arch bridges, Arch Bridges, Proceedings of the First International Conference on Arch Bridges, Bolton, UK, pp. 205–214, Thomas Telford, London.

Choo, B. S., Coutie, M. G., & Gong, N. G. (1991). Finite element analysis of masonry arch bridges using tapered elements. *Proc. Inst. Civ. Engrs, Part 2, 91*, 755–770.

Chrisfield, M. A. (1985). Finite Element and Mechanism Methods for the Analysis of Masonry Arch Bridges, TRL research report 19.

Clark, L. A. (1983). *Concrete Bridge Design to BS 5400*. London: Construction Press.

CSA. (1990). Supplement No. 1, Existing bridge evaluation to CSA Standard CAN/CSA-S6–88 Design of Highway Bridges, Canadian Standards Association.

Das, P. C. (1993). The assessment of masonry arch bridges, Arch Bridges, Proceedings of the First International Conference on Arch Bridges, Bolton, UK, pp. 21–27, Thomas Telford, London.

Das, P. C. (1997). Development of bridge-specific assessment and strengthening criteria, Safety of Bridges, International Symposium, London, pp. 53–57, Thomas Telford, London.

Eurocode 1. (1993). Basis of Design and Actions on Structures, Vol. 3, Traffic Loads on Bridges, CEN Brussels.

Ghosen, M. (1996). Safety evaluation of highway bridge structural systems. In J. R. Casas, & F. W. Klaiber (eds.), *Recent Advances in Bridge Engineering*. Barcelona: CIMNE.

Ghosen, M., Moses, F., & Khedekar, N. (1994). Response functions and system reliability of bridges. In P. D. Spanos, & Y. T. Wu (eds.), *IUTAM Symposium, Probabilistic Structural Mechanics: Advances in Structural Reliability Methods*. Heidelberg: Springer-Verlag.

Gilbert, M. and Melbourne, C. (1995). Analysis of multi-ring brickwork arch bridges, Arch Bridges, Proceedings of the First International Conference on Arch Bridges, Bolton, UK, pp. 225–238, Thomas Telford, London.

Gong, N. G., & Choo, B. S. (1994). Effects of diagonal cracks in masonry arch bridges. In B. I. G. Barr, H. R. Evans, & J. E. Harding (eds.), *Bridge Assessment, Management and Design*. Elsevier.

Hambly, E. C. and Hambly E. A. (1992). Bridge Deck Behaviour, E & F N Spon.

Hambly, E. C. and Hambly E. A. (1994). Risk evaluation and realism, Proc. Inst. Civil Engineers, Civil Engineering 1 (2).

Harvey, W. J. (1988). Application of the mechanism analysis to masonry arches. *The Structural Engineer, 66*(5).

Harvey, W. J. (1995). Loaded ribs or complex systems – A personal view of our ability to model arch bridge behaviour, Proceedings of the First International Conference on Arch Bridges, Bolton, UK, pp. 29–35, Thomas Telford, London.

Henriques, A. A. R., Calheiros, F. J. L. C. and Figueiras, J. A. (1996). Methodology for the assessment of the structural behavior of concrete bridges, Third International Conference on Bridge Management, University of Surrey, Guildford, UK, pp. 138–146, E & FN Spon.

Heyman, J. (1982). *The Masonry Arch*. Ellis Horwood.

Hillerborg, A. (1962). Reinforcement of slabs and shells designed according to the theory of elasticity. *Betong, 38*(2), 101–109, 1953. Translation by G. N. Gibson, Building Research Station, Watford.

HMSO. (1996). *National Audit Office: 'Highways Agency: The Bridge Programme', HC282*. London: HMSO.

Hodgson, J. A. and Fashole-Luke, P. S. (1997). Finite element analysis of the multi-ring brickwork arch, Proceedings of the Seventh International Conference on Structural Faults and Repair, Edinburgh, pp. 135–142.

Hughes, T. G., & Blackler, M. J. (1997). A review of the UK masonry arch assessment methods. *Proc. Inst. Civ. Engrs. Structs. and Bldgs, 122*, 305–315.

Hughes, T. G., Davies, M. C. R., & Taunton, P. R. (1998). Small scale modelling of brickwork arch bridges using a centrifuge. *Proc. Inst. Civ. Engrs. Structs. and Bldgs, 128*, 49–58.

Ibell, T. J., & Middleton, C. R. (1998). Shear assessment of concrete beam-and-slab bridges. *Proc. Inst. Civ. Engrs. Structs. and Bldgs, 128*, 264–273.

Jackson, P. (1996). Analysis and assessment of bridges with minimal transverse reinforcement, Third International Conference on Bridge Management, University of Surrey, Guildford, UK, pp. 779–785, Spon.

Jackson, P. and Cope, R. J. (1990). Strength assessment methods for concrete bridges, First International Conference on Bridge Management, University of Surrey, Guildford, UK, pp. 429–438, Elsevier Press.

Johansen, K. W. (1962). *Yield Line Theory*. London: Cement and Concrete Association.

Kumar, A. (1995). Semiautomatic approaches for masonry arch assessment, Arch Bridges, Proceedings of the First International Conference on Arch Bridges, Bolton, UK, Thomas Telford, pp. 267–276.

Lark, R. J. and Mawson, B. R. (2000). Assessment at the Serviceability Limit State, Fourth International Conference on Bridge Management, University of Surrey, Guildford, UK, Thomas Telford Ltd, London, pp. 426–433.

Leeming, M. B. (1990). *Keeping water out of concrete, First International Conference on Bridge Management*. University of Surrey, Guildford, UK: Elsevier Press.

Melbourne, C. (1990). The assessment of masonry arch bridges – The effects of defects, First International Conference on Bridge Management, University of Surrey, Guildford, UK, pp. 523–531, Elsevier Press.

Melbourne, C., & Gilbert, M. (1995). The behaviour of multiring brickwork arch bridges. *The Structural Engineer, 73*, 39–47.

MEXE. (1963). Military Engineering Experimental Establishment, Classification of Civil Bridges the Reconnaissance and Correlation Methods, Christchurch, UK.

Micic, T. (1997). *Statistical Reliability Methods for Probabilistic Bridge Assessment: Part 1*. London: Highways Agency.

Middleton, C. R. (1997). Concrete bridge assessment – An alternative approach. *The Structural Engineer, 75*(23 & 24), 403–409.

Moses, F. and Verma, D. (1990). Load Carrying Evaluation of Existing Bridges, Final Report, NCHRP 12–28(1).

Nowak, A. S. (1997). Application of bridge reliability analysis to design and assessment codes, Safety of Bridges, International Symposium, London 42–49, Thomas Telford, London.

OHBDC. (1991). Ontario Highway Bridge Design Code, 3rd Edn, Ministry of Transportation, Downsview, Ontario, Canada.

Page, J. (ed.). (1993). *Masonry Arch Bridges, TRL State of the Art review.* London: HMSO.

Pucher, A. (1964). *Influence Surfaces for Elastic Plates.* Wien and New York: Springer-Verlag.

Ryall, M. J. (1992). Application of the D-type method of analysis for determining the longitudinal moments in bridge decks. *Proc. Inst. Civ. Engrs. Structs. and Bldgs, 94*, 157–169.

Shetty, N. K., Chubb, M. S., & Manzocchi, G. M. E. (1998). *Advanced methods of assessment for bridges, Symposium on The Management of Highway Structures*. London: ICE.

SIA 160. (1989). Actions on Structures, Societe suisse des ingenieurs et des architects, Zurich.

SIA 462. (1994) Evaluation de la Securite Structurale des Ouvrages Existantes, Societe suisse des ingenieurs et des architects, Zurich.

Sicilia, C., Hughes, T. G. & Pande, G. N. (2000). Three-dimensional centrifuge test of Pontypridd Bridge, Fourth International Conference on Bridge Management, University of Surrey, Guildford, UK, pp. 63–71, Thomas Telford, London.

Sinopoli, A., Corradi, M. & Foce, F. (1995). A modern evaluation of the historical theories about masonry or stone arches, Arch Bridges, Proceedings of the First International Conference on Arch Bridges, Bolton, UK, pp. 131–142, Thomas Telford, London.

Sumon, S. (2000). Analysis of bonded, ring separated and strengthened masonry arch bridges, Fourth International Conference on Bridge Management, University of Surrey, Guildford, UK, pp. 600–607, Thomas Telford, London.

Thoft-Christensen, P., Jensen, F. M., Middleton, C. R., & Blackmore, A. (1997). Revised rules for concrete bridges, Safety of Bridges, International Symposium, London, pp. 175–188, Thomas Telford, London.

Vrouvenvelder, A. C. W. M., & Waarts, P. H. (1993). Traffic Loads on Bridges. *Structural Engineering International, 3*(93), 169–177.

Westergaard, H. M. (1931). Computation of stresses in bridge slabs due to wheel loads. *Public Roads, 2*(1), 1–23.

White, K. R., Minor, J., & Derucher, K. N. (1992). Bridge Maintenance, Inspection and Evaluation, M. Dekker inc., New York.

Woolfenden, P. A. (1993). Modelling the masonry arch: improving modern bridge assessment using a nonlinear finite-element software package (MAFEA), Second International Conference on Bridge Management, University of Surrey, Guildford, UK, pp. 254–263, Thomas Telford, London.

FURTHER READING

BA 44. (1990). *Advice note – The assessment of concrete highway bridges and structures, Design Manual for Roads and Bridges*. London: HMSO.

BD 37-01. (1988). *Loads for highway bridges, Design Manual for Roads and Bridges*. London: HMSO.

Ditlevsen, O., & Madsen, H. O. (1995). *Structural Reliability Methods*. Wiley Interscience-Europe.

Micic, T. (1997). *Statistical Reliability Methods for Probabilistic Bridge Assessment: Part 2*. London: Highways Agency.

White, K. R., Minor, J., & Derucher, K. N. (1992). Bridge Maintenance, Inspection and Evaluation, M. Dekker Inc., New York.

Schneider, Jorg, (1997). Introduction to Safety and Reliability of Structures. IABSE Structural Engineering Documents 5.

APPENDIX 3A

Natural variability arises from the fact that the basic variable (load, strength, etc.) is random and not deterministic. This natural variation is represented by a probability distribution which enables the mean and standard deviation to be calculated which can then be used as a basis for the values of the partial safety factors for use in the assessment.

Estimation error stems from the fact that statistical data for a given variable is limited and so the probability of exceedance cannot be precisely calculated. The estimation error decreases with the number of samples or observations taken. By testing or sampling at the bridge being assessed it is possible to reduce uncertainty and so reduce the partial safety factor.

Model imperfection naturally arises since it is not possible to perfectly represent a real life bridge. Load testing can provide results which may indicate that, for example, the transverse stiffness characteristics can be improved for use in the model. Further modifications can also be made which take into account any additional strength characteristics that may be mobilised (see Chapter 5).

Human error can arise at the design (incorrect load specification), construction (poor quality or missing materials) and operation (traffic overload) stages all of which can be detected by close inspection and/or testing of the bridge.

APPENDIX 3B

The nomenclature adopted by the MEXE method is shown in Figure 3B.1

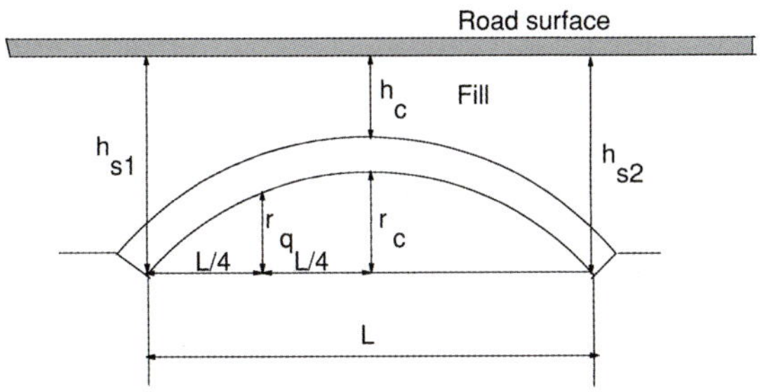

FIGURE 3B.1 *Arch idealisation for MEXE method.*

Example of use

A single span mass concrete arch has a span of 6.0 m, with a rise at the crown 'r_c' = 1.61 m and a rise at the quarter points of $r_q = 1.283$ m. The depth $d + h = 600$ mm (west) and 740 mm (east). It is to be assessed to see if it can support a 40-tonne vehicle having one, two or three axles.

Table 3B.1 MEXE Calculations

Ref: BA 16/97	Modifying factors	MEXE axle loads
$L = 6.0$ m	Span/rise = 6/1.61 = 3.73 < 4	From CL 3.1 of BA 16/97 Provisional Axle Load (PAL) = 49.039 tonne
$r_c = 1.61$ m $r_q = 1.2283$ m	Therefore span/rise factor = 1.00 $r_q/r_c = 1.283/1.61 = 0.80$	Modified Axle Load MAL = $F_{sr} \cdot F_p \cdot F_m \cdot F_j \cdot F_c \cdot$ PAL = 1 × 0.88 × 0.88 × 1 × 0.95 × 49.039 = 36.1 tonne
$d = 0.305$ m $h_c = 0.52$ m	Therefore Profile Factor $Fp = 0.88$ Barrel Factor = 1.2 Fill Factor = 0.7 $F_m = 0.88$ $F_w = F_{mo} = F_d = 1.0$ Therefore Material Factor $F_m = 1.0$ Joint Factor $F_j = 1.0$ Condition Factor $F_c = 0.95$	

(continued)

Table 3B.1 MEXE Calculations *(continued)*

Ref: BA 16/97	Modifying factors	MEXE axle loads
Maximum permitted gross axle and bogie weights (Assuming no axle lift-off)		
Single Axle	$A_f = 1.27$ per axle $= 1.27 \times 36.1 = 45.8$ tonne	per bogie = 45.8 tonne
Double Axle Bogie	$A_f = 1.00$ per axle $= 1.00 \times 36.1 = 36.1$ tonne	per bogie = 72.2 tonne
Triple Axle Bogie	$A_f = 1.00$ per axle $= 1.00 \times 36.1 = 36.1$ tonne	per bogie = 108.0 tonne
Weight restrictions		None

APPENDIX 3C

Modus operandi for UK levels of assessment I–IV

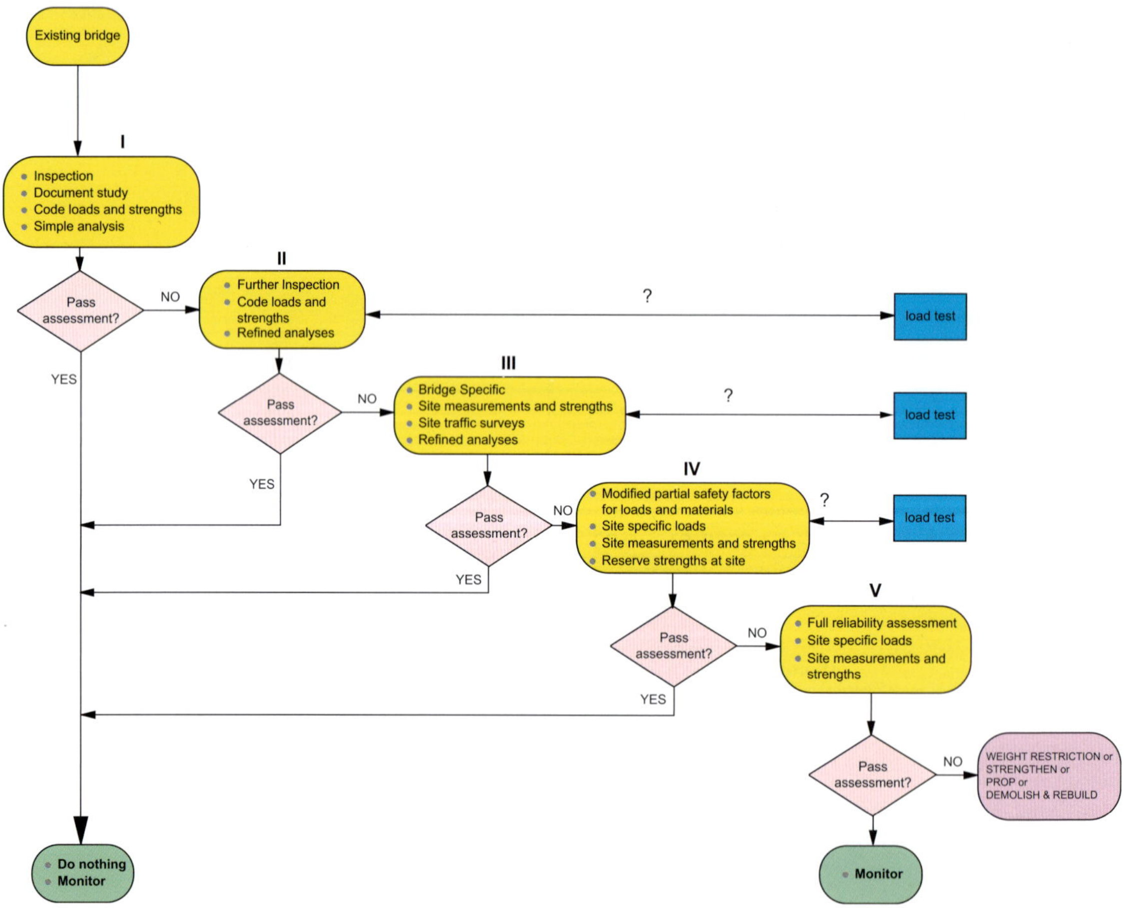

FIGURE 3C.1 *Flow chart for utilising assessment levels I–V.*

CHAPTER 4

Non-Destructive Testing

CONTENTS

4.1 INTRODUCTION

Once a bridge has been inspected for its overall visual condition, it is often necessary to carry out *non-destructive tests* to further extend the diagnostic process if it is suspected that the bridge has been weakened in some way. Non-destructive in this context means that the structure is not destroyed though parts of it may be broken away for testing or inspection.

The tests are normally ordered to determine:

- The physical quality of the materials.
- The position and extent of hidden defects, elements, and material boundaries.

The tests are carried out in situ and provide further information from which an improved diagnosis can be made to enable the bridge engineer to make decisions on the remedial work necessary. Some of the tests are carried out during the course of a normal inspection while others are only applicable during a special inspection.

The use of much of the equipment described in this chapter requires very specialised skills and is normally carried out by trained operatives. They record and report the data from test runs that can be professionally interpreted and translated into meaningful information by the bridge engineer to act upon.

4.2 CONCRETE ELEMENTS

By far the majority of non-destructive tests relate to concrete. This is to be expected as most bridges have some concrete element to them, and the use of concrete in conjunction with steel has proven (under certain conditions) to be the most volatile combination of materials in use today.

4.2.1 Corrosion-analysing equipment

The corrosion of reinforcing bars in concrete constitutes probably the single most expensive cause of local failure. The corrosion of steel is an electrochemical phenomenon that takes place in the presence of oxygen and moisture. It is a two-step procedure; the first is without the presence of chlorides where the high alkaline environment of the concrete prevents further corrosion by the formation of a thin, dense layer of oxide–known as passivation–and remains in place until disturbed by the presence of chlorides in the form of de-icing salts that attack the steel through the passive layer. Once this happens, a *half-cell* is formed with the concrete pore water as the electrolyte. Anodes (corroding areas) produce electrons that are consumed at cathodes (non-corroding areas). This causes a change in voltage along the reinforcing bar, which can be measured and used to identify the areas of corrosion. The name half-cell surveying (or potential mapping) was first coined in the United States by Stratfull (1973) and is derived from the fact that one half of the battery cell is considered to be the steel reinforcing bar and the surrounding concrete. Modern half-cells are of the silver/silver chloride type and have proven to be very stable, and the problems of electrode and electrolyte contamination are greatly reduced. A typical setup for measuring the potential field with a half-cell electrode is shown in Figure 4.1.

The electrode itself has a water-saturated tip and may consist of a rod or a wheel. The area of concrete to be investigated may be vertical or horizontal (either above or below) and is divided into suitable areas for investigation and further subdivided into an orthogonal grid pattern (Vassie, 1991). A scan is then made along the grid lines and readings taken at grid line intersections. A pixel print-out is then usually available as a grey numerical *map* that indicates the potentials in millimetres, and this can be converted to a colour isopotential map. A typical instrument and potential plot are shown in Figure 4.2.

By taking measurements over the whole surface, an appraisal can be made of the *probability* of corrosion. The *map* is then used in conjunction with a rebar location survey and an assessment can be made of the corrosion risk. The commonly accepted

Risk values are related to the potential reading value (E) (ASTM C 876–80) and are indicated in Table 4.1.

The figures in Table 4.1 offer only a rough guide, and are dependent upon the moisture content of the concrete. It is normal practice, therefore, to produce a three-dimensional isopotential contour plot, either of the whole

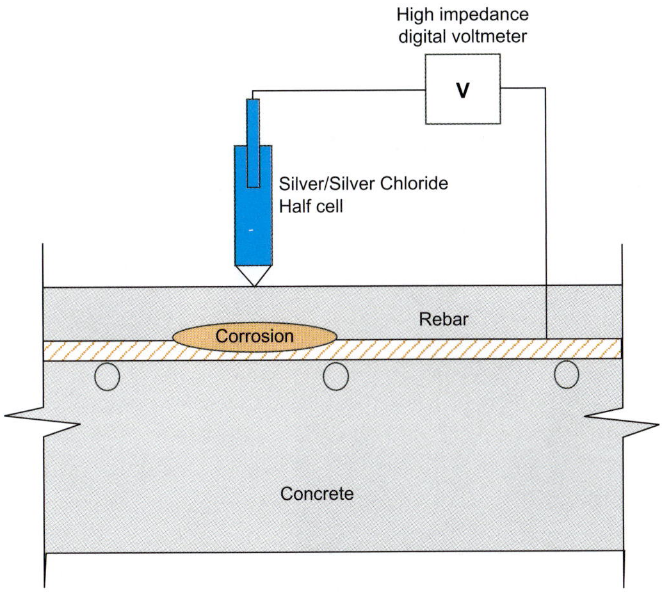

FIGURE 4.1
Typical half-cell setup.

area or along lengths of rebars, and to break out and examine a bar in the most highly negative areas with a view to correlating measured potentials with the actual reinforcement corrosion condition.

After break out, points to be considered are

- Brown or orange (rust) patches in the rebar indicate depassivation and the onset of corrosion.
- Delamination of the steel indicates severe corrosion.
- Cracks in the concrete parallel to the rebars indicate delamination.

There are many corrosion-analysing instruments (CAIs) on the market, and each must be judged according to its merits. It should be remembered that the isopotential method gives an indication only of whether corrosion of steel can occur, that is, whether it is thermodynamically possible, but it can give no information regarding the kinetics or rate of corrosion.

4.2.2 Resistivity measurements

The *rate* at which rebar corrosion occurs in concrete is generally controlled by the *resistivity* (resistance per unit length of unit cross section) of the concrete. The easier it is for the corrosion current to flow through the surrounding concrete, the greater the amount of metal loss from a corroding length of reinforcement. The electrical resistance of concrete is measured in Ohm

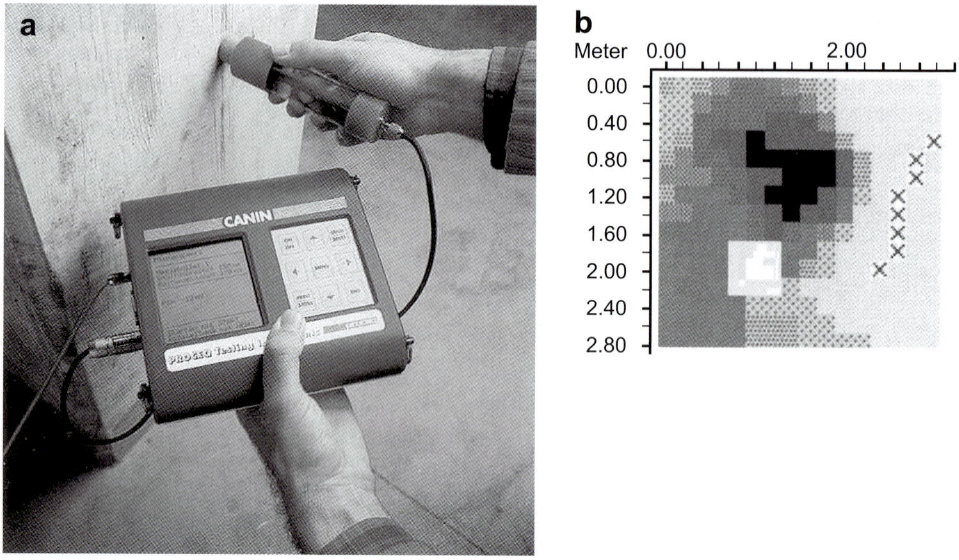

FIGURE 4.2 *(a) Typical corrosion meter and (b) potential plot (see colour plate section).*

centimetres (Ω cm). Increasing resistivity will result in a reduced corrosion content. Dry concrete is an insulator with a resistivity R of 10^{11} Ω cm and embedded rebars will corrode less quickly than in most concrete, which is a semiconductor with a resistivity of 10^4 cm. Resistivity measurements predict the probability of suffering significant levels of corrosion when half-cell tests show that corrosion is possible. The usual ranges are indicated in Table 4.2. Resistivity is measured using a method developed by Wenner (1915) for determining soil resistivity. It consists of four contact points placed at equal distances in a straight line. The contact resistance on each of the probes can differ and cause serious errors if sine wave current sources are used. It is better, therefore, to use a meter that has a flat-topped current wave form (C.N.S. Electronics Ltd, 1990). Also, before using an instrument for the first time, a controlled baseline survey should be carried out to become familiar with the instrument to ensure consistency of the readings.

Table 4.1 Corrosion Risk

Potential Value	Corrosion Risk
$E > -200$ mV	<5%
-200 mV $< E < -350$ mV	Uncertain
$E < -350$ mV	99%

Table 4.2 Resistivity Ranges

Resistivity	Probability of Corrosion
R > 12 kΩ cm	Insignificant
5 kΩ cm < R < 12 kΩ cm	Possible
R < 5 kΩ cm	Almost certain

A typical proprietary meter in use is shown in Figure 4.3. It measures both the current (I) and voltage (V) and converts the readings to resistance within the meter, and stores the gathered data in a semipermanent memory. The data can later be downloaded and analysed using suitable software.

Care should be exercised when interpreting the readings as they can vary with the season. In the United Kingdom, Darby (1996) showed that for long-term monitoring with permanently embedded sensors there was a clear

a

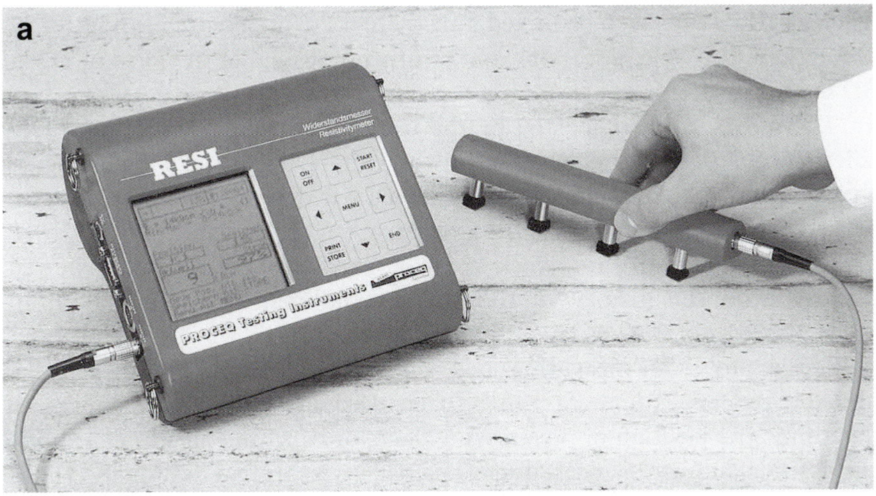

b

	A	B	C	D	E	F	G	H	I	J	..	..	..	AX
1	38	30	27	20	17	14	14	15	25	27				
2	34	29	24	18	12	12	9	6	13	25				
3	32	26	23	15	13			5	11	24				
4	32	28	23	16	13			4						
5	32	28	23	17	13			6						
6	34	29	24	16	14	11	9	7						
7	29	26	25	17	14	13	10	8						
8	28	28	26	20	18	16	14	11						
..														
..														
50														

Table of stored measured values for an extensive investigation of a component

```
RESI ResistivityMeter 1.10  0011.0208
100020   23  129   4  38
 38  30  27  20  17  14  14
 34  29  24  18  12  12   9
 32  26  23  15  13
 32  28  23  16  13
 32  28  23  17  13
 34  29  24  16  14  11   9
 29  26  25  17  14  13  10
 28  28  26  20  18  16  14
 28  29  26  25  19  18  17
```

Data transmission to PC under WINDOWS

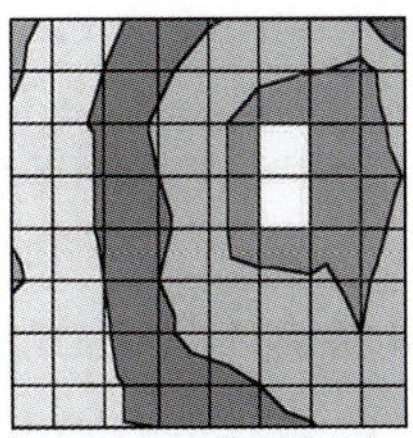

Processing on the PC with EXCEL

FIGURE 4.3

(a,b) Typical resistivity meter and plot.

seasonal variation with higher resistivity in the winter and lower resistivity in the summer. Changes in temperature were also shown to affect the readings. They urge extreme caution in the use of threshold values as the basis for decision making.

In the United Kingdom, if possible, it is better to carry out resistivity testing after a prolonged period of dry weather in a reasonably moderate temperature environment, usually May or June.

4.2.3 Rebar locator

Prior to any tests to determine the extent and rate of corrosion, it is necessary to gather information regarding the layout of rebars in a concrete structure. The current commercially available instruments operate on the principle that the presence of steel affects the field of an electromagnet. Most modern instruments are portable and capable of measuring the position, direction, depth, and diameter of bars within the concrete and of producing a visual display on the instrument itself with an *x–y* metre scale. They can, however, also be linked directly to a PC by the use of the RS 232 C interface to enable processing and printing of results in the form of 2D or 3D contour plots.

The rebars are first located by the use of a scanning probe that identifies the position of a rebar by means of a short 'beep' and a visual signal that displays the cover and the preselected bar diameter. The position of a bar is then automatically stored on an *x–y* grid. Rebars that are to be exposed or where drilling work is to be carried out can be transferred directly from the display to the surface and marked on a maximum measured area of 2 m × 2m.

Special probes for measuring steel depths of up to 300 mm are available, and the general equipment can also be used in some circumstances to determine the position of prestressing ducts or bars, and also to determine the position and size of encased steel beams. A typical rebar locator and visual display are shown in Figure 4.4

To verify the results, it is advisable to check random locations by breaking out the concrete, physically measuring the cover and bar diameter, and comparing with the results obtained from the instrument. For the complications regarding multiple bar situations, see Alldred (1995).

4.2.4 Ultrasonic testing

The velocity of ultrasonic pulses travelling in a solid medium depends upon the density and elastic properties of the medium. The transmission of such

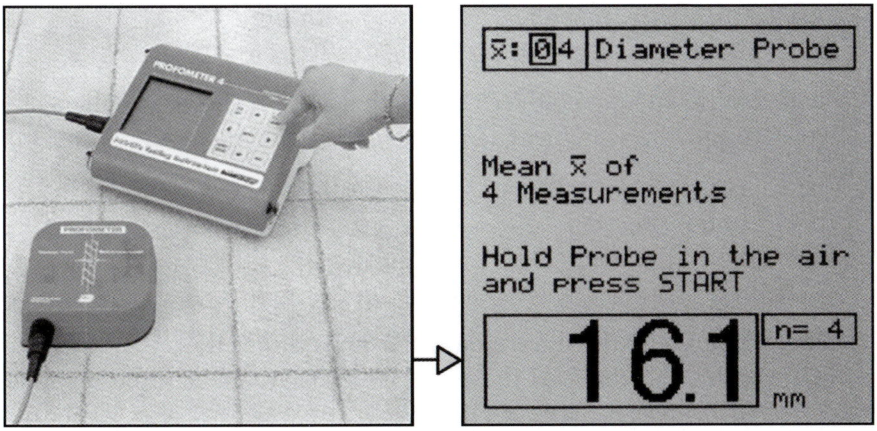

FIGURE 4.4
Typical rebar locator and visual display.

pulses can, therefore, provide information on the *integrity* of plain, reinforced, or prestressed concrete bridges.

The ultrasonic technique measures the transit time of sound waves passing from an *emitter* transducer through the concrete to a *receiver* transducer as shown in Figure 4.5. The *pulse velocity* can be calculated if the length of the path taken by the pulse is known. It is then possible to assess the *quality* of the concrete in relation to:

1. The homogeneity of the concrete.
2. The presence of voids, cracks, or other imperfections.

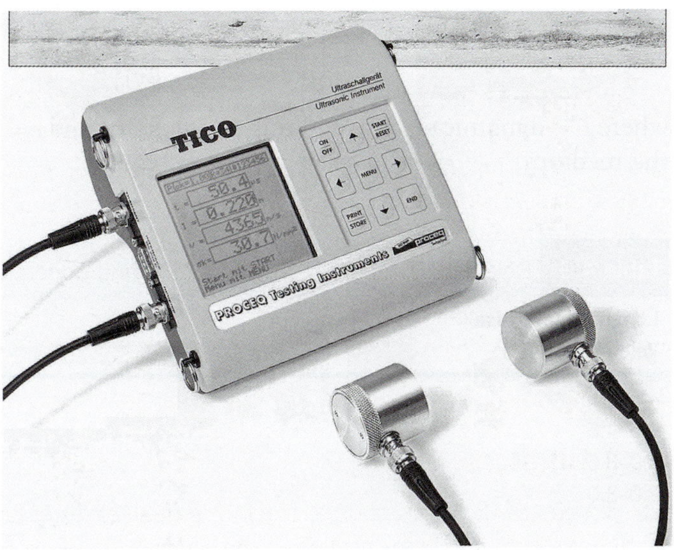

FIGURE 4.5
Typical ultrasonic testing equipment.

3. Changes in the concrete, which occur with time.
4. The quality of the concrete relating to strength.

Typical correlation of pulse velocity with quality is shown in Table 4.3.

Pulse velocity measurement can be made in three modes as indicated in Figure 4.6:

1. Direct transmission.
2. Semidirect transmission.
3. Indirect or surface transmission.

The principle and the three modes are shown in Figure 4.6. The instrument measures the time for the pulse to travel from the emitter to the receiver. Where opposite faces of a member are exposed such as a slab, beam, or column, the direct mode is usual and most satisfactory because the path is well defined. This is the preferred path for *void* detection. In some cases, all faces of bridge elements are not accessible and the indirect modes are used. The path is less easily defined but it is the preferred arrangement when measuring the presence and depth of surface *cracks*. In both cases, the pulse takes longer to reach the receiver than if the concrete was sound.

In metals, which are relatively homogeneous compared with concrete, the equipment is used to detect flaws and inclusions. The ultrasonic beam is reflected back off the flaw and the time taken enables its position to be determined.

The pulse velocity (V) of the longitudinal ultrasonic vibrations travelling in an elastic medium is given by:

$$V = \left[\frac{E(1-\nu)}{p(1+\nu)(1-2\nu)}\right]^{1/2}$$

where E = dynamic elastic modulus, p = density, and ν = Poisson's ratio of the medium.

Table 4.3 Relationship Between Pulse Velocity and Concrete Quality

Longitudinal Pulse Velocity, km/h	Quality of Concrete
4.5	Excellent
3.5–4.5	Good
3.0–3.5	Doubtful
2.0–3.0	Poor
2	Very poor

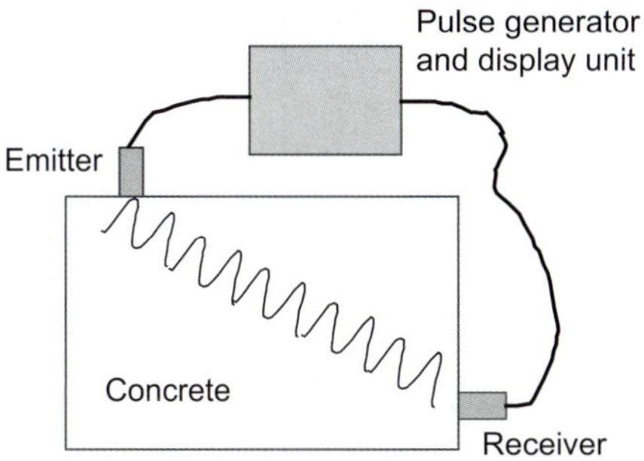

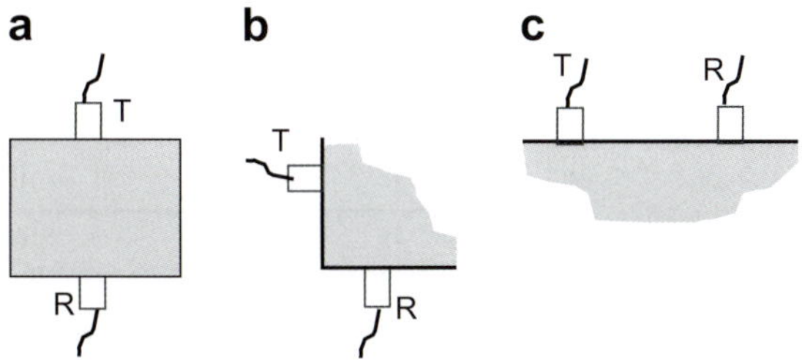

FIGURE 4.6

Transmission modes:
(a) direct;
(b) semidirect; and
(c) indirect.

In concrete, the pulse vibrations are usually low, typically 50 kHz; and providing the least lateral dimension is not less than the wavelength of the pulse vibration (about 200 mm), the above equation is valid. It is clear that measurement of the pulse velocity enables an assessment to be made of the condition of the concrete.

The pulse velocity measurement depends on accurate measurement of the path length and the transit time of the pulse, and usually an accuracy of 1% is required. This demands that there is good contact between the flat surface of the metal transducer and the concrete, and so the concrete surface may need to be prepared by rubbing down with a carborundum block prior to testing. A coupling agent is also often used to ensure good pulse transmittance.

The presence of steel reinforcement considerably affects the pulse velocity, and this has to be taken into account according to the manufacturer's literature. Although the equipment is primarily to establish the quality of the concrete in general terms, some manufacturers provide information about the cube strength, age, and the Modulus of Rupture. Typical relationships are shown in Figure 4.7a–c. The information gathered can help to provide a quick diagnostic picture for the bridge engineer and can enable him to plan a more focused testing programme if necessary. Bungey and Millard (1996) describe the method very thoroughly and although they give detailed information on how to apply the method, they do urge caution in the interpretation of results.

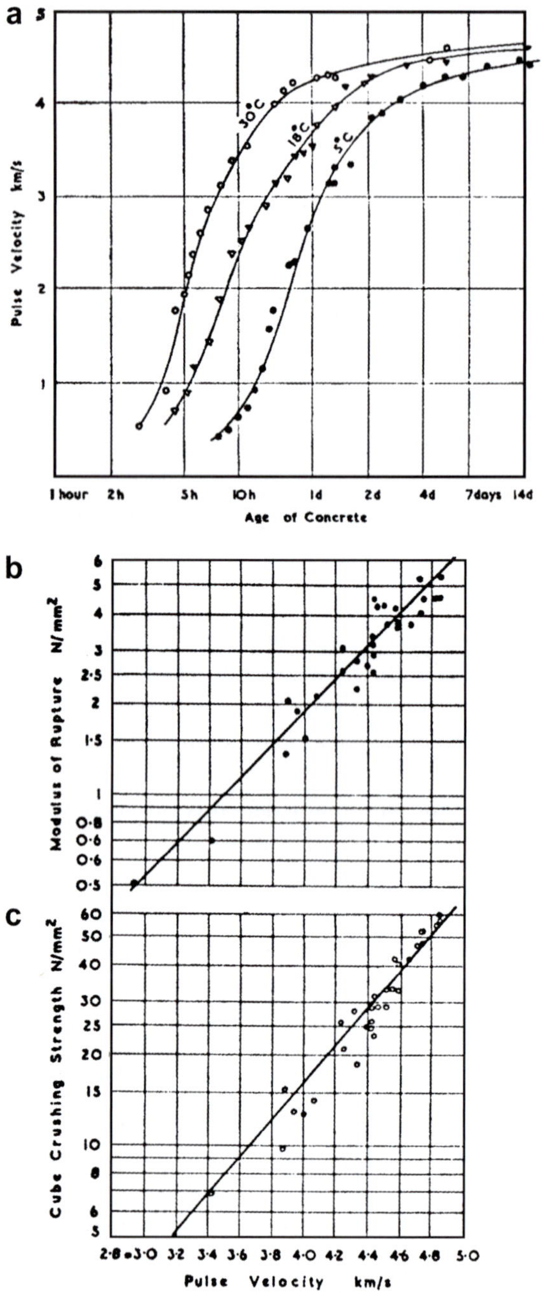

FIGURE 4.7 *(a, b, c) Correlation charts for cube strength, age, and Modulus of Rupture.*

Modern equipment is lightweight and portable, simple to operate, and has a high order of accuracy and stability. A typical instrument in use is shown in Figure 4.8a.

4.2.5 Rebound hammer

One of the primary measures of the quality of concrete is its *compressive strength*. This cannot be measured directly in the field, but an assessment can be made by first measuring its *surface hardness*, by means of a *rebound hammer*.

The hammer is a self-contained unit and a typical instrument is shown in Figure 4.9. The tip of the unit, the impact plunger, is held vertically or horizontally against the smooth surface of the concrete and pushed. This compresses an internal spring that automatically lifts and releases a hammer mass on to the impact plunger and then to the concrete. The impact energy is well defined, and the rebound of the hammer mass is dependent on the hardness of the concrete. Rebound values are indicated on a gauge built into the instrument. It is mostly a comparative technique because the absolute value depends on the local variations in the surface properties due to the presence of voids or aggregate particles. A number of measurements are therefore required in the same location from which the mean and standard deviations values can be determined.

Recently developed instruments incorporate a digital readout with a printer, which can store readings for subsequent statistical processing in a PC. This means that absolute values of compressive strength can be determined with greater accuracy from rebound values.

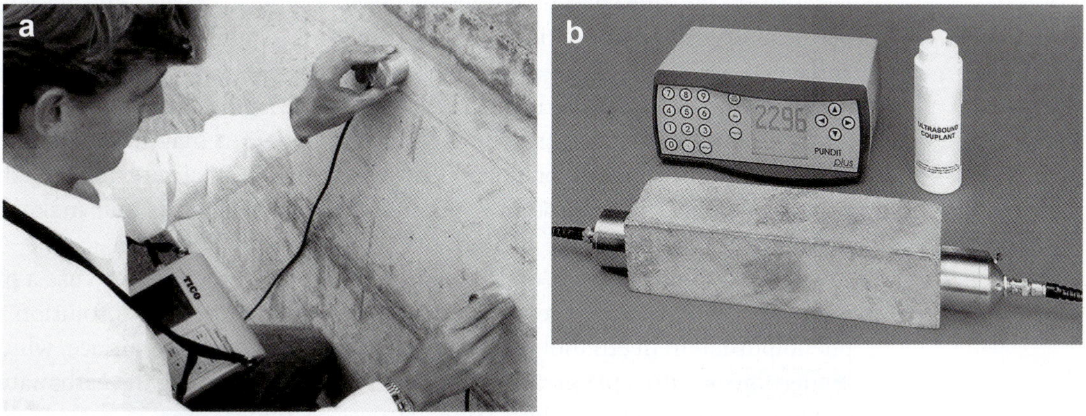

FIGURE 4.8 *(a, b) Portable ultrasonic tester (in situ and in the laboratory).*

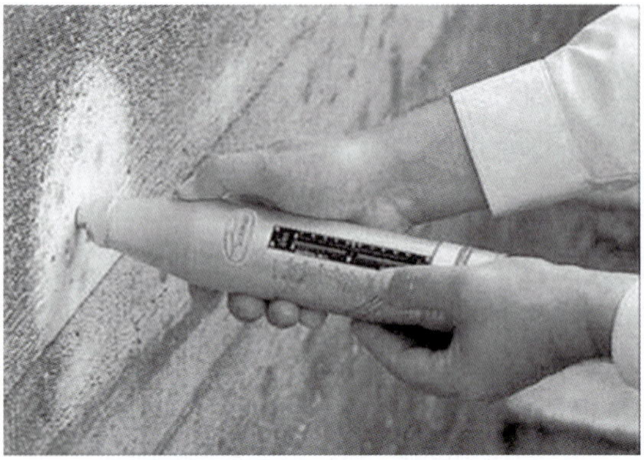

FIGURE 4.9 A rebound hammer.

Because calibration of the most well-known rebound hammer is based on conventional testing on 200 mm cubes, it is necessary to multiply hardness values by a form factor to take account of different-sized specimens. Typical form factors are normally supplied by the manufacturer.

Great care is necessary when using the compressive strength derived in this manner for use in strength assessment of bridges because the values are limited to the surface of the concrete only, and cannot automatically be extrapolated to the body of the concrete.

4.2.6 Carbonation tests

Carbonation is a naturally occurring phenomenon that occurs as the result of absorption of carbon dioxide from the atmosphere into the pore water. This results in the formation of *carbonic acid* near to the surface of the concrete, which reacts with compounds in the hydrated cement paste and increases the strength of the concrete and reduces its permeability.

The rate and extent of carbonation depends on the concentration of carbon dioxide, the permeability of the cement paste, the moisture content of the concrete, the relative humidity of the environment, and the temperature. Thus, permeable concrete in a warm, humid environment (25–90%) is the most vulnerable.

Under normal conditions, the depth of carbonation in a good quality concrete does not exceed 10–20 mm even after many years of exposure. Because of the presence of porous aggregates, voids, cracks, and variable permeability, the carbonation front is rarely a straight line parallel to the surface.

Of itself, carbonation is not necessarily damaging, but in porous or cracked reinforced or prestressed concrete, carbonation may reach as far as the steel (reinforcement or ducts) and the reduced alkalinity renders the naturally passive layer around the reinforcement ineffective and makes it susceptible to corrosion.

To determine the depth of carbonation, the most usual test is to use a pH indicator. To determine the extent of *surface* carbonation, a 2% solution of phenolphthalein in ethanol (ppe) is sprayed on to the concrete surface, which renders areas with a pH greater than 10.5 pink, while leaving the carbonated concrete (with a pH less than 10) colourless. A typical area after testing is shown in Figure 4.10. To determine the *depth* of penetration, a portion of concrete is broken out or cored and the curved surface area coated with ppe. The depth to the pink/clear boundary is then measured and recorded.

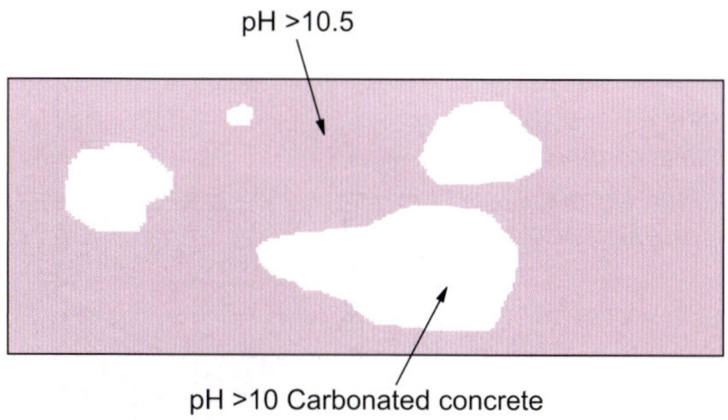

FIGURE 4.10 *Results of a carbonation test.*

4.2.7 Permeability testing

The durability of concrete is directly related to its *permeability in the cover zone (covercrete)* between the surface of the concrete and the nearest reinforcing bar (or duct) and can be measured in a variety of ways (Basheer, 1993). The CEB-FIP Model Code (1991) recognises that the permeability of the cover layer of concrete is directly related to the 'potential durability of a particular concrete'.

4.2.7.1 Air permeability

There are several systems on the market, but only one is described here, which is simple to use, of proven reliability (Torrent & Frenzer, 1995), and one that can take account of the initial moisture content of the concrete.

The permeability coefficient kT (m^2) can be calculated by measuring the flow of air through the concrete by use of a two-chamber vacuum cell and a pressure regulator. The depth of penetration of the vacuum can also be determined. The equipment is simply placed on the concrete, a vacuum is applied, and flow of air commences from the surrounding concrete and through into the vacuum chambers. The data are collected automatically and converted to a value of kT and the depth of penetration, which is then displayed visually.

The measurement takes from 2 to 12 minutes depending on the permeability of the concrete. For dry concrete (not subject to water for about 2 weeks), the quality class of the concrete cover can be read from a table using the kT values similar to those shown in Table 4.4.

Table 4.4 Quality Classes of Cover Concrete Based on Air Permeability (Torrent & Frenzer, 1995)

kT (10^{-16} m^2)	Index	Quality of Cover Concrete
0.01	1	Very good
0.01–0.1	2	Good
0.10–1.0	3	Normal
1.0–10	4	Bad
10	5	Very bad

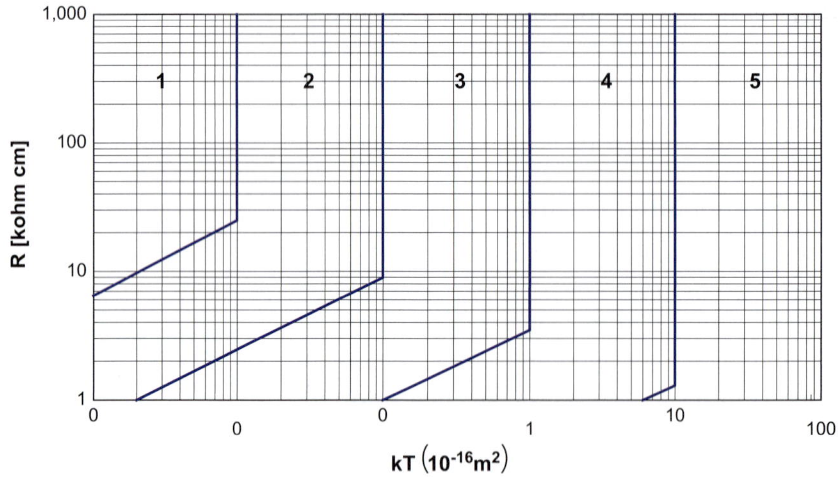

FIGURE 4.11 *Nomogram for determining quality class of moist concrete (after Torrent & Frenzer, 1995).*

In the case of moist concrete, *kT* is combined with the electrical concrete resistance ρ (kΩ cm) and the quality class can be determined from a nomogram similar to that shown in Figure 4.11. In this case (providing the concrete is young), for every single measurement of *kT* there should be three to six measurements of ρ from which the mean value of ρ can be calculated and the quality index read from the nomogram.

The equipment is portable, simple to use, and results are in good agreement with the traditional laboratory methods such as oxygen permeability, capillary suction, and chloride penetration (Torrent, Ebensburger, & Gebauer, 1994; Torrent & Frenzer, 1995). It can be used on concrete surfaces inclined at any angle.

4.2.7.2 Air and water permeability

Equipment to carry out a combined air and water test was first put forward by Figg (1973). The current application is based on his work. The method is based on the time it takes for a small volume of water to escape from a blind hole drilled into the concrete.

A 10 mm diameter by 40 mm deep hole is drilled into the horizontal surface of the concrete and thoroughly cleaned using compressed air. A 20-mm long moulded rubber plug is then inserted into the hole and a steel hypodermic needle is pushed through the plug into the void. This is connected to a hand vacuum pump, and the time taken in seconds for the vacuum to fall from –55 kPa to 50 kPa is defined as the Figg number. The correlation between this and the quality of the concrete is shown in Table 4.5.

In the water test, water is inserted into the void using a hypodermic syringe and then forced out under pressure. The time taken for the water to

Table 4.5 Quality Classes of Cover Concrete Based on Air Permeability (Figg, 1973)

Figg Number (Time in Seconds)	Index	Quality of Cover Concrete
Less than 30	0	Poor
30–100	1	Moderate
100–300	2	Fair
300–1000	3	Good
Greater than 1000	4	Excellent

travel a fixed distance is sensed by the flow measuring instrument or manometer. The time in seconds is the Figg number for water permeability. The steps are shown in Figure 4.12 and the equipment in use is shown in Figure 4.13.

4.2.8 Tensile strength of concrete

The surface strength of concrete in *tension* can be a good indicator of its quality. It can be measured in situ by securing a 50 mm diameter aluminium test disc directly onto the concrete surface by means of a standard commercial epoxy-based, rapid hardening adhesive, which is then pulled with a force of up to 50 kN. A typical device is shown in Figure 4.14.

The rate of loading can be pre-set, and a digital readout displays the highest tensile stress to the nearest 0.01 N/mm^2. The compressive strength can then be determined *very approximately* by multiplying the pull-off tensile stress by 10. In some cases where the concrete is particularly strong, failure occurs at the adhesive/concrete interface, in which case the results refer to the tensile strength of the adhesive.

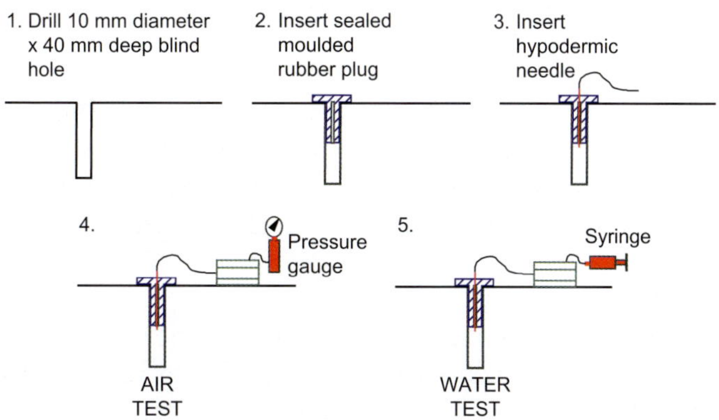

FIGURE 4.12 *Steps in the application of the modified Figg method.*

FIGURE 4.13
Typical Figg equipment in use.

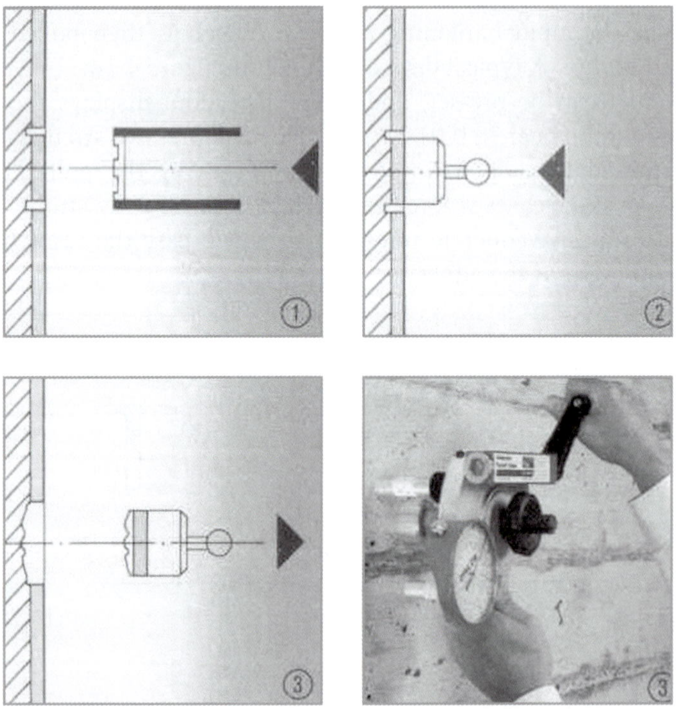

FIGURE 4.14
Pull-off tester.

4.2.9 Internal fracture tester

The most accurate means of testing the *in situ compressive strength* of concrete is by means of the internal fracture test developed by Chabowski and Bryden-Smith (1980a).

A 6 mm diameter Parabolt is placed into a pre-drilled 6 mm diameter hole so that the sleeve is 20 mm below the surface of the concrete. Load is applied in increments every 10 s until the concrete fails and a cone of concrete is pulled off. The bolt is first-hand tightened and then loaded by means of a calibrated torque meter. This method suffers from the fact that some torque is applied to the bolt and the meter is relatively insensitive. Bungey (1981) developed a loading method using a proving ring, which has the advantage of providing a direct pull to the bolt, but so far this does not seem to be in popular use. A typical setup using a torque meter is shown in Figure 4.15.

Torque results are compared with calibration curves and the compressive strength read directly. In the United Kingdom, calibration curves have been produced by BRE (Chabowski & Bryden-Smith, 1980b) to give the compressive strength equivalent to those expected from compressive tests on 150 mm concrete cubes. These are reproduced in Figure 4.16. Six tests are recommended, and the average compressive stress calculated from $f = 3.116T^{1.69}$ N/mm^2.

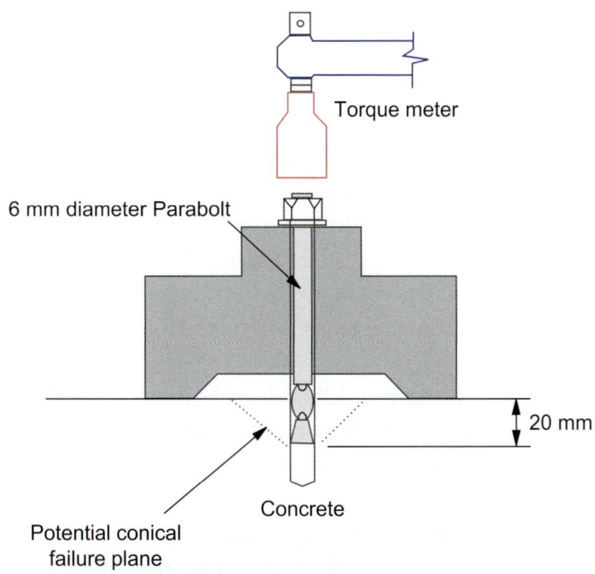

FIGURE 4.15 *Fracture test arrangement.*

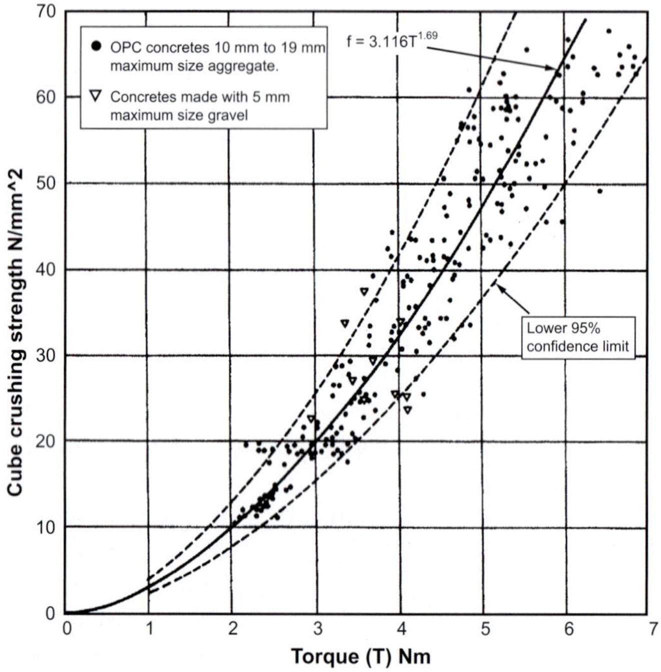

FIGURE 4.16
Calibration curves for natural aggregates.

4.2.10 Impulse radar

Impulse radar is a relatively old technique that has been used for many years for geophysical surveys, but is now currently in use for the non-destructive testing of bridge decks. It is sometimes referred to as surface penetrating radar (SPR) or ground penetrating radar (GPR) and is based upon an electromagnetic signal that is emitted and received by an antenna. When material features with different properties are encountered, part of the energy is reflected back to the surface and part is transmitted through the interface as shown in Figure 4.17. It therefore maps boundaries and can reveal the construction arrangement beneath the surface of the bridge deck. The evaluation of the data is based upon the expert analysis of these reflections, but very good results can be achieved as reported by Mesher, Dawley, Davis, and Rossiter (1995).

Testing takes place along pre-defined grid lines to obtain what are, in effect, *electromagnetic borehole logs*. Additional information is sometimes required from a traditional borehole or coring operation to supplement radar data.

The antenna equipment is very rugged and portable and can be traversed over a bridge either manually or by the use of a motor vehicle. Collection and processing of the data are rapid and offers real-time viewing if required.

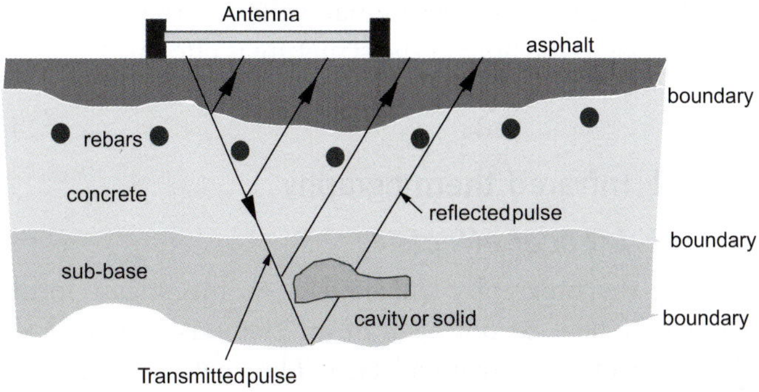

FIGURE 4.17
Principle of impulse radar.

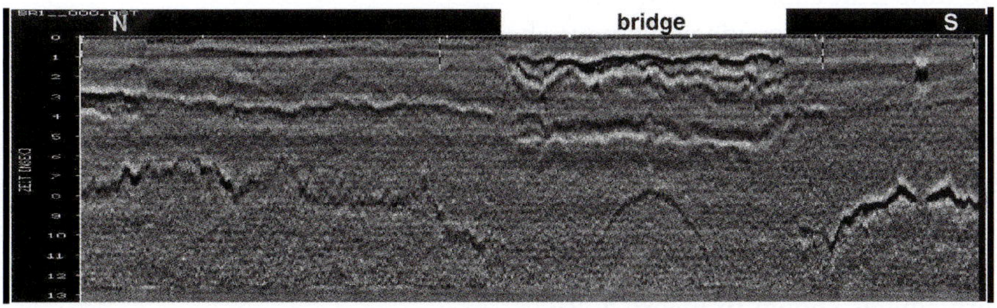

FIGURE 4.18
Typical impulse radar plot (courtesy EMPA, Switzerland).

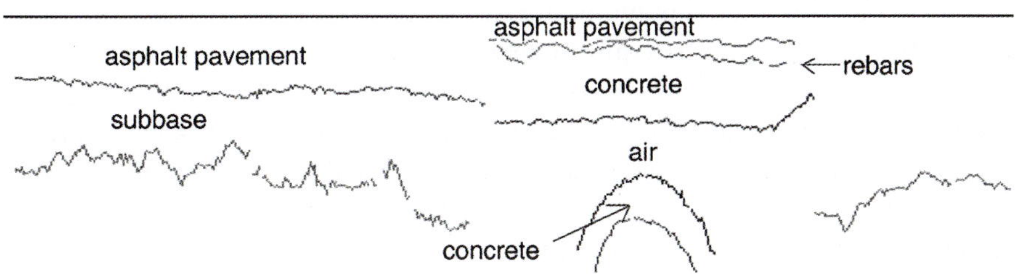

FIGURE 4.19
Interpretation of plot (courtesy EMPA, Switzerland).

A typical plot is shown in Figure 4.18 with interpretation shown in Figure 4.19 (Hugenschmidt, 1997).

Impulse radar has also been used successfully to *locate prestressing tendons* in post-tensioned concrete bridges, and to determine the *presence and extent of voids within a duct* where the injection of grout has not been successful.

The difficulty with this method as far as bridge engineers are concerned is that knowledge of the properties of the materials, which affect the electromagnetic behaviour such as conductivity, permittivity, magnetic permeability, and polarisation, is not generally understood. A very good definition of these terms along with basic electromagnetic theory can be found in a paper by Loulizi and Al-Quadi (1997).

The method is of particular value in the inspection of masonry arch bridges where the bulk of the structure is effectively hidden (Millar & Ballard, 1996).

4.2.11 Infrared thermography

4.2.11.1 Local effects

Infrared thermography can be used to investigate local effects by applying a pulse of heat to one side of an element, which is diffused through the element to the opposite surface. The spatial distribution is affected by the presence of any internal defects, diffusivity (related to conductivity and volume heat capacity), and time.

View on underside of element showing image contrast where CFRP strip has not bonded.

The area of element under investigation is photographed once immediately after the heat flash using a thermographic camera, and again after a short period of time, say 30 min. The difference in temperatures on the surface facing the camera causes an image contrast, which can be related to an internal defect or delamination (causing a near surface void). The basic principle is shown in Figure 4.20.

The method is particularly useful for testing the integrity of the bonding of carbon fibre reinforced polymer (CFRP) strips to concrete or delaminations on concrete surfaces. The equipment is light and portable and requires no physical contact between the recording camera and the element. If necessary, the test can be recorded on a video cassette, and with the advance of digital technology, the data will be able to be recorded within the camera and downloaded to a PC for processing.

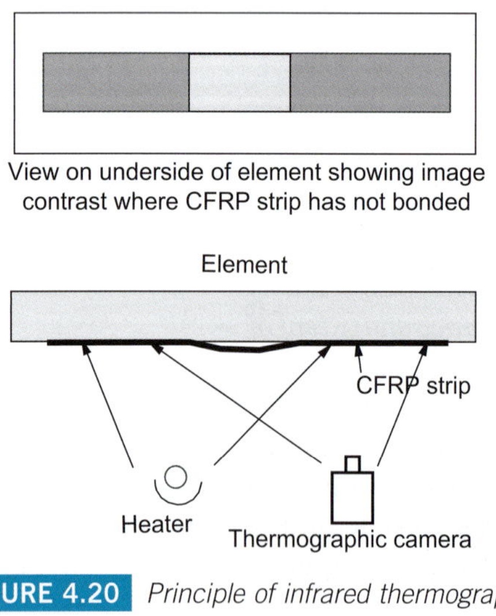

FIGURE 4.20 *Principle of infrared thermography.*

4.2.11.2 Bridge decks

Infrared thermography can be used to detect delamination of embedded reinforcement in concrete bridge decks, the breakdown of surface concrete due to freezing/thawing cycles, and debonding between asphalt surfacing and concrete (Manning & Masliwec, 1990).

When a surface is heated it emits energy that is directly proportional to the temperature. A remote sensing method is used to detect differences in surface temperature that can indicate different areas of sound and unsound concrete. On an asphalt-covered deck, any delamination or debonding will impede the transfer of heat and this will affect the surface temperature. On a thermographic display (thermogram) delaminations appear as white areas against the dark concrete areas; freeze-thaw areas are indicated as mottled grey-white tones, while debonding is barely detectable.

The data acquisition for both impulse radar and infrared thermography when applied to bridge decks is usually via equipment housed in a dedicated motor vehicle, which crosses the deck at a speed of around 5 mph (for a full description of the operational characteristics of such a vehicle, refer to Manning & Masliwec, 1990).

Naik, Singh, and Zachar (1992) have studied in detail the use of infrared thermography on a number of structures and conclude that 'although the equipment is costly, the test can be performed quickly over a large area'.

4.2.12 Endoscopy

A device used traditionally to inspect wall ties in cavity wall construction can also be used to inspect the presence of voids in prestressing ducts. The basic equipment consists of a highly flexible 6 mm diameter probe about 1000 mm long connected to an inspection tip at one end and the endoscope at the other and is shown in Figure 4.21. Other attachments are possible to improve the light source and to interface with TV or photographic equipment.

The inspection locations have to be chosen with care in places where possible lack of grouting and water accumulation can occur such as at high points over supports, low points in mid-span areas, and at vent points. The inspection is intrusive in that the concrete is first broken out locally using a 25 mm diameter overcore drill. The concrete core is removed and a small hole is drilled through the duct liner and the probe is then inserted. If voids are discovered, it is possible to measure the volume of the void by the use of a 'void volume estimator' (VVE), a simple device based on Boyles Law, $P1 \times V1 = P2 \times V2$. Any leakage of air from the duct can distort results and so vacuum pressure tests are often resorted to.

4.2.13 Impact echo

The impact (or pulse) echo is a technique originally developed to measure the integrity of concrete piles. A hammer blow is applied to the surface of concrete via a velocity or impacting transducer and the resulting shock wave travels through the concrete and is reflected back off a discontinuity to a signal analyser via a receiving transducer placed on the surface of the concrete after the surfacing has been cut away locally (Figure 4.22). The signals are processed in a portable analyser and yield a plot of distance versus

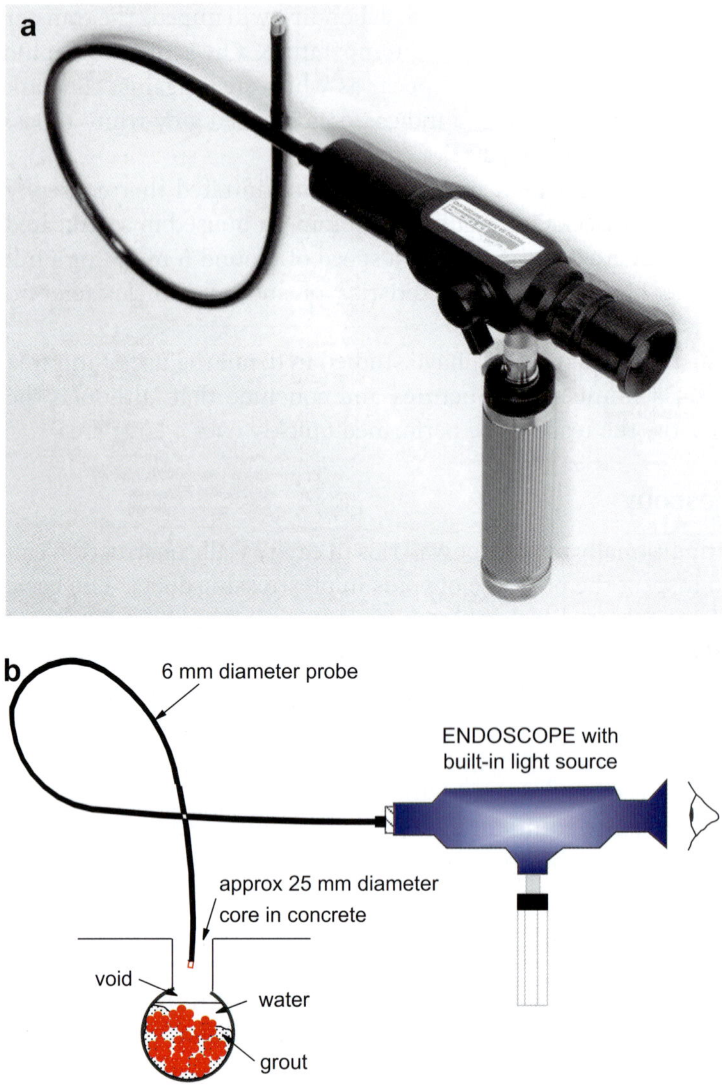

FIGURE 4.21
(a, b) Typical endoscope and setup.

frequency for each impact point. The results are then stored on a disk for further processing using an 'on-board' PC to give a plot across the bridge. The depths of any discontinuities can be found if the wave velocity is known. Its application to concrete bridge decks is relatively new, and can be operated in the time domain or frequency domain to detect voids and delaminations around reinforcing bars (Cheng & Sansalone, 1993).

Its most promising use, however, is the *detection of voids in prestressing ducts*. Martin, Hardy, Usmani, and Forde (1997) carried out exhaustive tests

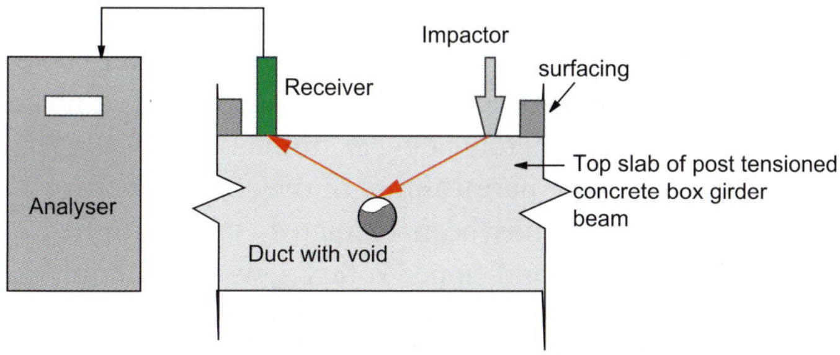

FIGURE 4.22
Typical impact echo setup.

to determine the advantages and limitations of the method as applied to detect flaws in the cylindrical metal ducts of grouted post-tensioned concrete bridge beams. Good results have also been reported by tests carried out by Carino and Sansalone (1992), particularly in the detection of voids.

Because of the way concrete disperses the sound waves, Martin et al. (1997) have observed that defects can only normally be detected 'provided that a sufficiently high frequency is used', namely, greater than 2 × wave length (λ).

4.2.14 Radiography

The principles of radiography are described in section 4.3.5. Its primary use on bridges is to scan grouted post-tensioned tendons for voids. The French have developed a lorry-mounted system called Scorpion II (Parker, 1994; Winney, 1992). The integrity of post-tensioned bridge beams is achieved by tracking the X-ray source along one side of the beam and detecting the output on the other. Inspection is carried out as the lorry moves along the bridge and can be viewed as a real-time video display inside the vehicle, sometimes called radioscopy. Any suspect areas can be identified and re-inspected while the lorry is stationary by making a radiograph that can be minutely studied later. The method is not foolproof but seems to be the most reliable at the present time (Woodward, Hill, & Cullington, 1996). According to Martin et al. (1997) results can be enhanced by the use of tomography that provides information in three dimensions, that is, the presence and the position of the defect.

4.2.15 Coring

Cores of concrete are very often extracted from in situ structures to determine such properties as *density, tensile strength, compressive strength, carbonation, and permeability*. The diameter of the core should be as large as possible to ensure that the local effects of the aggregates do not adversely affect the results. In the United Kingdom, it is usual to cut a 150 mm diameter core, although 100 mm is becoming more common, and the ratio of length to diameter (*L*/*D*) is normally 2.0. A water-cooled diamond-tipped overcoring drill bit is used, which is mounted on a stand and can be bolted to the surface of the concrete being examined. A typical arrangement is shown in Figure 4.23. If the permeability of the covercrete is not being determined, then after the core has been extracted, it is trimmed at each end using a water-cooled diamond-tipped rotary saw. Care should be taken not to cut through any steel reinforcement, or if this cannot be avoided, then a judgement must be made on whether such local damage will affect the strength of the bridge.

4.2.15.1 Density

The density can be determined either in the traditional manner from its mass and overall dimensions or by following a procedure suggested by the Concrete Society (1987) in the United Kingdom involving water displacement.

4.2.15.2 Compressive strength

On new bridge building projects, quality control by measuring the 28-day strength of 150 mm concrete cubes in compression is a well-established

FIGURE 4.23
Typical coring setup.

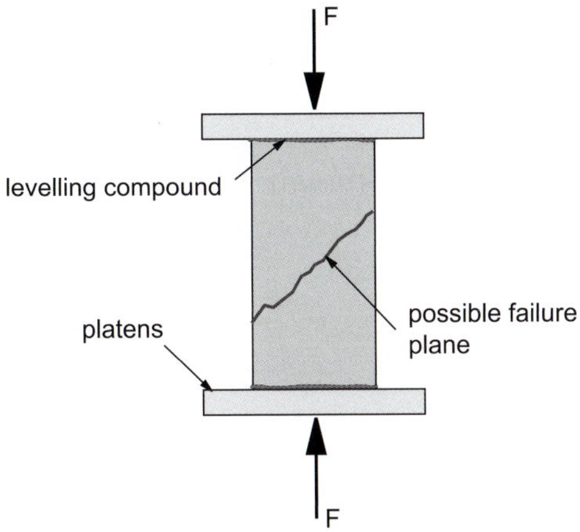

FIGURE 4.24 *Typical setup for compressive testing of a concrete cylinder.*

procedure throughout the world. The number of cubes tested is related to the quantity of concrete to be poured. With in situ concrete in existing bridges only a relatively small number of *cores* can be taken, and these must be chosen with care to be representative of the area being tested.

Compression testing of a concrete core is carried out in a suitable compression testing machine at a controlled speed of about 5 kN/s for a 150 mm diameter core. The ends can either be trimmed square with a disc cutting saw or levelled using a suitable epoxy resin compound. With the precision drilling equipment available the diameter can usually be assumed, but if there is any doubt then the average measured diameter in two orthogonal directions can be used to calculate the area, and after crushing failure the maximum strength can be determined. Cubes and cylinders of the same concrete have different crushing strengths; the cylinder failing at about 80% of the cube strength. Which value is to be taken should be decided by the engineer based on guidance given in national Codes of Practice. The cylinder strength probably gives a more realistic estimate of the uniaxial strength of the concrete as it fails in shear at a natural angle whereas the cube is restrained by the platens on the testing machine and so is in a confined state. A typical cylinder test setup is shown in Figure 4.24.

4.2.15.3 Direct tensile strength

A way of measuring the tensile strength on the surface of a concrete element has been described in section 4.2.8. Concrete cores can be tested, however, in such a way as to give an estimate of the indirect tensile strength of the heart concrete. For this test, the cylinder is laid on its 'side' and subjected to a line loading along its length and perpendicular to its diameter until failure occurs by splitting of the concrete along a vertical plane as indicated in Figure 4.25. The distribution of the horizontal stresses along a vertical diameter is as shown in Figure 4.25. (The distribution local to the top and bottom of the diameter is very similar to that induced in a pre-stressed concrete member at an anchorage position.)

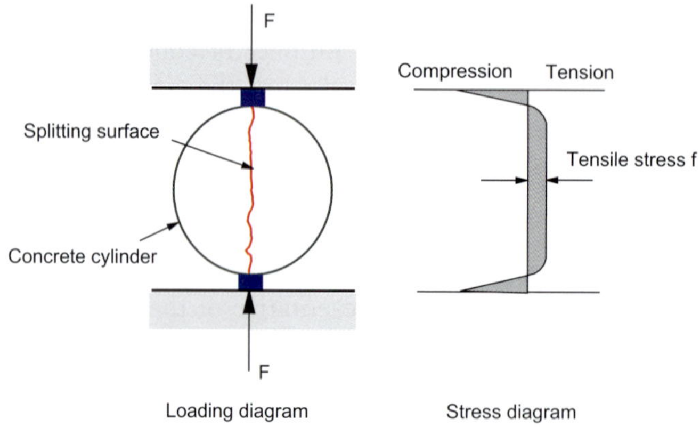

FIGURE 4.25
Typical setup for tensile testing of a concrete cylinder.

The splitting or tensile strength (N/mm^2) is then given approximately as:

$$f_t = 2F/\pi DL \tag{4.1}$$

where D and L are the diameter and length of the cylinder in mm. The value of f_t is generally assumed to be about 0.1 of the 150 mm cube strength, or 0.08 of the cylinder strength. If the flexural tensile strength is required for assessment purposes, this can be found by multiplying the cylinder strength by 1.5.

4.3 STEEL

4.3.1 Introduction

Structural steel is one of the most common materials used for the construction of medium to long span bridges of the truss, cable-stayed, and suspension types. Being a metal, it tends to corrode if left exposed to air and water, and undergoes a conversion from the metallic to the oxidised state. The *rate* of such corrosion varies depending upon the composition of the metal and its environment. It also suffers from the presence of *defects* embodied during manufacture such as *voids and inclusions*. It can *crack* at low temperatures, and *fracture* due to in-service fatigue.

Surface treatment and attention to detail by bridge engineers are attempts to control the rate of deterioration but inevitably corrosion will take place. Non-destructive tests play an important part in detecting both the *presence* and the *rate* of corrosion and are also employed to seek evidence of *imperfections and section loss*. Figure 4.26 illustrates typical defects that can be detected.

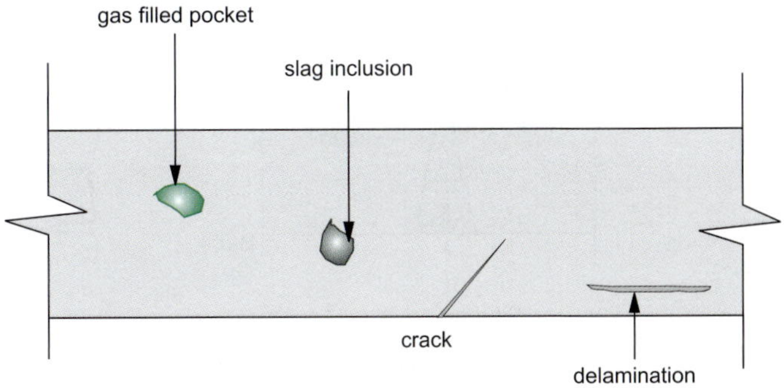

FIGURE 4.26
Imperfections in steel.

4.3.2 Ultrasonic testing

This method of testing concrete has already been described (section 4.2.4). It can also be used to test steel elements by passing a very high frequency (2–5 MHz) through the metal. The ultrasonic waves that travel through the metal are reflected back by either a defect or from the far surface of the member. The degree of reflection produced by a defect depends upon its shape and size. A void, for example, will reflect almost 100% of the waves, while a defect-free element will reflect 0% of the waves.

There are two types of probe that can be used: the *single* crystal or the *twin* crystal. In the former, the probe both transmits and receives the waves; and in the latter, one probe transmits and the other receives. Both can be used to enter a wave at any angle to the surface of the element. For example, when a weld is under examination, both types of probe would be used depending upon the nature of the weld. Typical probe arrangements for butt and fillet-type welds are shown in Figure 4.27.

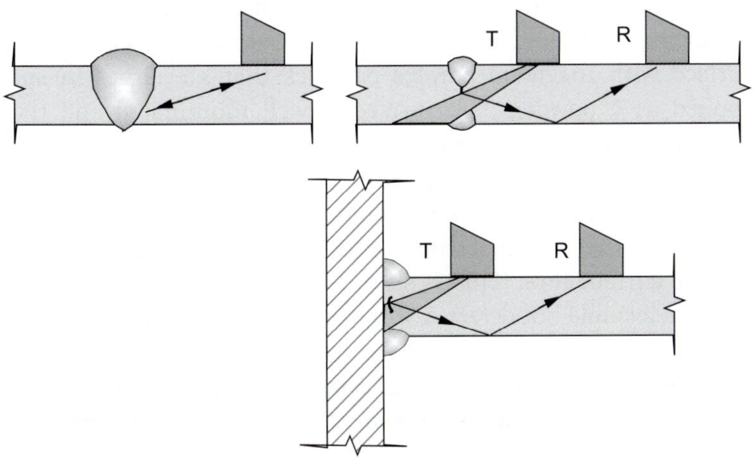

FIGURE 4.27
Typical probe arrangements when testing welds.

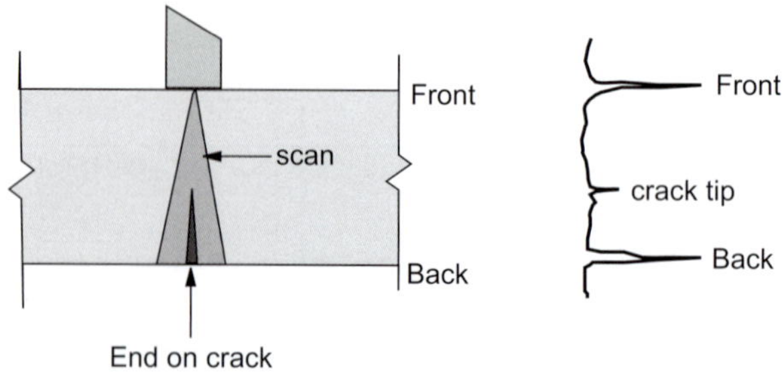

FIGURE 4.28 *Use of a low-noise probe.*

New developments in ultrasonic testing now make it possible to use high-power, low-noise compressed wave probe to detect the tips of a crack 'head-on' and can sometimes provide information on the crack size (Figure 4.28).

Ultrasonic diffraction (as opposed to deflection) is now being developed as what is called a 'time-of-flight' technique. This may prove to be a useful way of determining the location, position, and size of a crack. Another development is the 'Delta-scan' that uses ultrasonic mode conversion and can also provide information about head-on cracks (Cartz, 1995).

Ultrasonic technology is probably the most widely developed and used method of flaw detection at the present time and because of its relative simplicity will most certainly remain ahead of the field for some time to come.

4.3.3 Magnetic particle inspection (MPI)

This method *can only detect surface or near surface flaws*. It consists of producing a magnetic field around the area to be tested and then coating the surface with magnetic powder particles that are either in suspension or applied as a powder. The surface is well illuminated and then a visible inspection of the powder is carried out (Halmshaw, 1988).

Flaws give rise to magnetic leakage fields and are best detected when the magnetic field direction is perpendicular to the direction of the suspected flaws – in other words, a defect is best revealed if parallel to the direction of electric current flow. This is shown in Figure 4.29 where in (a) defect B is easily detectable but defect A will be missed. (If it is possible to remagnetise in an orthogonal direction, then the converse is true.) There are several ways of magnetising the area to be investigated (Halmshaw, 1988) and the most appropriate will depend on the site conditions. The two most common are the current flow techniques where two contact heads are placed on the

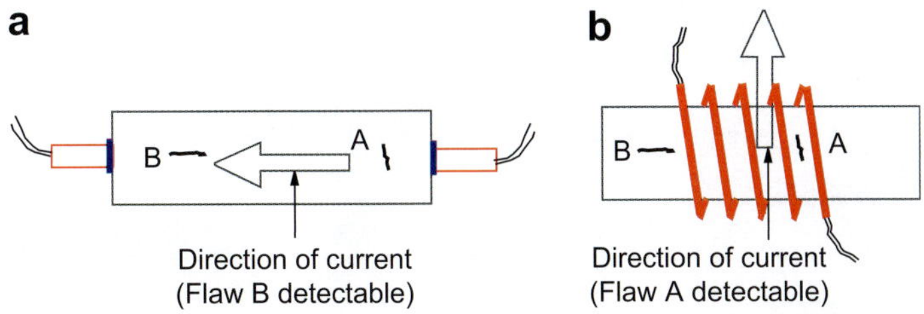

FIGURE 4.29
Flaw detection using MPI.

specimen, which becomes part of the magnetic circuit, or electromagnetic induction using a coil wrapped around the specimen.

For example, current flow could be used in Figure 4.29(a) with contact heads at each end, and electromagnetic induction could be used in Figure 4.29(b) with the coil wrapped around the specimen or element.

Flaws will only be detected when the magnetic flux (which is perpendicular to the current) is at right angles (or nearly so) to the flaw.

If magnetic 'ink' is used, the defects attract the ink, which remains in the defects when the magnetic field is removed. The defect is then visible as either lines or spots on the surface. There is no limit to the size or shape of element, which can be tested provided that a magnetic field can be set up. Figure 4.30 shows a typical MPI setup on a steel bridge.

4.3.4 Liquid penetration inspection (LPI)

This is a very extensively used method for testing defects that are *open to the surface* (Cartz, 1995). The results are heavily dependent on the experience and skills of the operator, and to that extent they are somewhat subjective.

The area to be tested, for example, near to a weld, is first cleaned and then coated with a *liquid penetrant* containing a coloured dye (usually red) for use in normal light, or a fluorescent dye for use under UV light. As a result of capillary action, dye is drawn into any defects on the surface trapped. After a few minutes, the dye is wiped off and a developer is sprayed over the surface, which reduces the surface tension and allows the dye in the defect to seep to the surface thus highlighting the defect. After use, the penetrant should be cleaned off. The steps are shown in Figure 4.31.

The technique is simple and requires no equipment, but it is limited to defects that are visible on the surface, and interpretation must be by a person experienced in the process. This ensures that false LPI indications are identified and eliminated from the study. If more information is required

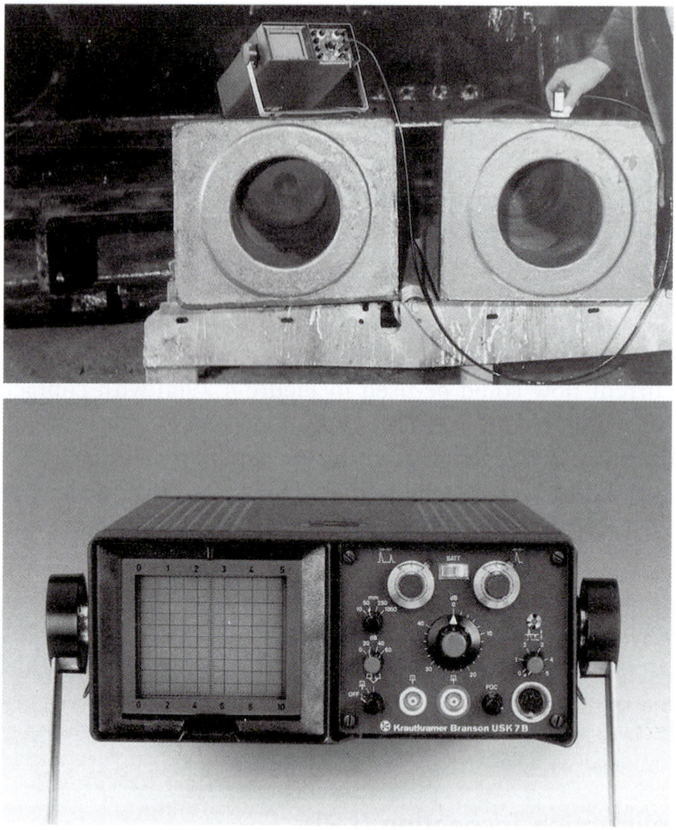

FIGURE 4.30
Typical MPI setup on a bridge (courtesy of M. Neave).

about flaw depth and width, then it must be supplemented with a UT or MPI investigation.

4.3.5 Radiography

Radiography is a technique that uses penetrating radiation such as X-rays or gamma γ-rays to obtain an image of a solid or void. Access to both sides of the element under test is required and images are recorded on film (radiographs) placed behind the element.

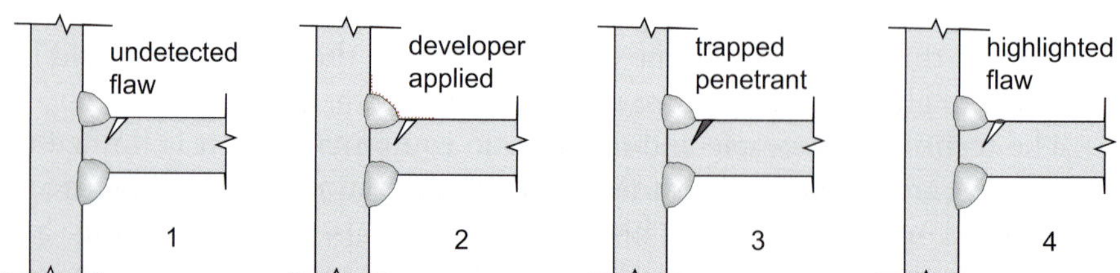

FIGURE 4.31
The application of LPI.

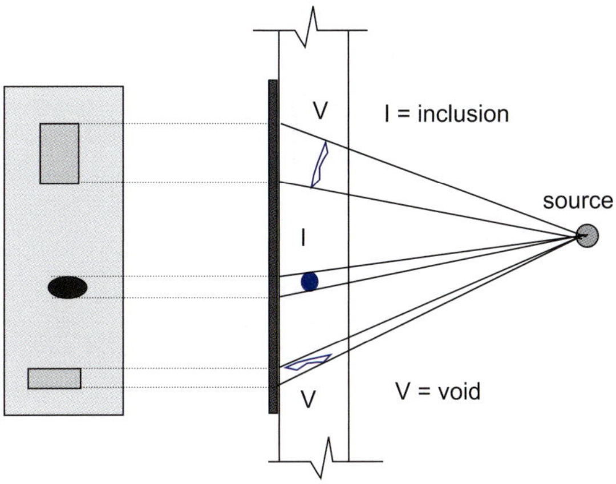

FIGURE 4.32 *Principle of radiography.*

The contrast in a radiograph is controlled largely by the energy of the ray, and can be enhanced by the use of two or more films. No details are possible apart from its size perpendicular to the source of the rays. The basic principle is shown in Figure 4.32. Generally, defects that are less dense than the parent metal appear lighter, and defects which are more dense appear darker. The experience of the operator and engineer in charge plays an important role in the interpretation of the results.

Radiography is a hazardous technique and due precautions have to be taken to ensure safety of all personnel involved. It is costly and involves the use of heavy equipment, and for that reason it is not a popular choice among bridge engineers. It can be quite useful, however, for examining components of a bridge that can be temporarily removed (such as bearing plates, trunions, bolts) and brought into a laboratory environment for examination.

4.3.6 Metal thickness gauges

4.3.6.1 Ultrasonic thickness gauge

This type of instrument is based on the same principles outlined in section 4.3.2. A small emitter and receiver are built into a single probe that is simply placed onto one face of the structural element and a beam of ultrasound is transmitted through it and reflected off the back of the element (Figure 4.33). The time taken is used to calculate the thickness knowing the velocity of the ultrasound in the material from the formula:

$$t = v \cdot T/2 \qquad (4.2)$$

where t is the thickness of metal part, T is the transit time, and v is the velocity of the ultrasound in the material.

It can measure from 1.2 mm to 200 mm, and the thickness is presented on a digital display. The instrument is very small, lightweight, portable, and can be used in the field with ease. For reliable, accurate measurements, it is

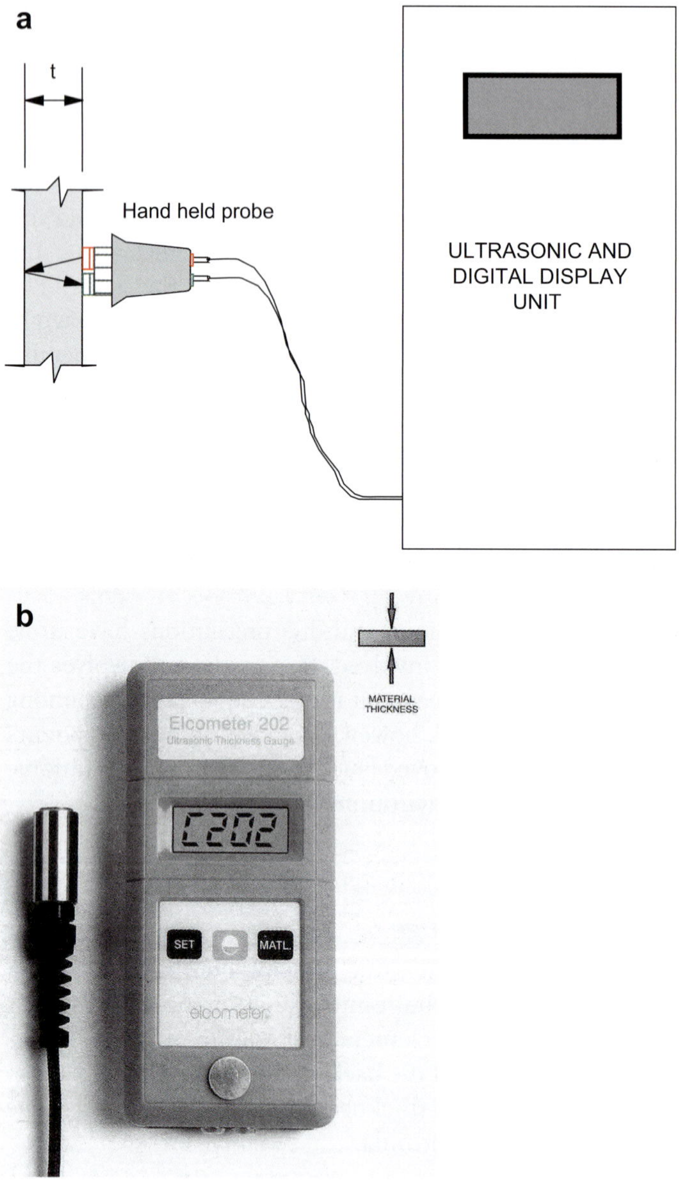

FIGURE 4.33
(a) Principle and (b) ultrasonic thickness gauge (courtesy of Elcometer Ltd. UK).

important to ensure that the surface of the element is first cleaned and a couplant applied to ensure that there is a suitable sound path between transducer and test surface and that there is no air layer. It is best to use thick oil or grease but water may also be used, in fact almost anything will do as long as it is liquid.

4.3.6.2 *Vernier scale*

The vernier scale is a well-known traditional device and has the advantage that the measuring points will settle easily into the deepest pits of corroding metal thus measuring the thinnest parts of the material. Modern versions have a digital readout and can measure thicknesses from 1 mm to 50 mm.

4.3.7 Other tests

There are many other well-known tests that can be applied to steel elements to establish their properties of *toughness, hardness, brittleness, elasticity, ductility, and fatigue limit* and these would normally be investigated after the onset of any visible defect such as cracking or plastic deformation, or in the event of a bridge failure as part of a diagnostic investigation. Apart from the hardness test that can be implemented in situ, small coupons of steel would need to be cut out for laboratory testing.

Chemical analysis is carried out if it is suspected that the steel does not have the properties assumed in the original design. This will include measurements of the amounts of *silicon, phosphorus, sulphur, and manganese*, each with a different effect on the properties listed above.

Iffland and Birnstiel (1996) provide a very useful list of NDT methods for steel giving their applications and limitations together with an exhaustive list of references.

4.4 MASONRY

Masonry bridges are not usually subject to the problems of corrosion, but instead can deteriorate by weathering of the exposed masonry, arch ring separation, deformation of the spandrel walls and parapets, break-up and settlement of the fill materials, and fracture of the transverse ties.

4.4.1 Hammer tap test

This is the 'traditional' way of testing the *integrity of masonry arches* by trying to locate separation of the arch rings. It is acoustic in nature and involves striking the masonry with a hammer and listening to the acoustic response. A 'hollow' sound indicates possible ring separation since the mechanical sound wave does not travel well through air, and a 'solid ring' implies that the integrity of the arch rings is sound. This works well for separations near to the intrados but does not always detect deeper or multiple separations.

4.4.2 Coring

If separation is suspected at a particular location, then the same coring procedures detailed in section 4.2.16 can be applied to the arch ring. The core will confirm the presence or otherwise of the separation(s) but will not provide information about its thickness. Compression testing of the core can be carried out as for concrete (see section 4.2.16.2), which will provide information about the strength and the quality (class) of the brickwork or masonry, and (if dry coring is carried out) the extent of water saturation.

Cores can also be taken through the surface to discover the nature and extent of the layered materials in the 'fill'. If unbound materials are discovered, then a borehole needs to be sunk to ensure a reasonably undisturbed sample.

4.4.3 Impulse radar

This method has already been described in section 4.2.11. Not only is it non-invasive (unlike trial holes and slit trenches), but a large amount of information can be gathered in a relatively short space of time with a minimum interference to traffic and at a relatively low cost.

4.5 CONCLUDING REMARKS

The wide ranging non-destructive (semi-invasive) tests described in this chapter are well worth considering when planning a visual inspection. They can add considerably to the information obtained from the inspection and are relatively inexpensive. The results of well-chosen tests can help to determine the quality of the construction materials for assessment purposes. They are also able to measure the extent of a *visible* defect, and to reveal the presence of any *hidden* defects that might otherwise go undetected and result in a collapse – witness the ill-fated Ynys-y-Gwas bridge (section 5.2.6 in Chapter 5).

REFERENCES

Alldred, J. C. (1995). Quantifying losses in cover meter accuracy due to congestion of reinforcement. *International Journal of Construction, Maintenance and Repair, 9*(Jan/Feb), 41–47.

ASTM C 876–80. *Half cell potentials of uncoated reinforcing steel in concrete*. Philadelphia: American Society for Testing and Materials.

Basheer, P. A. M. (1993). A brief review of methods for measuring the permeation properties of concrete in-situ. *Proc. ICE, Structs and Bldgs, 99*, 74–83.

Bungey, J. H. (1981). Concrete strength determination by pull-out tests on wedge anchor bolts. *Proceedings of the Institution of Civil Engineering, Pt. 2, 71*, 379–394.

Bungey, J. H., & Millard, S. G. (1996). *Testing of concrete in structures*. Blackie.

Carino, N. J., & Sansalone, M. (1992). Detecting voids in metal tendon ducts using the Impact-Echo method. *Materials Journal of the American Concrete Institute, 89*(3).

Cartz, L. (1995). *Nondestructive testing*. ASM International.

CEB-FIP Model Code. (1991). Final Draft, Section d.5.3: 'Classification by Durability'. CEB Bulletin d'Information No. 205, July, Lausanne.

Chabowski, A. J., & Bryden-Smith, D. W. (1980a). Assessing the strength of in-situ Portland cement concrete by internal fracture tests. *Magazine of Concrete Research*, 32(11), 164–172.

Chabowski, A. J., & Bryden-Smith, D. W. (1980b). *Internal fracture testing of in situ concrete: a method of assessing compressive strength*. BRE Information Paper IP 22/80.

Cheng, C., & Sansalone, M. (1993). Effect on impact echo signals caused by steel reinforcing bars and voids around bars. *ACI Materials Journal, 90*, 421–434.

C & S. Electronics Ltd. (1990). *Description and Specification of the Resistivity Meter*. Concrete Society (1987).

Concrete Society. (1987). *Concrete core testing for strength*. Technical Report No.11. London.

Darby, J. J. (1996). The effectiveness of Silane for extending the life of chloride contaminated reinforced concrete (pp. 838–848). *Third International Conference on Bridge Management, Inspection, Maintenance, Assessment and Repair*, University of Surrey, Guildford, UK. E&FN Spon.

Figg, J. W. (1973). Methods of measuring air and water permeability of concrete. *Magazine of Concrete Research, 25*(85), 213–219.

Halmshaw, R. (1988). *Introduction to the non-destructive testing of welded joints*. Cambridge: The Welding Institute.

Hugenschmidt, J. (1997). Non-destructive evaluation of bridge decks using GPR – benefits and limits. *Recent Advances in Bridge Engineering* (pp. 360–365). Dubendorf, Switzerland.

Iffland, J. S. B., & Birnstiel, C. (1996). Guide to non-destructive tests (pp. 921–930). *Third International Conference on Bridge Management, Inspection, Maintenance, Assessment and Repair*, University of Surrey, Guildford, UK. London: Thomas Telford.

Loulizi, A., & Al-Quadi, I. L. (1997). Use of electromagnetic waves to evaluate Civil Infrastructure. *Recent Advances in Bridge Engineering* (pp. 333–343). Dubendorf, Switzerland.

Manning, D. G., & Masliwec, T. (1990). Application of radar and thermography to bridge deck condition surveys (pp. 305–317). *First International Conference on Bridge Management, Inspection, Maintenance, Assessment and Repair*, University of Surrey, Guildford, UK. Elsevier.

Martin, J., Hardy, M. S. A., Usmani, A. S., & Forde, M. C. (1997). NDE of post tensioned concrete bridge beams. *Recent Advances in Bridge Engineering* (pp. 352–359). Dubendorf, Switzerland.

Mesher, D. E., Dawley, C. B., Davis, J. L., & Rossiter, J. R. (1995). Evaluation of new GPR technology to quantify pavement structures. *Transportation Research Record, 1505*(July), 17–26.

Millar, R. J., & Ballard, G. S. (1996). Non-destructive assessment of masonry arch bridges – The price:value ratio. *Construction Repair, 10*(1), 32–34.

Naik, T. R. S., Singh, S. S. R., & Zachar, J. A. (1992). Applications of the infrared thermography technique for existing concrete structures (pp. 539–548). In: *Proceedings of the seventh international conference on structural faults and repairs*, Edinburgh.

Parker, D. (1994). X-Rated video. *New Civil Engineer*, 21 April, 10.

Stratfull, R. F. (1973). Half cell potentials and the corrosion of steel in concrete. *Highway Research Record, 433*, 12–21.

Torrent, R., Ebensburger, L., & Gebauer, J. (1994). On site evaluation of the permeability of the 'covercrete'. *Third CANMET/ACI International Conference on Durability of Concrete*, Nice, France.

Torrent, R., & Frenzer, G. (1995). A method for the rapid determination of the coefficient of permeability of the 'covercrete'. *International Symposium on Non-Destructive Testing in Civil Engineering (NDT-CE)*, 26–28 September.

Vassie, P. R. (1991). The half cell potential method for locating corroding reinforcement in concrete structures. *TRRL (now TRL) Application Guide 9*.

Wenner, F. (1915). A method of measuring earth resistivity. *Bulletin of the Bureau of Standards (USA), 12*, 469–478.

Winney, M. (1992). *Grouted duct tendon plan poses problems. New Civil Engineer*, October, 4.

Woodward, R. J., Hill, M. E., & Cullington, D. W. (1996). Non-destructive methods for inspection of post tensioned concrete bridges. *FIP Symposium on Post-Tensioned Concrete Structures* 295–304.

CHAPTER 5

Load Testing

CONTENTS

5.1 INTRODUCTION

The load testing of bridges can be of enormous benefit to engineers in aiding their understanding of the behaviour of bridges at both the serviceability and the ultimate limit states under the action of *live loads*. After a bridge has been constructed it is quite usual to carry out a live load test to measure the response of the structure by the use of appropriate strain and displacement devices. This will confirm (or otherwise) the assumptions made in the original analysis regarding load distribution, stress levels and deflections. In France and Switzerland it is common practice to load test all new bridges with a static load ($DL + LL$) equivalent to 80% of the maximum design life bending moments (Markey, 1991; Favre, Hassan, & Markey, 1992). A good example of this was the recent test carried out on the new Pont de Normandie cable-stayed bridge at Le Havre (New Civil Engineer, 1995) when 80 lorries weighing 15 tonnes each were positioned on the deck and on the approach spans. The measured midspan deflection values were close to those predicted.

In the bridge management domain, however, where an existing bridge has been in service for many years, the situation is not so simple. The problem arises at the post-assessment stage where, from a *theoretical* standpoint, a bridge has not proved strong enough to carry the specified live load (i.e. it has 'failed' its assessment) but where there may be good reason to suspect that its load-carrying capacity is greater than calculated. In spite of a satisfactory load-carrying history, and even though it may be showing no signs of distress, a failed assessment means that the bridge has to be *weight restricted, propped, strengthened or replaced* (zero-rated). Each of these solutions causes traffic disruption and costs money and so if a way can be found of proving that the bridge can, in fact, carry the required live load, then it can remain in service. One way to establish this is by means of *load testing* of the actual bridge superstructure under investigation.

A suitable load test can help to bridge the gap between theory and reality. This test is relatively fast and can remove some of the doubts about assumptions made in the initial assessment.

5.2 LOAD TESTS

There are three types of *static* load tests that can be used on bridge superstructures:

- Supplementary load testing.
- Proving load testing.
- Collapse load testing.

5.2.1 Supplementary load tests

This type of test is *supplementary* to the assessment process (see Chapter 3) and is preferred by most engineers because it involves applying a known load to the bridge which is no greater than the existing traffic load and therefore should be safe with no damage or risk of collapse. It is therefore essentially a diagnostic performance test. Instrumentation is extensive with both strain gauges and transducers strategically placed to provide an overall 'map' of the structures elastic behaviour. This is compared with the output from the mathematical model and adjustments made as necessary. For example, the load test may indicate smaller deflections than initially obtained from the model and so the deck stiffnesses can be adjusted and the analysis re-run. It also enables a 'what-if' scenario to be developed, whereby various strengthening options could be considered and the computer model modified accordingly.

Care should be exercised, however, in the interpretation of the test data to ensure that what is being observed is, in fact, due to the 'normal' response of the structure and not some abnormal factor such as bearings which have seized or expansion joints that have locked. This reinforces the need for a thorough visual examination prior to testing. Bahkt and Jaeger (1990) have shown that tests on a wide variety of bridge types reveal some aspects of bridge behaviour not considered in design and which, if considered, have resulted in higher calculated load-carrying capacities. They conclude with the statement that 'no amount of refinement in the analysis can yield the true load-carrying capacity of the bridge'.

It is argued that the ultimate load for a bridge can be deduced by extrapolating the effects of the load test to the point of failure. Care should be exercised, however, since the appropriate type of extrapolation can only be known if results are available for a similar type of bridge up to failure, and even then no two bridges are alike and so assumptions will have to be made.

The decision on whether to undertake a supplementary test is dependent on factors such as:

- The type of bridge
- The condition of the bridge
- The presence of structural actions not normally considered in analysis and which provide a degree of *hidden strength*

5.2.1.1 Type of bridge

5.2.1.1.1 Shallow deck construction

It is generally the older, single span bridges of *shallow deck construction* which fail their assessments, primarily in bending and shear. This includes bridge types which exhibit a certain degree of ductility shown in Figure 5.1, such as (a) jack arch construction, (b) filler beam deck, (c) steel trough and concrete, (d) singly reinforced concrete slab or (e) beam and (concrete or timber) slab decks. These are likely to give warning of impending failure and the test can be stopped.

Bridges with a brittle nature such as cast iron with a concrete infill may not be appropriate because failure could be sudden and catastrophic.

Arch bridges can also be load-tested to understand their behaviour due to live loads provided that certain criteria are observed (Davey, 1953). They usually fail due to the formation of hinges in the arch ring and fracture. The main structural element is the *arch barrel* (comprising separate rings of brick or masonry) which supports both the *fill* and the roadway above, and which is contained by the *spandrel walls*. Abutments at each end transfer the reaction from the arch ring to the ground. A typical arch is shown in Figure 5.2.

5.2.1.2 Condition of bridge

A visual inspection of a bridge scheduled for a load test is mandatory. If there is extensive corrosion of reinforcement or steel beams, cracked and broken concrete, damaged bearings, missing bolts or any significant damage which reduces its basic structural worthiness, then it would be imprudent to proceed with a test. Likewise, if there are large permanent deflections or if the

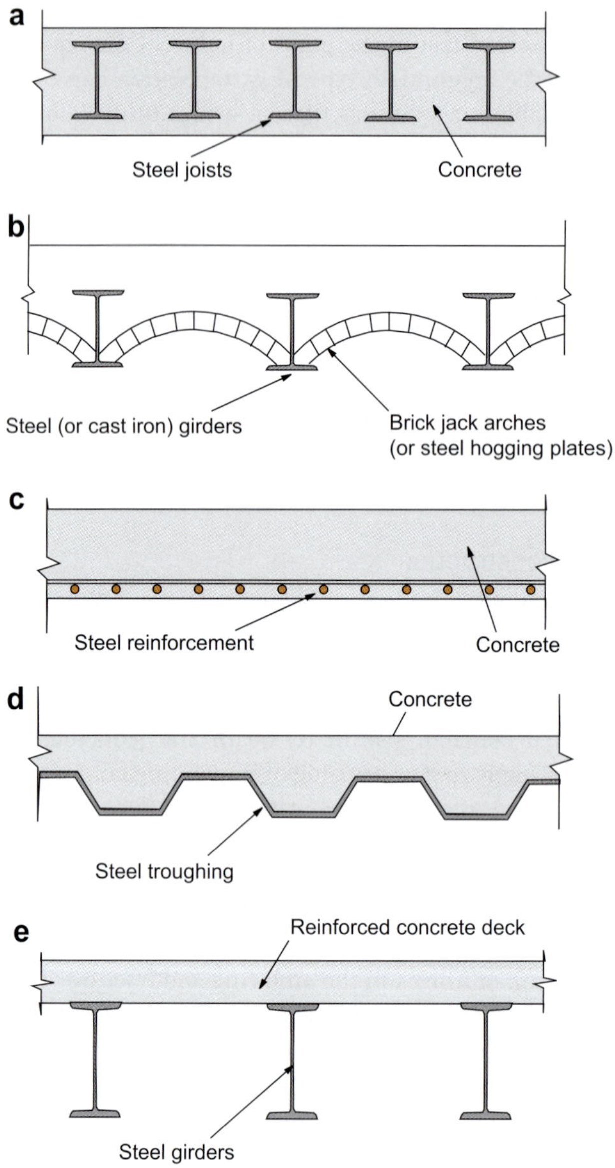

FIGURE 5.1
Bridges suitable for supplementary testing.

bridge appears to be 'lively', then testing should not proceed. The final decision on whether to test a bridge with a poor or good condition rating is straightforward enough. Where the rating is between average and low, then sound engineering judgement is required from specialist engineers experienced in testing.

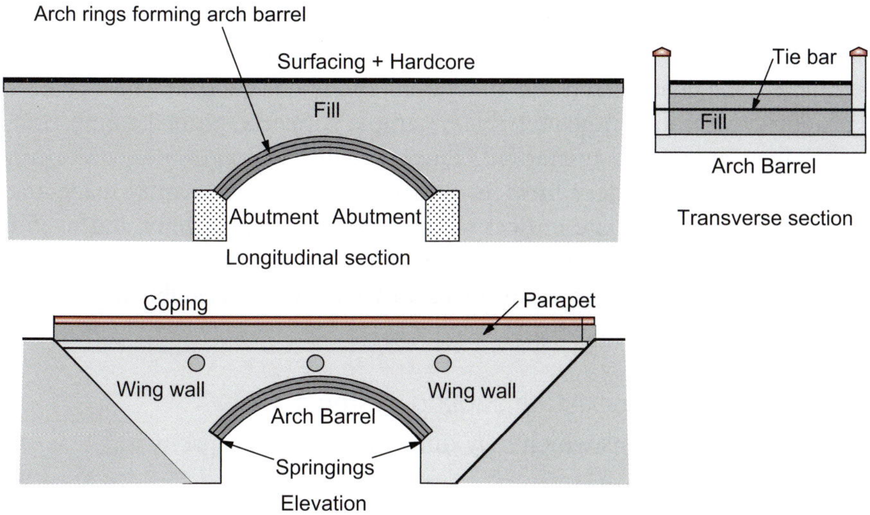

FIGURE 5.2
Typical arch bridge construction.

5.2.1.3 Hidden strength

The existence of factors which provide a degree of *hidden strength* in shallow bridge decks is now well recognised (Chalkley, 1994; Bahkt & Jaeger, 1993; Mehrkar-Asl, 1994).

Chalkley (1994) lists the three main contributory factors as:

- Material strengths
- Design assumptions
- The presence of other coexisting support systems

5.2.1.3.1 Material strengths

Material strengths are usually assumed to be low for the purposes of assessment and are referred to as the *worst credible strengths*. This assumption, although conservative, is not realistic even though it is obtained from testing representative samples of the materials actually used in the bridge. For example, there may be a wide variation of compressive strengths resulting from core testing, and careful interpretation and engineering judgement is needed to ensure that the correct value is used in the assessment (see Cullington & Beales, 1994). If areas of weak concrete can be identified, and they are not near the critical flexural sections, then it may be possible to eliminate them from the calculation of the worst credible strength and thus obtain a higher value. Of course, the tests may in fact indicate *lesser* values, and these must be taken into account.

5.2.1.3.2 Design assumptions

Design assumptions regarding the structural action of bridge decks can influence the load effects on individual members such as the degree of composite (or non-composite) action, simple supports, pinned connections, continuity, cracked or uncracked concrete, bending action, nonconforming details, absence of shear links in prestressed concrete beams, inadequate anchorage, etc. They are critical to the load-carrying capacity and need to be correctly allowed for. This may mean carrying out model tests on complete bridge decks or laboratory testing of a detail or element to establish the effects of defects.

5.2.1.3.3 Other coexisting support systems

Other support systems which may influence the load capacity are:

5.2.1.3.3.1 Membrane action. This has been shown to considerably enhance the strength of concrete beam-and-slab bridges (Long, Kirkpatrick, & Rankin, 1994; Jackson, 1995) and exists where the surrounding concrete offers an in-plane restraint as shown in Figure 5.3. This effectively alters the behaviour of the slab to one of arching rather than *bending*, resulting in an increase in strength well above that calculated assuming flexural action only. The actual increase will depend upon the degree of restraint related to the value of the spring constant (k). A perfectly rigid restraint means $k = \infty$, a condition not possible in practice.

A very notable conclusion drawn by Long et al. (1994) after an exhaustive testing programme at Queens University, Belfast, was that: 'Compressive membrane action enhances the strength and serviceability of laterally restrained slabs and can be utilised in the assessment of concrete bridge decks'. The degree of enhancement depends upon the level of restraint and is generally illustrated in Figure 5.4.

Jackson (1995) has shown that for zero reinforcement the arching action is dependent on the compressive and tensile strengths of the concrete and the stiffness of the restraints, and he has attempted to quantify the enhancement using both simple and numerical analyses. He concludes that

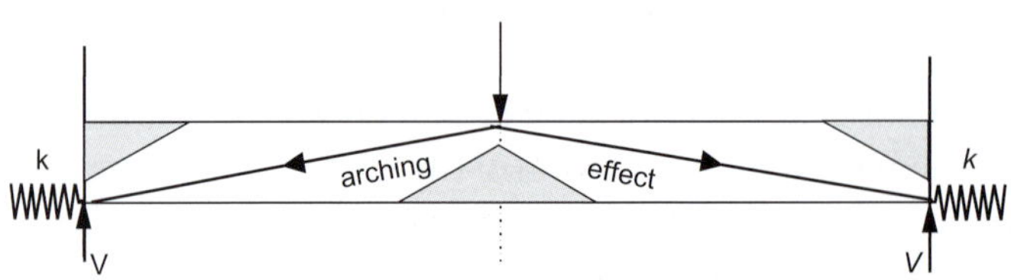

FIGURE 5.3
Restrained slab (after Jackson, 1995).

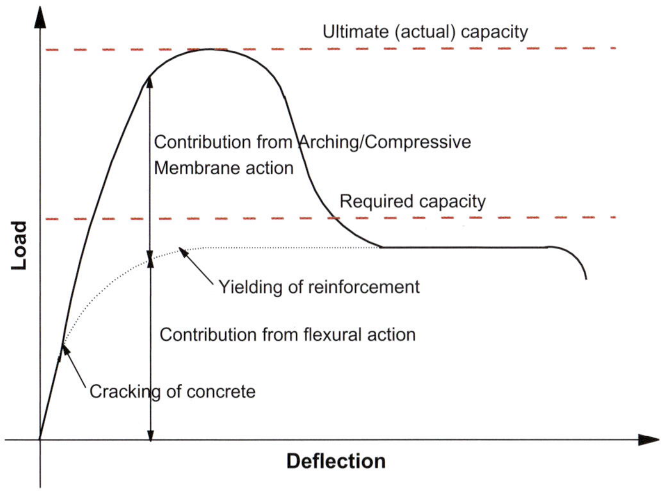

FIGURE 5.4 *Interaction between flexural and compressive membrane action (after Long* et al., *1994).*

actual testing of bridges indicates that strengths greater than those suggested by analysis can be achieved, and so further research is necessary.

An interesting point raised by Das (1993) is that in *narrow* bridge decks, even if there was a perfect distribution of load between beams, the assessed capacity would not be greatly increased above that for minimal distribution, and since the longitudinal reinforcement is so much greater than in the transverse direction, a considerable restraint in the longitudinal direction is necessary to enhance the strength due to membrane action.

5.2.1.3.3.2 End restraints. End restraint at bearing points is possible in an otherwise simply supported deck due to induced end moments arising from the horizontal resistance of the bearings themselves, and the vertical load from the road surface 'arching' through the fill and deck construction as in Figure 5.5. The bearing may be elastomeric, steel/concrete, sliding or roller and will provide some horizontal resistance (F) to the movement of the bridge deck when it tries to deflect under load. In very old bridges where the bearings have completely seized up the effect will be greater than, for example, a new sliding bearing with a low coefficient of friction (Table 5.1).

Table 5.1 Horizontal Restraint Forces at Simply Supported Bearings

Type of Bearing	Horizontal Restraint Force (kN)
Elastomeric	$F = V \times$ Shear stiffness
Sliding	$F = V \times$ Dynamic coefficient of sliding friction
Rolling	$F = V \times$ Coefficient of rolling friction

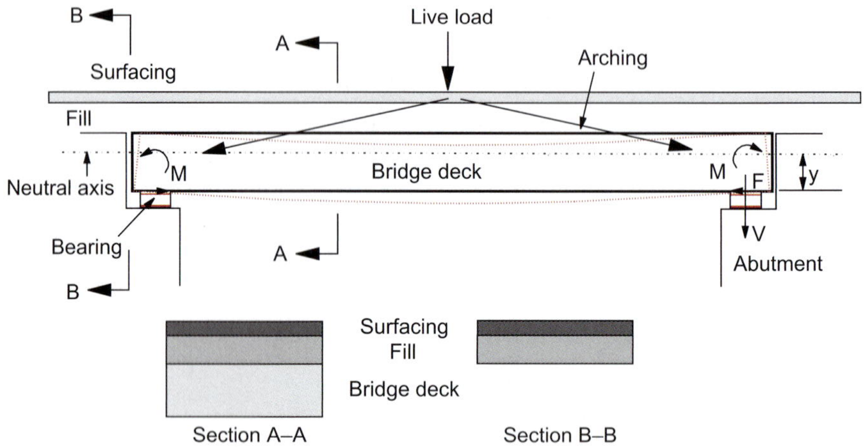

FIGURE 5.5
Effect of end restraint.

The induced end restraint moment M is equal to $F \times y$, where y is the distance from the neutral axis to the plane of the resisting force. The applied moments along the span are therefore effectively reduced by M. It should be noted that the value of the restraint moment is affected by the time of year. When the ambient temperature is causing the bridge to expand the restraint moment is fully active, whilst when the ambient temperature is causing the bridge to contract, the opposite is true, and the restraint moment *reduced*.

A good example illustrating the restraint effects of the bearings is that of the North Muskoka river bridge in Canada (Bakht and Jaeger, 1993). Five steel girders act compositely with a 14.6 m wide reinforced concrete deck slab over a span of 45.7 m. Each girder was supported on elastomeric bearings, each with a shear stiffness of 30 kN/mm. Supplementary load tests were carried out with the load in five different positions and the strains on the bottom flanges of all girders were measured at three transverse locations (one at midspan and the others adjacent to abutments). The test loads induced *compressive strains* in the girders adjacent to the bearings. For a simple pinned analysis these should have been zero, and so the results prove the fact that the bearings were providing restraint. The maximum bearing restraint force F was calculated as 175 kN equivalent to a reduction in the maximum bending moment of nearly 20%. Birnstiel (1999) suggests caution when interpreting such data because it is not known whether the bearings will continue to give the same level of restraint in the future, or indeed, whether movement of the abutments might negate the positive effect of the bearings. It would be irresponsible to allow for such strength enhancement when designing a *new* bridge, and engineering judgement must decide whether it should be taken into account during an assessment.

There is also some additional strength due to the surfacing and fill acting compositely with the deck as in Section A–A of Figure 5.5, and also (to a lesser degree), some restraint is offered by the stiffness of the composition of the surfacing and the fill at the ends of the deck as in Section B–B. These effects are more difficult to quantify, because they rely on the nature of the fill and surfacing materials as to strength and behaviour (linear or nonlinear elastic/plastic) and also the degree of interfacial shear achieved by a combination of chemical and frictional bond.

5.2.1.3.3.3 Soil structure interaction. This effect is more marked in arch bridges. Early tests on masonry arch bridges by Davey (1953) indicated that the superimposed fill material had consolidated to such a degree that *the fill, the arch barrel and the abutment were acting as one integral unit*. This is not surprising as this is exactly what would be expected in reality. (The disadvantage in breaking a bridge down into its individual parts, such as, superstructure/supports/foundations/soil for the purpose of strength calculations is that the total effect is often considered as the sum of its parts instead of the synergy of one integral unit.) Recent load tests by Page (1987, 1993) and soil structure analysis by Mann and Gunn (1995) have confirmed this view.

5.2.1.4 Supplementary testing – case histories

The benefits of supplementary testing can be readily seen from the following case histories.

5.2.1.4.1 Clockhouse bridge

Daly (1996) reports on a load test carried out on the Clockhouse bridge in the London Borough of Bromley, UK. The bridge is described as having 'longitudinal cast-iron beams with transverse ties and brick jack arches between them.' The deck spans 9 m onto brick abutments with a skew of 29°, and the beams are variously spaced at between 830 mm and 920 mm. It was originally assessed, using simple transverse distribution factors, to have a capacity of 7.5 tonnes. It was felt that there were reserves of strength due to end fixity and composite action which could be mobilised to increase this value and so a load test was carried out. The specific aims of the load test were '*to obtain data on the behaviour of the bridge under load and to use this data to determine a better analytical model for the deck*'. Data from the tests indicated that the bridge was considerably stiffer and had better load distribution characteristics than that suggested by the simple method and that strengthening of just one beam would increase the capacity to 38 tonnes! It was decided, therefore, to run a grillage analysis to confirm this using model parameters based on the test results. The analysis indicated that a final capacity of 25 tonnes was permissible. In this case the test was fully justified and proved that not only was the bridge stronger than first thought but that, with some modest strengthening works to three of the beams, the capacity rating could be increased to 40 tonnes. As fatigue endurance is critical to the integrity of the bridge, regular inspections will be undertaken to ensure that any damage to the cast iron beams is recorded and appropriate action taken.

5.2.1.4.2 Cadnum Green Bridge

This bridge has a span of 10.5 m over the River Cadnum in Hampshire, UK. It was built in the mid 1950s and is comprised of 305 mm deep trapezoidal precast post-tensioned beams with mortar packing between them. The width varies from 425 mm at the top to 45 mm at the bottom with the surfacing placed directly on to the beams as in Figure 5.6.

Mehrkar-Asl (1994) reports that due to the absence of a waterproof membrane over the deck and the evidence of leaking 'the effectiveness of the transverse prestressing cables was in doubt' with the resulting possibility of a loss of load distribution in the transverse direction meaning that some beams could be overloaded. A supplementary load test was therefore scheduled to enable transverse and longitudinal strain measurements to be made on the underside of the bridge. Vibrating wire (VW) gauges were placed transversely at midspan and longitudinally along two beams to see if there was any restraint at the abutments. The loading regime was chosen to ensure that measurable strains were induced and that no permanent damage would result.

The measured results confirmed that the transverse distribution characteristics were much greater than initially thought in spite of the complete

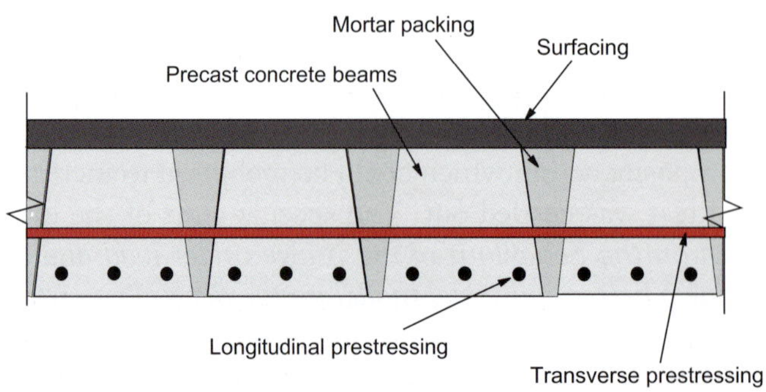

FIGURE 5.6
Part section of the Cadnum Green Bridge.

loss of the transverse prestress, and this enabled an *improved* assessment of its capacity. Other unquantifiable mechanisms were thought to be present to help the distribution of the loads due to the 'surfacing, shear transfer across the mortar packing, dowel action of the transverse prestressing and in-plane forces due to lateral restraints'.

5.2.1.4.3 Wormbridge

Low and Ricketts (1993) describe the testing of a highway bridge at Wormbridge near Hereford, UK. It comprises a filler beam deck with 305 mm rolled steel joists (RSJs) at 457 mm centres cast centrally into a 380 mm concrete slab without *transverse reinforcement*. The deck had a span of 5.8 m and was cast directly onto concrete abutments.

Because of unfavourable transverse distribution properties the bridge failed its structural assessment and was scheduled for replacement. Just prior to demolition it was decided to carry out load tests. The tests indicated a more favourable pattern of strains and displacements than had been expected, and when these were back-substituted into the analysis the deck 'passed' its assessment (Ricketts, 1991). The test was eventually continued right up to collapse and it was found that the failure load was 5 times that calculated from the assessment calculation. Low and Ricketts (1993) conclude that such decks have favourable distribution characteristics arising from a combination of 'composite action, longitudinal end restraint and transverse distribution by a raking strut and tie model'. They conclude that it is safe to assess these decks using an elastic analysis assuming a shear key deck with uncracked properties.

5.2.1.4.4 Egremont Bridge

This sandstone ashlar masonry arch bridge was built in 1822. It has two spans each approximately 12 m on a 6° skew, and each with a rise of approximately 2.3 m. The arch ring varies from approximately 770 mm at the springings to 540 mm at the crown.

One span of the bridge was loaded with a moving 32 t, four axled articulated lorry and a twelve axled vehicle carrying 153 t (Page, 1995). It was found that the measured strains and deflections were less than those calculated and it was concluded that analytical models need to be refined to improve their suitability.

5.2.1.5 A warning

In the early hours of 4 December 1985 the single span Ynys-y-Gwas Bridge over the River Afan in South Wales collapsed without warning. Fortunately there was no traffic on it at the time and so there were no casualties. The bridge was built in 1953 and had a cross-section similar to that shown in Figure 5.6. In the longitudinal direction, however, the beams were not continuous but of segmental construction. The reasons for the collapse have been well documented (Woodward and Williams, 1988), but it was primarily due to a combination of poor detailing, unusual design incorporating segmental post-tensioned segments and the absence of a waterproof membrane on the deck. The ingress of water (both clear and salt-laden) to both the longitudinal and transverse prestressing tendons resulted in severe and widespread corrosion of the prestressing steel. As the steel corroded, the cross-section diminished and the stress in the wires increased to the point of sudden fracture. *Since the contribution of the steel wires to the overall stiffness of the section was small, the loss was not detected during the routine monitoring of the bridge deck deflections and so it was assumed that all was well and that the bridge could continue in service.*

The relevance of this incident to load testing is a salutary one. Had the bridge been scheduled for an assessment and found theoretically to have 'failed' then a supplementary load test could have been recommended. Nothing about the condition of the bridge or its stiffness gave cause for alarm, and it would probably have gone ahead. Unknown to anyone, however, was that the bridge was fragile and on the point of incipient failure and thus the result of carrying out a load test would have been catastrophic, possibly leading to serious injury or loss of life.

Accurate information on the construction of the bridge and the possibility of *hidden defects* is therefore vital prior to testing. Middleton (1997) has commented that: 'if a structure has been in service for many years and there is no visual evidence of distress, the assessing engineer can be reasonably confident that the structure is capable of sustaining significantly higher loads than those already experienced by the structure'. This may be true in 99% of cases, but in the case of the Ynys-y-Gwas bridge it was not!

It is important not to confuse *stiffness* with *strength*. This point is brought out forcefully by Jackson (1996) who argues that, for instance, in the case of a filler beam bridge deck with a mixture of cast iron and steel girders (as may be the case where a deck has been widened in the past) 'improved' distribution characteristics, taken from the results of a supplementary load test, may result in load being attracted to the stiffer cast iron beams which they could not resist. This could result in 'a sudden failure at a *lower* level than a conventional assessment would suggest'. In the case of the Ynys-y-Gwas bridge, the loss of section of the prestressing tendons had a negligible effect on the stiffness of the bridge and therefore could not have been detected by taking deflection measurements.

Another more recent failure illustrating the point is that of one 25 m span of a four span, prestressed concrete footbridge at Lode's Motor Speedway in Concorde, North Carolina, USA (Oliver, 2000) which collapsed without warning as spectators were leaving the ground after a day's racing. The probable cause was corrosion of the tendons due to water leaking through holes running down from the deck to the tendons. The holes had been formed during construction by steel rods used to deflect the tendons. The rods were subsequently removed and the holes grouted. According to witnesses, the failure was that expected from a classic under-reinforced design with several loud 'rifle cracks' as the deck slowly sagged and fractured at midspan. Fortunately no one was killed but there were several serious injuries.

5.2.2 Proving load testing

This type of test is the ideal one from an assessment point of view since it involves applying loads *incrementally* up to a value equivalent to the assessment loads factored for the ultimate limit state and, in some ways, is an extension of the supplementary load test. The problem is that if the full design factors are used, the bridge would *theoretically* have collapsed. If it is suspected that there are hidden reserves of strength then a test may be sanctioned. If not, then a lower load factor will have to be assumed and the results extrapolated to failure. The safe load capacity is then calculated from the maximum test load reduced by an appropriate factor.

There is always the danger that the bridge may be damaged during such a test. Because the bridge response is monitored by measurement of stresses and deflections as the test progresses, it is possible to plot a trend. Any unacceptably high values, unexpected sudden increases or non-linearities may indicate imminent failure and the test can be halted, thereby preventing damage. Chalkley (1994) quotes a UK Task Force run by LoBEG (London Bridge Engineering Group) who are of the opinion that 'it may be preferable to accept a small probability of damage during *controlled* proving load testing in order to prove conclusively that a bridge has the required strength than to accept doubts about the actual capacity of the bridge based on a mathematical extrapolation of results obtained from [supplementary] testing with insignificant loads'. What is a 'small probability', and what if unseen local damage has occurred thus weakening the bridge? These are questions that need addressing before carrying out a proving load (or proof load) test.

5.2.3 Collapse load testing

This type of test is normally carried out on older bridges which are to be taken out of service or on more modern ones which have deteriorated beyond economic repair. In effect, the early stages of the test are essentially a supplementary test and the main advantage of such tests is not only to observe the failure load and collapse mechanism, but to enable back-analysis into established computer models. This could mean for example, allowing composite action where it was thought none existed, allowing for movement or fixity of abutments, modifying the end fixity of individual elements, considering soil/structure interaction, load dispersion, etc. to the point where the computed values converge to the actual applied loads and measured strains and displacements.

Both Page (1994) and Cullington (1994) give examples of such tests on arches (Page) and other older forms of bridge deck (Cullington). Almost exclusively, they found that the tested bridges were both stronger and stiffer than the calculated values.

Observations made on the strains and displacements during testing up to collapse can be used to indicate what loads could be safely applied to similar bridges in proving load tests without causing damage.

5.2.4 Load application

5.2.4.1 Supplementary load testing

Before the test load is specified, it is essential to carry out a pre-test analysis based on a full material and dimensional survey of the bridge in order to establish the allowable maximum load which can safely be applied. Reliance should not be placed on drawings since the 'as-built' structure may differ from them due to alterations either during construction or during later repair works.

The amount of instrumentation should be kept to a minimum, commensurate with that necessary to interpret the effects of the load test. It is largely a matter of experience and involves the measurement of *strains and displacements*.

A popular way of measuring strains is in the form of demountable vibrating wire (VW) gauges. They are robust and can be surface-mounted on uneven surfaces using a filled polyester resin adhesive (see Figure 5.7). The gauges utilise a high-tensile steel wire in tension which is enclosed in a tube between two end mounting blocks. A coil is mounted on the tube and, by means of a remote signal, the coil 'plucks' the wire causing the wire to vibrate. The frequency of the vibration is sensed remotely and this is related to the tension in the wire which is related to the surface strain measured over the gauge length. When testing has been completed the gauges can be demounted for reuse.

FIGURE 5.7 *Typical vibrating wire gauges in use.*

One way of measuring *displacements* is by the use of displacement transducers mounted off a scaffold tower. A very accurate form described as a *position transducer* uses a precision potentiometer to sense displacement driven by a stainless steel cable attached to a spring retraction device. The cable is attached to a hook glued to the soffit of the bridge directly above the transducer. Another type is the *linear variable displacement transducer* (LVDT), which can be mounted on poles or telescopic tubes (Packham, 1993) extending from a scaffold tower or fixed directly into the ground [see Figure 5.8 (RHS)]. The most common forms are the 'resistance' and 'induction' ones which are accurate to 0.01 mm.

All instrumentation is then checked for any malfunction and 'zeroed'.

5.2.4.1.1 Mobile vehicles

The preferred way of loading is by use of *lorries (trucks) with or without trailers*, as traffic disruption is kept to a minimum especially if carried out in the small hours of the morning when traffic is light. The vehicles are mobile and can be positioned at chosen stations in each lane along the span, usually the centre, quarter span and support positions. Each axle load can be measured accurately at a weighbridge, and further increments of load can be added if necessary to the lorries/trailers by means of a mobile crane lifting precast concrete blocks (kentledge) stored adjacent to the bridge.

Where the test is purely to determine the transverse load distribution characteristics, a standard vehicle which can provide a single *fixed axle loading* is all that is required. In the UK this is normally by means of a standard HB truck which is hired out by the Transport Research Laboratory. This vehicle can also be incrementally loaded up to a maximum axle load of 450 kN. For arch bridges it is usual to apply the load via a single, double or triple axled trailer. The most severe effect is when two single axle trailers are placed side by side.

Common practice is to load the bridge lane by lane and then (assuming a linear response) use the principle of superposition to establish results for all lanes loaded.

FIGURE 5.8
Typical VW gauges and displacement transducer in use (New Civil Engineer, 1996).

In some cases, if it is required to obtain a gradually applied uniformly distributed load, to prevent local overstress and simplify analysis (Yeoell, Blakelock, & Munson, 1996), then *water filled army pillow trucks* can be utilised.

5.2.4.2 Kentledge

Kentledge is the name given to *dead weights* in the form of precast concrete or steel blocks. They are lifted into position using a mobile crane to form either concentrated or distributed loads. Care needs to be taken when placing them so as not to damage the bridge deck surfacing.

5.2.4.3 Flexible water containers

These are another form of dead weight and are ideal for simulating patch or uniformly distributed loads. They are relatively light and so can be placed manually and can be filled gradually thus obviating the risk of sudden load application. The different types are described in the ICE guidelines on Supplementary Load Testing (ICE, 1998) as (i) *Pillow or mattress tanks* which are square in plan and about 500 mm high when filled, (ii) *Flexible (circular or rectangular) tanks* open at the top with a capacity of up to 100 tonnes, (iii) *Water bags* designed to be hung from the bridge deck. A major disadvantage is that they need a ready supply of water and they take a long time to fill and then empty.

5.2.4.4 Collapse load testing

5.2.4.4.1 Jacking against ground anchors

This is a particularly good system for generating the very high loads necessary for loading to collapse. A full ground survey in the form of a borehole log and soil test information is required in order for the anchors to be designed.

In the case of *arch bridges*, ground anchors are installed via holes drilled through the deck, fill and arch barrel to a steel loading frame and the load applied in increments up to failure. There may be several frames across the width of the bridge (similar to that shown in Figure 5.9) depending on the expected failure load. The load is normally taken up to a point where failure has deemed to have occurred although the bridge may not have actually collapsed, in which case the testing frames remain in place (albeit having dropped and tilted a little!). If the load is taken to collapse, the frames may fall through the arch along with the debris to be salvaged later for reuse. If the arch is over water then the frames can be connected via a spine beam at the top which allows longitudinal beams to be inserted underneath to support the frames after failure (see Figure 5.9). After each increment, relevant information such as the displacement of the arch ring, damage in the form of cracks, the formation of hinges and the final mode of failure are recorded. Visual records of the actual failure can be made using colour still photography and video equipment, the latter providing excellent information on collapse behaviour (Figure 5.10).

Because any displacement instrumentation is in danger of being lost if positioned under the arch, it is better to use a method such as *precision surveying* with reading stations marked on the edge of the arch barrel.

In the case of single span *shallow bridge decks*, similar loading frames can be used to simulate either a strip load or patch loads representing the wheels of heavy vehicles (see Figure 5.11.)

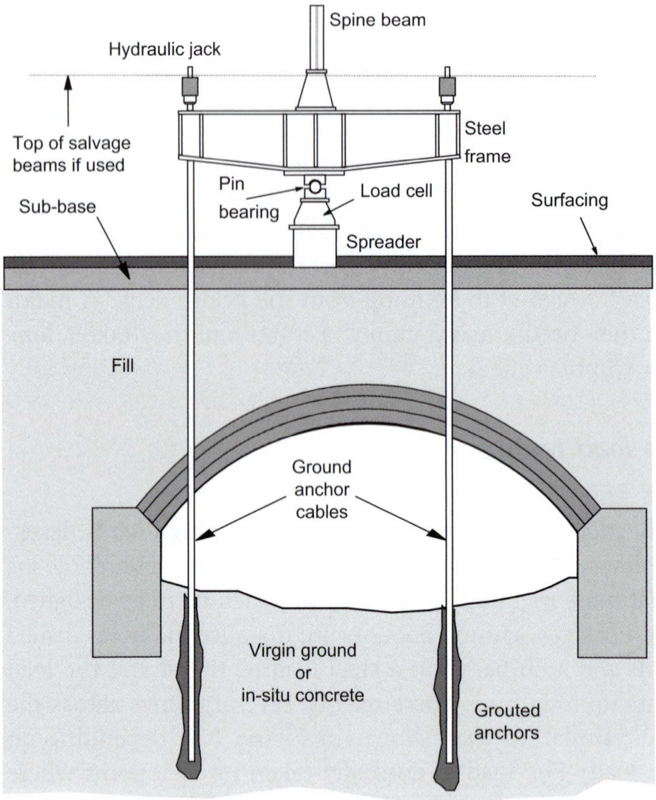

FIGURE 5.9
Typical jacking arrangement to an arch bridge.

FIGURE 5.10
Collapse of the Torksey Bridge, UK (after Page, 1995) (photograph courtesy of TRL).

FIGURE 5.11
Typical loading frame for testing a shallow deck (courtesy of Ricketts, 1991).

Loading is normally carried to failure defined as excessive cracking, yielding (of steel) or displacement. If the loading is taken beyond failure to collapse, then the same precautions need to be taken to recover the rigs as for arch bridges.

5.2.5 Traffic management

The supplementary load testing of a bridge will inevitably result in some disruption to traffic – both pedestrian and motor. This can prove costly to commercial traffic in terms of journey times and can cause frustration in casual users, and so the test must be planned so as to keep disruption to a minimum. The best time would be late at night or very early morning. The ideal situation is to keep one or more lanes open during the test. A traffic light sequence can then be arranged so that there are short time periods when no traffic is on the bridge and readings can be taken.

5.2.6 Risk assessment

Load testing is a potentially hazardous operation with many uncertainties and so it is essential that a *risk assessment* is prepared by experienced engineers prior to the test. There are many forms that such an assessment can take, but the aim is to calculate the probability of a certain event occurring and to identify the probable consequences, for example, loss of equipment or life. The risk can then be assessed by the formula:

$$\text{Risk} = \text{Probability of} \times \text{Consequence of failure} \qquad (5.1)$$

If the risk is low then the procedure can be accepted; if the risk is high then ways of reducing the risk must be sought. If none can be found then adequate safety precautions must be put in place.

A typical list of uncertainties can be prepared similar to the following:

- Identification of the hazards, such as dangers inherent when instrumenting the bridge, the expected type of failure (explosive; brittle or ductile), the possibility of worsening of existing damage to a structural element under normal traffic flows, a limited ground investigation, lack of hidden strength, possibility of collapse, use of personnel to control traffic by hand as opposed to using traffic lights, management of equipment and personnel, etc.
- An estimation of the frequency of occurrence of these risky events.
- Estimation of the level of each risk, i.e. is it low, medium or high?
- Production of method statements for events with a medium to high level of risk, and examination of events with a low level of risk with the aim of possibly eliminating those events.

In Appendix C of the ICE (1998) document on supplementary load testing there is a risk analysis procedure which enables the risk of an event occurring to be determined, calculating a normalised probability score (1–10) and a normalised consequences score (1–10) and substituting into equation 5.1 – the higher the score the greater the risk, minimal risk = 1.0 and maximum risk = $10 \times 10 = 100$.

Contingency plans are always required in the case of an unforeseen event occurring such as instrumentation malfunction, greater time required to complete the test than expected, partial damage requiring temporary propping, lighting failure or injury to personnel, etc. This reinforces the need for proper planning of the pre-test; test and post-test phases to develop a safe detailed testing programme.

5.2.7 Supplementary-proof load testing

In section 5.2.1 the dangers of extrapolating test data when the 'normal assessment' loads had been reached were noted. Indeed it is better not to extrapolate at this point but to make a judgement on whether to increase the test loads. For example, it is theoretically possible to calculate the critical deflection of a bridge deck under live load and to determine that deflection just as the failure load is reached. If a supplementary load is applied incrementally and it consistently deflects at levels less than the theoretical values up to the load corresponding to normal traffic conditions then it can be

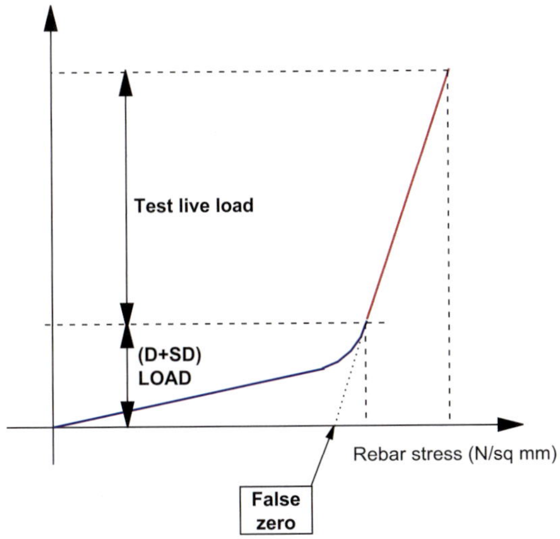

FIGURE 5.12 *Load–stress plot under applied test load.*

concluded that the bridge is *stiffer* or has *better load distribution properties* than expected.

If no damage is observed, the test could continue with loads up to a value equivalent to the assessment loads with slightly lower factors of safety than normally used in design to take it close but not up to the ultimate limit state. If the load deflection plots for the structure indicate a linear elastic response and the structure is ductile, then they can be extrapolated up to failure and the bridge can reasonably be assumed to be safe. For example, if the bridge is reinforced concrete and the *dead load stresses* can be fairly accurately estimated by using drawings and site measurements then the test load stresses can be plotted against applied load. If the final combined stress is less than (or equal to) the yield stress then the bridge is safe (see Figure 5.12). In effect a *Supplementary-proof* load test has been carried out (see section 5.2.1.4).

5.2.8 Post-test analysis

5.2.8.1 Supplementary load test

On completion of the test, an analysis of the resulting stresses, strains and deflections is made and compared to those obtained from the analytical model (see section 5.2.1). Refinement of the model is then carried out with the aim of simulating as closely as possible what was measured. This may mean modifying longitudinal and transverse stiffnesses, providing the bearings with a measure of rotational stiffness or improving the load distribution.

Caution is required in the interpretation of the test results. Lower midspan deflections may indicate that the bridge is stiffer than expected but care should be taken, for example, not to confuse the better stiffness characteristics with better load distribution to the supports due to membrane action. A plot showing the trend of the measured quantities at a given location compared with the calculated linear response can be of help in this respect.

Generally though, a better distribution of loads means that structural elements carry smaller loads, which means that the load-carrying capacity is improved. Whether the structure will fail in the expected manner is another matter. There is no way of knowing this, and so an engineering judgement has to be made as to whether the results viewed in a qualitative manner indicate a failure mechanism that was implicitly assumed in the analytical assessment. This is notoriously difficult with arches, but is not such a problem with shallow deck structures. If published research is available relative to the type of bridge being tested, this should be examined to see if there is any information which will help decision making.

The end result is to demonstrate with a high degree of confidence that the bridge can or cannot carry the applied traffic loads factored up to the ultimate limit state.

If the bridge is marginally safe then it may be permissible to allow it to remain in service for a limited number of years and then to re-inspect it. If there is little or negligible deterioration in its condition, it can be re-tested.

5.2.8.2 Collapse load test

The value of a collapse load test can be enormous since data in an incremental form are available right up to collapse. This enables existing analytical models to be refined to reflect observed behaviour. Whether the data from one test can be considered as representative of all bridges in that class cannot be answered with certainty and, again, engineering judgement has to be exercised in order to reach a 'safe' decision.

5.3 CONCLUSION

Before embarking on either a proof load or supplementary load test every other analytical option should first be explored. Load testing can be time-consuming and expensive and is therefore a last resort. The costs can vary from as little as £5000.00 for a small span bridge to as much as £100,000.00 for a larger, more complex structure. A lot will depend on the degree of instrumentation required, access, the amount of traffic crossing the bridge, for how long the bridge has to remain closed, and the degree of traffic management required to keep users on the move. A good pointer as to the need for a test is to carry out a cost-benefit analysis comparing a load test with strengthening works. After all, a load test *may* show the bridge to be inadequate, and so the total cost includes the cost of the test *plus* the cost of inevitable strengthening works.

A summary of the factors to consider when load testing is given in Appendix 5A.

REFERENCES

Bahkt, B., & Jaeger, L. G. (1990). Bridge testing – a surprise every time. *ASCE Journal of Structural Engineering, 116*(5), 1370–1383.

Bahkt, B., & Jaeger, L. G. (1993). *Analytical and observed responses of steel girder bridges, IABSE Colloquium, Remaining Structural Capacity.* Denmark: Copenhagen.

Birnstiel, C. (1999). Private communication.

Chalkley, C. (1994). *Modelling problems in short-span bridges and the basics of load testing, Surveyor.* London, UK: Hidden Strength Seminar.

Cullington, D. W. (1994). *Supplementary load testing and ultimate load testing: part 1: TRL experiences*. London, UK: Surveyor, Hidden Strength Seminar.

Cullington, D. W., & Beales, C. (1994). *Is your strengthening really necessary? Bridge Modification*. London: Thomas Telford. (pp. 270–284).

Daly, A. F. (1996). Load test on a cast iron Jack Arch bridge. In: *Proceedings of the third international conference on bridge management*. Surrey, UK: E & F N Spon. (pp. 287–294).

Davey, N. (1953). Tests on Road Bridges, National Building Studies Research Paper No. 16, TRL.

Favre, R., Hassan, M., & Markey, I. (1992). *Bridge behaviour drawn from load tests, 3rd. Int. Workshop on Bridge Rehabilitation*. Darmstadt, Germany: Ernst & Sohn Pub.

Institution of Civil Engineers. (1998). *Supplementary load testing of bridges.* London: The Institution of Civil Engineers National Steering Committee for the Load Testing of Bridges, Thomas Telford.

Jackson, P. (1995). *Flat arch action. Proceedings of the first international conference on arch bridges*. UK: Bolton: Thomas Telford. (pp. 407–415).

Jackson, P. (1996). Analysis and assessment of bridges with minimal transverse reinforcement. *In: Proceedings of the third international conference on bridge management*. Surrey, UK: E & F N Spon. (pp. 779–785).

Long, A. E., Kirkpatrick, J., & Rankin, G. I. B. (1994). Enhancing influences of compressive membrane action in bridge decks. *Bridge Modification*. London: Thomas Telford. (pp. 217–227).

Low, A., & Ricketts, N.J. (1993). The assessment of filler beam bridge decks without transverse reinforcement, TRL, Crowthorne, Report RRS383.

Mann, P., & Gunn, M. (1995). Computer modelling of the construction and load testing of a masonry arch bridge, *Arch Bridges*. In: *Proceedings of the first international conference on arch bridges*. Bolton, UK, London: Thomas Telford. (pp. 277–286).

Markey, I. (1991). Load testing of Swiss bridges. *Steel Construction Today, 5*(1), 15–20.

Mehrkar-Asl, S. (1994). *Testing and assessment of post-tensioned concrete bridges in Hampshire, Bridge Assessment Management and Design*. Elsevier. (pp. 347–352).

Middleton, C. R. (1997). Concrete bridge assessment: an alternative approach. *The Structural Engineer,* (23 & 24), 403–409.

New Civil Engineer. (1995). Weighing the evidence, p. 5.

New Civil Engineer. (1996). 117, October 18.

Oliver, A. (2000). Rusty tendons are likely culprit in US bridge collapse. *New Civil Engineer 25 May* 5–7.

Packham, A. J. (1993). Cost-effective load testing of bridges on British Railways. In: *Proceedings of the second international conference on bridge management*. Surrey, UK, London: Thomas Telford. (pp. 450–457).

Page, J. (1987). *Load testing to collapse of masonry arch bridges, Structural assessment – the use of full and large scale testing*. Butterworths. (pp. 265–271).

Page, J. (1993). *Masonry arch bridges,* TRL State of the Art Review. London: HMSO.

Page, J. (1994). *Supplementary load testing and ultimate load testing: part 2: masonry arch bridges*. London, UK: Surveyor, Hidden Strength Seminar.

Page, J. (1995). Load tests for assessment of in-service arch bridges. In: *Proceedings of the first international conference on arch bridges,* Bolton, UK (pp. 299–308).

Ricketts, N. (1991). *Assessment and load test to failure of a filler beam bridge deck, MSc Dissertation submitted to the Department of Civil Engineering*. UK: University of Surrey.

Woodward, R. J., & Williams, F. W. (1988). The collapse of the Ynys-y-Gwas Bridge, West Glamorgan. *Proc. Inst. Civ. Engrs., Part, 1*(84), 634–669.

Yeoell, D. A., Blakelock, R., & Munson, S. R. (1996). The assessment, load-testing and strengthening of Westminster Bridge. *In: Proceedings of the second international conference on bridge management*. Surrey, UK, London: Thomas Telford. (pp. 307–315).

FURTHER READING

(1993). In: *Proceedings of the second international conference on bridge management*. Guildford, Surrey, UK. London: Thomas Telford.

(1996). In: Proceedings of the third international conference on bridge management. Guildford, Surrey, UK: E & F N Spon.

Surveyor. (1995). Testing times, Hidden Strength Seminar, London, UK.

APPENDIX 5A FACTORS TO CONSIDER WHEN LOAD TESTING

Objectives for the Load Testing of Bridges	Guidance on Load Testing	Dead Load Tests	Pre-Load Test Requirements	Supplementary Load Tests	Ways of Providing Load	Mathematical Models for Bridges
To develop design formulae ▪ Tests to destruction ▪ Appropriate partial factors	Technical guidance available ▪ Canada – MoT Ontario ▪ BS 5400 – no advice ▪ BA 54/94 ▪ ICE – National steering committee on bridge testing	Stress relief methods ▪ Holes ▪ Slots ▪ Rebar cutting	Pre-test analysis ▪ Evaluation of possible modes of failure ▪ Estimates of expected strains, stress and deflections	To improve knowledge on bridge response ▪ Transverse load distribution ▪ Composite action ▪ Contribution of parapets ▪ Membrane action	Jacking from static kentledge ▪ Jacks give good control of load; high loads; load can be quickly removed if any distress ▪ Support structure required; long setup time; cannot vary load position; road closure	Model for the structure ▪ Based on the section and material properties ▪ Simple models ▪ Grillage models ▪ Finite element models
To monitor performance ▪ Obtain real response to loads ▪ Produce a signature for the bridge	The risks involved ▪ Overloading ▪ Extrapolation ▪ Brittle types of failure	Jacking methods ▪ Materials in compression – Abdunur	Pre test planning ▪ Materials required ▪ Risks analysis ▪ Method statement ▪ Detailed test programme	To Supplement mathematical calculations ▪ Low load model ▪ High load model	Jacking from mobile kentledge ▪ Jacks give good control of load; ▪ Load can be moved quickly if signs of structural distress ▪ Load can be put at different locations ▪ Loads limited by size of transporter	Model for the bridge ▪ Based on the response of the bridge to loads ▪ Simple models ▪ Complicated models ▪ Multiple models

(continued)

Objectives for the Load Testing of Bridges	Guidance on Load Testing	Dead Load Tests	Pre-Load Test Requirements	Supplementary Load Tests	Ways of Providing Load	Mathematical Models for Bridges
To aid assessment ■ Improve the mathematical model ■ Prove the bridge capable of taking the load ■ Examine critical aspects of assessment ■ Investigate hidden strength reserves	Level of load to apply ■ BA 54/94 ■ BS 8110 Part 2	Coring methods ■ Concrete stresses	Loading limits ■ Pre test analysis ■ History of previous loads ■ Loading survey prior to test (daily maximum loading) ■ Lateral distribution	Test loading ■ No greater than already carried by the bridge ■ Usually a lot less ■ Loads not usually incremented ■ Actual loads may not be known ■ No risk of damage to the bridge	Axle loading from trailers ■ Easy to move around ■ Road can remain open	The perfect model ■ Not achievable ■ More accurate model possible for long span bridges than short span ■ Must understand bridge behaviour ■ Lateral distribution ■ Stiffness model varies across deck
To prove the performance of a new bridge	Reports on past tests ■ Difficulty to obtain more than a summary	Rebar stresses ■ LoBEG method	Safety ■ Traffic above the bridge ■ Traffic below the bridge ■ Testing staff ■ Damage/ collapse of the bridge ■ Contingency plans	Advantages ■ Safety of bridge not compromised ■ Sufficient load is easily obtained ■ Many load locations can be incorporated ■ No need to close the road ■ Tests are quick (less than 10 hrs)	Kentledge applied directly to bridge ■ Easy to change load/load patterns ■ Road remains open ■ Large loads possible ■ Cannot unload quickly if distress seen ■ Time consuming to increment load	

Limitations of load testing
- Hidden strength
- Hidden weaknesses
- Shear failures

Metal sections (increbars)
- Blind hole technique
- Through hole technique

Disadvantages
- Low loading does not sufficiently test the bridge
- Only short term effects are measured
- Small effects difficult to measure
- Low load model
- Extrapolation is a problem with supp. load testing

Water bags
- Transportation to site easy
- Increment of load easy
- Cannot move load
- Loads are low and of UDL
- Long time required to increment load
- Difficult to reduce load in an emergency

Types of bridge to be wary of testing
- Buckling
- Brittle failure

Locked in (residual) stresses
- Overcoring
- Flat jack

CHAPTER 6

Repair

CONTENTS

6.1 INTRODUCTION

In this chapter the term *'repair'* is taken to mean an activity which involves the removing or cleaning of defective building materials which are then replaced and protected in a suitable way. It may also include some decorative enhancement to restore the appearance of the bridge. *Repair works generally add no extra strength to the bridge structure*, but are merely curative measures taken to restore the serviceability of the bridge – rather like taking out decayed matter in a tooth and replacing it with an amalgam filling followed by a 'brush and clean'.

To avoid unnecessary repetition of the phrase 'masonry or bricks', the generic term 'masonry' will be used to refer to both stonework and brickwork.

6.2 MASONRY

6.2.1 Introduction

Of the total UK bridge stock of about 150,000 bridges, approximately half of these are constructed of masonry built before 1922 (Smith & Obaide, 1993) and are of the arch form. These can very roughly be divided into those built before the invention of the railway to carry light traffic consisting of horse-drawn vehicles, livestock and pedestrians (they are small, usually built of stonework and anything up to 300 years old), and those built after about 1830, when brickwork began to replace stone as the dominant building material for construction of the many arch bridges and viaducts required to carry railway traffic. It is not surprising, therefore, that the bulk of bridge repair work underway is directed at masonry bridges.

6.2.2 Common defects

Some of the more common defects found in masonry bridges are:

- Rutting and distortion of carriageway.
- Leaning or missing parapets.
- Spreading of spandrel walls due to earth pressure from the fill.
- Disintegration and lamination of masonry or brickwork from severe weather conditions such as driving rain and wind, alternate freezing and thawing and vegetation growth.
- Breakdown of bedding mortar and pointing and poor quality pointing.
- Leaching of fine material through joints causing staining and the growth of stalactites on the intrados.
- Soaking of the arch barrel stonework due to water from the fill, causing breakdown due to freeze-thaw cycles and wind erosion.
- Cracking in the arch barrels.
- Arch ring separation.
- Pier corrosion and damage to cutwaters.

The sorry state of affairs that can ensue when a bridge is neglected can be seen in Figure 6.1 depicting the Laigh Milton Viaduct over the River Irvine, Kilmarnock, Scotland (Meagher, 1997). It has since been restored and is shown in Figure 6.1.

6.2.3 Repair and replacement

Once enough of the access platforms and suspended scaffolding are in place the work can commence. Surface blemishes and graffiti can be removed partly by hand and partly by the use of a proprietary saturated wet abrasive low pressure blast cleaning system, which has the advantage of eliminating dry, airborne dust. Any damaged bricks and stone blocks are cut back individually by hand (using a hammer and chisel, an electric drill or disc saw) to a sufficient depth to ensure that all of the decayed, deleterious and damaged material is removed (see Figure 6.2 after Coonie, 1992).

It is good practice for damaged or missing masonry to be replaced, where possible, with masonry that matches – or at least is compatible with – the original and then repointed with a lime-based (matching) mortar. The lime

FIGURE 6.1 *Laigh Milton Viaduct (a) Before repair; (b) After repair.*

allows a degree of movement often not afforded by modern cement mortars and there is, therefore, less likelihood of cracking due to thermal movements.

Due to the age of many of the bridges, finding matching masonry often requires a good deal of research and ingenuity. Old imperial-sized bricks may be found at an old local brickworks or farther afield at a builders merchants or even from a local crumbling mansion! Stonework may be discovered at an old quarry or local builders merchants or from the demolition of a nearby bridge

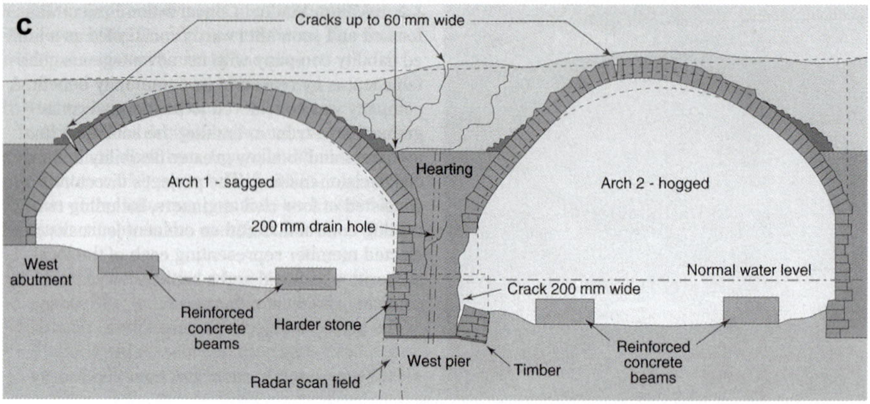

FIGURE 6.1(c) *Extent of internal damage (Paxton, 1998) (see colour plate section).*

(Coonie, 1992). Sand is usually less difficult to find as there was often a supply local to the bridge when it was originally constructed.

There is a case to be made for the use of *lime* mortars when repairing old masonry because they are *permeable and more plastic* than cement-based mortars. Cement mortar can often be stronger than the masonry it is binding and can cause the masonry to crack if there is any movement due to temperature changes. Lime mortar is slow to set and allows movement within the joints (Allen & McDonald, 1993).

New bricks are cemented into place, whilst new stones are cut and dressed on site by a stonemason and cemented into place by a stone fixer.

6.2.4 Pinning

To save on materials it is possible to pin damaged stones together in a manner indicated in Figure 6.3. A 50 mm diameter hole is drilled approximately 50 mm deep using a diamond-tipped overcoring tool and the core removed. A further hole, 25 mm in diameter, is drilled to the required depth and injected with a low-viscosity epoxy resin. The 16 mm diameter pin is inserted, which then displaces the resin into surrounding cracks, delaminations or voids. This is then capped with a plug cut from the masonry core. The technique can be used on either the spandrel walls, or the frontages or intrados of the arch ring.

Pinning is especially useful for repairing longitudinal cracks in a face arch ring where it has been detached from the main arch. The pins are put in normal to the spandrel and in the centres of the alternate voussoirs (wedge-shaped stones) to a depth of 500 mm or more, beyond the crack to be tied. The stainless steel bar usually has an enclosing, expandable nylon sock and the injected ultra-fine grout fills the sock and grout milk is forced through into all

FIGURE 6.2(a) *Suspended scaffolding to Viaduct Pier.*

of the crevices and cracks along the length (Figure 6.4). The plugs are then inserted and the cracks pressure-pointed.

6.2.5 Arch ring separation

When one arch ring moves relative to another arch ring leaving a gap, this weakens the structure (because composite action is lost between rings) and is referred to as *arch ring separation* (see Figure 6.4). This can be repaired using the same technique as described in 6.2.4 (but with the pins inserted

FIGURE 6.2(b) *Removing damaged brickwork.*

perpendicular to the intrados) or, if the damage is local, by the use of a crack injection system (see 6.2.10).

6.2.6 Spandrel walls

Spandrel walls tend to distort with time if the bridge has been subjected to sustained passage of heavy traffic, and/or if tie rods have disintegrated

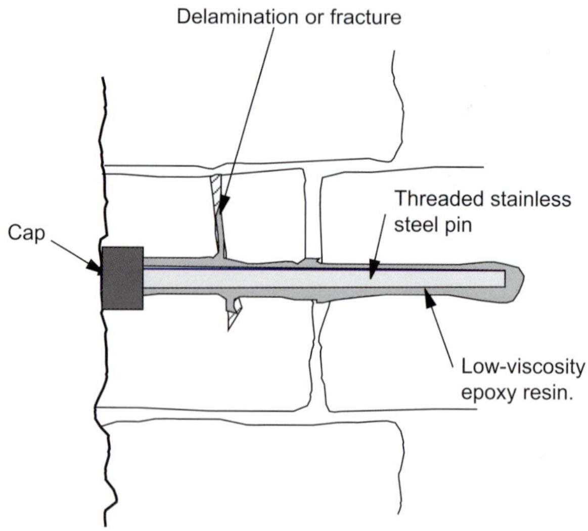

FIGURE 6.3 *Pinning technique.*

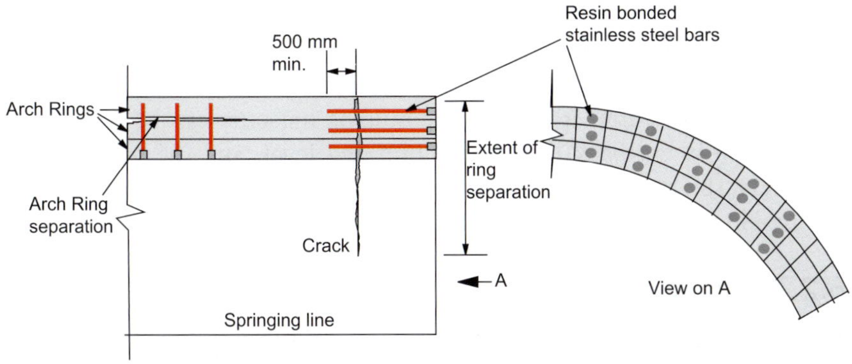

FIGURE 6.4

Tying of longitudinal cracks in arch ring.

through. Rebuilding the bulging walls is best accomplished by excavating behind them and rebuilding conventionally. This condition may require partial demolition and realignment, or complete removal and rebuilding. The wall can then be backed with 300 mm–450 mm of weak mix concrete to provide extra edge stiffening, and to avoid any local damage to the edge voussoirs it may be advisable to use lightweight concrete as in Figure 6.5a.

An unconventional approach is suggested by Welch (1995) to remove lateral pressure completely from the spandrel walls by using the reinforced earth system to support the fill. The principle is shown in Figure 6.5b. Basically, an inner wall is constructed from filled sandbags and tied with tensile webbing to a similar wall on the opposite side of the bridge, and the gap between this and the spandrel wall is then filled with a granular drainage material. This method assumes that there are few services to be interfered with and also great care must be taken if future works involve digging up the roadway by utility companies so as not to damage the webbing!

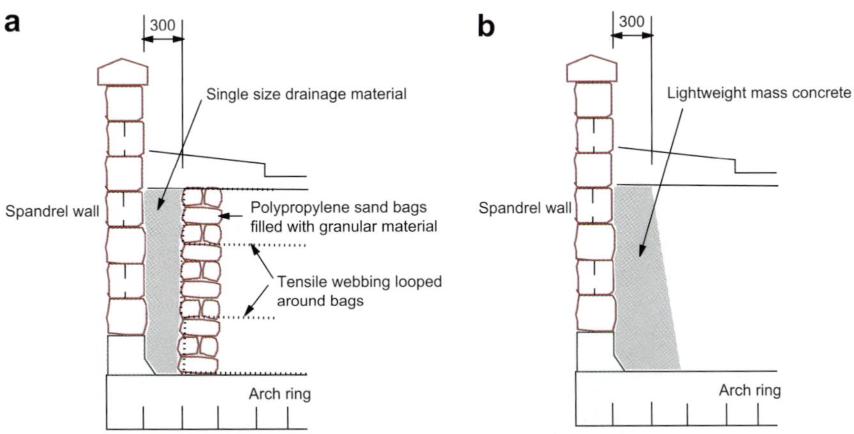

FIGURE 6.5

Mass concrete fill and reinforced earth repair options for bulging or displaced spandrel walls.

If the situation means that the bridge cannot be closed entirely, Welch (1995) has suggested a solution as shown in Figure 6.6. Here the contained ground is grouted over the zone shown using a weak mix (1:6 or 1:8) of cement:sand and after it has set, the bulging masonry can be removed and the weak mix cut back to the true line and level, and the wall rebuilt. If necessary, the reconstructed wall could be tied back with stainless steel pins at suitable intervals.

6.2.7 Parapets

Masonry parapets are very strong (Middleton, 1995), but they do occasionally get damaged and need repair. For masonry arches (unless otherwise specified by a Highways agency) there is generally no requirement to upgrade and so repairs follow the conventional way of rebuilding broken areas in materials, similar to the existing. Damaged wrought iron or cast iron railings can be repaired compatibly with the existing and, if necessary, the posts can be replaced and reinstated in holes drilled into the masonry which are then filled with molten lead. Finally, they can be wire brushed (in-situ) or shot blasted (in a workshop), and given two (125 μm DFT) coats of zinc phosphate primer followed by a top coat to match the original.

6.2.8 Piers

Pier corrosion and damage to cutwaters usually consists of loss of facing to the masonry with open joints, missing and cracked masonry. These can be repaired above low water level in the traditional way using a quick-setting cementitious mortar. Ideally the mortar should be capable of application above and below the water line. Older, rubble-filled piers very often suffer from voiding near the inner surface of the masonry where open joints have

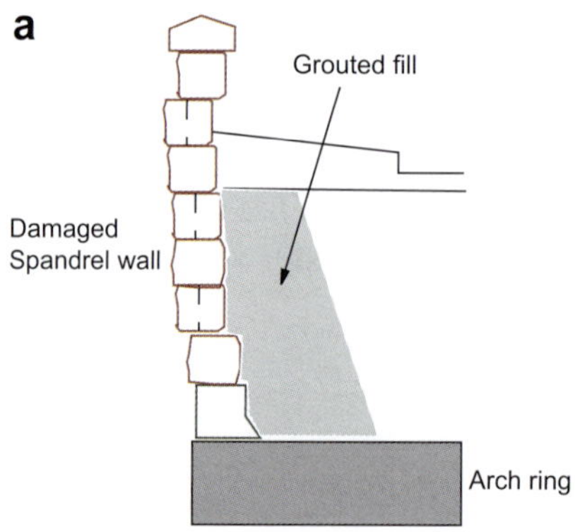

FIGURE 6.6
Grouted fill repair options for bulging or displaced spandrel walls.

allowed water through, which then begins to wash out the weak binding materials. These can be filled by the use of a suitable grout which should, according to Ball (1997) have the following characteristics:

- Capable of displacing water.
- Seek and fill both large and fine fissures behind the masonry.
- Capable of being pumped through a 10 mm diameter tube.
- Remain cohesive while being poured underwater.
- Have an open time of two hours to provide adequate mixing and pumping time.
- Have a strength similar to the infill material already existing in the pier core.
- Be non-shrinkable.

Once repairs to piers and cutwaters are complete, it is prudent to provide further protection on the upstream and downstream sides by the use of bagging or loose rocks to just above high water level to limit damage from floating debris and possible scour.

6.2.9 Carriageway, fill and waterproofing

Normally, it is possible to remove all of the infill and hearting materials below the carriageway without providing supplementary support to the arch barrel. In the case of badly damaged and fragile arches, however, support must be provided from beneath in the form of timber or steel centring frameworks founded on independent, temporary foundations at bed level.

Once removed, and provided that no strengthening works are required, the extrados of the arch barrel and longitudinal *sleeper* walls are repaired as necessary using techniques described in section 6.2.3. If drainage outlets are in place over the piers, these should be inspected and cleared (or alternatively grouted up) and new graded infill material ('no-fines' or lightweight concrete) placed over them up to formation level. Alternatively, the old outlets can be rendered redundant by filling them with grout and an effective waterproofing barrier provided above by installing either a concrete *saddle* directly on to the arch barrel or a horizontal *concrete slab* just below the carriageway surfacing layer (these will be discussed fully in Chapter 7).

Sprayed membranes are being increasingly used on older bridges as they are able to deal with awkward profiles and have no seams which can be troublesome in sheet systems. Also, because they consist of methacrylate monomers, they can be applied in virtually all weathers. Modern systems have a proven track record and are applied cold, have rapid curing characteristics, and form a tough flexible membrane. It is usual to apply a methyl methacrylate reactive primer, followed by two membrane layers, each colour-coded to ensure proper coverage, followed by a high-strength methacrylate copolymer hot melt adhesive tack coat which helps to provide a good bond between the membrane and the surfacing.

6.2.10 Crack injection

There are two types of crack injection system: pressure and vacuum. Both use a low modulus (3.5 kN/mm^2) epoxy resin compound with tensile and compression strengths of about 20 N/mm^2. In the former case the resin is injected under pressure, and in the latter it is drawn in under the action of a vacuum. They can be utilised to repair small cracks in individual masonry blocks or to fill larger continuous cracks. Vacuum injection has a typical fill level of 95% and can fill cracks as small as 0.025 mm, whilst the figures for pressure injection are 75% and 0.15 mm respectively. The integrity of such systems depends upon the development of a sound bond between the parent material and the resin, and so it is vital that the brick or masonry is dry and dust free, achieved by the use of a vacuum appliance. Individual stones which show signs of cracking can be treated in this way. Each stone is core drilled, vacuum cleaned and injection fittings bonded to the surface. A vacuum is established by attaching a vacuum pump to the appropriate fittings and, as the air and water are drawn out, the resin is introduced under slight pressure. When the resin stops flowing the fitting is plugged, and once the resin has set the fittings themselves are removed. Surplus resin is then removed by grit blasting and pointing carried out as necessary.

Cracks in brickwork can be treated in a similar fashion. For cracks in mortar joints, the joint is first routed or drilled out to a depth of about 3/4 of the depth of the brickwork. The inner surfaces are then vacuum cleaned using an extension tube, and the resin is injected using a skeleton gun to fill the hole but leaving a gap of 10 mm at the surface. The joint is finally repointed in matching mortar.

For cracks running through bricks or blocks, a hammer drill is used to drill 13 mm holes at approximately 20 mm centres along the length of the crack, which is then vacuum cleaned to remove all dust. The injection nozzle is held tight to the surface, and the thixotropic resin is injected to fill the hole, leaving a 10 mm recess at the surface which is then filled with matching material.

6.3 CONCRETE

6.3.1 Introduction

Concrete by itself is a relatively inert material, but if it is in contact with embedded steelwork (be it reinforcement bars or steel beams) or made from reactive aggregates, decay can ensue. The concrete can crack, spall, delaminate and crumble depending on the aggressiveness of the environment, the quality of the concrete and the proximity of corroding reinforcement. Repairs can range from simple crack injection to wholesale replacement (Figure 6.7).

6.3.2 Preparation for patch repairs

Preparation of affected areas of concrete is necessary before any repair is attempted. It is also important that the ambient temperature at the time of repair is moderate, and that the concrete surfaces are not too hot or cold. The first step is to *clean the concrete surface* by removing dirt, debris and other deleterious materials. The area is then *marked out* and the *decayed concrete broken out* using one (or a combination) of three methods, namely:

- Hand held hammer and chisel
- Pneumatic hammers
- Hydrodemolition

The last method is basically a high pressure water jetting system for the selective removal of concrete, and is gaining in popularity where large areas of concrete need to be removed. Unlike pneumatic hammers, it tends to be less violent and there is, therefore, less chance of damage to the existing concrete substrate. It also provides a good surface for keying the new concrete to the old (Medeot, 1990; Baldo & Medeot, 1994). The concrete removal is continued behind the corroded reinforcing steel, and the edges of the breakout area should be cut square to ensure success of the repair. Once the carbonated concrete has been completely removed down to at least 15 mm behind reinforcing bars, then repairs can commence. *The reinforcement is then cleaned of corrosion products by grit blasting and primed with a zinc-rich primer or a wet cementitious powder mix*. The actual repair method employed

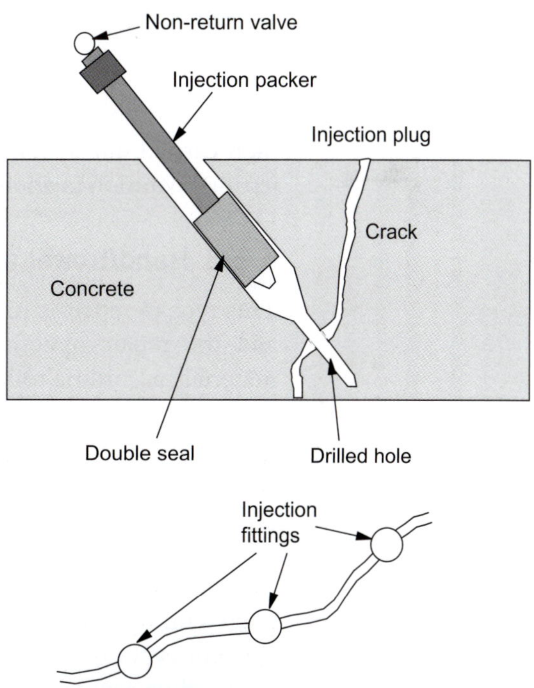

FIGURE 6.7 *Crack injection system.*

FIGURE 6.8 *Water jetting on the Silverton Viaduct (courtesy Cons. Main & Repair).*

will depend upon the size of the affected area and its location. The repair is then carried out by *hand/trowel, fluid* or *spray* application. A typical water jetting operation is shown in Figure 6.8.

6.3.3 Hand/trowel application

This type of repair is most suitable for small areas of approximately 0.1 m^3 and the repair materials are available pre-packed. The common repair materials according to Mallet (1994) page 106, are:

- Plain cementitious material.
- Polyester and epoxy resins.
- Polymer-modified cementitious material.
- Cement/pozzolanic-modified material.

There are many proprietary brands of material on the market, with advice offered on what to use in given circumstances. It is prudent to obtain independent evaluation of the manufacturers claims, however, if at all possible.

Bonding agents are normally specified by specialist repair pack suppliers, and these must be faithfully applied prior to concreting; alternatively the surface of the exposed concrete (the substrate) should be dampened to prevent suction of water from the applied concrete which can prevent proper hydration of the cement and failure of the repair.

During application of the repair works it is vital to ensure that proper protective clothing is worn, including gloves, goggles and dust masks. These precautions will help to prevent burns, dermatitis, blindness and respiratory problems.

A typical example of the hand-applied types of repair (Liquid Plastics Ltd., 1990) are shown in Table 6.1.

Various *surface* preparations are indicated by items 1,3,4 and 5 in Figure 6.9.

Item 1 is a polymer-modified cementitious screed which gives a fair faced finish in either a thin film application or as a filler for surface defects. It has good waterproofing characteristics and provides increased protection from acid gases, chloride ions and freeze-thaw cycles.

Item 2 is a waterproof screed and pore filler to provide protection against pollution, airborne chemicals and carbon dioxide.

Item 3 indicates the use of a surface stabiliser which is a water-based material with high penetration in concrete. It binds together powdery surfaces and evens out different absorptions.

Item 4 is a substance used to provide protection against aggressive atmospheric pollution, and because of its self-cleaning finish, remains in a 'just applied' condition.

Item 5 illustrates a smooth finish to the concrete surface. Once the repair has been levelled or smoothed and the concrete has cured, any small irregularities can be covered with a 'fairing coat' and, finally, the surface of the concrete can be protected by the use of water-based products such as a *copolymer or a resin-rich, elastomeric copolymer – sometimes known as 'anti-carbonation paint'*. This gives a very pleasing finish and should last for 10–15 years.

Item 6 has a proven track record of over 28 years (Liquid Plastics Ltd., 1990) and is 'highly resistant to carbon dioxide diffusion. It can facilitate movement and allows the emittance of substrate moisture. It provides protection against aggressive atmospheric pollution and is self-cleaning. It can be fibre reinforced and is available in various colours'. The latter has a 'high water vapour transmission rating and excellent resistance to pollution and acid rain. It also contains a fungicidal system to prevent growths of mould and fungi'.

An example of the types of *hand* repairs for more serious repairs illustrated by the tee beam section shown in Figure 6.9, Items 6, 7, 8 and 10, (Liquid Plastics Ltd., 1990).

Table 6.1 Different Types of Hand-applied Repair

No.	Defect	Repair	Properties
1	Minor surface defect	A two-pack polymer-modified, cementitious screed.	Gives a fair faced finish in either a thin film application or as a filler. Good waterproofing characteristics; gives protection from acid gases, chloride ions and freeze-thaw cycles.
2	Surface cavities and honeycombed concrete	Highly adhesive, thixotropic mortar.	Waterproof with anti-carbonation finish to both concrete and repairs. Good resistance to pollution and airborne chemicals.
3	Powdery surfaces	A two-component surface stabiliser.	Binds together powdery surfaces and evens out different absorptions.
4	Surface protection	A water-based copolymer.	Highly resistant to carbon dioxide diffusion and is self-cleaning.
5	Surface protection	Resin rich, water-based, elastomeric copolymer.	One coat application. High water vapour transmission rating and excellent resistance to pollution and acid rain. Contains a fungicidal system.
6	Non-structural cracks and gaps	Non-shrinking polyol flexible filler.	Tough, easily applied, elastic compound with excellent adhesive and chemical resistance. Cures at temperatures down to freezing.
7	Minor voids (100 mm × 100 mm × 50 mm deep)	Rapid curing, polymer-modified cementitious mortar.	High strength – compacted in layers.
8	Major voids (200 mm × 200 mm × 150 mm deep)	Heavy duty, thixotropic, fibre reinforced, polymer-modified, cementitious mortar.	Can be applied up to 100 mm thick without slump or sag. Excellent adhesion and easy to mould and profile.
9	Bonding agent for 7 & 8	Polymer-modified, cementitious surface impregnant.	High penetration, creating enhanced adhesion between porous concrete and repair concretes.
10	Protection of steel reinforcement	A highly alkaline, two-component product of cementitious powder and a polymer dispersion which react chemically to passivates with the aid of anodic inhibitors and protects steel reinforcement.	High penetration, creating enhanced adhesion between porous concrete and repair concretes.

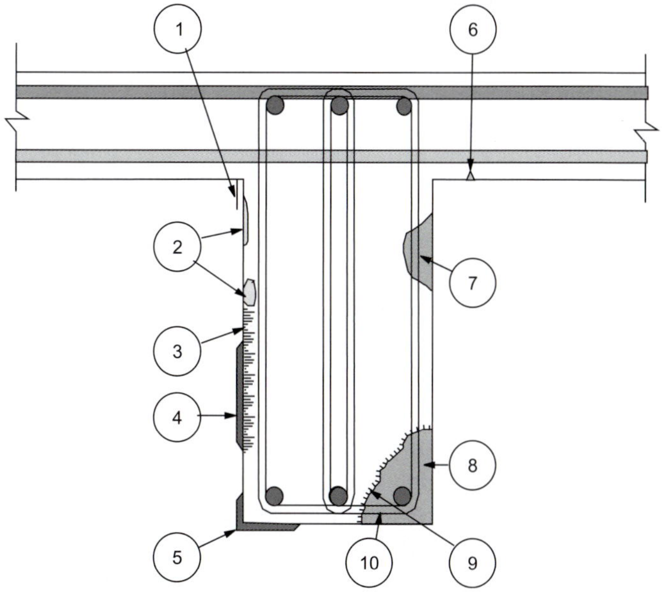

FIGURE 6.9 *Typical hand-applied concrete repairs.*

Any exposed steel reinforcement is cleaned and protected with a highly alkaline two-component product consisting of a cementitious powder and a polymer dispersion which react chemically together to form a passive protection to the steel (Figure 6.9, Item 10). Small voids about 30 mm–50 mm deep or wide are usually filled with a rapid curing, polymcr-modified, high-strength cementitious mortar (Figure 6.9, Item 7). For small gaps or local cracks, a non-shrinking polyol is used. It is tough, easy to apply and cures at temperatures down to freezing (Figure 6.9, Item 6). For surface cavities or honeycombed concrete, a highly adhesive thixotropic mortar is used which acts as a weatherproof screed and provides an engineering anti-carbonation finish to both concrete and repairs. It has excellent water-proofing properties and good resistance to pollution and airborne chemicals (Figure 6.9, Item 2).

6.3.3.1 Loading effects

In places where the repaired concrete areas are stressed in tension when the live load is applied, it may be necessary to temporarily load the bridge with an equivalent live load, repair, and then remove the load. This ensures that the repaired areas are in compression when in service. Alternatively, if the tensile stresses induced under live load are low, then it is sufficient to add an expanding agent to the concrete mix.

If no pre-load is applied, then the new concrete, unlike the parent concrete, is stressed under the action of live load effects only.

6.3.4 Fluid application

In large areas where it is not possible to apply a stiff mortar mix, a pourable or 'fluid' concrete is used. For example, where there is congested reinforcement, where the shape of the section is not compatible with hand-applied techniques or where access is difficult.

Preparation is carried out as previously described, shuttering is put in place and the concrete poured. A typical letter box arrangement is shown in Figure 6.10 where the concrete can be (a) *poured* into the void, or (b) *pumped* under pressure from below. A cementitious free-flowing micro-concrete is normally used which is self-compacting (so that no vibration is necessary), shrinkage compensated, has high strength, has high chloride/carbon dioxide resistance and has low alkalinity to minimise the risk of alkali-silica reaction or ASR (this is achieved in the UK by compliance with the stringent requirements of the Department of Transport Specification for Highway Works, Clause 1704.6: Control of Alkali-Silica Reaction). After the concrete has set, the shuttering is removed and the small volume remaining above the final level of fluid concrete can then be hand-filled. Where decayed concrete to the sides, soffit and top of reinforced beams, or completely around a pier column has been cut back, the fluid concrete can be poured from the top as shown in Figure 6.10.

The large surface areas can be protected using a three-part protection system comprising, (i) a 'Silane' primer (see Chapter 10, section 3.8) which

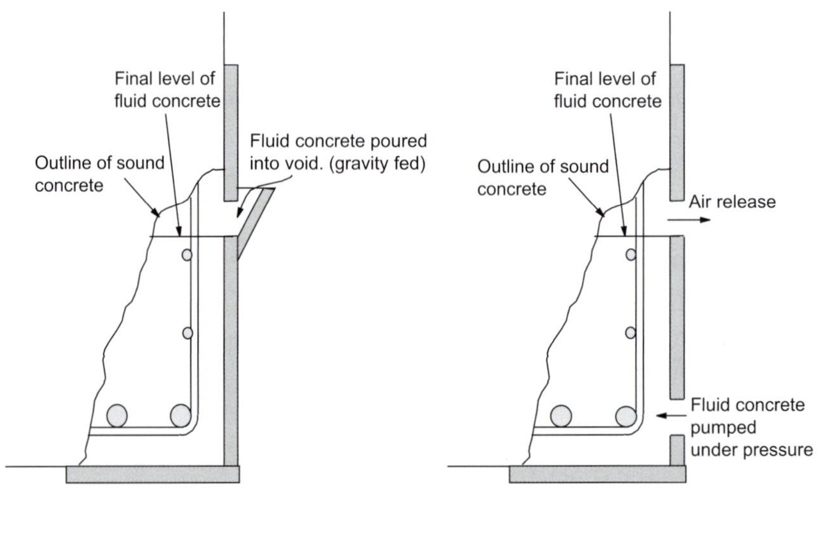

FIGURE 6.10
Traditional 'letter box' arrangements for the application of fluid concrete to a beam element.

searches deep into the concrete pores, lining their surfaces with a chemically bound water-repellent layer, (ii) a topcoat barrier to carbon dioxide and other acidic gases from the atmosphere, but which permits the passage of water vapour, thus, allowing the concrete to breathe, and (iii), a final pigmented topcoat. This system has proven very reliable in the UK and possesses outstanding resistance to scaling.

In *deck* areas where the salt has penetrated the surface layers of concrete, it is often necessary to completely strip the bridge surfacing, the waterproof protection layer and the old waterproof membrane. The whole area is then water-jetted to remove the infected concrete, and repairs carried out as in section 6.3.3, or new concrete is sprayed on as in section 6.3.5. New waterproofing is then normally applied using a three-part sprayed technique which is now a proven and effective system, and has the advantage that it can cover every irregularity and surface orientation (Figure 6.11).

The *top members and columns to concrete piers or trestles* can be subject to severe corrosion due to water containing chlorides leaking from the superstructure through imperfect joints. In such cases it is usual to jack up the beams forming the deck to remove the load and improve access for repair. Contaminated concrete is then removed as normal and, once the reinforcement has been coated with an inhibiting primer, the new concrete is applied. If extensive areas of the members are damaged, then formwork is added to

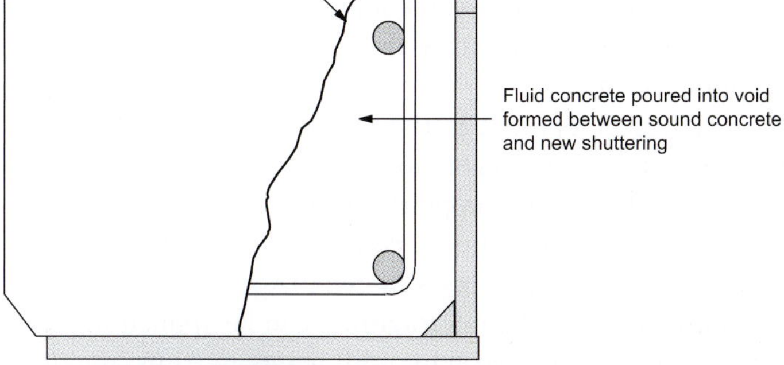

FIGURE 6.11 *Plan on column to be repaired.*

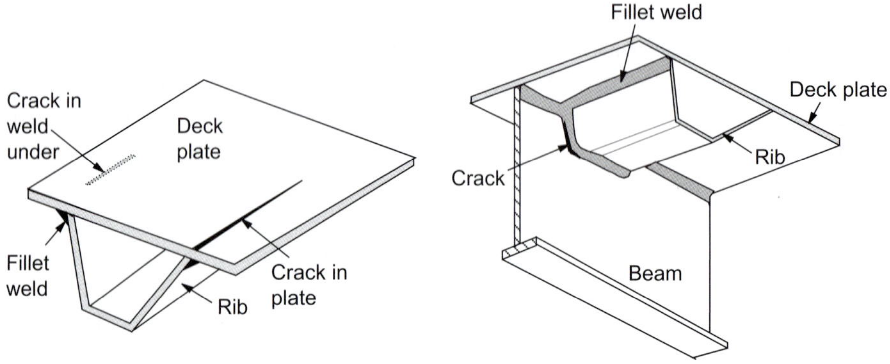

FIGURE 6.12
Cracking in orthotropic plate decks (after Mehue, 1990).

the original profile and pourable concrete added. In order to be sure of the flow characteristics of the repair materials, it may be necessary (subject to quality assurance procedures) to carry out a test on a model shutter.

A typical repair is shown in Figure 6.12.

6.3.5 Spray application

The concept of spraying a cementitious mix at high velocity from a nozzle onto a prepared concrete surface where 'infected' concrete has been removed is often referred to as '*gunite*'. The concrete is made with a small aggregate size – 10 mm or less – and has been found to be useful in repairing large, relatively shallow (50 mm–125 mm depth) surface areas where the reinforcement is uncongested. It can be applied as either a *dry* or *wet* spray. The dry process was pioneered in America and dates back to around 1911 (Bolton, 1998). It consists of mortar powder forced by the use of compressed air to a delivery nozzle where it is mixed with water and applied. The latter method consists of premixed mortar being pumped directly to the delivery nozzle and forced out with compressed air. The application rates are much higher than hand-applied techniques, and several suppliers claim better bond strength, compaction and lower permeability due, presumably, to the effect of the pressurised application.

A fairly recent development is the use of micro-gunite (Aguilar, 1994) produced by Sika which consists of 'non-reactive 3 mm aggregate and a blend of admixtures to enhance the properties of the repair material and provide superior qualities to that of ordinary concrete'. The enhanced performance has been proven by exhaustive tests by Sika (Aguilar, 1994) and is used in practice on many bridges. The micro-gunite can be applied in layers as thin as 10 mm and eliminates the need for reinforcement, thus, making it a cost-effective alternative to conventional gunite especially in bridges where soffit repairs are being carried out. By virtue of its reduced thickness it also reduces the dead load. It can also be used for thicknesses of up to 150 mm. The makers also claim that the micro-gunite has low creep, low shrinkage, good interface bond characteristics and can be designed to have a modulus of elasticity (E) value compatible with the original concrete, thus minimising differential strains between the micro-gunite and the parent concrete. Once completed, the repairs can be hand-floated to a specified smooth finish.

6.4 STEEL

6.4.1 Introduction

The most common cause of deterioration of steel bridges is corrosion. This may be due to chloride solutions when the steel is surrounded in concrete or due to the effects of the environment when the steel is exposed.

A reminder of the importance of the environment was described recently in an American publication (Bettigole, 1994) in which a renowned bridge engineer, commenting on the Williamsburg Bridge at its opening in 1903, said, 'So far as engineering science can foretell with confidence, this colossal structure, if protected against corrosion, its only deadly enemy, will stand hundreds of years in unimpaired strength.' It never occurred to him that people would be foolish enough to let it rust almost to oblivion'.

The Forth Railway Bridge, at the opposite end of the scale, demonstrates that steel bridges can be maintained virtually intact for many years, and represent good value for money. Since its opening in 1890, there have been no major structural repairs. This has been partly due to the sound design principles employed, but primarily, it is due to the foresight of the original design engineers who recognised that prevention was better (and cheaper) than cure. The continual painting of the structure since its opening has ensured that repair works have been kept to the absolute minimum, consisting only of the replacement of some bracings and angle plates corroded by steam and smoke.

Below water level, apart from the normal corrosion mechanisms, a new form of bacterial corrosion has been found affecting steel sheet piles around bridge piers (NCE, 1998). This is a very aggressive form of corrosion found in warm climates, both tropical and temperate. The bacteria can consume up to 1 mm per year and is an example of 'accelerated low water corrosion' (ALWC). It is a little understood phenomenon and detection is extremely difficult. The current repair regime recommended by British Steel in the UK is to strip off the corrosion products and to coat with a glass flake-based paint formulation.

6.4.2 Damage and repair

Accidental damage and that due to wear and tear result in numerous repairs to steel bridges each year, and it is, clearly, not possible to provide a comprehensive list. Damage is due to corrosion, collision, fatigue and natural disasters (earthquakes and floods), and all associated repairs are of an individual nature.

Classification of the *damage* needs to be carefully recorded and can be broadly classified as shown in Table 6.2.

6.4.2.1 Corrosion

Corrosion damage can range from as little as surface rusting to complete erosion of the section. The most severe cause is that from salt-laden water which is washed or splashed from the roadway onto the various parts. This is especially the case where no proper drainage has been provided at the joints (principally in truss bridges) and the solution (along with other deleterious matter) has collected to form a pool of corrosive matter continually eating away at the steel. This is, in effect, 'stress-corrosion'. As the section corrodes, the cross-sectional area decreases, thus increasing the direct stress in the member which results in increased deformation, the section corrodes again and the cycles continue. Many decades in such an environment can cause complete loss of section of both the structural element (tie, strut or beam) and the fastening, whether welded or bolted (Lichtenstein, 1990).

The most effective methods of repair where access is possible has been to thoroughly clean the affected areas by means of grit or shot blasting, and, providing that there has been no significant loss of section, to apply at least a three-part paint protection barrier. Two priming coats are normally applied with either micaceous iron oxide or zinc phosphate-based primers, with a total thickness not less than 75 μm DFT followed by a topcoat applied in two layers of 75 μm DFT each.

Table 6.2 Classification of Damage to Steel Elements

Parent Material	Fastenings
Corrosion	Corrosion
Cracking	Fracture
Tearing	Bending
Buckling	Cracking
Distortion	Slipping

Epoxy-based paints are being used as a long term repair and protection measure on the Forth Railway Bridge (NCE, 1997). After grit blasting, the original hot-dipped linseed oil protection is stripped down to base metal, a zinc-rich priming coat is applied and then, finally, it is covered with two coats of two-part epoxy paint enriched with tiny glass flakes which will act like the scales of a fish to repel the water. The system is forecast to last 20 or between 25 and 40 years, but has not been proven in practice. It is forecast to cost £74 m and to be completed in 2012.

6.4.2.2 Cracking/fracture

Cracking in the parent material usually occurs adjacent to a stress raiser such as a hole or notch. If in the vicinity of a bolt, the bolt has to be removed and the area thoroughly cleaned locally (see 6.4.2.1). A small hole is first drilled at the tip of the crack to prevent further propagation, the steel on each side of the crack is then dressed by grinding down with a power grinder, and then the crack is flush welded. Finally, an MPI (see 4.3.3) is carried out to test the integrity of the weld, and if satisfactory, the section is rebolted. If cracked bolts are found, these are usually taken out, and if there is no damage to the surrounding such as deformation due to bearing, then the bolts are simply replaced.

Fatigue cracking can sometimes occur in the orthotropic decks of steel bridges at either the stiffener/rib-to-deck plate junction or the (discontinuous) stiffener/rib-to-floor beam junction (see Figure 6.12). In each case, *providing the cause of the cracking has been identified and eradicated*, the crack can be ground out and welded. In actual fact, Mehue (1990) suggests that in some orthotropic bridges inspected and monitored in France, where such cracks 'are few in number, rather short and fine, and grow at a slow or gradual rate' they should be monitored but not repaired.

Other weak points also exist in steel truss or beam bridges at positions of halved (half) joints (Matthews et al., 1996; Kitada and Mizobuchi, 1993) where cracks appear in the joints themselves and at the main bearing points, and other cases have been reported in steel arch bridges (Shimada, Yamada, Aoki, & Fukushima, 1996) where such cracks appear at the ends of stringers, floor beams and struts.

6.4.2.3 Tearing/buckling

Tearing of steel plates can occur due to an accident, where, for example, a collision takes place between a motor vehicle or river traffic and the bridge. Such tearing causes local bending of the plate(s) away from the tear. The damage can be repaired by either bending the plate(s) back to their original position and welding/bolting a cover plate over the damaged area, or by cutting away the distorted plates and installing new plate(s). The latter method is usually the more practical, since to try and jack 16 mm–25 mm plates back into position is often not possible because of the lack of a suitable jacking position.

6.4.2.4 Bending/distortion

Bending or distortion of steel plates after an accident does not necessarily result in any loss of strength. For example, in the case of the lateral distortion of the bottom flange/web area of a steel beam and slab bridge which is in tension, or where the distortion is in an area at or close to a point of contraflexure, the stresses are low and there is usually a reserve of strength. In such areas a decision can be made to do nothing at all other than to touch up the paintwork, and monitor on a regular (annual) basis.

In the case of steel crash barriers or parapets where holding down bolts have pulled out or suffered severe bending, or where welds have bent or cracked, the only solution is to remove the damaged sections and replace with new ones.

6.4.2.5 Slipping

At splice connections incorporating high-strength friction grip bolts, damage can result in slippage (but not failure) of the joints. This may cause displacements greater than the design serviceabilty limits. Provided that no damage such as cracking or spalling has resulted to concrete parts then no repair is needed. If the slip is beyond acceptable limits, then the beam can be jacked up locally and propped, the nuts are then loosened and if the faying surfaces are sound, the beam can be jacked up to its original position and the nuts tightened. A typical case study which illustrates some of these repair techniques is that of the Erskine Bridge.

6.4.3 Erskine Bridge

The Erskine Road Bridge carries the A898 over the River Clyde in Scotland. It is a cable-stayed steel bridge and has a main span of 305 m. The deck of both the approach viaduct and the main span is formed of a trapezoidal steel box section. On 4 August 1996 the soffit of the main span section was struck by the top of an oil topside module as it was being transported out to sea (Wadsworth, 1997). The resulting damage consisted of tearing of the bottom flange, the lower webs of the steel box, and the lower part of one of the diaphragms. It had also resulted in local buckling of the transverse bulb stiffeners (see Figure 6.14). The damage was not severe and calculations indicated that the decrease in load capacity of the bridge was localised to the area of the damage (Wadsworth, 1997).

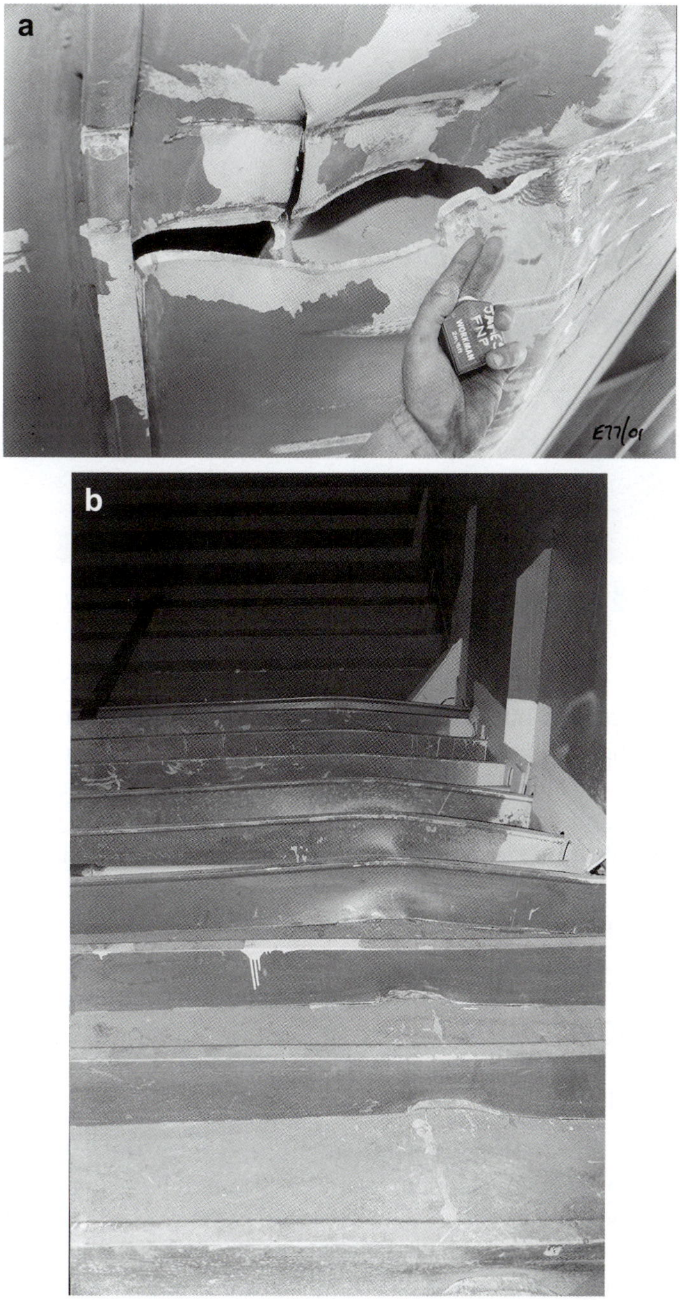

FIGURE 6.13 *(a) Damage to bottom flange and (b) Transverse stiffeners (courtesy Construction Repair).*

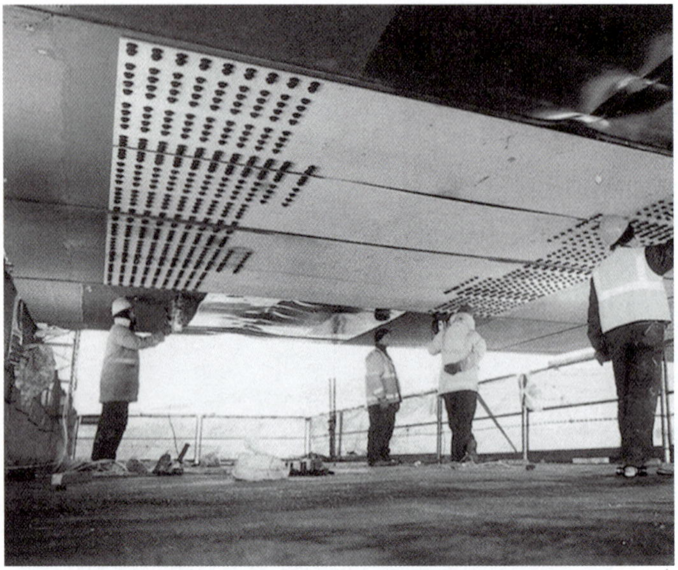

FIGURE 6.14 *Fitting of doubler plates to the bottom flange (Construction Repair, 1997).*

The repair work consisted initially of preventing further propagation of the tears by dressing out the exposed edges using disc grinders and drilling holes at the ends of the tears (or cracks) to act as crack arresters. Plates were straightened out wherever possible, and new steelwork was then added to the bottom plate and the transverse diaphragms secured with high-strength friction grip bolts (see Figure 6.15). The diaphragms have been made to act as a truss rather than a plate. Badly torn plates were cut out and new sections bolted in where necessary. Welding was kept to a minimum.

6.5 WROUGHT IRON

6.5.1 Introduction

Wrought iron (WI) became popular in the middle of the 19th century in Europe and especially in America. Its tensile strength is greater than cast iron but less than steel, and, as its name suggests, it is worked (wrought) when white hot to remove impurities making it highly fibrous. It has a characteristic yield strength of about 230 N/mm^2 in tension and good mechanical properties.

Bridges with WI elements built in the UK date from about the early 1800's where WI links were used to form chains for suspension bridges.

Riveted WI plates were later used to form waterways for aqueducts and the tubes of the Conway Bridge and then the famous tubular Britannia Bridge built in 1850. Smaller (15 m) single span bridges and continuous bridges were built up from prefabricated WI girders supporting WI buckle plate decks, or, on occasions, in-filled with mass concrete, though lattice girder bridges were also popular with their distinctive pattern of riveted cross bracing in the webs and high-level lattice cross beams.

6.5.2 Damage repair

Chafing of WI links in suspension bridge chains is a common form of damage (Day, 1993) and can be repaired by first blast cleaning the affected areas and then building up with weld material to the original profile. Because of the nature of WI and the possibility of slag inclusions Day (1993) found, it was necessary to develop a welding procedure by experimenting on some links that were beyond repair. These links were also used to supply the material for welding, and were themselves replaced with mild steel links.

Missing, broken or corroded rivets can be replaced with high-strength friction grip bolts (HSFG) which ensures compatibility with existing pre-loaded rivets. Tension control (TC) bolts are a viable alternative to HSFG bolts, which are designed with a unique 'tension control tip' that twists off when the bolt has achieved its minimum tension value, as specified by the designer. It is dubbed 'the intelligent bolt' because of its direct tension installation, controlled clamping force, one-man operation, no tool calibration, no compressor, no noise, reduced labour costs, visual inspection and less operator fatigue (IStruct E, 1996). Corroded plates or lattice bridge members can be replaced with sections of steel plate connected to the original with either HSFG or TC bolts.

Severe acid damage from smoke discharged from steam trains to the high cross beams of the lattice girders or top plates of the tubular structures results in pitting and corrosion. A combination of welding and replacement with steel parts is the usual solution (Mathews et al., 1993).

6.6 CAST IRON

6.6.1 Introduction

Cast iron (CI) bridges date as far back as 1779 when the Ironbridge was constructed over the River Severn at Coalbrookdale in Shropshire as a showpiece of CI engineering. The open spandrel arch form was used at the time as engineers were familiar with the structural action and confident of assessing its compressive strength. It is well known that CI is generally twice as strong in compression than it is in tension (Walton, 1981) and, for that reason, the arch remained the most popular form of CI bridge well into the 19th century when the material was superseded by wrought iron and eventually by steel.

A popular form of construction was to form the main span of the bridge with a series of solid cast iron arch ribs at a spacing of around 1.5 m. Open CI spandrels were often set on top of the ribs with CI road plates set on top to carry the roadway. Alternatively, the transverse construction was formed of jack arches (see Chapter 5, Figure 5.1) supporting granular fill on which the surfacing was placed. The ribs themselves were restrained with CI cross braces at intervals along their length.

Such bridges carry loads far in excess of those originally assumed, and, as such, they are subject to higher levels of stress and vibration than when originally constructed. This often causes tensile stresses to be set up resulting in minor cracking to the arch ribs which are revealed during inspection and general refurbishment works. Cast iron is a brittle material, very crystalline in nature, and susceptible to fatigue cracking. It has a characteristic tensile strength of only 120 N/mm^2, and with present traffic flows, it does not take long for critical members to reach their fatigue endurance limit.

Grit blasting of the ribs is usual to remove layers of existing paint and surface dirt, and it is during these operations that cracks, failed minor components and casting defects are detected. The full extent of cracking can be determined by the use of MIP (see 4.3.3).

6.6.2 Cold stitching

The method of cold stitching was pioneered by Metalock Industrial Services in 1949 (Metalock, 1998) and is used for repairing cracked and broken CI castings with ductile high tensile nickel alloy *keys and studs*. The process is heat-free, and can be carried out in-situ or in a workshop. Figure 6.15a shows the process and Figure 6.15b shows the application to Bures Bridge in Suffolk, UK comprising five arched cast iron ribs spanning 20 m, with brick jack arches in between, supporting a layer of fill and topped with 75 mm of asphalt surfacing (Neave, 1992). Several of the cast iron springing plates had suffered extensive cracking over the years since 1810 and required extensive stitching to restore their integrity.

There are four steps involved in the process (see Figure 6.15a):

- Apertures are formed up to 100 mm deep at right angles to the crack by jig drilling to accept the nickel keys.

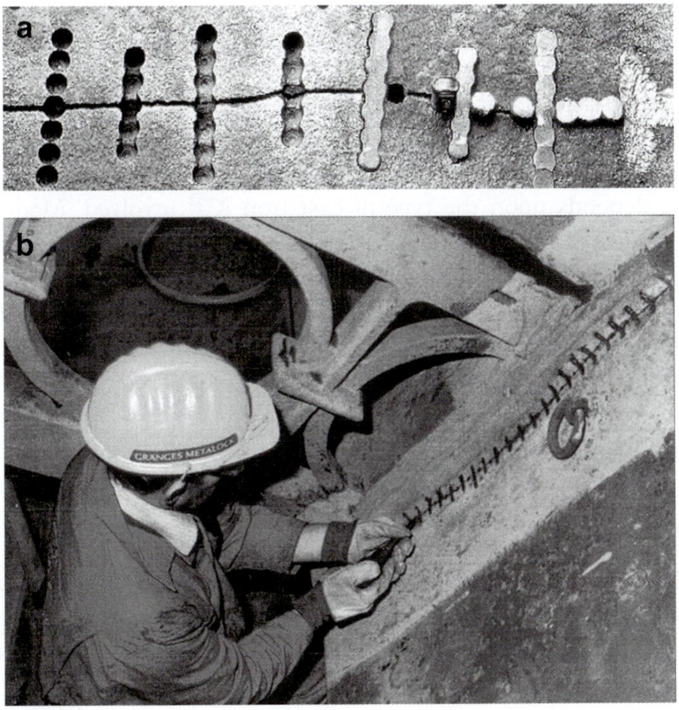

FIGURE 6.15 *(a) Cold stitching system and (b) Repair to Bures Bridge (courtesy of M. Neave).*

- The keys are tightly fitted and peened into the apertures.
- Holes are drilled and tapped along the line of the fracture and fitted with overlapping studs.
- The repair is peened and hand-dressed to blend with the parent metal. The resulting combination of keys and studs produces a robust repair, restoring strength to the member and preventing further propagation of the cracks under continued loading. In the case of the High Bridge in Staffordshire (Addison, 1997), the method was also used 'to repair the diaphragms and to stitch sections of bolt washers and housings back into position'. Once completed, the repairs are all but invisible and can be protected by a proprietary paint system consisting of a 'two-pack high build epoxy-containing micaceous iron oxide (MIO) with a final coat of two-pack recoatable polyurethane'.

6.6.3 Casting defects

Casting defects such as blow holes, pits and surface bumps are repaired in a variety of ways. After cleaning the area by wire brush or blast-cleaning, the blow holes and pits can be repaired by the non-fusion welding process, such as, powder welding, using a filler with a low melting point such as copper. The ease of welding depends upon the type of cast iron (CI) involved. Modern *decarbonised malleable* irons are the best, followed by *nodular CI* and then *flake graphite CI*. For small pits it may be more appropriate to use brazing or soldering.

Once filled, the weld material is ground down and painted using a zinc phosphate epoxy ester primer, with either a phenolic micaceous iron oxide (MIO) or silicone alkyd undercoat followed by a finish coat.

6.7 TIMBER

6.7.1 Introduction

Timber bridges in Europe date from the 12th century, but few, if any, still exist. They became popular again in Europe in the 18th century, and in America during the middle period of the 19th century when thousands were constructed to carry both pedestrian and road traffic.

The most popular form was that of the truss inspired by timber roof construction during the latter half of the 18th century. The addition of a timber barnlike structure over the main load-bearing trusses to protect them from the weather resulted in the well known 'covered bridges' of North America and Scandinavia with spans up to about 100 m.

Because of the difficulty of jointing timber members, development of timber bridges has been slow during the 20th century but they now seem to be enjoying somewhat of a renaissance (Mettem, 1998) due to 'sustainability, aesthetic appeal, new materials and jointing techniques, prefabrication and the availability of good practice guidance to ensure durability and value for money'.

As well as small-span footbridges in rural areas, some notable long-span road bridges have been built, such as the 200 m span Mur arch bridge in Germany (Mettem, 1998). According to many other bridge types such as cable-stayed, tension ribboned, twin-span lifting, Warren girder and bowed beam are also being built (Mettem Bainbridge, & Jayanetti, 1998).

6.7.2 Repair methods

Timber members in bridges can be weakened due to the normal ageing process which produces 'shakes' along the length of the members, wet rot and

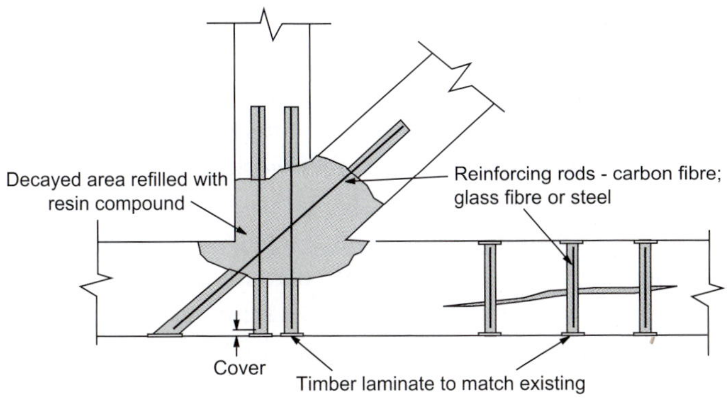

FIGURE 6.16 *Timber repairs using the rod-and-resin technique.*

decay due to beetle infestation, or fire. TRADA (1992) have produced a two-page general information sheet on the repair of structural timber mainly addressed to the conservation of timber buildings. There are some established techniques, however, which are appropriate to bridges.

6.7.2.1 Reinforced resin repair

Shakes in timber can be repaired by first cleaning out the area and then injecting with an epoxy resin compound. Epoxy resin is to be preferred as it is relatively unaffected by moisture and does not shrink or expand. The resin can also be coloured to match the surrounding timber.

Decayed areas of timber usually occur in the bottom chord members of truss bridges because, not only are they subject to tensile forces which tend to pull the joints apart, but they are more exposed to the weather (this is especially true in the case of covered bridges).

With small localised repairs, at joints for example, or along the length of a member, it is not necessary to provide temporary support, and as with all works of this nature, the repaired section will be called upon to resist live loads only.

Surgery is performed on the decayed areas either by carefully removing them using a water jetting or sand blasting system, or more drastically, by sawing and gouging out the decayed material down to sound wood. Reinforcement in the form of steel bars, glass fibre rods or carbon fibre rods is inserted into predrilled holes and bonded in with a suitable resin compound. Cover is provided at the ends of the rod for protection and, if required, timber laminates can be bonded at the ends to match the existing. The reinforcement is of the order of fifty times as strong as the parent timber and so the repaired areas become the strongest parts of the bridge.

The area is then shuttered and filled with resin and allowed to cure. The exact formulation of the resin compound will depend on the particular application, but polyester, epoxy and acrylic resins have and are being developed that can be used in a variety of forms to consolidate, impregnate, fill, coat and repair almost any substrate. For timber work, the resin must have good fluidity in order to obtain maximum penetration and impregnation of the wood fibres to ensure a sound bond. Typical repairs are shown in Figure 6.17.

If a member is cracked, split or rotted right through, a repair can be effected by filling the void with resin and then reinforcing with rods and resin in two planes from opposite surfaces as shown in Figure 6.17 or in four planes from all sides. Alternatively, carbon fibre strips can be bonded on to the surface as shown, or buried and covered with a timber laminate if desired, rendering the repair almost invisible. If bending resistance is required, then surface-mounted carbon fibre strips are the only solution. The latter form of repair is a very popular form of strengthening cast iron, steel, concrete and timber bridges and is referred to in detail in Chapter 7.

If strength calculations show that timber members damaged by fire still have adequate residual strength, then it is possible to repair them. The charred timber is stripped off using a water jetting or sand blasting system. Two coats of a suitable resin compound are then applied followed by an intumescent coat to provide the required fire protection rating (in hours). Severely weakened timbers have to be removed and replaced and secured in place using one of the foregoing methods.

Composite wood textiles have recently been introduced into the design process of timber joints (Haller & Chen, 1999) as a way of strengthening conventionally bolted joints, and producing a *ductile* failure at a higher load as compared with the normal *brittle* failure in shear and tension perpendicular to

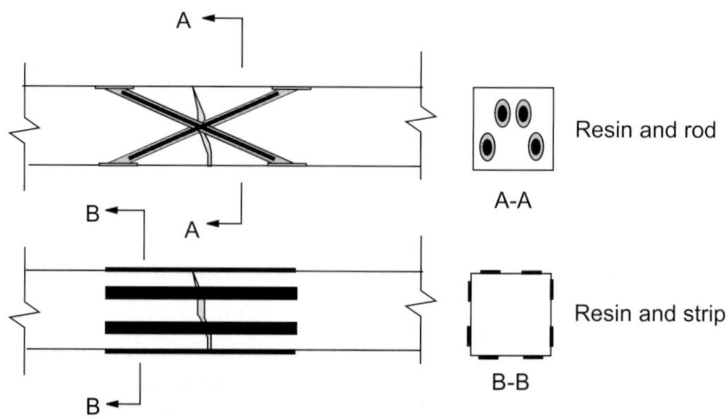

FIGURE 6.17 *Timber repair of a complete fracture.*

the grain. The same technique can be used for the rehabilitation of wooden bridges which involves gluing a *glass fibre fabric* on to the surface of the wood in the area of the joint. It not only improves the strength but provides a barrier against environmental attack.

6.7.2.2 Double shear joints using GRP Dowels

A recent innovation in the design of joints in prefabricated laminated veneer lumber (LVL) and glue-laminated timber (GLT) is the use of pultruded glass reinforced polymer (GRP) dowels in double shear (Drake & Ansell, 2000). There is no reason, however, why they should not be used in the repair of timber joints/members with traditional metal fasteners in areas where either the metal bolts have corroded, or the timber has rotted. They are light, strong, durable, fire-resistant, easily manufactured and can be machined on site. The repair would be in the form of a splice as shown in Figure 6.18b.

Equations in Eurocode (EC5) are not applicable and so Drake and Ansell, (2000) have proposed a design methodology based on a series of laboratory tests and have modified the EC5 equations. The tests also indicate that the use of GRP dowels provides a ductility not possible using steel dowels (or bolts). For further details reference should be made to (Drake & Ansell, 2000).

6.7.2.3 Steel bar/epoxy resin repair

The particular problem of decaying *ends* of timber members such as stringers or joists has been addressed by the proposal of a novel system using steel bars in conjunction with epoxy resin (Rotafix, 1997). The decayed section is cut

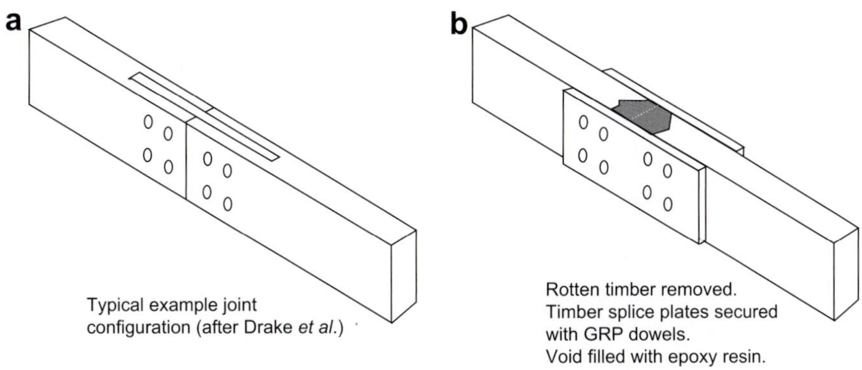

FIGURE 6.18
(a, b) Typical joint configuration.

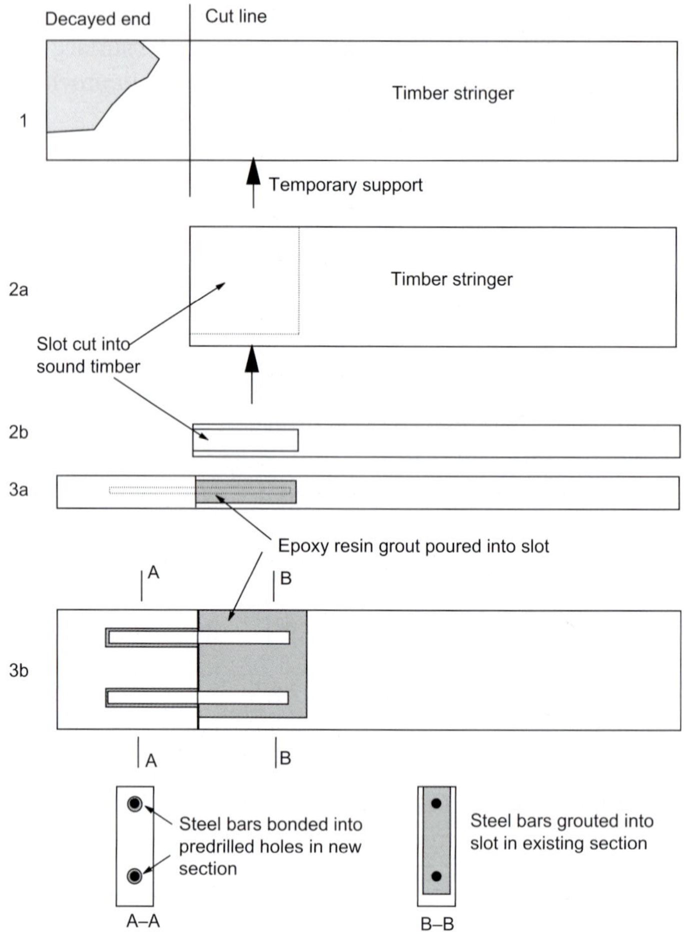

FIGURE 6.19 *General procedure for steel bar/epoxy resin repair.*

off and replaced with a new preformed section with protruding epoxy-bonded steel reinforcement bars which are placed in a slot cut in the sound timber adjacent to the cut end. Epoxy mortar is then poured into the slot to bond the bars to the timber (see Figure 6.19).

Limited testing has been carried out by Jones and Smedley (2000) and the results are promising. The general configuration is shown in Figure 6.20.

An alternative to the solution in Figure 6.20 for the repair of the decaying ends of timber beams carried out by J. Jasienko (2001) is shown in Figure 6.21. Clearly the cross-sections' dimensions of the new timber are chosen to match those of the old timber.

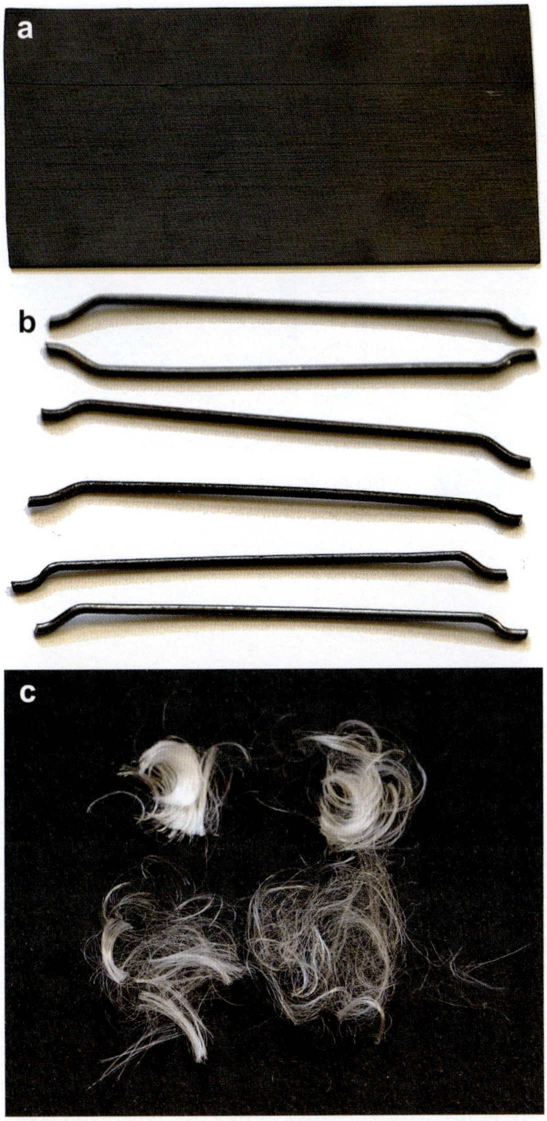

FIGURE 6.20 *Details of Glass Fibre bar/epoxy resin repair. (a) CarboDur S 512.50 × 1.0 (carbon), (b) Hooked-end Steel Fibre 50 × 1.0 (steel), (c) Stealth 6922 12 mm (polypropylene).*

6.8 GRAFFITI

Graffiti is the scourge of the present day. Buildings and bridges everywhere are afflicted with the words, patterns and pictures of artists (sic) armed with tins of spray paint. Figure 6.21 is typical of the vandalising that takes place.

FIGURE 6.21 *Graffiti on the abutment of an old arch bridge.*

FIGURE 6.22 *The needle eye bridge (courtesy concrete repair).*

One of the most effective ways of cleaning such defaced surfaces is by the use of an *abrasive blast cleaning system*. Modern systems are wet (fully saturated) and so there is no airborne dust. They are very effective and operate at very low pressures (typically 5 – 100 psi) which means that cleaning is 'gentle' and that no damage is done to the bridge material, be it concrete, brick, stone, steel or cast iron (Figure 6.22).

Once cleaned, decorative and protective coatings (such as UV-cured acrylic copolymers) can be applied, to enhance the appearance of the bridge. A fine example of such a typically restored bridge with an intelligent use of colours for the finishing coating, is the Guildford Town Bridge across the River Wey built in 1902 and shown in Figure 6.23.

FIGURE 6.23 *The Guildford Town Bridge.*

6.9 CONCLUSIONS

Repair technology is a fast-growing area, with new techniques being developed annually for all types of situation, very often by specialist companies with a vested interest in selling them. It is up to the bridge engineer to assess any claims made by these companies for their particular 'new' technique or system by reading the back-up research behind them, witnessing field trials first hand and checking out any contracts undertaken to date. Only then can he be confident in recommending the repair method for a given situation.

REFERENCES

Addison, S. (1997). High bridge rehabilitation contract. *Construction Repair, 11*(1), 14–17.

Aguilar, L. A. (1994). Remedial works to the Cardiff Bridge over the River Taff including the application of Sprayed Concrete, Bridge Assessment, Management and Design. In: *Proceedings of the centenary year bridge conference*, Cardiff, UK, 26–30 September 1994, Elsevier.

Allen, W. J., & McDonald, L. A. (1993). Lime as a building material. *The Structural Engineer, 71*(17), 317–318.

Baldo, P., & Medeot, R. (1994). The use of high water pressure water jet for bridge deck repair and widening, Bridge Assessment, Management and Design. In: *Proceedings of the centenary year bridge conference*, Cardiff, UK, 26–30 September 1994, Elsevier.

Ball, D. (1997). Waking a tired old lady. *Construction Maintenance and Repair, 11*(2), 22.

Bettigole, N. H. (1994). Rebuilding our bridges: why and how. *ASTM Standardisation News*.

Bolton, A. (1998). Lundy landing. *New Civil Engineer*. 20/27 August.

Coonie, R. (1992). Leaderfoot viaduct restored. *Construction Maintenance and Repair, 6*(1), 2–4.

Day, W. (1993). Gattonside suspension footbridge refurbishment. In: *Second international conference on bridge management, inspection, maintenance, assessment and repair*, University of Surrey, Guildford, UK (pp. 376–385). Thomas Telford.

Drake, R. D., & Ansell, M. P. (2000). Evaluation of double shear joints for timber structures using pultruded GRP dowels, and implications for EC5-based strength. *The Structural Engineer, 78*(12), 28–32.

Haller, P., & Chen, C. (1999). Textile reinforced joints in timber construction. *Structural Engineering International, 9*(4), 259–261.

Jasienko, J. (2001). Glue joints used for reinforcing the damaged ends of wooden beams. *Structural Engineering International, 4*.

Jones, R., & Smedley, D. (2000). Replacement of decayed beam ends with epoxy-bonded timber composites: structural testing. *The Structural Engineer, 78*(15), 19–22.

Kitada, T., & Mizobuchi, S. (1993). Design and construction of steel bridges considering maintenance and durability. In: *Second international conference on bridge management, inspection, maintenance, assessment and repair*, University of Surrey, Guildford, UK (pp. 170–179). Thomas Telford.

Lichtenstein, A. G. (1990). Inspection and rehabilitation of steel trusses for highway bridges. In: *First international conference on bridge management, inspection, maintenance, assessment and repair*, University of Surrey, Guildford, UK (pp. 695–704). Elsevier.

Liquid Plastics Ltd. (1990). The Flexcrete concrete repair & protection system.

Meagher, J. (1997). The Laigh Milton Viaduct restoration project. *Construction Maintenance and Repair, 11*(1), 30–32.

Medeot, R. (1990). Hydrodemolition - a modern technique of concrete removal in bridge repair. In: *First international conference on bridge management, inspection, maintenance, assessment and repair*, University of Surrey, Guildford, UK (pp. 765–776). Elsevier.

Mehue, P. (1990). Cracks in steel orthotropic decks. In: *First international conference on bridge management, inspection, maintenance, assessment and repair*, University of Surrey, Guildford, UK (pp. 633–642). Elsevier.

Mettem, C. J. (1998). Turning to timber. *New Civil Engineer*, 20–27 August, 39–40.

Mettem, C. J., Bainbridge, R. J., & Jayanetti, D. L. (1998). Timber bridges - design and appropriate use of materials. *The Structural Engineer, 76*(16), 313–318.

Middleton, W. G. (1995). Research project into the upgrading of unreinforced masonry parapets. In: *First international conference on arch bridges*, Bolton Institute, Bolton, UK (p. 527). Thomas Telford.

NCE. (1997). At last, an end to the never-ending job. *New Civil Engineer.* 11/25 December.

Neave, M. (1992). The construction, maintenance and repair of cast iron bridges. *MSc Dissertation*. Guildford, UK: University of Surrey.

Paxton, R. A. (1998). Conservation of Laigh Milton Viaduct, Ayrshire. *Proc. Instn. Civ. Engrs, 126*, 73–85.

Rotafix. (1997). Timber repair. *The Structural Engineer, 75*(19), A16.

Shimada, S., Yamada, K., Aoki, H., & Fukushima, I. (1996). Structural modification of the two-hinged arch Semimaru Bridge. In: *Third international conference on bridge management, inspection, maintenance, assessment and repair,* University of Surrey, Guildford, UK (pp. 279–286). Spon.

Smith, N. J., & Obaide, T. (1993). Extending the life of existing bridges by rational use of economic criteria in maintenance management. In: *Second international conference on bridge management, inspection, maintenance, assessment and repair.* University of Surrey, Guildford, UK (p. 519). Thomas Telford.

TRADA. (1992). *Assessment and repair of structural timber,* Wood information sheet 34.

Wadsworth, S. (1997). Erskine bridge repairs. *Construction Repair, 11*(1), 18–22.

Walton, C. F. (1981). *Iron Castings Handbook*, Ed., Iron Castings Society, Inc., p. 235.

Welch, P. J. (1995). Renovation of masonry bridges. In: *First international conference on arch bridges*, Bolton Institute, Bolton, UK (pp. 601–610). Thomas Telford.

FURTHER READING

Blockley, D. (2000). Puente Durate bridge: an unusual challenge. *The Structural Engineer, 78*(5), 11.

Broomfield, J. P. (1997). *Corrosion of steel in concrete*. London: Spon.

DD ENV 1995-1-1:1994, Eurocode 5, Design of Timber Structures, Part 1.1: General rules and rules for building, London British Standards Institution, 1994.

Holland, R. (1997). *Appraisal & repair of reinforced concrete*. Thomas Telford.

IStructE. (1996). The intelligent bolt. *The Structural Engineer, 74*, 15.

Mallett, G. P. (1994). *Repair of concrete bridges - (A Transport Research Laboratory sponsored state of the art review)*. UK: Thomas Telford.

Mathews, R. A., & Paterson, I. A. LMR to LRT: restoration of the Cornbrook viaduct. In: *Second international conference on bridge management, inspection, maintenance, assessment and repair,* University of Surrey, Guildford, UK (pp. 418–427). Thomas Telford.

Matthews, S. J., & Ogle, M. H. (1996). Investigation and load testing of a steel latticed truss viaduct. In: *Third international conference on bridge management, inspection, maintenance, assessment and repair*, University of Surrey, Guildford, UK (pp. 544–555). Spon.

Metalock Services Ltd. (1998). *The Metalock Process - patented cold repairs for cracked and broken castings*, Information sheet.

Parker, D. (1998). Steel eating bugs spread worldwide. *New Civil Engineer*. 5th March.

CHAPTER 7

Strengthening

CONTENTS

7.1 INTRODUCTION

The term strengthening is taken to mean the structural enhancement of existing weak members in such a way as to restore or increase their ultimate strength in bending, shear or direct tension and compression. In some cases it may mean removal of major structural members and the addition of new ones (in the same or different structural arrangement).

A bridge may be strengthened to allow the passage of a single exceptionally heavy vehicle, to allow for increasingly heavy traffic, to counteract effects of vehicular damage or to rectify deficiencies in design or construction. Strengthening generally has less of an impact, both environmentally and economically, and is to be preferred to demolishing and rebuilding. The means of strengthening chosen should ensure minimal disruption to traffic and pedestrians, be cost-effective, robust and reliable, and should, where possible, not impair the visual appearance of the bridge.

Strengthening options vary from bridge to bridge, depending on the problems which need to be addressed, and much individual attention is needed at the assessment stage. Some of the more common forms of strengthening are described in this chapter.

7.2 STEEL PLATE BONDING

7.2.1 Introduction

This technique is most commonly used on concrete bridges where cracking may indicate a deficiency in the amount of reinforcement necessary to carry the traffic. Application of bonded steel plates can increase the flexural strength and load-carrying capacity, improve stiffness, which reduces deflection and cracking, and enhance the shear capacity. The process is not designed to carry dead load but only live load due to traffic.

7.2.2 Principle

Shaw (1993) defines steel plate bonding as 'a method of strengthening achieved by bonding additional reinforcement to the external faces of a structure'. The reinforcement is normally in the form of steel plates secured to the structure by means of an epoxy resin compound. The method depends on the ability of epoxy to transmit forces between the structure and plates by means of shear stress.

Bolts are often used in conjunction with the adhesive to provide a mechanical anchor for the plates at the ends to prevent premature debonding due to peeling, and also to help support the plates whilst the adhesive cures. Plates can be placed in areas where either the bending, shear or compression capacity needs to be increased. A typical application is shown diagrammatically in Figure 7.1.

7.2.3 Design

In bending the plate acts as an extra layer of tensile bending reinforcement. The tensile forces have to be transferred from the concrete to the steel plate, and it is the function of the adhesive to transfer the tensile forces by means of shear between the steel plate and the concrete surface and to produce a continuous bond to ensure that full composite action is achieved.

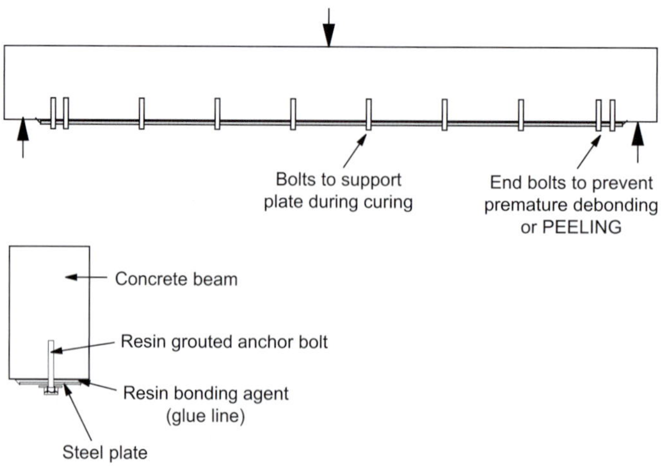

FIGURE 7.1
Typical plate bonding application.

7.2.3.1 Peeling

Mays (1993) carried out a finite element analysis of a simply supported reinforced concrete beam strengthened with a steel plate on the tension side and subjected to a four-point loading regime representing the working load, and obtained a plot of the distribution of the shear stress in the adhesive. There was a very marked increase in the shear stress in the region at the end of the plate over a length approximately equal to the width of the plate. The peak shear stress at the end of the plate was about four times the mean longitudinal stress calculated using simple beam theory. This was also confirmed by Oehlers and Moran (1990). The normal stresses also peaked in this area to a value of about 3.5 times the mean. Details of the test set-up and results are shown in Figure 7.2. These results suggest that there is a tendency for the plate to peel away from the concrete in the manner shown in Figure 7.3, and, to prevent this, the plate should be bolted over an effective anchorage zone at the end of the plate of at least 1.5 times the width of the plate plus a nominal amount to allow for curtailment. The bolts are normally designed to resist the longitudinal shear stress ($\sigma_{ö}$) at the ultimate limit state (ULS) times a load factor (typically 3). Raoof and Zhang (1996) have very carefully examined the mechanism of plate peeling and have proposed a way of predicting such failures.

As would be expected, this type of failure is not dissimilar to that of a conventional reinforced concrete beam without plating, which eventually fails in bending and exhibits bond failure near the supports, see Figure 7.3(b).

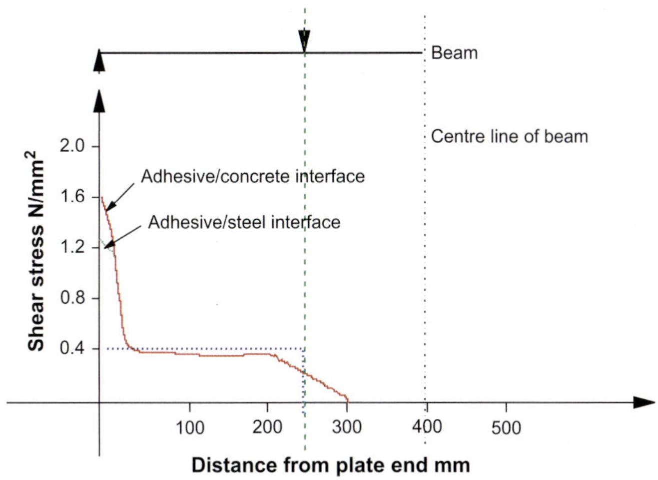

FIGURE 7.2
Shear stress distribution in a steel-plated beam (Mays 1993, p. 674).

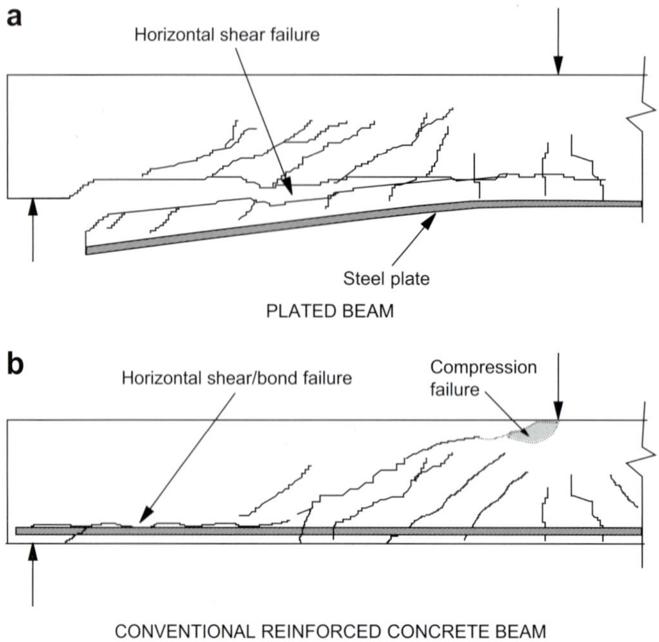

FIGURE 7.3
Typical peeling failure for (a) a plated beam, compared with (b) the bending failure of a conventional RC beam.

7.2.3.2 Strength and stiffness

Steel plates are normally added to improve the live load-carrying capacity and to improve the stiffness. The steps in design are, therefore, as follows:

- Determine the depth to the neutral axis and the second moment of area of the existing section for the long term.
- Calculate the dead load stresses in the steel and concrete.
- Determine the depth to the neutral axis and the second moment of area of the *strengthened* section for the short term.
- Calculate the live load stresses in the steel and concrete.
- Add the stresses from steps 2 and 4 and compare with allowable serviceability values.
- Check the section capacity of the strengthened section at the ultimate limit state and compare with the design value.
- Design anchorage bolts to resist three times the ultimate shear stress in the plates (see 7.3.2.1).

Deflection calculations are also necessary to ensure that the stiffness has improved sufficiently to limit cracking under live loads.

7.2.4 Application to slab, or beam-and-slab concrete bridges

Many bridges have been strengthened by this means since the mid 1970s (Ramsey, 1993) and, so far, no failures have been recorded. The Transport Research Laboratory in the UK have been closely monitoring the interchange at Quinton strengthened by this technique in 1975 and the only degradation recorded has been limited debonding due to corrosion of the steel and not breakdown of the bonding material (Ramsey, 1993; Hutchinson, 1996).

The three key stages in the procedure are:

- Surface preparation.
- Application of the adhesive.
- Placing of the steel plates.

7.2.4.1 Surface preparation

The concrete surfaces should be thoroughly cleaned by grit blasting to remove organic growths, cement laitance, surface coatings, oil, dirt, grease etc. and to produce a rough key for the adhesive. Any cracks or weak areas of concrete are then repaired (see Chapter 6).

Finally the surface is checked for profile, any dips are filled with adhesive, and any projections ground down using a carborundum block, or removed by scabbling or the use of a needle gun. The surfaces are then vacuum-cleaned and the centre lines of the plate positions marked on the concrete.

It is essential that the bolt positions do not coincide with the existing reinforcement/prestressing tendons. To ensure this, a cover meter survey is carried out and the positions of the bolts in the plates are designed accordingly.

For beams, the plate would be positioned on the soffit within the width of the web or bottom flange. For slabs, in order to prevent congestion and ensure that all of the slab is strengthened, the effective width of the slab portion is assumed to be encompassed within a 45° spread zone from each side of the plate to the top of the slab as shown in Figure 7.4. The overlap (L) is to allow for a small margin of error in fixing and is to the order of 100 mm.

After fabrication, the *surfaces* of the steel plates are cleaned before arrival at the site by grit blasting (normally grade Sa $2\frac{1}{2}$ to SIS 055900) to obtain a surface peak-to-trough amplitude between 50 and 100 microns (μm). The plates are then degreased and vacuumed to remove any dust or debris and primed with a spray of two-part micaceous iron oxide within four hours to prevent re-oxidisation of the surface. The plates are then delivered to site wrapped in a protective covering, such as bubble wrap or polythene, and are handled by operatives wearing clean rubber gloves. To prevent any moisture

FIGURE 7.4
Effective width of slabs strengthened with steel plates.

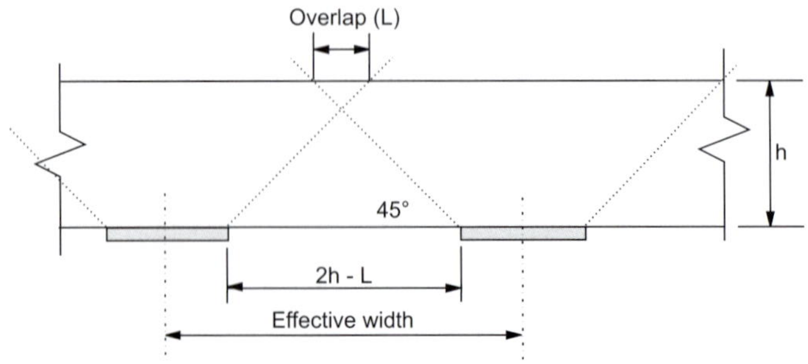

build up prior to placement, a silicon gel desiccant is placed inside the wrapping.

7.2.4.2 Adhesive application

The usual method of application is by 'buttering' the adhesive onto the respective surfaces using a stiff plate. The adhesive – such as Sikadur 31 (Sika, 1997) – is a two-part, epoxy-based compound, with the base coloured white and the hardener coloured grey. This helps to ensure on-site quality control when the mixing takes place to produce a uniform consistency and mid-grey colour. Other adhesives which have also been used in practice are Ciba Polymers XD 800 and Shell Epikote 828 (Hutchinson, 1996).

The coatings of the adhesive are applied to both the steel and concrete surfaces simultaneously (usually 1 mm to the steel and 2 mm to the concrete). A specially shaped trowel can be used to ensure that the distribution of the adhesive is thicker at the centre of the plate than the edge – typically 3 mm to 1 mm (Henwood & O'Connell, 1993). When the plate is then applied to the concrete 'the adhesive is pushed from the centre to the edge and reduces the risk of air entrapment'. The finished thickness of the adhesive can be ensured by placing 2 mm plastic individual spacers down the length of the plate.

7.2.4.3 Placing of the steel plates

The placing of the steel plates has to be completed within the pot life of the adhesive – typically 40 minutes. If there is plenty of working space and the plates are relatively light and short, they can be lifted manually into place and secured. For ease of handling, plates are normally limited to 6 mm thickness and 6–8 m in length. The full continuous length of plating is then achieved either by splicing or by using a staggered overlapping system as in Figure 7.5.

Very often, however, the plates can be both long and heavy, and a lifting and support system has to be devised to aid placement. For soffit or side plates, the support system can often take the form of a working platform

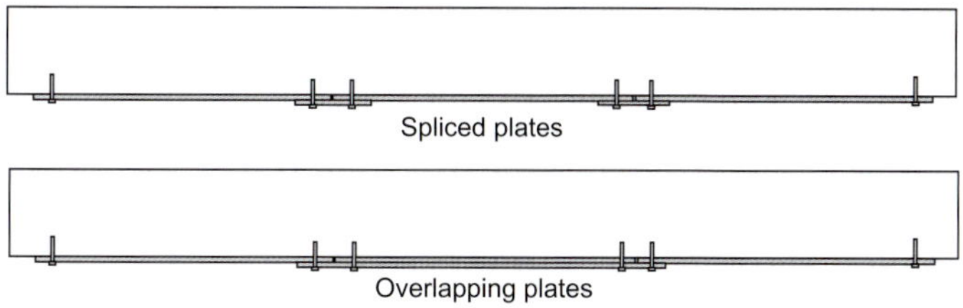

FIGURE 7.5
Means of providing continuity in plates (for clarity, only anchor bolts shown).

which can be either hung from the deck or supported from below. Its precise form depends upon the nature of the site. Once the plates are in place on the platform they can be offered up to the concrete by mechanical means utilising jacks (or lifting beams) and then kept in place until curing is complete by a system of folding wedges. Figure 7.6 shows a typical arrangement. Once in place, the plates are then protected with a two-part micaceous iron oxide primer followed by two top coats of high build PVC acrylic.

7.2.5 Plates secured to the sides of concrete beams

Because of the problem associated with access, bearing restrictions, the congestion of shear links and tensile reinforcement at the ends of concrete beams, and the danger of premature failure by shear peeling (see Figure 7.7a), plates bolted to the soffit of beams can prove impractical.

An alternative solution proposed by Oehlers and Ahmed (1996) is to provide plates secured to the sides of the beams. These can be provided in addition to

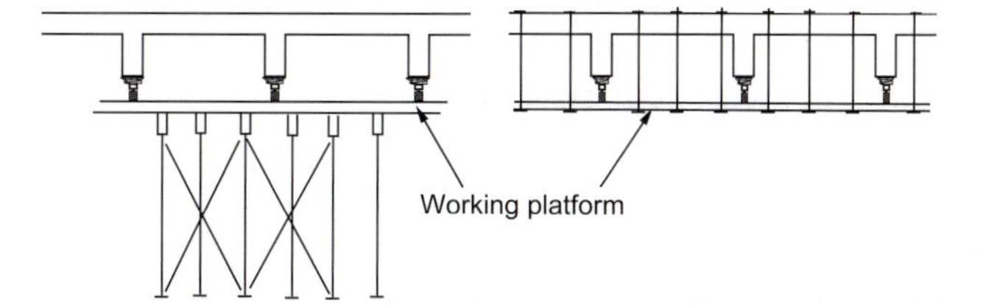

FIGURE 7.6
Typical working platform and support system for plate bonding.

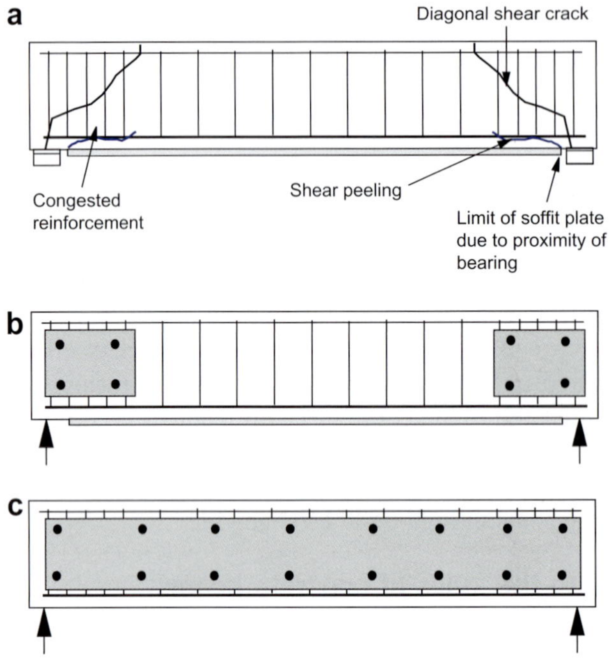

FIGURE 7.7 *Alternative plate bonding solutions.*

soffit plates (Figure 7.7(b)) or as an independent system (Figure 7.7(c)). The plates can be fixed using adhesive to satisfy serviceability requirements, and bolted to ensure safety should the adhesive fail. The theory is outlined by Oehlers and Ahmed (1996) and an analysis presented using a computer model, but no test data is given to corroborate the method.

7.2.6 Testing

Henwood and O'Connell (1993) recommend that three tests be carried out to check the integrity and quality of the operations, namely:

- Flexural modulus of elasticity – to ensure the consistency of the adhesive mix – usually one per batch.
- Bond to steel substrate – by the use of five double lap shear tests on every batch of adhesive.
- Bond to concrete substrate – by means of a minimum of six pull-off tests on selected areas.

In addition to these, Henwood and O'Connell (1993) suggest that a trial be carried out on a part of the actual bridge, rather than a trial panel, to demonstrate the suitability of the installation proposals and procedures.

7.2.7 Bures Bridge

An unusual application of steel plate bonding is that carried out to a cast iron bridge. Bures Bridge (Neave, 1992) located on the B1508 at the Suffolk/Essex border over the River Stour is comprised of five arched cast iron ribs with a clear span of 20.8 m. Jack arch construction supports the deck. During an inspection in 1990 cracks were discovered in one of the ribs. After closer inspection and assessment it was found that a single arch rib was only capable of carrying its own self weight if permissible stress levels were not to be exceeded. After looking at various strengthening options it was decided to provide steel plates bonded to the underside of the arch ribs. The plates were successfully secured with epoxy resin and anchored at the ends with bolts, and the bridge was reopened to traffic in October, 1992.

Similar repairs were carried out to two cast iron (CI) bridges in Birmingham, UK, namely, Girder Bridge (18 m span) and Top Lock Bridge (9 m span) (Haynes, 1995). The former bridge was stripped down to its bare components (two CI beams) and cleaned by needle gunning and grit blasting. The 200 mm wide 6 m long steel plates were similarly cleaned and primed before being secured to the soffits of the CI beams with a specially produced epoxy resin: Sikadur 31 PBA. The ends of the plates were secured with high strength friction grip bolts to prevent peeling failure. Finally a new reinforced concrete deck was cast on top to provide lateral restraint and a walkway for pedestrians.

The Top Lock Bridge was of a similar construction and was strengthened in a similar manner but with the addition of a crack repair in one of the beams using the Metalock technique referred to in Chapter 6.

7.3 COMPOSITE PLATE BONDING

7.3.1 Introduction

Steel plates are both heavy and bulky and pose problems in transport, delivery and handling on site. There is also the need for an expensive working platform and support system (see Figure 7.5). There are potential problems with the corrosion of the plates, and there is the need to provide an expensive auxiliary bolting system, both to prevent end anchorage failure and to support the self-weight of the plates during the curing process. Development of comparatively lightweight carbon fibre reinforced polymer (CFRP) materials began in the mid 1980s in Switzerland and their subsequent use in strengthening concrete bridge structures dating from 1991 onwards (Meier, 1992; Sika, 1997) has provided a means of overcoming these problems.

Extensive testing in the UK under the Government LINK Structural Composites programme ROBUST (strengthening of bridges using polymeric composite materials) both in the laboratory (Leeming, 1996; Peshkam & Leeming, 1996) and on full-scale prestressed concrete beams (Leeming, Darby, Lane, & Fashole-Luke, 1997) has indicated that laminates offer a viable alternative to steel plating for the flexural strengthening of concrete beams. A whole range of applications is quoted in a book dedicated to the subject of CFRP strengthening (Hollaway et al., 1999). The book is a very useful reference manual, and considers in detail all aspects of manufacture, design and construction.

7.3.2 Carbon fibre reinforced polymers

CFRP materials consist of strong and stiff carbon fibres (approx. 7 μm in diameter) embedded in an epoxy resin matrix, and possess high strength-to-weight ratios and corrosion resistance, thus helping to reduce maintenance costs. They can be manufactured in long lengths by the pultrusion process (Garden, Hollaway, Thorne, & Parke, 1996) to form laminates (or strips) with a unidirectional structure ranging from 50 mm–150 mm wide and 1.2 mm or 1.4 mm thick, with a fibre volumetric content greater than 65%. They are black in colour and are about ten times as strong as mild steel and one-fifth the weight. Some typical properties are shown in Table 7.1.

By comparison, mild steel has properties of $E = 200$ kN/mm^2, UTS $= 450$ N/mm^2 and an elongation at fracture of 20%. In addition, steel weighs 7850 kg/m^3 compared with 1530 kg/m^3 for laminates.

Luyckx, Lacroix, Fuzier, and Chabert (1997) lists further advantages of laminates as having low density, high fatigue strength, high wear resistance, vibration absorption, high dimensional stability, high thermal stability, and high chemical and corrosion resistance against acids, alkalis, salts and organic solvents. Farmer (1997) states that 'tests in Switzerland and Japan

Table 7.1 Mechanical Properties of Laminates

Grade	*E*-Modulus kN/mm^2	Ultimate Tensile Strength (Mean)* N/mm^2	Design Tensile Strength Based on $\gamma m = 1.5$ N/mm^2	Elongation at Break*
Sika Carbodur S	165	3050	2033	1.7%
Sika Carbodur M	210	2900	1933	1.2%
Sika Carbodur H	300	1450	967	0.45%

* Mechanical value obtained from longitudinal direction of fibres.

have confirmed that neither long term creep nor surface degradation are problematical and fatigue characteristics are much better than for steel'.

7.3.2.1 Strength

There is no doubt that laminates are sufficiently strong to enhance the live load-carrying capacity of bridges significantly. However, carbon fibres do not yield gradually like steel. They exhibit linear elastic behaviour up to failure when they rupture suddenly at strains of up to 1.7%. The general consensus seems to be that a suitable material safety factor to deal with this would be about 1.5 (similar to that for concrete). Typical stress–strain curves are shown in Figure 7.8.

7.3.2.2 Stiffness

CFRP laminates have, strength-for-strength, much smaller cross-sectional areas than steel and, therefore, do not contribute significantly to the flexural stiffness of the section. Their effect on reducing deflections, therefore, is minimal. For example a 100 mm × 200 mm deep concrete beam strengthened with a 90 mm × 1.4 mm laminate would increase its stiffness by about 10%. An equivalent 90 mm × 8 mm steel plate, on the other hand, would stiffen the beam by about 40%.

7.3.2.3 Failure

Some very interesting laboratory tests carried out on small identically reinforced concrete beams by Juvandes, Figueiras, and Marques (1997) indicate

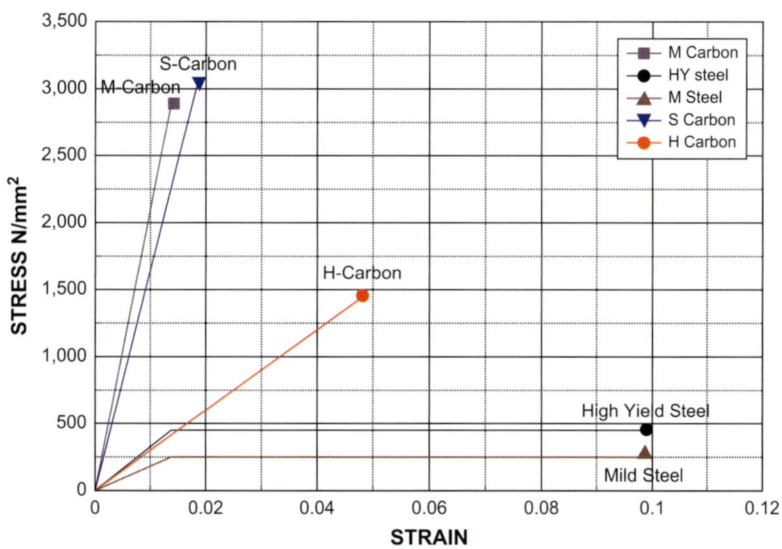

FIGURE 7.8
Comparison of stress–strain characteristics of steel and carbon.

that the mode and level of failure of strengthened beams depend critically on the extent of the laminate, that is, whether the end of the laminate stops short of the supports (or points of zero bending moment) or whether it is carried under the supports. Their results are shown in Table 7.2 where in beams B.1 to B.3 the laminate was under the supports, and in beams B.4 and B.5 the laminate stopped short by 10 mm and 200 mm respectively.

These results indicate the importance of the effect of end anchorage, and it can be seen from the low failure load for beam B.5, that the effect is not inconsiderable. This is an issue that needs to be carefully addressed by researchers, and underlines the need for caution before adopting laminates, and for careful consideration of the details of a particular application. Beams B.1 and B.3 indicate the preferred modes of failure which are primarily flexural rather than shear-like in nature.

Similar tests by Garden (1997) with very small end distances of 20 mm and 40 mm resulted in failure modes similar to those for B.4 and B.5. Buyukozturk and Hearing (1997) investigated failure modes of pre-cracked beams strengthened with laminates and reported similar failure modes.

Table 7.2 Results of Tests by Juvandes et al. (1997)

Beam	Maximum Failure Load (kN)	Maximum Strain in CFRP (mm)	Mode of Failure	Pictorial Mode of Failure
B.1	31.2	–	Flexural + bond failure	
B.3	32	531	Concrete shear + strip peeling with delamination	
B.8	26.8	360	Concrete crushing + bond failure + delamination	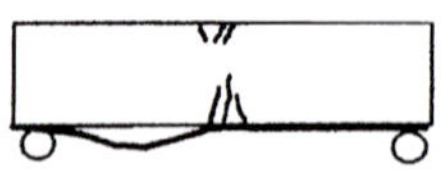
B.9	25	309	Concrete shear + bond failure with concrete cover separation	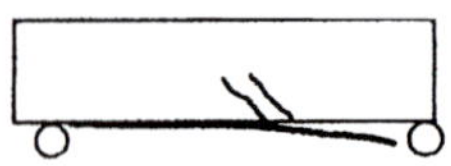
B.11	13.4	143	Concrete shear	

Using test data from other sources, Raoof and Hassanen (2000) have developed a semi-empirical model for premature brittle plate peeling (similar to B.11 in Table 7.2) of RC beams strengthened with CFRP on the tension face because the plate peeling failure moment can be significantly lower than the ultimate failure moment of the original unplated RC beam. They warn, therefore, that it is absolutely necessary to guard against such premature failure with some sort of end anchorage.

Beber, Campos Filho, Campagnolo, and Silva Filho (2000) tested ten simply supported reinforced concrete beams strengthened with CFRP plates and found that they all failed by horizontal shearing at the concrete/internal steel interface, whilst the CFRP was still intact and that strains in the carbon sheets at failure was around 1.0%, considerably less than its breaking strain.

In practice, the only situation where it would be possible to provide an end anchorage would be in the sagging regions of continuous members where the strips could be continued past the points of contraflexure. In simply supported slab bridge decks it may be possible to extend the strips beyond the bearing line by placing them between bearings.

7.3.2.4 Design

The design procedure for strengthening of reinforced or prestressed concrete beams using CFRP laminates is the same as that for steel plates as outlined in section 7.2.3.2. Walser and Steiner (1997) have noted that, according to general practice, structural elements strengthened with bonded reinforcements should have a residual safety (factor) of 1.0 after failure of the added strengthening. This may mean – as in the case of the Oberriet-Meiningen Bridge (Walser & Steiner, 1997) – that some concrete should be added to the top of the deck, in addition to the added reinforcement to the underside of the deck. This has the advantage of significantly increasing the stiffness as well as the strength of the section. The bridge crosses the Rhine River and is a two-beam composite bridge, continuous over three spans of 35 m, 45 m and 35 m. The steel beams are at 5 m centres and the deck slab cantilevers from each side of the beams by 2.3 m. The concrete deck slab between the longitudinal steel beams was strengthened transversely by the use of 80 mm × 1.2 mm strips at 750 mm centres, and 80 mm of concrete was added to the top over the full width of the slab. A typical section (after Walser & Steiner, 1997), together with the values of the ultimate bending moment at the centre of the slab, is shown in Figure 7.9.

In any event, the design should be such that failure of the tensile steel and the strip occurs prior to failure of the concrete in compression or premature peeling.

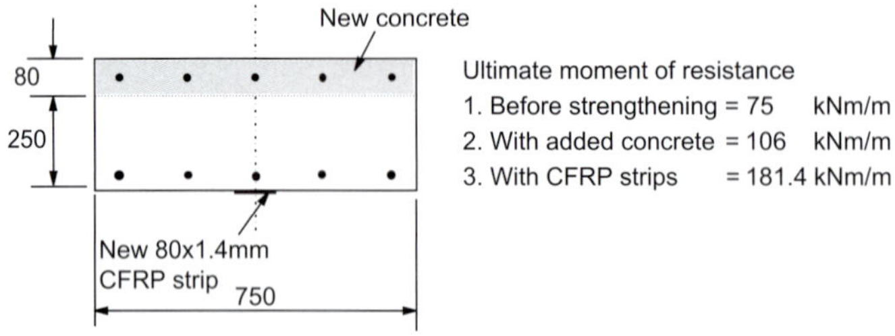

FIGURE 7.9
Typical section of a strengthened bridge deck using CFRP.

The Concrete Society (2000) has published a design guide for CFRP strengthened concrete beams which includes design equations to enable the strength of a given section to be calculated.

7.3.2.5 Numerical modelling

Extensive finite element modelling has been carried out using three-dimensional brick elements with non-linear material properties. Peshkam and Leeming (1996) and Lane, Leeming, and Fashole-Luke (1997) quote the use of such elements to model test beams with and without reinforcement. Their results for load versus deflection and plate strain versus distance from beam support, were within a few per cent of each other and the authors have been able to predict the results from experiments accurately. Peshkam and Leeming (1996) further report that their work suggests that plane sections may not remain plane and they are currently investigating the method further.

The main conclusions are that FE analysis has an important role to play in developing design rules and, possibly, for the prediction of load-deflection history, load-strain history and the ultimate capacity of strengthened concrete beams.

7.3.2.6 Fire

Behaviour under fire conditions is exemplified by tests carried out in 1994 at the Swiss Federal Laboratories for Materials Testing and Research (EMPA) which indicated that laminates can withstand temperatures of up to 925°C for 1 hour (Sika, 1997) before failure due to debonding. Steel plates in the same environment debonded and fell to the floor after only eight minutes. If necessary, fire resistant boarding can be applied to the laminates to give added fire resistance.

7.3.2.7 Installation

The installation of strips is less labour-intensive than that of steel plates. A two-man team is generally all that is required. The strips are easy to handle and can be manufactured to any length and delivered in a coil at site. No auxiliary bolting is required.

Once the surface of the concrete along the lines of the final laminate positions has been blast-cleaned, brushed and vacuumed, the width of the laminate is defined with adhesive tape. The procedure is then as follows:

1. Reprofile the substrate between the lines of tape by applying the adhesive by trowel to the concrete and level by scraping (this provides an even surface).
2. Remove the lines of tape.
3. Clean the contact face of the laminate (or remove the 'peel-off' protective covering).
4. Apply the epoxy adhesive to the laminate with a triangular profile by means of a shaped spatula.

5/6. Offer the strip to the concrete soffit and lightly fix in position with hand pressure.

7. Press the strips firmly into the substrate place by means of a hard rubber roller, at the same time squeezing out any excess adhesive (this ensures a good contact and the removal of any entrapped air).
8. Remove excess adhesive using a painters knife.
9. Clean the strip.

These steps are shown pictorially in Figure 7.10.

7.3.2.8 Quality assurance

The first stage in quality assurance is to carry out a visual inspection of the concrete surfaces after grit blasting, but prior to installation of the laminates, for the presence of voids, cracks or any foreign bodies. Appropriate repairs are then carried out and the surface levelled.

Pull-off tests are then carried out to determine, first, the tensile bond strength of the concrete and, second (on a test strip), the quality of the bond between the laminate and the concrete. Values greater than 1.5 N/mm^2 are acceptable. The uniformity and integrity of the bond is determined using infrared thermography or tapping. The compressive and flexural strengths of the adhesive are determined from small rectangular prisms primarily to ensure uniformity of the mix.

Finally the evenness of the laminates themselves is examined. If everything is satisfactory, the laminates may be painted over or left as they are.

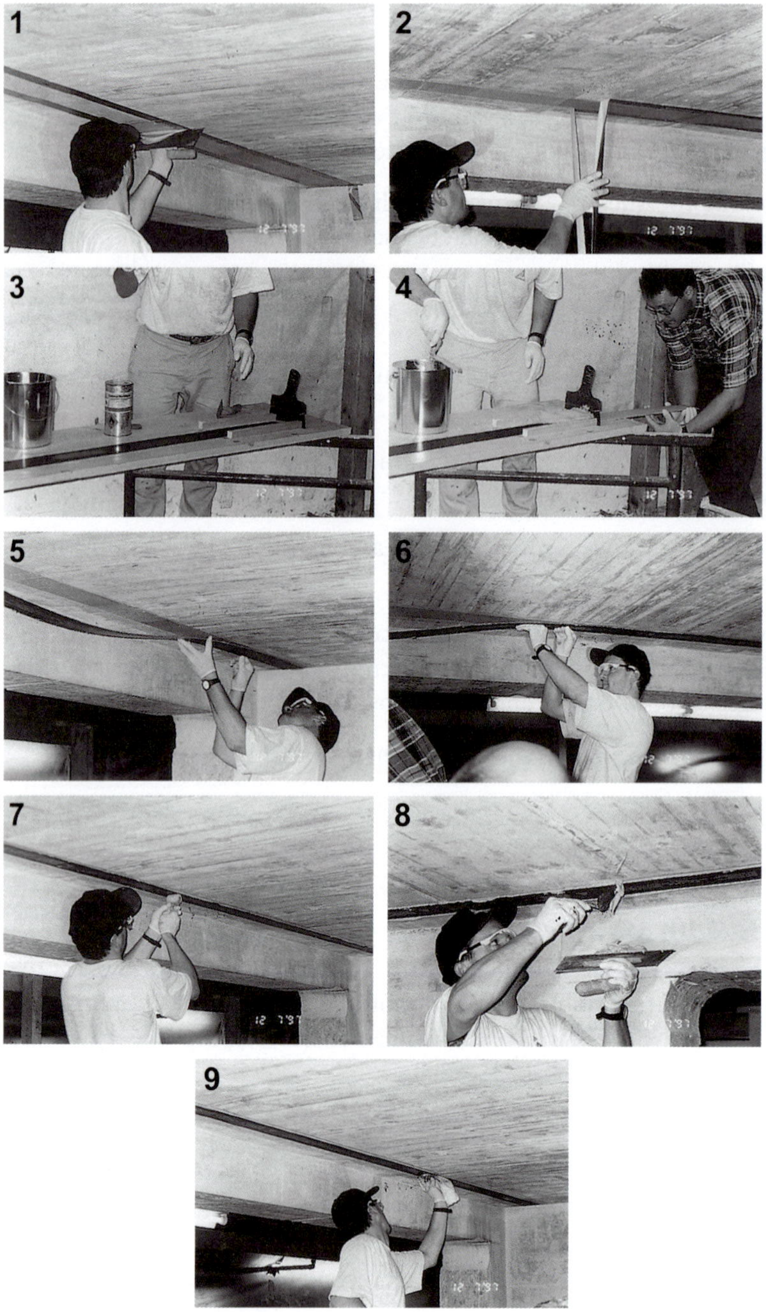

FIGURE 7.10
Methodology for the installation of laminates to the soffit of concrete bridge decks.

7.3.2.9 Prestressed CFRP

A recent development in advanced FRP is the development of prestressed carbon fibre reinforced polymers (PCFRP) developed under the ROBUST project (Darby, Kuke, & Collins, 2000) who are convinced that there is a potential for strengthening structures providing that it can be justified in terms of cost and durability. Such a system was used on the Hythe Bridge in Oxfordshire. Originally built in 1874, it is a jack arch bridge over two clear spans of 7.8 m each. The bridge failed its 40 t assessment and three of the cast iron beams were severely cracked and were repaired using the Metalock system referred to in Chapter 6, section 6.6.2. After extensive testing which proved the feasibility of prestressing, it was decided to provide four thin PCFRP strips at the bottom of each of the eight cast iron inverted tee beams forming the deck, each stressed to a total of 12–18 tonnes (Darby et al., 2000). They were attached to anchorages and bonded to the soffit of each beam. The load-carrying capacity of the bridge was increased and the project was deemed a success.

7.3.3 CFRP wraps

For many years, since the 1971 San Fernando earthquake in the USA revealed deficiencies in reinforcement details, existing concrete bridge columns and piers have since been retrofitted with concrete or steel jackets. Recently however, the use of CFRP continuous fibre jackets (or wraps) have proved to be a viable alternative (Seible, Priestly, and Innamorato, 1995) and detailed guidelines show how such systems can be designed.

Laboratory experimental verification of the new seismic retrofit system on large-scale models is described by Seible (1997) and shown to be at least as effective as conventional steel jackets. He identifies three possible modes of failure due to a seismic event as shear, plastic hinge confinement and lap splice debonding. The design procedures employed to convert a potential brittle shear failure to a flexural ductile one enable the extent and thickness (t) of the fibre wraps to be determined for application in the field. A typical example described by Seible (1997) is shown in Figure 7.11.

Tests on the retrofitted jacket indicated that the fibre wrap system was slightly less strong than a comparable steel one, but that its structural performance characteristics were similar to steel and, in both cases, the ductility was increased dramatically. The higher strength of the steel jacket was attributed to its isotropic nature. This could be solved in the future by use of a woven fabric having similar characteristics in two mutually perpendicular directions (Sika, 1997).

The conventional way of providing a seismic refit to a reinforced concrete bridge column involves the use of heavy steel sections to enclose the column,

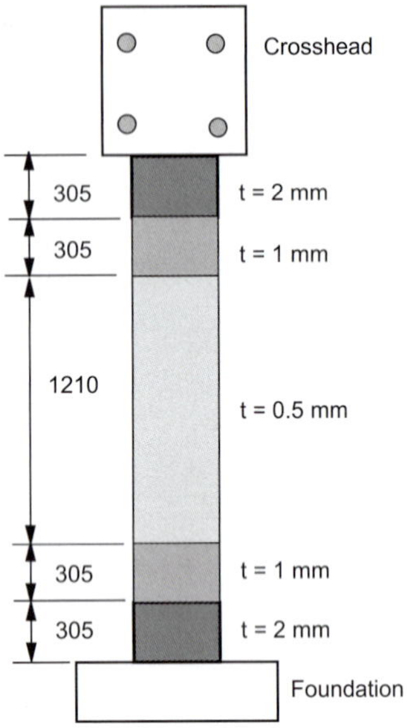

FIGURE 7.11 *Design jacket arrangement for a rectangular column with 2.5% longitudinal reinforcement (after Seible, 1997).*

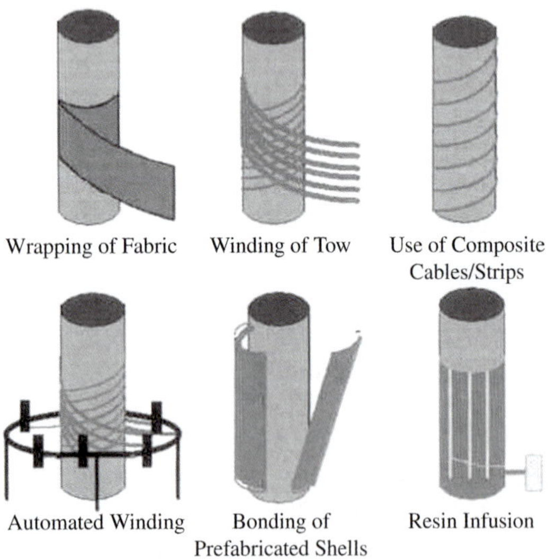

FIGURE 7.12 *(a) Application of CFRP wraps to columns (Karbhari & Seible, 1999).*

which are then filled with concrete thus providing a protective and strengthening 'jacket'. CFRP wraps offer an attractive cost-effective alternative as they can be rapidly fabricated, speedily applied and result in minimal traffic disruption. Karbhari and Seible (1999) describe six ways in which CFRP wraps can be applied (see Figure 7.12(a) and quote 'Wrapping of fabric', 'Winding of Tow', and 'Bonding of Prefabricated shells', as the most extensively used in the USA.

The typical procedure for 'wrapping' is as follows (see Figure 7.12(b)):

- Clean the surface of the concrete, fill if necessary.
- Apply resin.
- Check the resin for uniformity.
- Cut the FR sheet to size.
- Wrap the FR sheet around the column.
- Overlap the FR sheets and clean off excess resin.
- Bed down the FR wrap with a roller.

Once in place, the wrap can then be painted if required, or additional layers added.

7.3.4 Shear strengthening using CFRP

Relatively little research has been carried out in the area of shear strengthening of concrete beams using CFRP. There are many reasons for this, the most likely being that shear weakness is far less prevalent than flexural weakness.

Taljston (1997) carried out laboratory tests on ten rectangular reinforced concrete beams with CFRP tape strips secured to each of the vertical sides laid by hand, pre-preg and vacuum injection at an angle of 45°. He achieved a strengthening effect of almost 300%. The work was very limited in scope and no recommendations regarding design were presented.

Hutchinson et al. (1997) carried out tests with various configurations of CFRP laminates bonded

FIGURE 7.12

(b) Wrapping of a CFRP fabric to a circular column.

to the sides of scale models of I-shaped precast, prestressed concrete beams used in a highway bridge in Manitoba, which needed upgrading in shear to cope with increased truck loads. A typical cross-section of the beams tested, and details of the CFRP reinforcement are shown in Figure 7.13. The sheets were 250 mm wide, and in Configuration (1) the vertical sheets were placed with 100 mm gaps between each sheet, whilst, in Configuration (2), the diagonal sheets were placed with only 20 mm between each. The beams (including an unstrengthened control beam) were tested to failure and it was found that the ratio of the Ultimate shear for the strengthened beam/Ultimate shear of the control beam [Vu/Vu control] was a maximum of 1.36 for Configuration (2) and 1.29 for Configuration (1). This method shows considerable promise and, hopefully, further research will result in a sound design procedure and guidelines for its application.

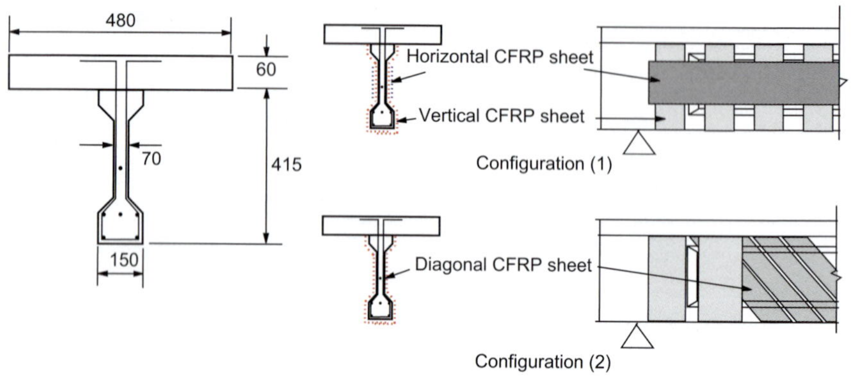

FIGURE 7.13
Details of CFRP shear reinforcement.

7.3.5 Unconventional strengthening

Very often, it is necessary to find a solution to a potentially serious problem which is identified during an inspection. It may not require major works but it is evident that some interim 'temporary' measure is needed in order to prevent further deterioration and danger to the public. Instead of utilising a conventional 'off-the-shelf technique' some lateral thinking can lead to an ingenious solution. Figure 7A.1 illustrates the point. This reinforced concrete column (supporting an over bridge on the M4, UK) was showing signs of spalling at the top due to bursting forces induced by the vertical load from the bearing (similar to the bursting forces which develop in the vicinity of the anchorage of post tensioned concrete bridges). To contain these forces, the top section of the column was provided with five steel hoops or belts, which are tensioned using a device similar to that used in the packaging industry to secure the contents of wooden boxes. This 'temporary' measure may, if proven successful, become 'permanent' with time.

7.4 EXTERNAL POST-TENSIONING

7.4.1 Introduction

The general purpose of providing external post-tensioning to an existing bridge is to restore its serviceability by relieving the dead load-bending effects, thus, resulting in reduced deflections and/or elimination of cracking. The additional post-tensioning material will also increase the ultimate limit state capacity in bending and shear. It has been used on a number of bridges in the UK (Daly, 1998) including Rakewood Viaduct in Yorkshire,

Friarton Bridge in Scotland (Murray, 1996), Kingston Bridge in Glasgow (Carruthers & Coutts, 1993), Clifton Bridge in Nottingham and the A3/A31 Flyover in Surrey (Robson and Craig, 1996). At the present time, high strength steel wire or rods are used exclusively for the post-tensioning tendons, but it is possible that the use of tendons made from CFRP strands will be accepted. Testing is presently underway in Switzerland (Maissen & de Smet, 1997) on the use of such tendons for conventional prestressed beams but the results will also be applicable to bridges strengthened using the same material.

7.4.1.1 Horizontal tendons

An example of the novel use of horizontal tendons is the strengthening of Rakewood Viaduct in Yorkshire (Pritchard, 1990). The 256 m long, six-span continuous deck, completed in 1969, consists of ten 3 m-deep steel plate girders composite with a reinforced deck. Upgrading was required to cope with increased traffic loading which would result in a 40% increase in the compression force in the bottom flanges where they passed over the piers (see Figure 7.14).

The strengthening was achieved by providing three overlapping pairs of Macalloy bars under each girder attached to steel anchors fixed to the bottom flanges with high strength friction grip bolts together with intermediate supports to prevent vibration of the bars in high winds. Particular attention was paid to anti-corrosion protection of the steel. The arrangement is shown in Figure 7.14 and a diagram of the additional bending moments shows how the unloading over the piers is affected. The end anchorage positions were chosen to coincide with points where local bending and shear were not critical and the existing structure could accommodate the high local stresses

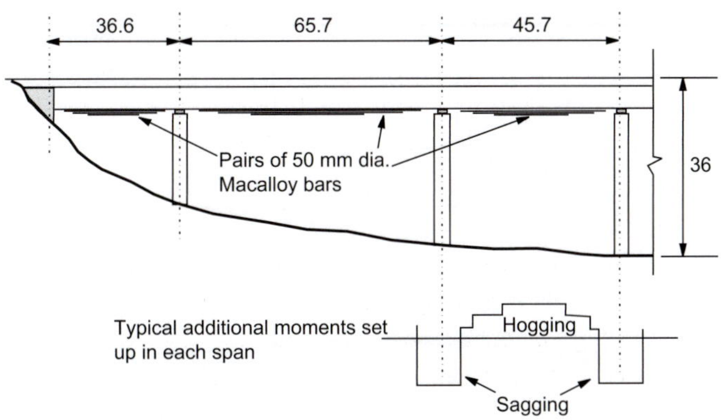

FIGURE 7.14
Strengthening of Rakewood Viaduct (after Pritchard, 1990).

induced. Finite element analysis was used to aid design. The whole system proved to be economical and was installed with a minimum of traffic disruption. A similar load relieving system has also been successfully employed on a three-span composite girder bridge in Iowa State, USA (Klaiber, Dunker, & Sanders, 1990).

Horizontal tendons were also used in the Sachsengraben Bridge in Germany (Ivanyi and Buschmeyer, 1996). The 97 m long, three-span continuous deck consists of a single post-tensioned concrete box section, with a uniform depth of 2.8 m from end to end (see Figure 7.15). During inspection it was found to have transverse cracks up to 0.6 mm wide in the top and bottom flanges, mainly caused by temperature fluctuations. To improve durability, it was decided to provide horizontal external prestressing to close the cracks. Any large cracks found not to close sufficiently were filled with an epoxy resin compound. The tendons were secured to new reinforced concrete anchor blocks placed inside the box section, which in turn were prestressed back to the main box. The prestressing force was higher in the centre span where the cracking was worst.

7.4.1.2 Deflected tendons

Small reinforced concrete bridges can be very effectively strengthened by the use of externally-placed, deflected tendons. An example of this type is given by Pritchard (1994), Cepela, Moravcik, Vican, and Zemko (1996) and Bavetta, Cardinale, and Spinelli (1996). Tendons can be anchored to the sides of the longitudinal beams near the supports (Figure 7.16(a)) or behind the end diaphragms (Figure 7.16(b)) and inclined downwards to deviators placed at cross-beam positions (if they exist), or by means of deviator blocks secured to the sides of the main beams. The actual forces transferred to the structure are shown diagrammatically (for condition b)

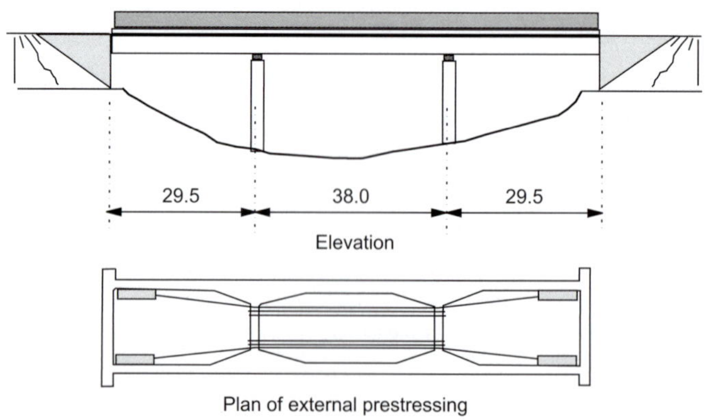

FIGURE 7.15
Strengthening of Sachsengraben Bridge (after Ivanyi and Buschmeyer, 1996).

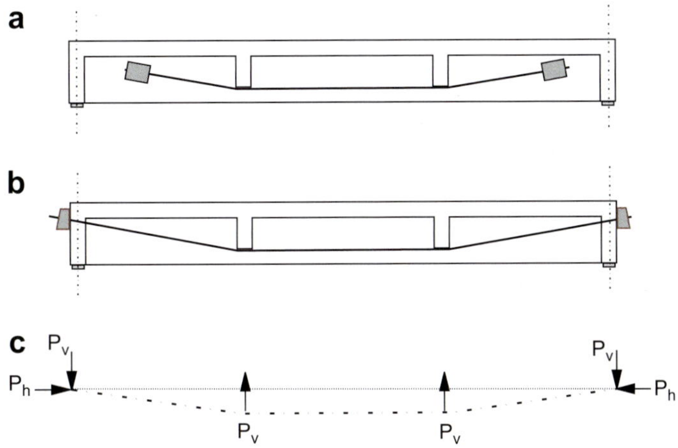

FIGURE 7.16
Bridge strengthening using externally-applied inclined tendons.

in Figure 7.16(c), resulting in a bending moment diagram (BMD) that will approximately balance the dead load BMD.

7.4.1.3 Detailing

Care must be taken with any system to ensure that the anchorage points and deviators are sound and that the additional prestressing forces can be safely transferred to the existing structure. They can be of steel or concrete, secured to the existing structure by means of chemically anchored bolts or local prestressing. In either case, the deviators should be shaped to accommodate the curvature of the tendon sheathing. The local stresses have to be carefully assessed to ensure that there is no danger of crushing or bursting of the concrete.

The external prestressing tendons should be well protected against corrosion. Strand systems should be contained in ungrouted plastic sheathing filled with grease, and bars should be protected with an epoxy based paint system.

The whole system should be designed so that, once the anchorages and deviators are in place, it is relatively easy to install the tendons. It is a good idea to build in some redundant unstressed tendons for possible further upgrades, or to be used in the event of a tendon failure.

The tendon system should, if possible, produce a bending moment diagram (BMD) that is the reverse of the dead load BMD, thus, a deflected system is the best. Straight tendons, if anchored at the very ends (Figure 7.16(b)) result in an unwanted negative moment at those positions. They are, however, the easiest to install as they require end anchorages only, and there can be a substantial saving in both material and construction costs.

If anchored at an intermediate point (Figure 7.16(a)), some of the prestressing moment is neutralised by the dead load moment. Careful judgement has to made for the particular bridge under review.

7.4.1.4 Trusses

The remaining life of truss bridges can be considerably enhanced by the addition of external post-tensioning. It can be applied to individual members, or it can be applied to the bridge as a whole where it provides a means of creating redundancy within the structural system by providing alternate load paths (Ayyub, 1990). Failure of a component in a statical system is tantamount to failure of the whole structure and so redundancy, in fact, strengthens the truss. Xanathos (1996) suggests various tendon configurations for post-tensioned trusses and these are indicated in Figure 7.17.

Where prestressing is applied to individual members, a check of the member's ability to support the anchoring system and its ability to withstand a compressive force without local buckling will generally be sufficient. Where whole chords or groups of members are considered together for prestressing (such as a top or bottom chord), the overall effects of the force redistribution must be carefully considered.

Sliding bearings must be in good condition to allow the full effects of the prestressing to be realised, and connection details need to be checked to ensure that additional local forces can be resisted.

Transverse members (such as floor beams and decks) can be strengthened in the same way in order to increase load-carrying capacity and to improve lateral load distribution characteristics. Van Dalen, Taylor, and Batchelor (1984) report on a laminated timber bridge with transverse strengthening, and Troitsky (1990) reports on the transverse strengthening of a steel truss bridge using Dywidag Bars.

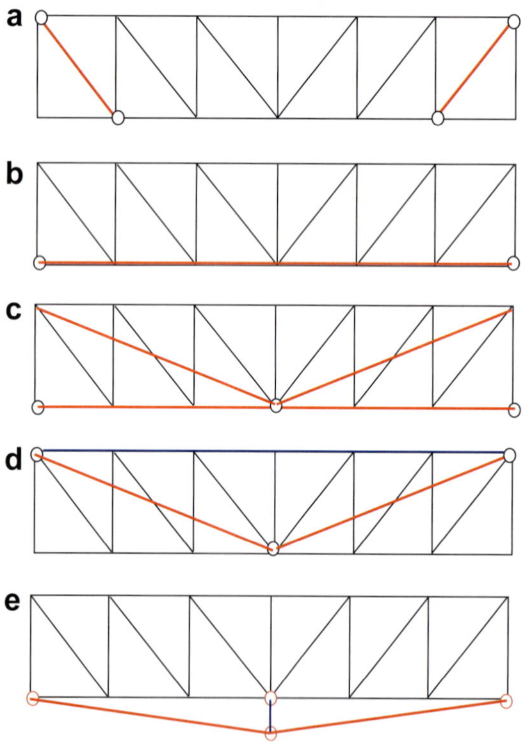

FIGURE 7.17 *Tendon configurations for post-tensioned trusses (after Xanathos, 1996).*

7.5 ARCHES

7.5.1 Introduction

The main load-carrying element in the superstructure of a masonry arch bridge is the barrel of

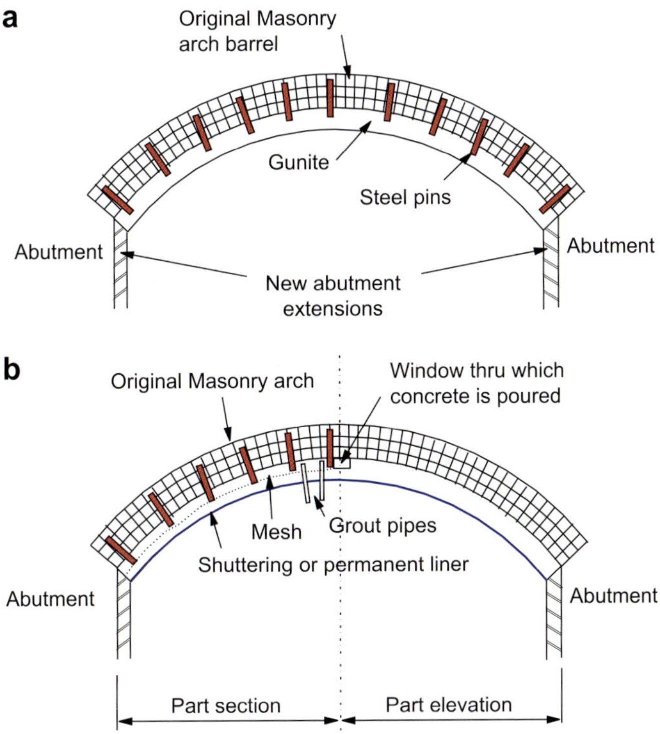

FIGURE 7.18

Strengthening of the Intrados.

the arch (see Figure 7.18). Major strengthening becomes necessary either where the bridge has to be upgraded to accommodate high traffic loads or when damage has been sustained by the barrel beyond the point where relatively simple repair techniques (Chapter 6) can provide sufficient rehabilitation. The solution is either to provide load relieving of the barrel or to strengthen the barrel itself. Some of the more common techniques are listed in the following sections. Waterproofing of arches has been overlooked in the past, presumably because it was assumed that the often small amount of iron in the form of anchor ties and balustrade fixings could be individually protected from water and that the masonry was inert. However, it is now recognised that leaking water and chloride-laden water can cause serious harm to the arch barrel, and for that reason waterproofing is carried out, as a matter of course, in all arch-strengthening operations.

7.5.2 Guniting

This method of spraying concrete under pressure is briefly described in Chapter 6 and a fuller explanation is given by Neville and Brooks (1991). It provides a concrete jacket to the intrados of the barrel. The concrete is built up in layers onto the precleaned masonry which when cured acts compositely with the masonry. If properly mixed and sprayed, the gunite stays in place but inevitably there is a loss of material due to rebound from the surface. This can be as much as 50% of the total. The significance of rebound is not so much the waste of material as the danger of build-up in other areas where it is not required and the loss of aggregate leading to increased shrinkage and possible cracking. The insertion of ribbed steel pins into the full depth of the barrel can help to support the concrete, and either the abutment has to be built out in masonry or the sprayed concrete continued down to the existing footings in order to provide a springing for the new concrete (see Figure 7.18(a)). A typical period for applying gunite can vary from between two to four days, depending on the area to be sprayed and the thickness required. For bridges with road traffic over only (such as river bridges) it is an ideal solution as it is completely non-disruptive. For bridges with limited headroom, or with a roadway under, it may not be appropriate.

The level of increase in strength that can be achieved has been demonstrated by Sumon and Ricketts (1995) who carried out tests on two model arches. Both consisted of a three-ring brick thickness for the arch barrel and one was strengthened with a sprayed concrete jacket to the intrados. The arches were identical in every other way. The gunite was allowed to cure for 28 days and the strengthened arch tested to failure. They found that the addition of the gunite dramatically increased the barrel stiffness, and the strength was increased nearly four-fold.

Choo, Pecston, and Gong (1995) carried out similar tests, and they found that the most significant factor affecting performance was the condition of the interface between the masonry and the concrete. Assuming perfect bond, they found that by doubling the barrel thickness using concrete the strength increased ten-fold, and for a 33% increase in thickness the strength doubled.

7.5.3 Jacketing

This is similar in principle to guniting. The concrete is placed by hand or preferably by pumping into a space created by placing shuttering beneath the intrados. The surface of the masonry is first cleaned by grit blasting, dowel bars are then inserted and grouted. Mesh reinforcement is secured to the projecting ends of the dowels and the shuttering is put in place. The concrete is poured through 'windows' in the vertical sides of the shuttering near the crown and external vibrators can be used to aid compaction of the concrete. Once the concrete has cured, grout is pumped through grout pipes previously cast into the concrete and placed at regular intervals across the arch near the crown (see Figure 7.18(b)). This is to take up the space left between the concrete and the masonry due to the settling of the concrete.

The shuttering may be conventional timber, which is stripped after the work is complete, or a permanent liner comprised of glass fibre-reinforced polymer (GFRP) or corrugated steel sheeting (Ricketts et al., 1995).

7.5.4 Saddling

Both guniting and jacketing have three potential drawbacks:

- Lack of durability due to water ingress from above, which, in time, causes debonding of the concrete from the arch barrel.
- Reduction of headroom.
- Unacceptability if the bridge is a listed structure.

Saddling is probably the most popular form of strengthening and takes the form of a concrete saddle or relieving arch placed directly on top of the existing extrados as indicated in Figure 7.19. The saddle can be made composite with the existing extrados, which not only increases the strength but helps to restrain cracked sections of the original arch, or it can act independently by introducing a debonding layer (such as polythene sheet) between the old and the new. The thickness is usually kept constant over the internal spans of multi-span bridges and thickened at the end spans where a haunch is formed. The top of the saddle is waterproofed with a sprayed membrane and drainage pipes laid at the low points of multiple arch bridges over the piers.

Welch (1995) states that 'saddles are typically 150 mm–200 mm thick, of relatively weak concrete and can, if judged necessary, be articulated to harmonise with the arch rings structural action, either by 'bands' of transverse brickwork or an inert transverse debonding lamina'. Sound judgement must be exercised if articulation is incorporated to ensure that the 'bands' are

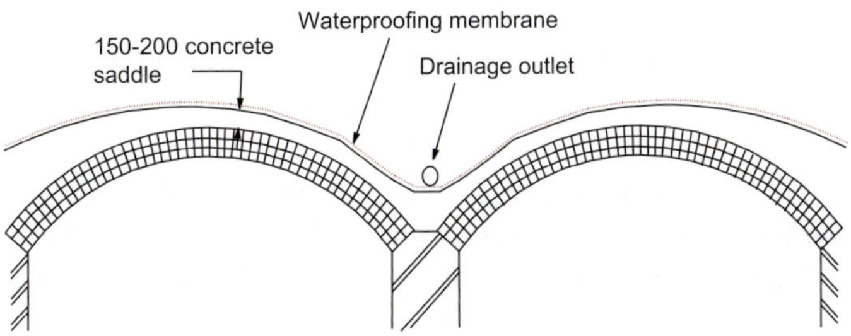

FIGURE 7.19 *Typical saddle strengthening.*

FIGURE 7.20
Wilderwick Bridge, Dormansland, East Grinstead.

positioned correctly and, unless there is already cracking to the existing arch defining the position(s), then articulation is best avoided.

To limit micro-cracking due to shrinkage and temperature stresses, the concrete can be reinforced with polypropylene or steel fibres. Protruding fibres have to be removed in order to prevent puncturing of sprayed water-proofing layers, but may be left in place if a hand-laid system is used. The effects of additional thrusts and vertical dead loads on the existing structure need to be assessed before the saddle option is employed.

A good example of the saddling technique used in practice is the Chertsey Bridge in the UK described by Palmer and Cogswell (1993) and the Wilderwick Bridge in the UK described by Welch (1995) (see Figure 7.20).

7.5.5 Overslabbing

Where additional thrusts cannot be resisted by the abutments, the solution is to provide an independent reinforced concrete (load relieving) slab over the existing fill, spanning between abutments, by partially excavating the fill above the arch as in Figure 7.21. The current concern of this type of solution is to assess exactly how the new composite bridge is behaving, since what is produced is essentially a slab/fill/arch configuration which has relative stiffness that will significantly alter the disposition of forces within the structure. In the case of very pointed arches, such as three-pinned gothic arches, the slab can be made to span between the crown and the springings (Welch, 1995). The load causing the bending effects in the arch ring is thus relieved, and the compressive force in the arch ring is increased.

The reader is referred to reference Kotze (1996) for a similar solution adopted for an old arch bridge in Virginia in the Free State in South Africa.

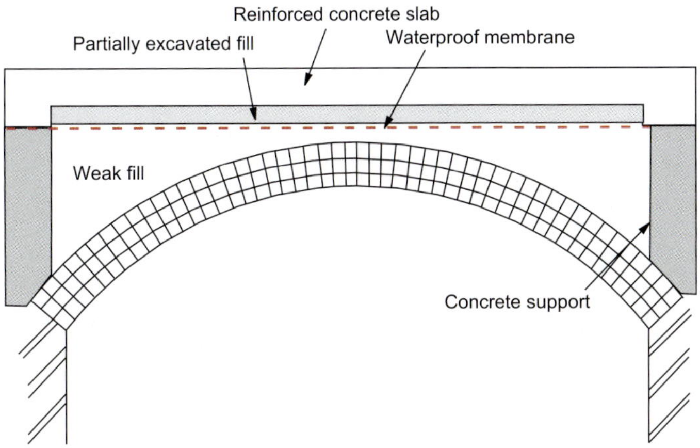

FIGURE 7.21
Overslabbing.

7.5.6 Retro-reinforcement

Garrity (1995) has proposed a technique known as retro-reinforcement for repairing masonry arch bridges. It has been successfully used to repair and strengthen masonry buildings (Garrity, 1994) but has yet to be used on a bridge.

The method involves the installation of stainless steel rods in the surface zone of existing masonry to provide continuity, tensile strength and ductility. It is claimed to be relatively easy to install, it creates little disruption to the users and is not visually obtrusive. A typical range of proposed strengthening measures appropriate to an arch bridge is shown in Figure 7.22. The basic philosophy adopted by Garrity is that the most appropriate method of strengthening masonry structures is to improve the tensile strength and the resistance to cracking without causing a significant change in the fundamental structural behaviour.

The term retro-reinforcement is described in detail by Garrity (1995) but, essentially, involves cutting small grooves in the masonry into which are placed stainless steel bars (typically 6 mm diameter), secured in place with a cementitious grout. Where bars are placed in bed joints, the outer 15 mm of the groove is left free of grout and repointed in matching mortar. Garrity maintains that 'the bars are small enough to be easily handled on site and can be readily bent to suit the actual profile of the masonry and the grooves cut into it.' Because of the highly targeted nature of the work, construction times can be kept to a minimum. Also, very little plant is required, and the work does not have to be carried out in one continuous operation – an attractive proposition when only limited possession times are available.

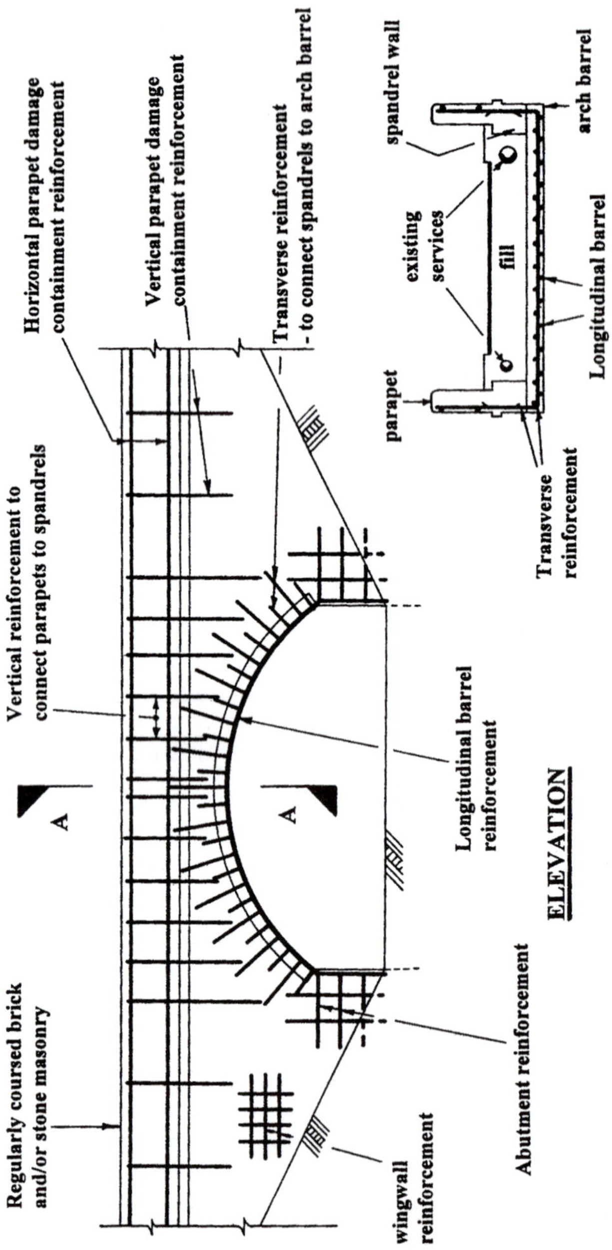

FIGURE 7.22 *Proposed retro-reinforcement for a typical single span masonry arch bridge (after Garrity, 1995, p. 561).*

Other structural benefits listed by Garrity are:

- There is no additional permanent load applied to the bridge as a result of the repair, hence the load on the foundations is kept to a minimum.
- There is no increase in the thickness of the arch ring. This avoids the potential problem arising with saddle repairs where the line of thrust is raised, thereby increasing the risk of abutment or pier movement which could lead to further structural damage (Melbourne, 1991).
- By installing the reinforcement in the outer zone of the masonry, large increases in the shear capacity and the moment of resistance are possible.

Garrity acknowledges that for various reasons quick-setting, high strength grouts will be necessary, having much higher stiffness properties than the original arch, and the long-term effects are not known. The long-term dynamic effects of highway loading are, similarly, not known. Further research is therefore necessary to establish soundness of the method.

7.5.7 Stitching

An innovation in the field of arch strengthening by the modification of an existing strengthening stitching technology for buildings has recently been unveiled in the UK, and is illustrated in Figure 7.23.

An oversized hole is first drilled in the masonry to the required depth, and a stainless steel hollow section wrapped in a fabric sock is inserted into the hole. Ultrafine concrete grout is then pumped through the middle of the anchor and the sock fills with grout from the end to the surface, thus eliminating the risk of pockets of air forming. The sock expands under pressure and takes up any irregularities in the hole. The grout milk is forced through the sock, thus creating a chemical and mechanical bond with the substrate. The hole is then made good at the entry point.

To test the system, a full scale bridge was tested at the Transport Research Laboratory (TRL). It had a span of 5 m, a width of 2 m and comprised a three-ring arch barrel with sand between the rings to simulate ring separation, and then strengthened with the new system. It was modelled by consultants Gifford and Partners using an advanced finite element package from Rockfield Software Ltd. to predict the failure load and collapse mechanism based on data obtained from the test to destruction of the Torksey Bridge (Page, 1995). Computer simulations of the progressive modes of collapse for the unstrengthened and strengthened arch are shown in Figure 7.24(a) and (b) (see also colour plate section) respectively.

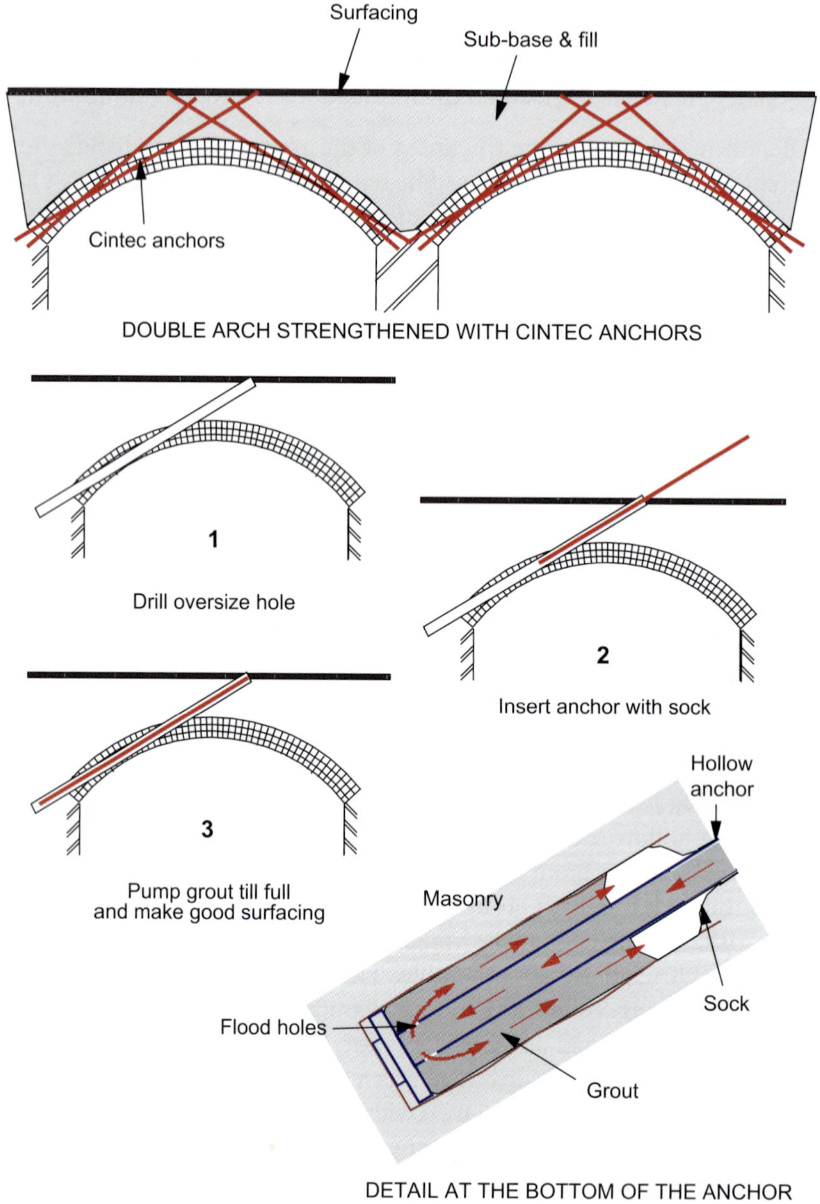

FIGURE 7.23
Typical stitching application.

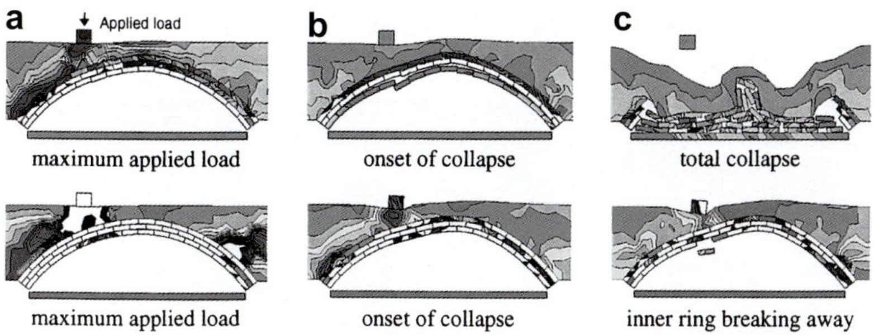

FIGURE 7.24
Predicted modes of collapse (see colour plate section).

The bridge was then tested by applying a load at the quarter span and load-deflection data recorded up to failure (Sumon & Ricketts, 1995). The load-carrying capacity was more than doubled, the elastic behaviour was improved and failure was of a ductile nature. The developed system is called ARCHTEC using steel anchors produced by Cintec.

This non-invasive solution to bridge strengthening is very appealing. The stitching operation can be applied from the top, bottom or sides of the bridge and causes minimal disturbance to traffic (which can continue to flow) and there is virtually no change to the appearance of the bridge. It has been proven to significantly increase the failure load (Sumon & Ricketts, 1995) and is claimed to be cost-effective. The method can also be used to strengthen the arch barrel transversely by inserting anchors into the voussoirs, or radially from the intrados; it can also restrain spandrel walls and stabilise parapets.

7.5.8 MARS strengthening system

Another arch-strengthening system which has been under development for many years is the MARS (masonry arch repair and strengthening) system developed by Protec Industrial Ltd. (Minnock, 1997).

An orthogonal grid of grooves is cut in the intrados of the arch barrel in both the transverse and longitudinal directions to a depth of about 75 mm and 40 mm wide. A prefabricated grid of 6 mm diameter 6 mm stainless steel bars is then fixed into the grooves and secured in place by pumping a specially formulated, low-modulus, structural adhesive into the grooves, thus forming a sort of 'singly reinforced masonry arch' (Figure 7.25). It is maintained (Minnock, 1997) that 'it is both cost-effective and quick to install'. Furthermore, it is a non-invasive technique, and does not significantly alter the appearance of the bridge. It has been used to strengthen the Well'i'th Lane Canal Bridge, Rochdale, UK, which has a span of 6.43 m, so that the 7.5 t weight restriction could be lifted and the bridge upgraded to sustain a live load of 40 t. The TRL tested the bridge in the elastic range up to a final

FIGURE 7.25
General and close-up views of MARS repairs to Well'i'th Lane Canal Bridge (Minnock, 1997).

applied load of 40.8 t and a maximum displacement of 0.7 mm, and it was found that this worked effectively.

7.6 INCREASED COMPOSITE ACTION

7.6.1 Introduction

Many beam and slab bridges with no mechanical connection between the beam and slab do, nevertheless, possess some limited composite action due to friction between the slab and the beam, residual bond between these parts

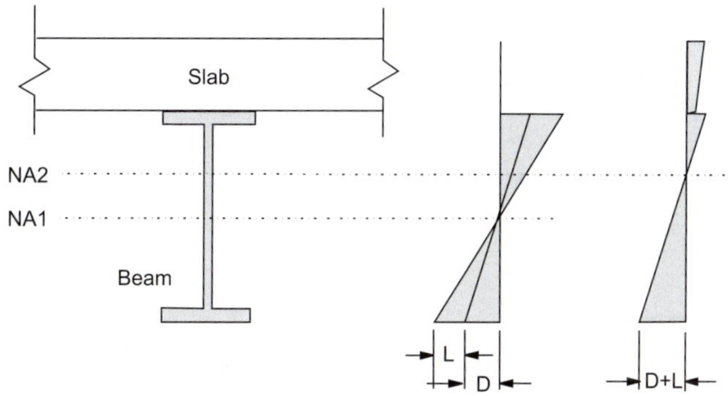

FIGURE 7.26
Stress distribution in non-composite beam and slab bridges.

and the presence of diaphragms. However, the introduction of full composite action between deck slab and beams is a way of significantly increasing the strength of a bridge. The load is shared between both materials allowing heavier live loads to be carried and also stiffening the bridge, thus reducing live load deflections. Figure 7.26 illustrates the principle. In existing bridges, the dead and live loads are carried by the beam section alone (a), but if the two parts are made to act together (b) then the live loads are carried by the composite beam-and-slab section. This will clearly increase the live load-carrying capacity by an amount which can easily be calculated.

7.6.2 Older bridges

7.6.2.1 Westminster Bridge, London, UK

Many road bridges built in the mid 19th Century in the UK and elsewhere consist of wrought iron and/or cast iron ribs (usually arched) with cast iron cross girders and spandrel webs supporting a complex arrangement of cast iron bearers, stringers and buckle plates all supporting a non-composite deck of concrete and timber. Such a bridge is the Westminster Bridge in London, UK constructed in 1862 (Yeoell & Munson, 1993; Yeoell, 1995).

Load tests carried out on the seven span bridge in 1991 led to the imposition of a 7.5 tonnes load limit. To determine whether the bridge could be strengthened, a feasibility study was ordered. Load tests were carried out and a post-test analysis suggested that if composite action between the ribs and the deck could be relied upon, then the capacity of the bridge could be dramatically enhanced to cope with 40 tonne vehicles. As the bridge is listed by English Heritage, care was taken to ensure that strengthening works did not interfere with the appearance of the bridge. It was decided that the most effective way of strengthening the bridge would be to remove the existing deck and replace it with a reinforced lightweight concrete deck on top of the

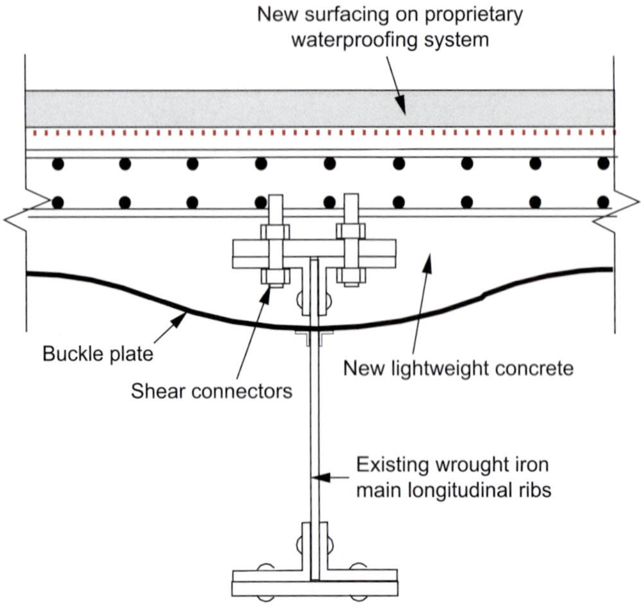

FIGURE 7.27
Deck reconstruction of the Westminster Bridge (Yeoell & Munson, 1993).

buckle plates and composite with the ribs (Yeoell, 1995). The shear connection between the deck and the ribs was achieved using a specially designed shear stud embedded in a new reinforced concrete deck slab cast over the buckle plates as shown in Figure 7.27. The change in load distribution meant that other parts of the bridge, particularly the existing cast-iron spandrels, had to be strengthened or replaced with new steel plates.

7.6.2.2 Woodbridge Old, Guildford, UK

This bridge, constructed in 1912, is one of the earliest examples of a reinforced concrete arch bridge built in the UK (Palmer & Cogswell, 1993). It has a single span of 21.4 m which is constructed monolithically of in situ concrete. The deck is supported on nine arch ribs spanning between cellular abutments. An assessment carried out in 1988 suggested that the top slab was inadequate in flexure by about 40% and the main rib beams were deficient in shear by between 600%–700%. After the consideration of many options, it was decided to strengthen the bridge.

The hogging moment of resistance at and near the supports was rectified by the provision of a new reinforced concrete top slab made to act integrally with the existing structure. Various methods of enhancing the shear strength were considered such as plating of the sides of the ribs and the use of external shear straps anchored to the new top slab. For various reasons these were

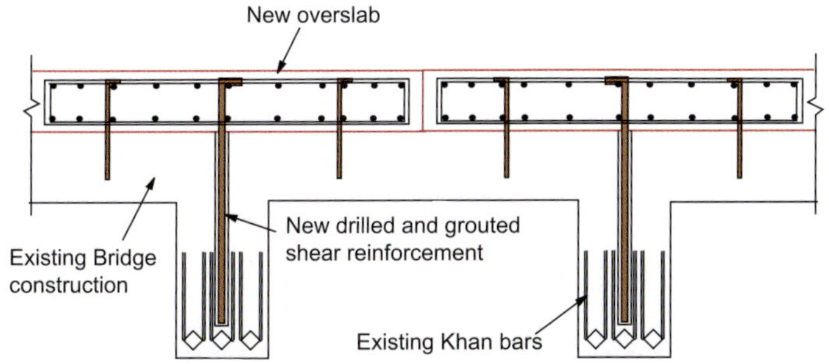

FIGURE 7.28
New overslab and shear reinforcement to the Woodbridge Old, Guildford.

rejected and it was decided to provide additional shear reinforcement by drilling centrally through the beams from the top at 500 mm centres and grouting-in 25 mm diameter 'L' bars. These acted in conjunction with the existing 'Khan' bars (shear reinforcement) by overlapping them and extending upwards to be anchored into the new top slab (see Figure 7.28).

Seible, Priestly, and Krishnan (1992) describe a similar method of strengthening an old, simply-supported reinforced concrete T-beam bridge, 3.5 m wide and 17 m long using a 150 mm reinforced concrete overlay. The top of the original deck was scarified and cleaned to ensure a sound bond between the old and the new to provide horizontal shear resistance and thus ensure composite action. Overload tests were carried out and no interface delamination was observed. The bridge has been removed from its location on Highway 41, California after 25 years of service and is now part of an ongoing research study at the California State Department of Transportation.

7.6.2.3 Jordanstown Railway Bridge, Belfast

This bridge, constructed in 1931, is a three-span, reinforced concrete structure of beam and slab construction (Lockwood, Whiteside, & Cleland, 1996). The beams are continuous and the slabs span transversely over them. An assessment showed that the deck beams were overstressed in both shear and bending. Various options for strengthening were considered and it was decided to add a 200 mm overslab to bring the bending capacity up to the required level. The shear capacity was increased by a combination of the overslab and external high-tensile link reinforcement in the form of U-bars secured to the beams. The main tensile reinforcement in the bottom of the beams was exposed, holes were drilled through the deck slab and the U-bars inserted through the holes and anchored in the new overslab. The vertical legs of the U-bars were then encased in a layer of gunite sprayed onto the sides of the deck beams as indicated in Figure 7.29.

Since no full-scale or laboratory tests on this form of shear strengthening was available, it was decided to construct a 1:10 scale model and test it in the

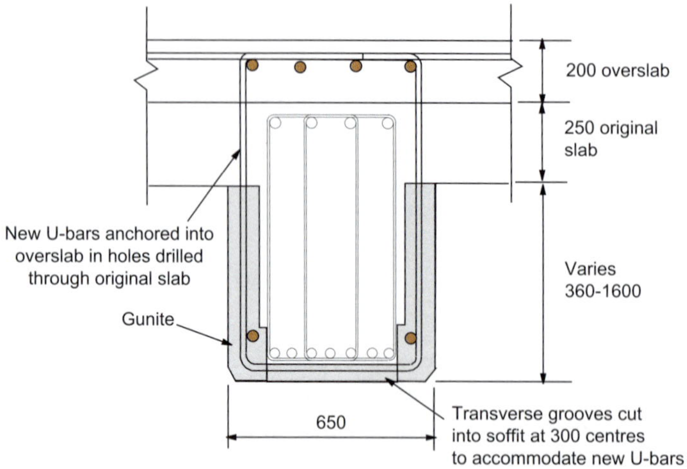

FIGURE 7.29
New shear reinforcement to the deck beams.

laboratory of the Civil Engineering Department at the Queens University of Belfast. Six models were constructed and tested under a four point loading regime. The beam with overslab and external shear reinforcement showed an increase in the ultimate shear load of approximately 100% over the basic unstrengthened beam. Code calculations indicated an increase of 45%. The method was therefore deemed satisfactory.

7.6.2.4 Langleybury Lane Bridge, Hertfordshire, UK

This bridge, comprising a three span, cast in situ, reinforced concrete structure with raking piers, was discovered during an inspection to have shear cracks at the interface of the voided and solid sections of the spine beams (Wall, Gardner, & Posner, 1996).

Strengthening was achieved by the introduction of a central pier which would permit the introduction of predetermined jacking loads in the bearings to overcome the shear deficiency, while, at the same time, keeping the central hogging moments within acceptable limits and also overcoming the anchorage problem that was discovered over the bank seats. The side span moments were reduced by replacing the raking piers with vertical ones placed 1 m closer to the end supports. The carriageway was relocated to be exactly in the centre of the deck thus removing the problem of accidental wheel loading. This was a novel solution which did not require any structural material to be added to the existing deck. The original and modified bridge are shown in Figure 7.30.

7.6.2.5 London Docklands Light Railway, London, UK

The London Docklands Light Railway, completed in 1987 to service commuters in the Docklands area of London, is of composite construction

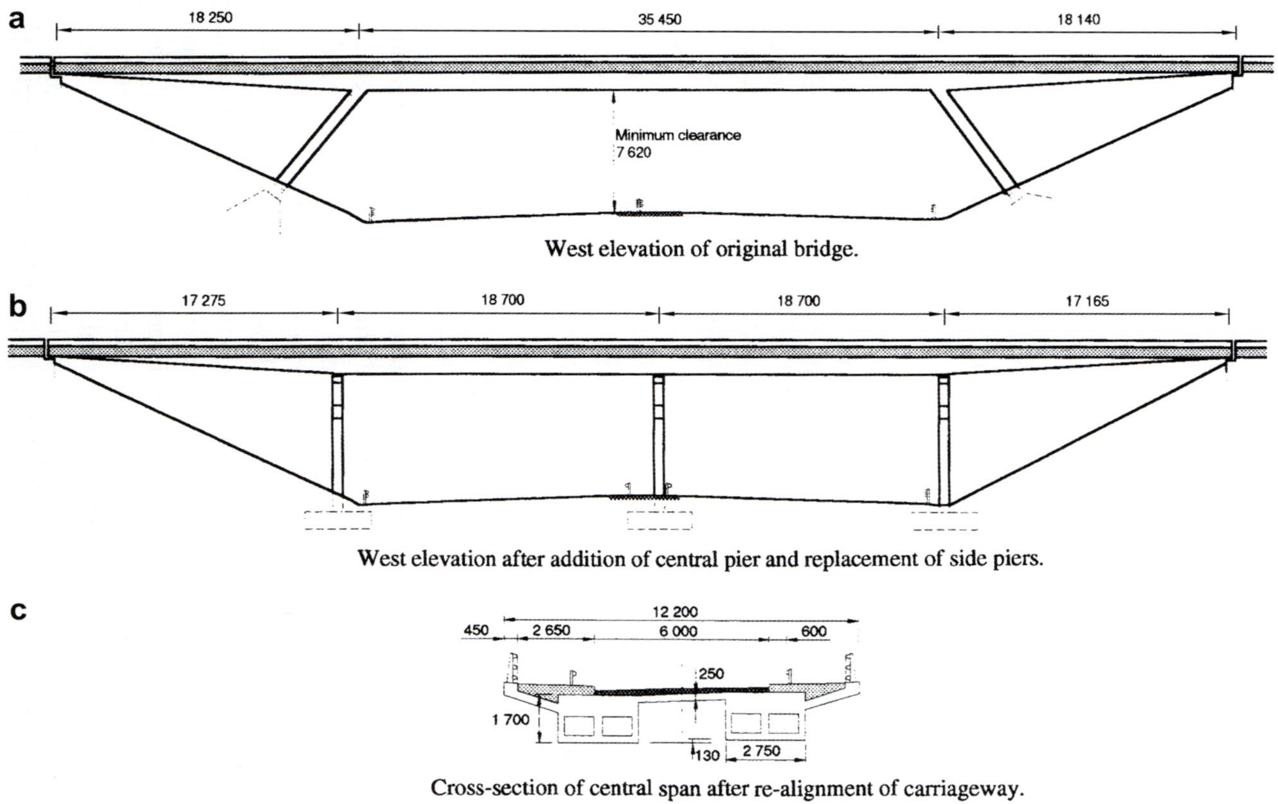

FIGURE 7.30 *(a) West elevation of original bridge and (b) West elevation after addition of a central pier and replacement of side piers and, (c) Cross-section of central span after re-alignment of carriageway (Wall* et al., *1996).*

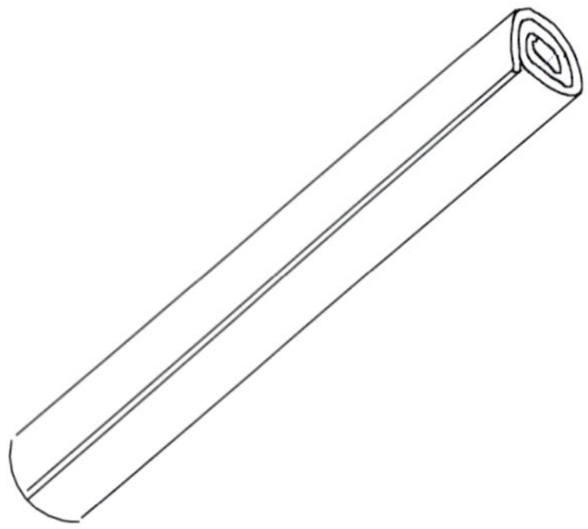

FIGURE 7.31 *Spring pin 'Spirol' Shear Connectors (Pritchard, 1994).*

(Pritchard, 1994). Fatigue considerations were a critical factor in the design of the stud shear connectors. A huge unforeseen development in the area meant that the weight and frequency of the trains had to increase dramatically in order to maintain the same level of service. This reduced the fatigue life of the connectors by up to a maximum of 75%. In order to re-establish the fatigue life of 120 years it was decided to install additional shear connectors between the original 19 mm welded stud connectors. This meant drilling in from under the top flange and inserting the connectors from below. Conventional studs were out of the question since, by virtue of their shape, they would not be in contact with the deck concrete. Eventually, 20 mm diameter spring steel pin fasteners were chosen (New Civil Engineer, 1990) as they could be force-driven into holes of slightly smaller diameter and, once inserted, they would spring out to be in contact with the concrete with no requirement for grouting, gluing or welding (Figure 7.31).

The pins were tested at The Welding Institute in Cambridge and found to be satisfactory, with good strength and fatigue properties and were installed with no interruption to train services.

7.6.3 Enforced frictional contact

Lin, Beaulieu, and Farad (1994) have proposed a way of providing composite action in existing non-composite bridges by the use of vertical prestressing through the concrete deck slab and into the top flanges of the steel beams below. The principle is shown in Figure 7.32.

Oversized holes are drilled and, once the prestressed bars are in place, the void is grouted. Partial composite action is thus achieved in the longitudinal direction due to the friction developed at the steel/concrete interface. Transverse load distribution is improved and the top flanges of the steel girders obtain a degree of lateral stabilisation. Tests are currently underway on third scale models to develop the method further. In general, results to date suggest that the method is viable and the load-carrying capacity of non-composite bridges can be significantly increased. The long-term efficiency might be impaired due to loss of prestress due to creep in the relatively short vertical bolts or bars.

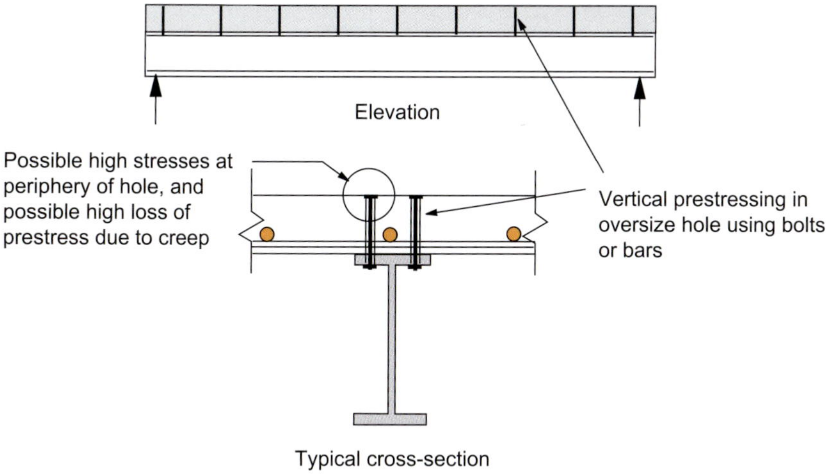

FIGURE 7.32
Enforced frictional contact using vertical prestressing.

7.6.4 Composite external steel reinforcement

An unusual application of composite action to strengthen concrete box girder spans on the approach to the George Washington Bridge is described by Englot (1996). The approach viaduct comprised simply-supported 18 m spans, separated by expansion joints. In 1995, after 35 years service, severe damage was noted. 'Chloride contaminated water had seeped through the joints onto the substructure and, in some cases, ran along the soffit of the inclined box girders'. This caused heavy spalling, particularly of the concrete cover on the underside of the box. Some of the post-tensioning cables in the 127 mm bottom flange had, in a few spans, completely corroded away resulting in a 100% loss of prestress, and there was evidence of delamination and active corrosion!

The repair scheme chosen was the use of external steel plates secured to the sides and soffits of the boxes. The design of the system was evaluated using half scale tests to failure and finite element analysis. These showed that the steel jacket behaved elastically up to two-thirds of the ultimate load, and that the steel alone could carry all of the imposed loads with all of the prestressing removed.

The construction involved the removal of all deteriorated concrete using an abrasive blast, and then side plates were secured in place using studs anchored with epoxy resin in drilled holes positioned to avoid the tendons in the webs. A bottom stiffened plate was then installed and welded to the bottom corners of the side plates. The void between the steel plates and the concrete box was then pumped full of fast-setting grout. A cross-section of the box showing the completed repairs is shown in Figure 7.33.

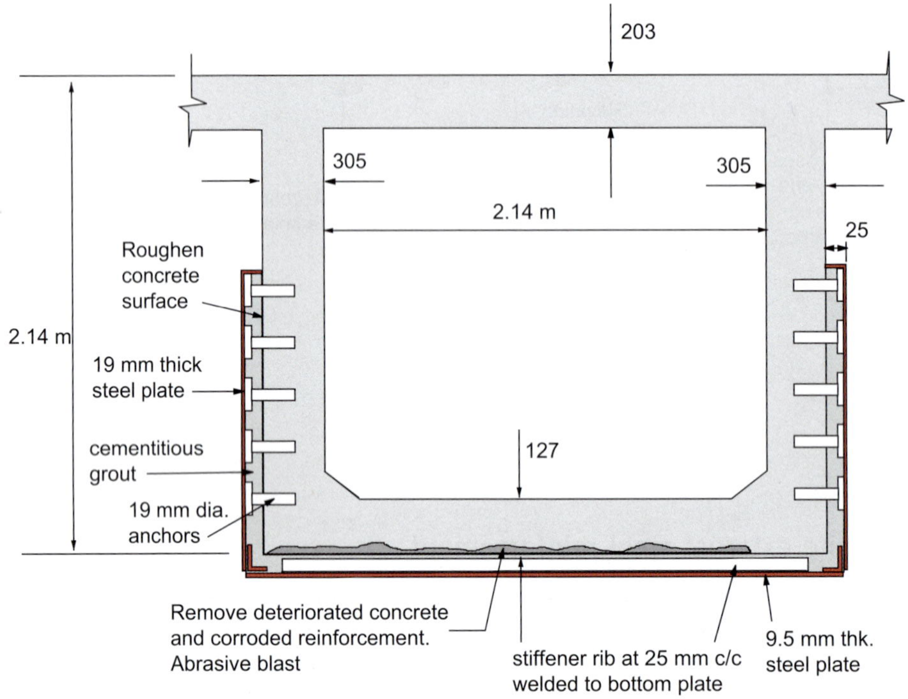

FIGURE 7.33 *Cross-section of strengthened concrete box girder (after Englot, 1996).*

7.7 PARTIAL END RESTRAINT

7.7.1 Introduction

The idea of strengthening bridge superstructures by the introduction of some partial restraint has been mooted by Klaiber et al. (1990) and Klaiber, Wipf, and Dunker (1992). When carrying out strengthening work on bridges by the use of post-tensioning, they found, as expected, that supports which had been assumed to be pinned, actually possessed a considerable amount of restraint, thus endowing the bridge with a reserve of strength not considered in design. Because of the difficulty of quantifying the amount of end restraint in actual bridges, it is neglected at the design stage. It was reasoned that it would be possible, however, to design end fixity details in previously 'simply supported' (steel or concrete) bridges, which would reduce the existing positive live load bending moments and thus effectively strengthen the bridge.

Tests were carried out on steel beam bridges with various cleat configurations (such as the one in Figure 7.34) to determine the amount of end restraint each one afforded. The beams tested were simply supported and the

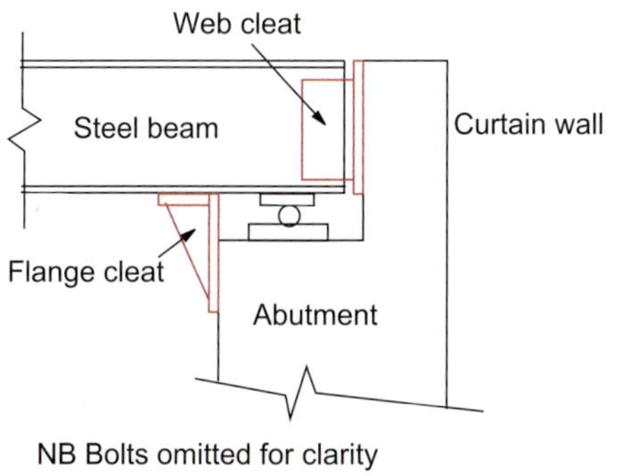

FIGURE 7.34 *Test beam used by Klaiber et al. (1990).*

supporting abutments were anchored to a strong floor. The beam was tested first without any cleats, then with a flange cleat only and finally with web and flange cleats. For each case, deflections and strains were recorded at several locations as well as end rotations. The effect of the imposed restraints is shown in Table 7.3 and it can be seen that the effect is quite dramatic.

Great care would have to be exercised in the implementation of such a scheme as the force regime on the abutment would change entirely and also the bearing would be prevented from either sliding or rotating.

So far there are no examples of this method being used in practice. Further research is necessary to consider the full technical and economical implications of the method but, until that time, it remains a relatively unnoticed and obscure technique.

7.8 PIERS, ABUTMENTS AND FOUNDATIONS

7.8.1 Introduction

Most piers, abutments and foundations are constructed in concrete. Strengthening is often required due to collision damage, corrosion damage, scour, the need to upgrade to resist more onerous vehicle collision forces, and to support extra loads that may be imposed from increased traffic loads and increased dead loads from superstructure strengthening.

Table 7.3 Effectiveness of Restraint Cleats (after Klaiber et al., 1990)

	Percentage Reductions	
Variables	**Flange Cleats**	**Flange and Web Cleats**
Mid-span strains	14.6	25.9
Mid-span deflection	19.9	30
Restrained end rotation	13	31.8

7.8.2 Piers

Probably the most effective way of strengthening the vertical parts of bridge piers is by the construction of a concrete jacket. A good example of this is given by Imam (1994) where the method was adopted for the piers of the Pease Pottage Interchange South Bridge over the M23 in the UK. The existing concrete faces were scabbled and cleaned and new reinforcement put in place around the pier to a predetermined height (which did not need to be the same as the existing pier) and the whole was shuttered and concreted.

Imam (1994) also describes the strengthening of the columns to the Horsham Road Bridge over the A23 which were hinged at their bases. The existing base was widened with mass concrete and a new prestressed concrete pedestal was constructed over. A reinforced concrete jacket was placed around the columns with a former placed at the base to maintain the 'pin' action (see Figure 7.35).

Similar strengthening by the use of a reinforced concrete jacket was carried out to the Kok Aquaduct in Maharashtra, India in 1991 (Phatak & Padalkar, 1994) where damage to the piers of the four-span structure was observed due to malfunctioning of the bearings causing extensive cracking in the masonry and dislocation of stones from the surface. A 300 mm concrete jacket was used, secured to the masonry, using grouted bolt anchors.

Steel piers, although quite numerous in the USA, are very unusual in Europe, but Mehue (1993) describes the strengthening of the steel piers to a temporary steel viaduct erected in 1977 in order to relieve traffic on a heavily damaged concrete bridge over the Paris–Lyon and Lyon–Geneve railway lines. The general arrangement of one of the piers is shown in Figure 7.36. In 1982, an inspection revealed that bolts were missing from the trunk/cantilever connection. These were replaced but it was found from calculations that some of the bolts could be over-stressed under certain loading conditions and so it was decided to strengthen by adding short steel corbels to the trunk/cantilever connection Figure 7.36(a). A routine inspection in 1985 revealed that bolts

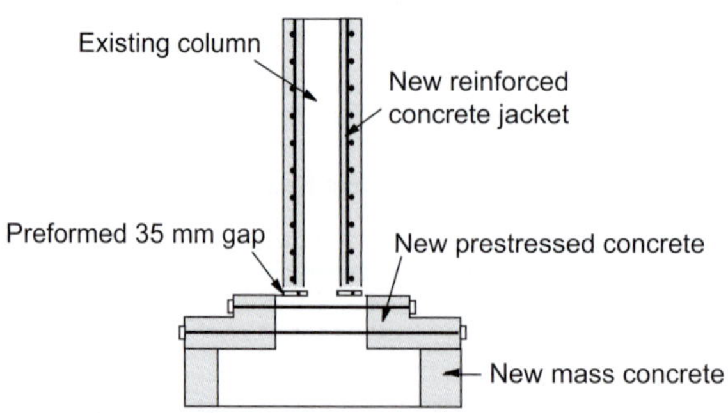

FIGURE 7.35
Strengthening of the columns to the Horsham Road Bridge.

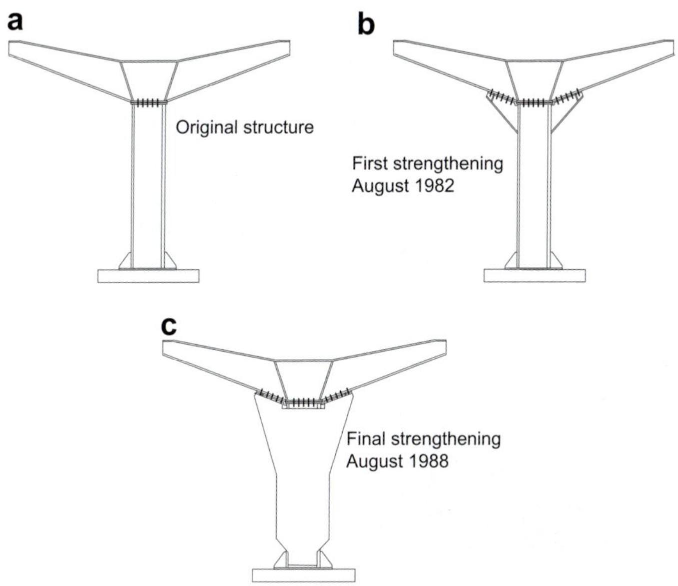

FIGURE 7.36
Strengthening of a steel pier.

were missing in the new cantilever brackets along with some cracks in the web of the corbels and in the trunk. After closer examination, it was decided to remove the corbels altogether and dress the trunk over the whole height with two wide plates welded to the flanges and connected on both sides to the cantilever with high strength bolts (Figure 7.36(c)). The strengthening work demonstrated the need to pay careful attention to details.

7.8.3 Abutments

Bridge abutments are usually massive concrete or masonry structures and there are no recorded cases of outright failure. Cracking can and does occur, however, in cantilever type abutments where the earth (and water) pressures may build up over a period of time to exceed original design values (see Figure 7.37).

Imam (1994) describes four methods to relieve these pressures which are illustrated in Figure 7.38(a)–(d). The first is the use of ground anchors, probably the simplest solution, but one which is very dependent upon the nature of the ground conditions. The second method involves considerable excavation but also gives the opportunity to examine the condition of the rear of the wall and to repair any perished waterproofing or drainage material. The final two are very similar, and again require extensive excavation. If possible, the level of the water table behind the abutment should be established. If it is high, and weep holes are in place, these should be checked and cleared. If

FIGURE 7.37
Cracking at abutment/ wing wall intersection due to high soil pressures.

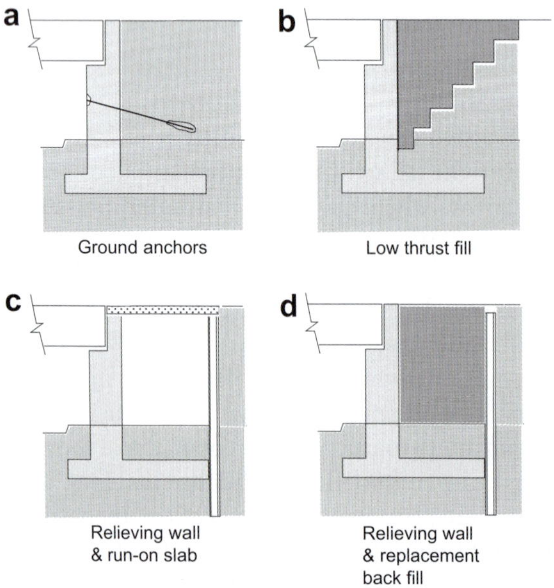

FIGURE 7.38
Pressure-relief methods to abutments.

there are no weep holes, then these should be provided using an overcoring drill. In either case, if water is released it may be that the cause of the problem has been rectified and no further works are necessary.

7.8.4 Foundations

Bridge foundations situated in river beds can be susceptible to erosion of the river bed during periods of flooding or simply by attrition over a long period of time under normal flow conditions (see Chapter 11, section 11.6). The resulting scour can undermine the foundations, thus increasing the bearing pressures under the undisturbed areas of the foundations. As scour continues, the bearing pressures exceed the strength of the soil and the foundation settles as loads are redistributed, resulting in both vertical and rotational displacement of the pier.

Such a situation was reported by Kittinger and Ashaber (1993) with regard to the Inn Bridge in Kufstein, Austria which partially collapsed in July 1990 during a flood. Displacements under one of the piers to the bridge were measured as 1.15 m on the downstream side and 0.22 m on the upstream side. The final strengthening solution is shown in Figure 7.39. After rehabilitation of the bearings, expansion joints and girders the damaged pier was strengthened by:

- Foundation grouting of the pier to partially restore it to its original position.
- Underpinning with 'Soilcrete' piles (columns).
- Surrounding with reinforced concrete apron.
- Scour protection with large boulders.

The piles are formed by a process called 'jet grouting' where the grout is injected into the ground to form a column which is load-bearing and acts just like a pile (see Figure 7.40).

Three other grouting methods described by Xanathos (1996) are hydrofracture, compaction and permeation.

Another form of strengthening using temporary steel 'needles' and steel 'soldiers' on jacks to support the superstructure prior to taking out and rebuilding a masonry pier is shown in Appendix 7A.

7.9 MOVABLE BRIDGES

Movable bridges are generally described generically as swing, vertical lift, bascule and retractile. They are a highly specialised form of bridge and differ

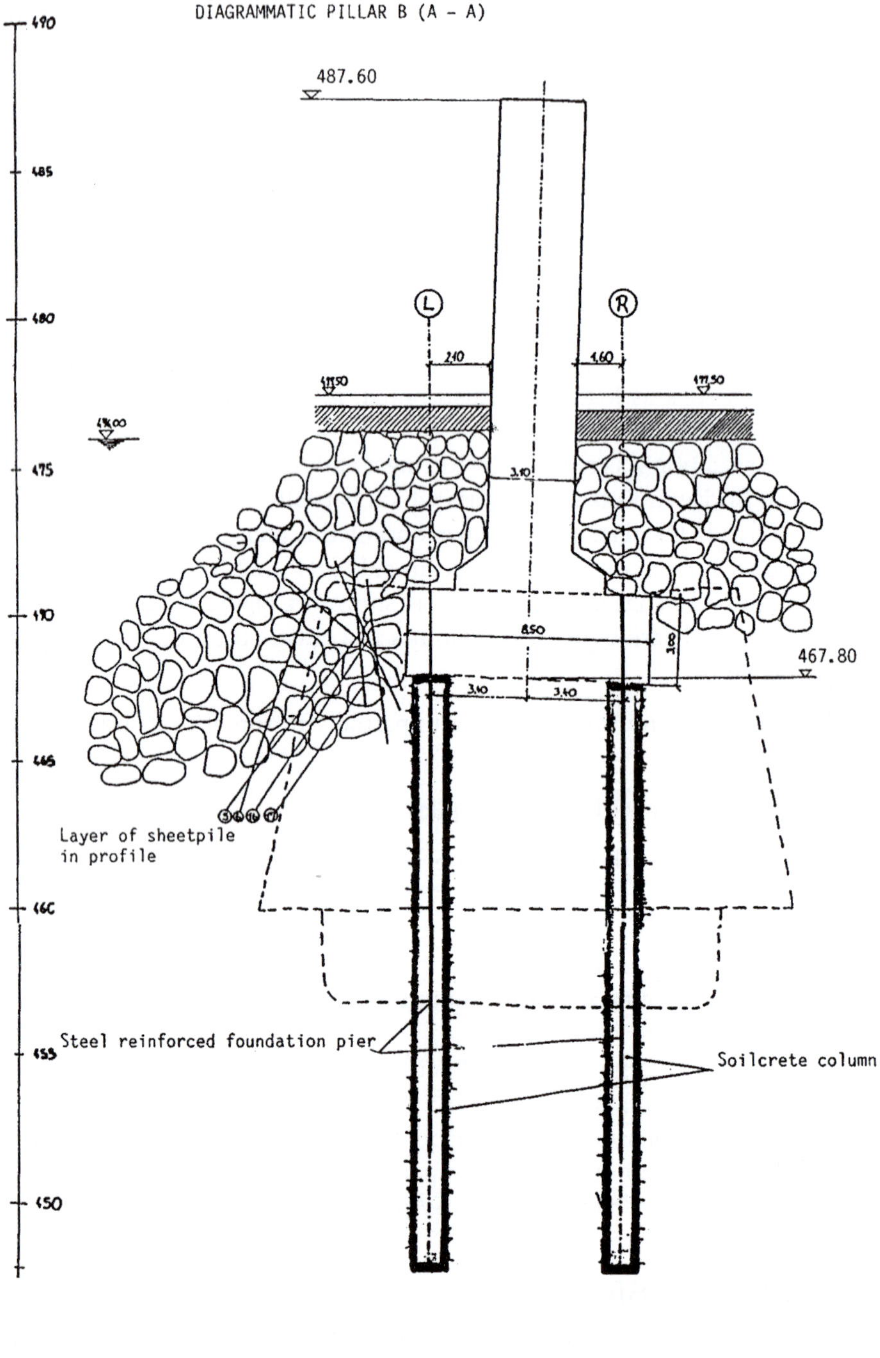

FIGURE 7.39 *Strengthening solution by (Kittinger & Ashaber, 1993).*

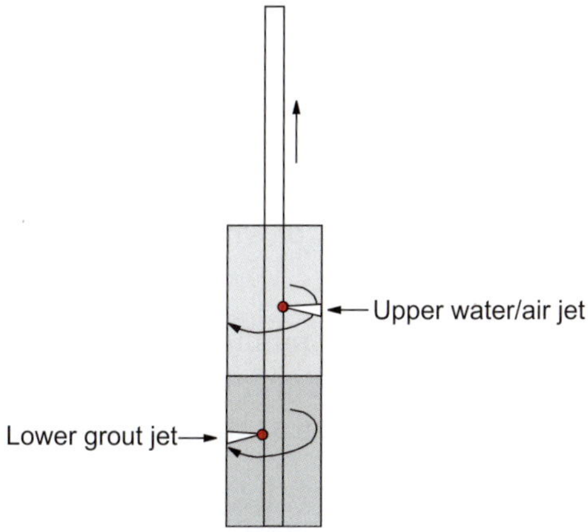

FIGURE 7.40 *Jet grouting.*

principally from all other types of bridge in that they have movable superstructures. This requires them to have extensive electrical and mechanical systems in order to function.

Birnstiel (1990) has observed that strengthening in the conventional sense is not usually required but that 'because of ageing, repeated stressing from traffic and the deleterious effect of marine environments, coupled with inadequate maintenance, the machinery deteriorates'.

Typical problems are exemplified by the rehabilitation of the 1940s Hutchinson River Parkway Bridge in The Bronx, New York (Birnstiel, 1996). The movable span of this seven span bridge is a double-leaf bascule and was found, after close inspection, to have suffered deterioration of the electro-mechanical system. Rehabilitation was required in order to extend its life by 30 years. Work involved the bascule electrical system, the span drive and stabilising machinery and traffic control devices.

Some very useful case Histories are listed by Birnstiel after the references at the end of this chapter.

7.10 LONG-SPAN STEEL BRIDGES

Experience has shown that long-span steel bridges of the suspension type are subject to fatigue damage, causing extensive cracking and, in some cases, buckling of members.

In the UK, one of the more spectacular strengthening and refurbishing contracts in recent times was the programme of the Severn Crossing consisting of the Severn suspension bridge, the Wye cable-stayed bridge and several viaduct sections described by Chaterjee (1992) and Evans (1993). At the time of its inception the Severn Bridge, with its new streamlined aerofoil cross-section in place of the traditional stiffening truss (as used on the Forth Road Bridge, opened in 1968), was hailed as revolutionary, as was the Wye Bridge with its single central stays. Since the Wye Bridge was built in 1996, it has suffered extensive damage to its superstructure due to an increase in both the volume and loading of traffic well above its design value. The effect of this manifested itself by the propagation of fatigue cracking in both the steel trough stiffeners, the deck surfacing material and in the outer wires of some of the hangers. Consultants were appointed to put forward a scheme for strengthening works, the most important elements, of which, were listed as follows:

- Reinforcing of the Severn Bridge towers by installing new tubular columns inside the tower boxes, grouting up the voids beneath the saddle towers and jacking the columns up against the saddles.
- Replacing all the hangers of the Severn Bridge with larger ones with an improved design and manufacturing specification for their sockets.
- Trough-to-deck plate fillet welds under the wheel tracks of the slow lanes of the entire crossing to be cut out, trough edge to be prepared and a deeper penetration and large fillet weld to be deposited, some butt welds on the backing flats to be repaired after removing the latter.
- Installing new sets of rocker bearings to suspend the ends of the deck boxes from the towers that would allow rotations about both longitudinal and transverse axes to cope with all movements of the deck.
- Additional stiffening inside and outside the deck boxes of the entire crossing.
- The Wye Bridge deck box girder to be supported by new twin stays in a harp fashion in place of the existing single stays which would require the heights of the two towers to be raised.
- Complete waterproofing and resurfacing of the deck of the entire crossing in such a way as to maintain structural composite action between the surfacing layer and the steel deck plate through the waterproofing membrane.

The works commenced in May 1985 and were completed in 1991 at a cost £40 m.

The Severn Crossing has proven to be a costly experiment and demonstrates the need for us all in the bridge fraternity to focus on two major aspects above all, namely, the design loading and the working environment of our bridges.

7.11 CONCLUDING REMARKS

Bridge strengthening is an art and cannot be learned from a textbook. It often requires great ingenuity to arrive at a viable and cost-effective solution. The effort is well worth it since a bridge can be given a new lease of life. None of the solutions in this chapter is prescriptive, but it is hoped that they will stimulate thought and provide a starting point for finding the 'custom' solution finally adopted.

REFERENCES

Ayyub, B. M. (1990). Post-tensioned trusses: reliability and redundancy. *Journal of Structural Engineering. ASCE, 116*(6), 1507–1521.

Bavetta, F., Cardinale, G., & Spinelli, P. (1996). Reinforced concrete bridge over Arno River: structural identification and repair. In *Proceedings of the seventh international conference on structural faults and repair* (pp. 521–530). Edinburgh: Engineering Technics Press.

Beber, A. J., Campos Filho, A., Campagnolo, J. L., & Silva Filho, L. C. P. (2000). Analysis of RC beams externally strengthened with carbon fibre sheets. In: *Proceedings of the fourth international conference on bridge management* (pp. 492–499). Surrey, UK, London: Thomas Telford.

Birnstiel, C. (1990). Movable bridge machinery inspection and rehabilitation. In: *Proceedings of the first international conference on bridge management* (pp. 295–304), Surrey: Elsevier.

Birnstiel, C. (1996). Bascule bridge machinery rehabilitation at Hutchinson River Parkway Bridge. In: *Proceedings of the third international conference on bridge management* (pp. 116–123). Surrey: E & F N Spon.

Buyukozturk, O., & Hearing, B. (1997). Failure investigation of concrete beams retrofitted with FRP laminate. *Recent advances in bridge engineering* (pp. 107–114). Switzerland: EMPA.

Carruthers, D., & Coutts, D. (1993). Refurbishmant aspects of Kingston Bridge, Glasgow. In: *Proceedings of the second international conference on bridge management* (pp. 604–613). Surrey, London: Thomas Telford.

Cepela, V., Moravcik, M., Vican, J., & Zemko, S. (1996). Reliability of reinforced existing bridges strengthened by external post-tensioning. *FIP symposium on post-tensioned concrete structures* (pp. 397–405, Vol. 1). London.

Chaterjee, S. (1992). Strengthening and refurbishment of the Severn Crossing part 1: introductions. *Proc. ICE. Structures and Buildings*.

Choo, B. S., Pecston, C. H., & Gong, N. G. (1995). Relative strength of repaired such bridges. *Arch Bridges* (pp. 579–588). London: Thomas Telford.

Concrete Society. (2000). *Design guide for strengthening concrete structures using fibre composite materials*.

Daly, A. F. (1998). External post-tensioning. *Concrete Engineering International, 2*(3), 27–29.

Darby, J., Kuke, S., & Collins, S. (2000). Stressed and unstressed advanced composite plates for the repair and strengthening of structures. In: *Proceedings of the fourth international conference on bridge management* (pp. 500–507). Surrey, UK, London: Thomas Telford.

Englot, T. (1996). Repair and strengthening of New York's infrastructure. *Structural Engineering International, 6*(2), 102–106.

Evans, J. E. (1993). The strengthening and improvement of the structures of the Severn crossing. In: *Proceedings of the second international conference on bridge management* (pp. 357–365). Surrey, London: Thomas Telford.

Farmer, N. (1997). Strengthening with CFRP laminates. *Construction Repair, 11*(1), 2–4.

Fowler, D. (1990). Glanrhyd report says BR had no warning. *New Civil Engineer, 5*, 14–15.

Garden, H. N. (1997). The strengthening of reinforced concrete members using externally bonded composite materials. Thesis presented for the award of PhD. Surrey University.

Garden, H. N., Hollaway, L. C., Thorne, A. M., & Parke, G. A. R. (1996). A parameter study of the strengthening of reinforced concrete beams with bonded composites. In: *Proceedings of the third international conference on bridge management* (pp. 400–408). Surrey: E & F N Spon.

Garrity, S. W. (1994). Retro-reinforcement of existing masonry structures. In: *Proceedings of the 10th international brick/block masonry conference*, Calgary, Canada.

Garrity, S. W. (1995). Retro-reinforcement – a proposed repair system for masonry arch bridges. *Arch bridges* (pp. 557–566). London: Thomas Telford.

Haynes, M. (1995). Repair and restoration of three cast iron footbridges in Birmingham. *Construction Repair, 9*(1), 16–17.

Henwood, A. M., & O'Connell, K. J. (1993). The use of externally bonded steel plates to strengthen Bolney Flyover. In: *Proceedings of the second international conference on bridge management* (pp. 660–671). Surrey, London: Thomas Telford.

Hutchinson, A. R. (1996). Strengthening of the Quinton Bridges with externally bonded steel plate reinforcement. In: *Proceedings of the third international conference on bridge management* (pp. 743–750). Surrey: E & F N Spon.

Imam, M. A. (1994). Strengthening and modification of bridge supports. *Bridge modification* (pp. 228–240). London: Thomas Telford.

Ivanyi, G., & Buschmeyer, W. (1996). Strengthening bridge superstructures due to external prestressing: experiences in design and construction. *FIP symposium on post-tensioned concrete structures*, (pp. 387–396, Vol. 1). London.

Juvandes, L., Figueiras, J. A., & Marques, A. T. (1997). Strengthening of concrete beams using CFRP laminates. *Recent advances in bridge engineering* (pp. 83–90). Switzerland: EMPA.

Karbhari, V. M., & Seible, F. (1999). Fiber-reinforced polymer composites for civil infrastructure in the USA. *Structural Engineering International, 9*(4), 275–277.

Kittinger, W., & Ashaber, M. (1993). Inn Bridge, Kufstein, Austria: bridge damage and repair. In: *Proceedings of the second international conference on bridge management* (pp. 787–793). Surrey, London: Thomas Telford.

Klaiber, F. W., Dunker, K. F., & Sanders, W. W., Jr. (1990). Strengthening of existing bridges (simple and continuous) by post tensioning – external prestressing in bridges (pp. 207–228). USA: American Concrete Institute.

Klaiber, F. W., Wipf, T. J., & Dunker, K. F. (1992). Recent bridge strengthening research. *Third international workshop on bridge rehabilitation* (pp. 89–98), Darmstadt.

Lane, J. S., Leeming, M. B., & Fashole-Luke, P. S. (1997). The role of 3-dimensional finite element analysis in plate bonding using advanced composite materials. In: *Proceedings of the seventh international conference on structural faults and repair* (pp. 271–275). Edinburgh: Engineering Technics Press.

Leeming, M. B. (1996). A ROBUST solution to strengthening RC and PC beams. *Construction Repair, 10*(1), 15–17.

Leeming, M. B., Darby, J. J., Lane, J. S., & Fashole-Luke, P. S. (1997). Field testing of 18m post-tensioned concrete beams strengthened with CFRP plates. In: *Proceedings of the seventh international conference on structural faults and repair* (pp. 209–214). Edinburgh: Engineering Technics Press.

Lin, J. J., Beaulieu, D., & Farad, M. (1994). Parametric study on non-composite slab-on-girder bridges with enforced frictional contact. *Canadian Journal of Civil Engineering, 21*(2), 237–250.

Lockwood, S. E., Whiteside, D. J. M., & Cleland, D. J. (1996). Strengthening concrete bridge decks – increasing shear capacity. In: *Proceedings of the third international conference on bridge management* (pp. 173–179). Surrey: E & F N Spon.

Luyckx, J., Lacroix, R., Fuzier, J. P., & Chabert, A. (1997). Bridge strengthening by carbon fibres: Recent advances in bridge engineering (pp. 144–148). Switzerland: EMPA.

Maissen, A., & de Smet, C. (1997). Prestressed concrete using carbon fibre reinforced plastic (CFRP) strands in statically indeterminate systems. *Recent advances in bridge engineering* (pp. 255–264). Switzerland: EMPA.

Mays, G. C. (1993). The use of external reinforcement in bridge strengthening: structural requirements of the adhesive. In: *Proceedings of the second international conference on bridge management* (pp. 672–680). Surrey, London: Thomas Telford.

Mehue, P. (1993). Investigation, repair and strengthening of a viaduct's steel piers. *Bridge management* (pp. 49–58, Vol. 2). London: Thomas Telford.

Meier, U. (1992). Carbon fibre reinforced polymers: modern materials in bridge engineering, IABSE. *Structural Engineering International, 1*(92), 7–12.

Melbourne, C. (1991). Conservation of masonry arches. In: *Proceedings of the ninth international brick/block masonry conference* (pp. 1563–1570). Berlin: DGfM.

Minnock, K. (1997). Masonry arch repair and strengthening. *Construction Repair, 11*(4), 45–46.

Murray, M. J. (1996). Friarton Bridge strengthening. In: *Proceedings of the third international conference on bridge management* (pp. 79–85). Surrey: E & F N Spon.

Neave, M. A. (1992). The construction, maintenance and repair of cast iron bridges. MSc thesis.

Neville, A. M., & Brooks, J. J. (1991). Concrete technology. Essex: Longman Scientific and Technical Press.

New Civil Engineer. (1990). Retrofit pins give longer bridge life, 5 April, 13.

Oehlers, D. J., & Moran, J. P. (1990). Premature failure of externally plated reinforced concrete beams. *Journal of the Structural Division of the American Society of Civil Engineers (USA), 116*(4), 978–995.

Oehlers, D. J., & Ahmed, M. (1996). Upgrading reinforced concrete beams by bolting steel side plates. In: *Proceedings of the third international conference on bridge management* (pp. 412–419). Surrey: E & F N Spon.

Page, J. (1995). Load tests to collapse on Masonry Arch Bridges. *Arch bridge* (pp. 289–298). London: Thomas Telford.

Palmer, J., & Cogswell, G. (1993). Bridge strengthening in practice. *Bridge management* (pp. 912–920, Vol. 2). London: Thomas Telford.

Peshkam, V., & Leeming, M. B. (1996). The use of advanced composite materials in strengthening and maintaining bridges. In: *Proceedings of the third international conference on bridge management* (pp. 732–742). Surrey: E & F N Spon.

Phatak, M. D., & Padalkar, J. S. (1994). Rehabilitation/repairs of Kok Aquaduct in Maharashtra, *International Seminar on Failures, Rehabilitation and Retrofitting of Bridges & Aqueducts* (pp. 207–219). Bombay, India.

Pritchard, B. (1990). Bridge deck strengthening using load relieving techniques. In: *Proceedings of the first international conference on bridge management* (pp. 667–676). Surrey: Elsevier Applied Science.

Pritchard, B. (1994). Bridge management with minimum traffic disruption. In: *Proceedings of the bridge modification conference* (pp. 185–283). ICE, London: Thomas Telford.

Ramsey, B. (1993). Steel plate bonding for concrete bridge strengthening. *Construction Repair, 7*(1), 14–16.

Raoof, M., & Zhang, S. (1996). Peeling failure of reinforced concrete beams with externally bonded steel plates. In: *Proceedings of the third international conference on bridge management* (pp. 420–428). Surrey: E & F N Spon.

Raoof, M., & Hassanen, M. A. H. (2000). Reinforced concrete beams upgraded with externally bonded steel or FRP plates. In: *Proceedings of the fourth international conference on bridge management*, Surrey, UK. London: Thomas Telford.

Robson, A., & Craig, J. M. (1996). The design of the strengthening of the A3/A31 flyover – Guildford. *FIP symposium on post-tensioned concrete structures* (pp. 406–415), London.

Seible, F. (1997). Bridge pier rehabilitation with continuous carbon fiber wraps, IABSE. *Structural Engineering International, 2*(97), 59–66.

Seible, F., Priestly, N., & Krishnan, K. (1992). Evaluation of strengthening techniques for reinforced concrete bridge superstructures. *Third international workshop on bridges rehabilitation* (pp. 39–48), Darmstadt.

Seible, F., Priestly, M. J. N., & Innamorato, D. (1995). *Earthquake retrofit of bridge columns with continuous fiber jackets, Vol. II, Design guidelines*, Advanced Composites Technology Transfer Consortium, Rpt. No. ACTT-95/08. San Diego: University of California.

Shaw, M. (1993). Strengthening bridges with externally bonded reinforcement. In: *Proceedings of the second international conference on bridge management* (pp. 651–659). Surrey, London: Thomas Telford.

Sika (1997). Technology and concepts for structural strengthening.

Sumon, S. K., & Ricketts, N. (1995). Repair and strengthening of masonry arch bridges. *Arch bridges* (pp. 501–508). London: Thomas Telford.

Taljston, B. (1997). Strengthening of concrete structures for steel with bonded CFRP-fabrics. *Recent advances in bridge engineering* (pp. 67–74). Zurich, Switzerland: EMPRA.

Troitsky, M. S. (1990). *Prestressed steel bridges – theory and design*. Van Nostrand Reinhold, Chap. 11, 373.

Van Dalen, K., Taylor, R. J., & Batchelor, B. D. (1984). *Transverse prestressing of prestressed laminated wood bridge decks*, Final report, IABSE 12th Congress, Vancouver, Canada, 3–7 Sept., p. 1118.

Wall, M. A., Gardner, G. D., & Posner, C. D. (1996). Strengthening of Langleybury Lane Bridge, Hertfordshire. In: *Proceedings of the third international conference on bridge management* (pp. 93–100). Surrey: Spon.

Walser, R., & Steiner, W. (1997). Strengthening a bridge with advanced materials, IABSE. *Structural Engineering International, 2*(97), 110–112.

Welch, P. J. (1995). Renovation of masonry bridges. *Arch Bridges* (pp. 601–610). London: Thomas Telford.

Xanathos, P. P. (1996). Bridge strengthening and rehabilitation. USA: Prentice Hall PTR. p. 392.

Yeoell, D. (1995). Westminster Bridge. In: *Proceedings of the third surveyor conference and exhibition – testing times.* London.

Yeoell, D., & Munson, S. R. (1993). The assessment, load testing and strengthening of Westminster Bridge, In: *Proceedings of the second international conference on bridge management* (pp. 307–315). Surrey, London: Thomas Telford.

FURTHER READING

Hollaway, L. & Leeming, M. B. *Strengthening of reinforced concrete structures using externally bonded FRP components in structural and civil engineering*, Woodhead Publishing Ltd., Cambridge.

Proceedings of the first international conference on bridge management (1990) Guildford, Surrey, UK, Elsevier.

Proceedings of the second international conference on bridge management (1993) Guildford, Surrey, UK, Thomas Telford, London.

Proceedings of the third international conference on bridge management (1996) Guildford, Surrey, UK, E & F N Spon.

CASE STUDIES

Bavetta, F., Cardinale, G., & Spinelli, P. (1996). Reinforced concrete bridge over Arno River: structural identification and repair. In: *Proceedings of the seventh international conference on structural faults and repair* (pp. 521–530). Edinburgh: Engineering Technics Press.

New Civil Engineer. (1998). Forth Road Bridge, upgrading of main towers, 28 January, pp. 16–17.

Hayward, D. (1998). Flow stopper. *New Civil Engineer* 5 February, 17–20.

Kelly, G., & Aston, C. (1996). Remedial strengthening works to Chappell Drive Bridge, Doncaster. *Construction Repair, 10*(1), 10–13.

Kotze, P. R. (1996). Rehabilitation of a historic bridge over the Sand River near Virginia, South Africa. In: *Proceedings of the third international conference on bridge management* (pp. 71–78). Surrey: E & F N Spon.

Mohindra, O. D., Narayan, D., Ram, A., & Gupta, V. D. (1994). Rehabilitation plan of Nizamuddin Bridge, *International seminar on failures, rehabilitation and retrofitting of bridges & aqueducts* (pp. 97–127), Bombay, India.

Parker, D. (1998). Hidden extras. *New Civil Engineer* 22 January, 16–17.

Structural Engineering International (SEI) 9(4) November. (Volume devoted to the use of FRP in design and strengthening.)

APPENDIX 7A

Strengthening of the Greatham Bridge near Pulborough, W. Sussex constructed in 1294 and pier repaired in 1987 with masonry filled with concrete and tied to foundation with stainless steel dowels.

FIGURE 7A-1

FIGURE 7A-2

CHAPTER 8

Bridge Modification

CONTENTS

8.1 INTRODUCTION

Bridges may need modification to accommodate widened motorways, to prevent deterioration from the environment, to distribute secondary (traction and braking) traffic and seismic forces, or strengthened in order to carry an increased volume and mass of traffic. In some cases, a bridge may have to be completely replaced. (Strengthening has been dealt with in Chapter 7; this chapter deals with the other issues.)

The worldwide economic growth that has taken place in the last 20 years or so has brought with it a concomitant growth in all forms of transportation (road, rail and air). This increase in mobility generally, and road traffic in particular, has been notoriously difficult to forecast and the increased traffic has put existing transport systems under an intolerable strain. Where trunk roads pass through city centres the problem is very acute, causing long delays and a build-up of exhaust emissions. The only resort is to provide a by-pass. However, the environmental impact of new road schemes is generally seen to be negative, and the obstacles placed in front of relatively straightforward by-pass schemes can be both frustrating and expensive. The recent construction of the Newbury by-pass in the UK is a case in point (Cole, 1996).

Between cities, on the other hand, one way of supplying the extra capacity on motorways without causing so much of an outcry is to *widen the carriageway in regions of congestion*, thereby relieving the pressure. However, unless a thorough assessment has been made to establish the location and reasons for the pressure points, then such measures will only provide a temporary solution locally, and simply push the congestion further on down the road. In the UK, at the moment (2001), the current Road Programme includes a substantial amount of new and widened roads.

8.2 WIDENING OPTIONS

There are three basic procedures for widening a carriageway, namely: *symmetrical, asymmetrical and parallel*. They reflect, in part, the way in which the traffic is *managed* and can be illustrated by an example of the conversion of a dual three-lane carriageway to a dual four-lane carriageway.

8.2.1 Symmetrical

In this option each of the carriageways is widened equally as in Figure 8.1. The amount of extra land required (land-take) is minimal and could even be

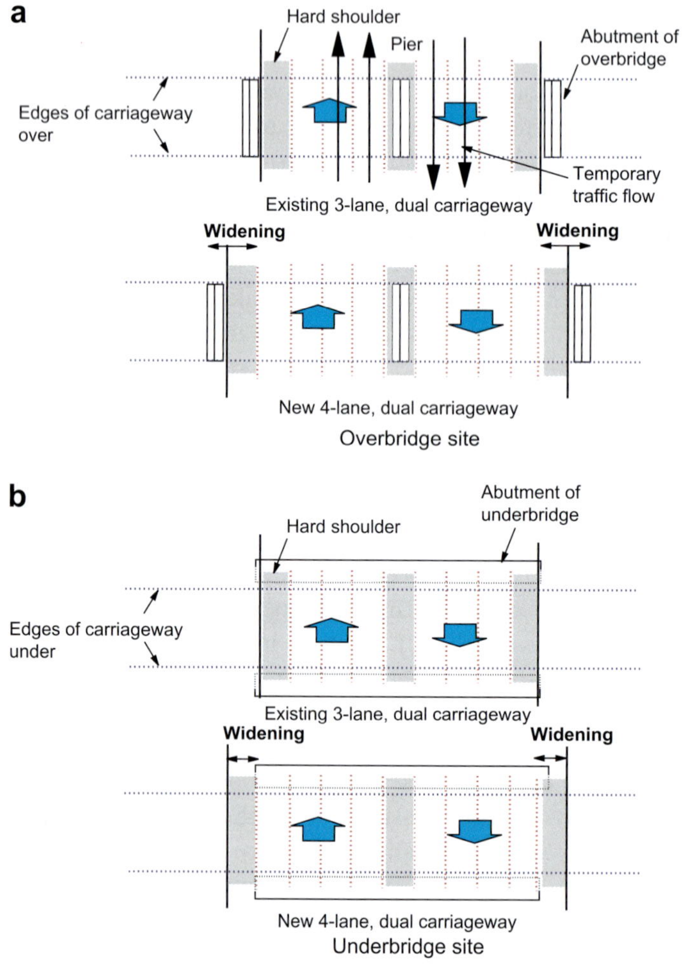

FIGURE 8.1 *Symmetrical widening.*

zero. The existing hard shoulders are upgraded and new hard shoulders built alongside.

To avoid demolition of the existing abutments of an *over bridge*, it may be possible to make the new hard shoulder discontinuous (or pinched) at the bridge site. This can, however, compromise user safety because a vehicle breakdown would cause a serious obstruction and reduce the four-lane flow to a three-lane one. Unless there is a margin of about 1 m between the edge of the new running lane and the abutment, it would be better to demolish the bridge and rebuild, even though that means a higher initial build cost. The existing central pier could remain in service for construction of a new over bridge on-line. The existing abutments could also remain in service, provided that the pinched hard shoulder solution is feasible.

Construction could prove difficult where the hard shoulder is continuous as there are essentially two very narrow sites – one each side of the carriageway. Traffic management is straightforward and only one arrangement is necessary for constructing both sides simultaneously in both the *over bridge and under bridge* situations. For separate construction each end, the traffic can be arranged in a 2 + 2 contraflow using three lanes of one carriageway plus the hard shoulder. The (simultaneous) construction sequence would be:

- Construct new hard shoulders.
- Convert existing hard shoulders to running lanes.
- Build abutments.
- Demolish and rebuild bridge.

At an *under bridge* site where the hard shoulders are *discontinuous*, the innermost lanes are very close to the edge of the deck and may need to be checked for strength. Where the hard shoulders are continuous, each side of the bridge will need to be widened.

8.2.2 Asymmetrical

This solution is generally favoured where one side of the carriageway is adjacent to an urban area. In this option one of the new carriageways is made coincident with the existing carriageway as shown in Figure 8.2, and the central reservation (median strip) is converted to a running lane. The carriageway is widened at one end only. The motorway will have effectively shifted sideways by two lanes. The existing central reservation would have to be upgraded to a running lane, and one pier could remain in service provided the new bridge stays on-line.

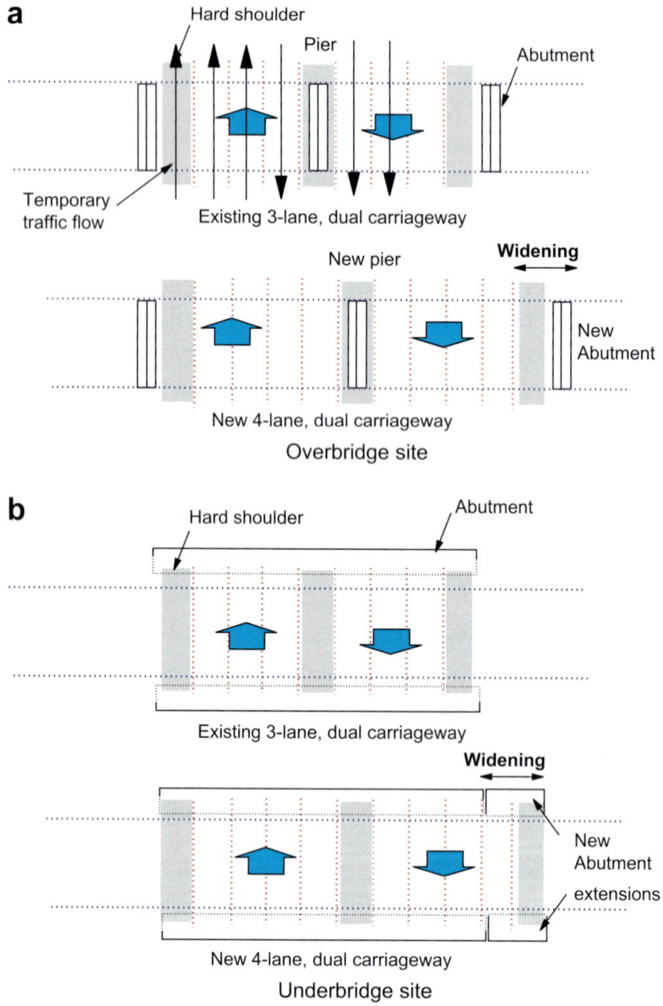

FIGURE 8.2 *Asymmetrical widening.*

In the *over bridge* situation, traffic management would be in two phases; the first involves the use of the hard shoulder and three lanes of one carriageway (with one lane in contraflow) and two lanes of the other carriageway as indicated in Figure 8.2(a). During this phase a new running lane, a hard shoulder and a new abutment are constructed. In the second phase when the existing central reserve is converted to a running lane, and a new central reserve pier are constructed, the traffic could be diverted to two running lanes plus the hard shoulders on each carriageway. If the contractor elects to complete the whole of the works from the new abutment side then it will only be possible to use two lanes of traffic each way on the other carriageway.

The sequence of operation would be:

- Demolish existing bridge, including existing pier.
- Build new running lane, hard shoulder and abutment.
- Convert existing central reserve to running lane, convert existing running lane to central reserve and build new pier.
- Construct new bridge.

The *under bridge* situation is easier in part, but may require the closure of the road under whilst the new abutments and deck are built. The traffic can be switched to two lanes and a hard shoulder on the permanent abutment side with one lane in contraflow, and two lanes adjacent to the pier in the other carriageway as shown in Figure 8.2(a).

8.2.3 Parallel

In this solution a completely new wider carriageway is built alongside the existing motorway, one of the existing carriageways is widened, and the other existing carriageway is converted to a single direction carriageway for maintenance, collection or distribution. It has the *disadvantage* that extra land has to be purchased for the new carriageway and the *advantage* that there is virtually no interference with traffic during construction. When the new carriageway is completed, all traffic can be diverted on to it, and work on the existing carriageway can commence. The principle is shown in Figure 8.3.

Traffic management is relatively simple (see Figure 8.3(a)) using the three lanes and hard shoulder of the carriageway adjacent to abutment B, and two lanes of the other existing carriageway.

The construction sequence is as follows:

- Convert abutment A to a pier (preferred option).
- Construct new carriageway and abutment.
- Construct new deck.

At an *under bridge* site where one hard shoulder has been upgraded to a lane very close to the edge of the deck, the existing deck may need to be checked for strength.

FIGURE 8.3
Parallel widening.

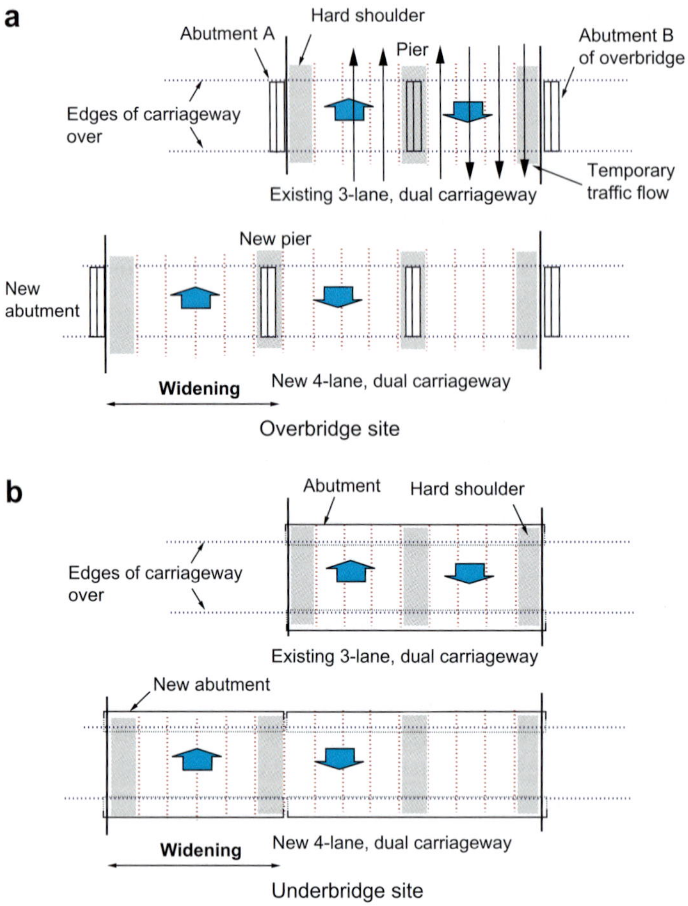

8.2.4 Alternative method

Each of the above methods illustrates that *management of the existing traffic dominates all other activities* and must be considered in detail before any works begin. An alternative to the traffic management described in the previous three sections has been proposed by Upstone (1996) which, so it is claimed, allows continuous uninterrupted movement of motorway traffic.

This is accomplished by directing motorway traffic above an over bridge by means of a temporary viaduct using, for example, Mabey and Johnson's Quick Bridge System (Mabey & Johnson, 1992), and re-routing the minor road traffic via slip roads (entry and exist lanes) and temporary roads at ground level.

The method is claimed to be cheaper than the traditional way and eight advantages of using the system are listed. As far as the author knows the method has not been tried in the UK.

8.3 BRIDGE OPTIONS

8.3.1 Over bridges

At existing *over bridge* locations, the improved capacity of a motorway is, of course, restricted and so works have to be put in hand to allow the widened road to continue under the bridge. The configuration of the *existing* bridge plays a major part in accommodating the widening process. Bridges built over motorways in the 1960s and 1970s were generally of the 'closed' abutment type, whilst many later designs were of the 'open' abutment type with the deck supported on bank seats at the ends (see Figure 8.4).

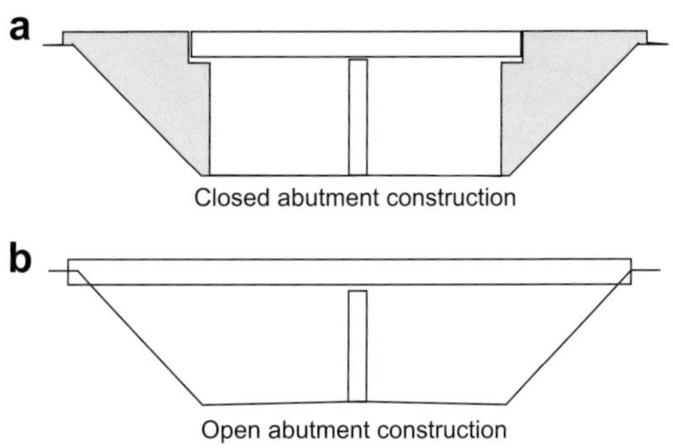

FIGURE 8.4 *Motorway bridge configurations.*

Should widening be required in the future at a closed abutment situation, then, clearly, more expensive works are necessary as both abutments would require demolition and rebuilding along with two additional piers and two outer spans as indicated in Figure 8.5(d). To avoid expensive demolition and replacement of bridges in the future it is better to accept an initially higher capital cost where possible by building a bridge with open abutments. This is the most amenable to motorway widening since it is possible to utilise the space between the bank seats and the toes of the embankment for extra lanes as in Figure 8.5(b). This requires considerable 'land-take' (Figure 8.5(a)) which may be expensive, and also the provision of a hidden abutment and wing walls (rather than a bank seat) so that widening can be carried out expeditiously in the future involving only reshaping of the embankments and

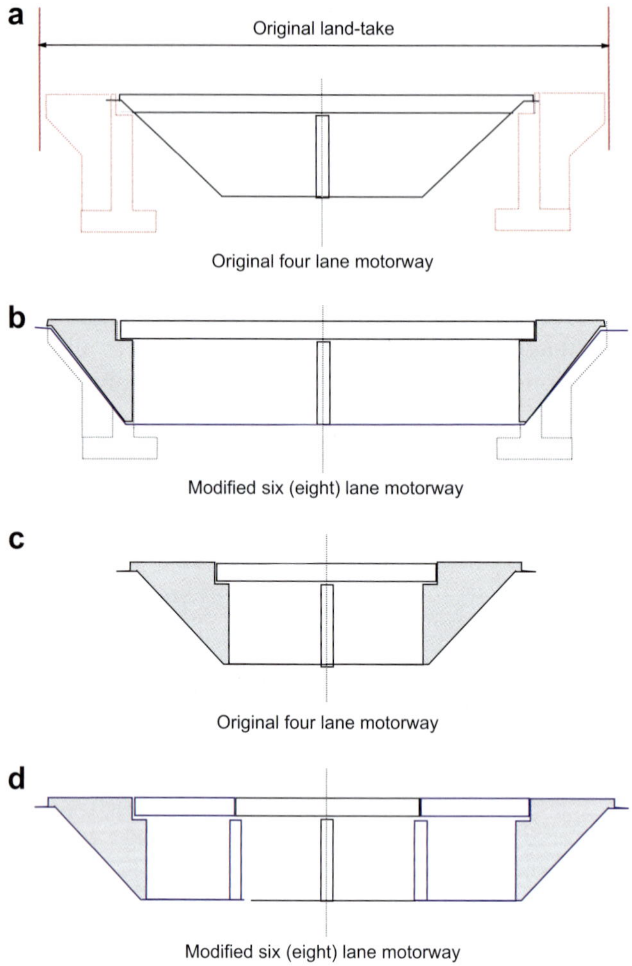

FIGURE 8.5 *Widening at over bridge locations with open and closed abutments.*

clearing of the once buried concrete abutments and wing walls. In planning for future widening, a whole life cost analysis (see Chapter 9) should be performed for the closed option and the open option (with buried abutments) as shown in Figure 8.5.

8.3.2 Underbridges

At *under bridge* locations the amount of construction works for small to medium spans is less and generally involves only the construction of abutment extensions and relatively narrow widths of new deck tied to the edge of the existing deck to provide some transverse distribution of the live load. This will depend on the original construction and the proposed new construction, which, if possible, should be the same. For beam and slab for example, this may be in the form of an in situ concrete stitch slab by lapping the reinforcement or provision of a shear joint utilising stainless steel dowels grouted into drilled holes in the existing deck (see Figure 8.6). For solid slabs and voided slabs a simple dowelled connection would suffice. If the carriageway is to be doubled in width then a joint is not necessary, and so the two bridges act independently.

Part of the forward planning must be concerned with what type of bridge deck would be appropriate in the new *over bridge* circumstances. New spans will probably be between 25 m and 45 m and so the choice would lie between an in situ concrete (voided) slab, or beam and slab (steel beams and concrete slab, or prestressed concrete beams and concrete) as in Figure 8.7. Experience has shown that new over bridges need to be built as fast as possible (Paul & Coke, 1994) and so the shallow decks utilising prefabricated sections are

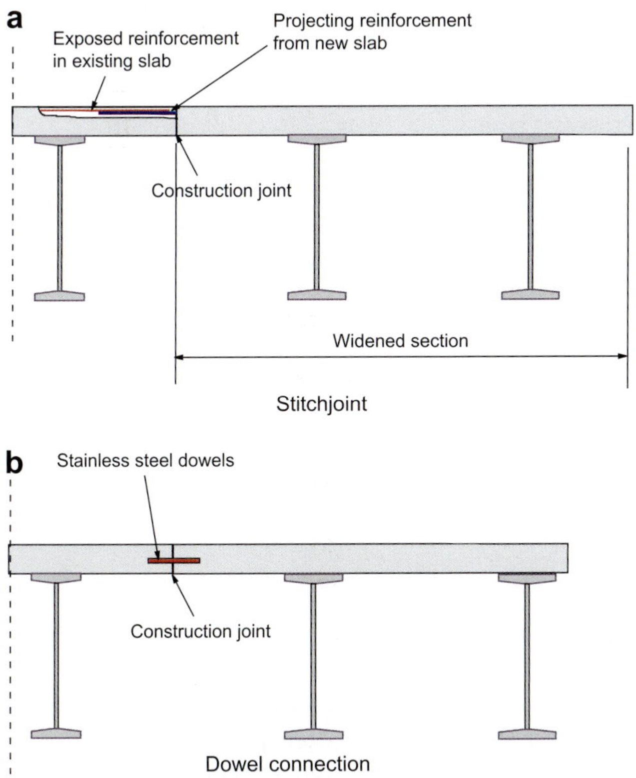

FIGURE 8.6 *Connection of new to old.*

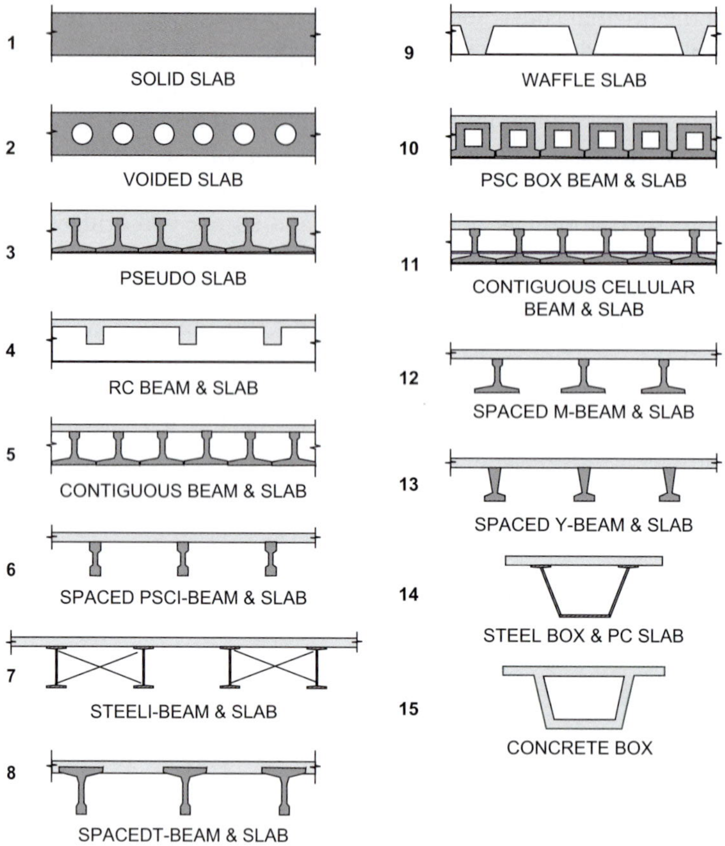

FIGURE 8.7
Bridge deck types.

generally preferred. For aesthetic reasons, it is desirable to match the original bridge type, or at least give an impression of similar construction by providing the same depth and side elevation details.

Spaced beam and slab construction is probably the most popular form of construction in the world, and the main decision is whether to opt for steel or concrete beams.

8.3.3 Steel beams

One of the great advantages of using steel beams is that they are readily manufactured in standard sections and lengths (up to 25 m in the UK) and are relatively light, which means that modest lifting equipment can be used to offload and erect them. With modern painting systems and weathering steel, they can be made very durable, but any corrosion (if it occurs) is readily seen at an early stage and can be rectified.

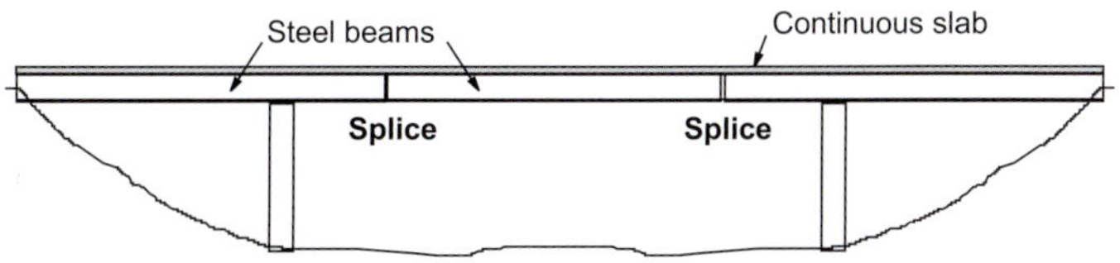

FIGURE 8.8 *Cantilever construction using steel beams.*

For longer spans, plate girders can be shop fabricated, and the lower flange can be profiled to form haunches (either straight or curved) to reflect the variation in bending moments along a continuous bridge (that is, deeper sections at internal supports and shallower depths in the mid-span regions). This is more aesthetically pleasing than parallel flanges, and adds interest to an otherwise banal appearance.

Long central spans can be provided by the use of cantilever construction, with splices near the points of contraflexure, thus obviating the need for costly intermediate piers (see Figure 8.8). In this case, negative end reactions may be present and the beams will need to be anchored. If headroom is a problem then the depth of construction can be reduced by introducing continuity over intermediate piers, or to employ half-through girder construction using either plate girders or trusses. Erection of the beams can be undertaken during a night-time possession. They are pre-assembled on site as braced pairs to ensure stability prior to constructing the deck slab, and then lifted into place by mobile (or tracked) crane. For this reason it is usual to have a finished bridge with an even number of beams.

To speed up erection, precast, transversely prestressed concrete deck panels may be used in lieu of the traditional in situ cast concrete on permanent formwork. Pockets are cast in the panels to match clusters of stud connectors pre-welded to the top flanges of the beams and the slabs lifted into place onto preformed bedding strips. Alternatively, the studs may be welded into place after the panels have been placed in position by *sliding* as shown in Figure 8.9 or by lifting. All joints between panels can be formed with epoxy resin, and the panels post-tensioned longitudinally if required. Finally the stud pockets are filled with an epoxy resin compound. The precast panels can be *full width* with the edge upstands in place and are ideal where a deck wide enough to carry a two or three lane carriageway is supported on two steel girders (or the top flanges of an 'open' steel box).

8.3.4 Concrete beams

Precast concrete (PC) beams have been developing in the UK since 1943 (Taylor, 1991, 1998; Discussion, 1999) and the latest are in the form of the Y

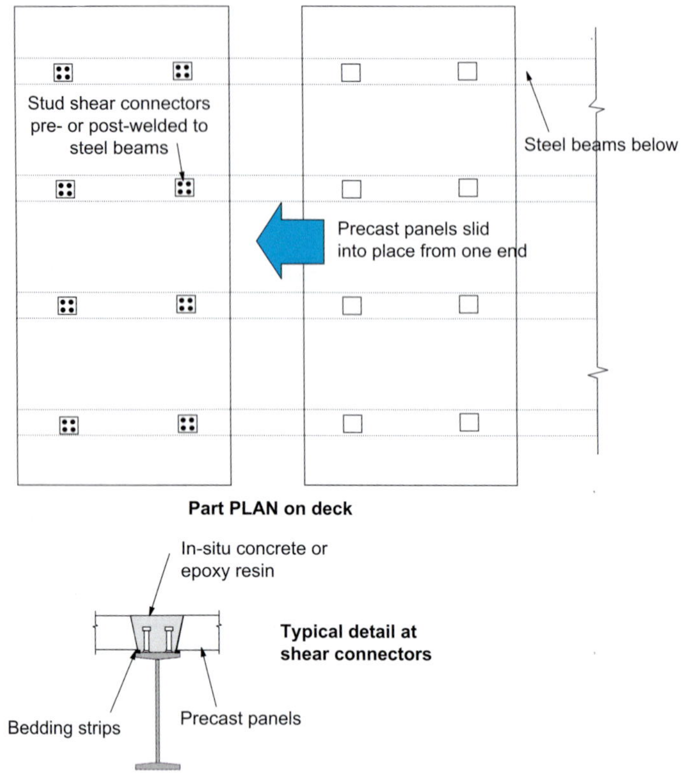

FIGURE 8.9 *Use of precast deck panels in composite construction.*

Beam (introduced in 1990), the SY Beam (introduced in 1992) and the TY Beam (introduced in 1994), with the SY able to span up to 45 m (see Figure 8.10). In the USA, PC beams were developed between 1958 and 1980, and the current ones are shown in Figure 8.10. The USA beams can be post-tensioned on site to increase the span range.

Traditionally, many multi-span bridges using PC beams have been constructed as a series of simply supported spans with bearings required at the ends of each beam. This is no longer desirable due to the corrosion caused to the expansion joints, the bearings, crossheads and piers by the infiltration of water through the joints, not to mention the unsightly appearance left on the concrete by white streaks of salts which have been leached out of the concrete by water.

Continuity can be introduced in a number of ways to eliminate intermediate joints, and also to increase both the span range and the span/depth ratio. Two well-proven methods (Pritchard, 1992) are shown in Figure 8.11. Type 1 is preferred in the UK and type 2 in the USA. For a type 1 detail with a width of 5 m and a 1 m embedment, an intermediate span can be increased by 3 m.

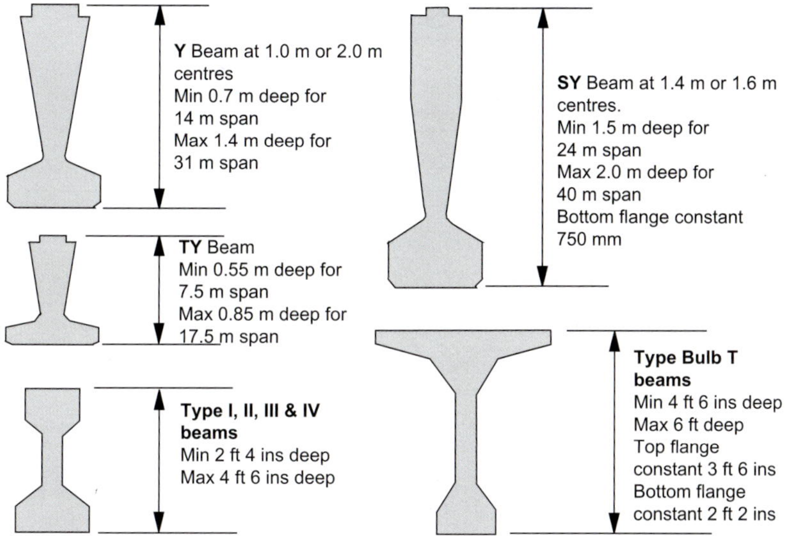

FIGURE 8.10
Y, SY and TY beams, and Types I–IV and Bulb T Beams.

Unlike steel beams, concrete beams are not amenable to decks curved in plan, as the beam lengths have to be varied across the width, with the curvature provided by the expedient of varying the span of the cantilever slab. In any event this is not a major problem with *bridge replacement* as most bridges crossing motorways are straight.

Concrete beams can be both *long and heavy* and, according to the Road Vehicles (Construction and Use) Regulations (1998) in the UK, the maximum allowable transportable length is 27.5 m. Above that limit a special order is required from the Department of Transport. For weights over 40 t the C&U regulations specify the number and configuration of axles required.

In spite of these limitations, concrete beams are, nevertheless, very popular and compete well with steel beams in terms of cost and durability. They are robust and have a high torsional resistance, thus providing good transverse load distribution properties across the width of the deck. They can also be speedily erected, are virtually maintenance-free and there is good access for inspection.

8.3.5 Aluminium system for bridge deck replacement

Aluminium has certain advantages over both steel and reinforced concrete in that it has:

- A high strength-to-weight ratio.
- A high resistance to corrosion.

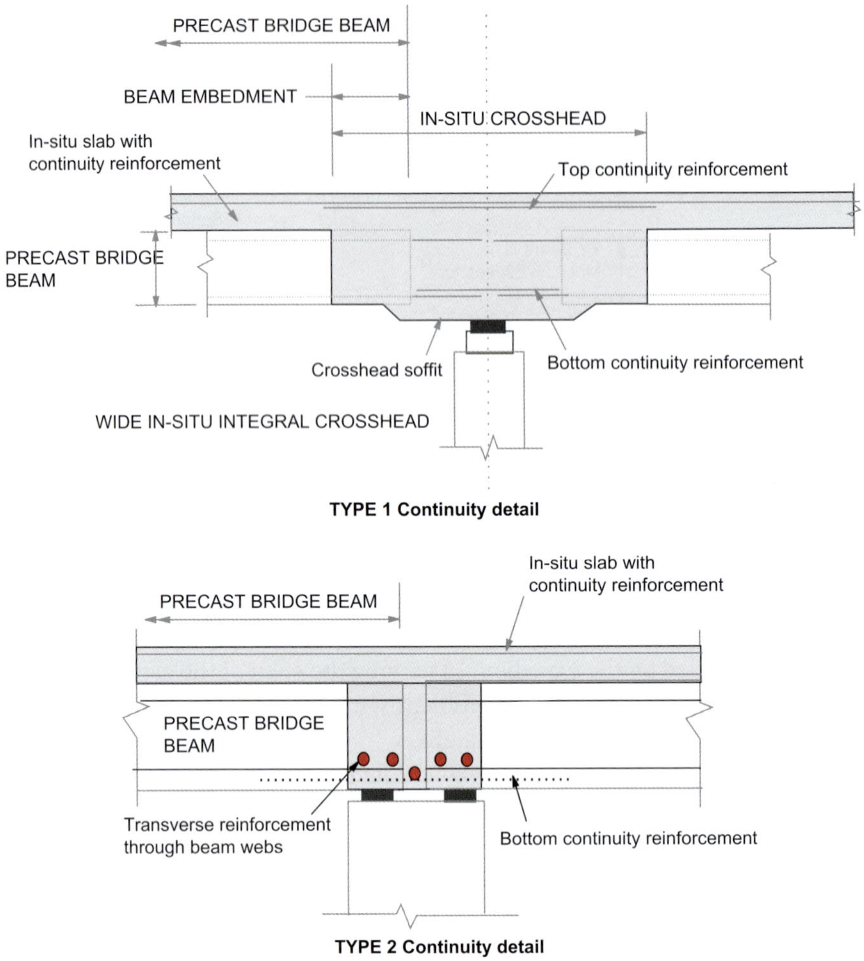

FIGURE 8.11
Deck continuity.

- Properties which enable it to be readily fabricated.
- The ability to be extruded into many different shapes.
- Resistance to salts (although salts do affect the appearance).

Its first use in the USA was to *redeck* the Smithfield Bridge, Pittsburgh, Pennsylvania in 1933 'with a lightweight aluminium deck plate with stringers and asphalt wearing surface to increase the live load capacity to H20 by reducing the dead load' (Trinidad, 1993). Later in 1967 this was replaced by a lighter weight orthotropic aluminium deck with a polyester and sand wearing surface to further increase its live load carrying capacity. Since then seven aluminium bridges have been constructed in the USA and all were still in service in 1993 (Trinidad, 1993).

The exploitation of aluminium as a potential material for the upgrading of steel bridges has been embraced and applied to many failing bridges in Sweden (Svensson & Pettersson, 1990). A lightweight truss-like system has been developed comprising an orthotropic plate built up from hollow aluminium extrusions, which are fitted together by a tongue and groove in the top flange as indicated in Figure 8.12. Each section can rotate freely whilst load distribution is by means of shear at the joints and Svensson and Pettersson (1990) claim that 'concentrated loads will be carried by at least seven extrusions'. Once the old concrete deck has been removed, the pre-assembled aluminium deck is secured to the existing steel structure via the bottom flange. The surface of the deck is covered with an acrylic-based material called Acrydur, a paving material that has proven successful in Sweden for many years. The combined deck system is only about one-tenth of the weight of the conventional concrete deck, a reduction which allows the live load to increase without the need to upgrade the foundations.

The system has been fully tested, both numerically and physically, and has been proven to function satisfactorily in use. Care has to be taken to prevent contact between the faces of steel and aluminium components to ensure that no galvanic corrosion of the aluminium takes place.

The cost of fabrication and erection of the aluminium deck is similar (in Sweden) to a traditional reinforced concrete deck, but there are a number of significant advantages (Svensson & Pettersson, 1990):

- Design time is minimal compared to concrete.
- Construction time drastically reduced.
- The reduced weight means that the existing (unmodified) foundations are satisfactory.
- Maintenance costs can be kept to a minimum because the Acrydur paving is very hard wearing, and the aluminium has a high resistance to corrosion.

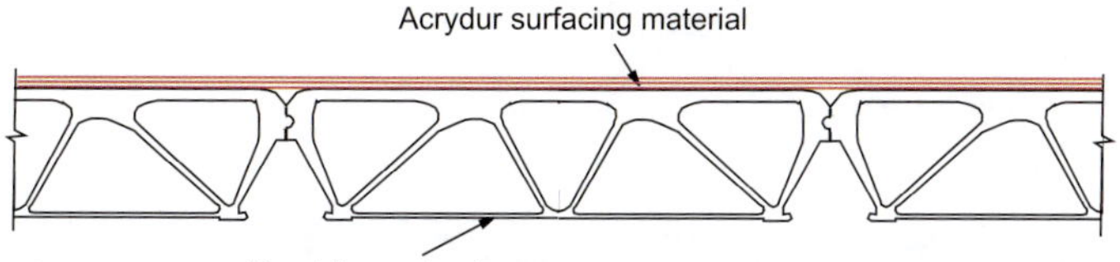

FIGURE 8.12 *Section through aluminium bridge deck extrusion (after Svensson and Pettersson, 1990).*

8.3.6 FRP system for bridge deck replacement

In the USA vigorous research has been conducted over many years in the development of bridge decks using composite materials (Henriquez, Plecnik, Munley, & Cooper, 1997). An initial study carried out on fibreglass-reinforced plastic (FRP) bridge decks highlighted the advantages over traditional systems.

- Relatively lightweight, which means that the live load carrying capacity can be maintained or increased.
- The modular units can be fabricated off-site, delivered and quickly assembled, which represents a potential cost saving in construction.
- Composites will resist a wide range of corrosive environments.

Five different configurations were considered before finally selecting the X-shape shown in Figure 8.13, which also shows the sequential build-up process.

FIGURE 8.13
Steps in the build-up of the FRP units.

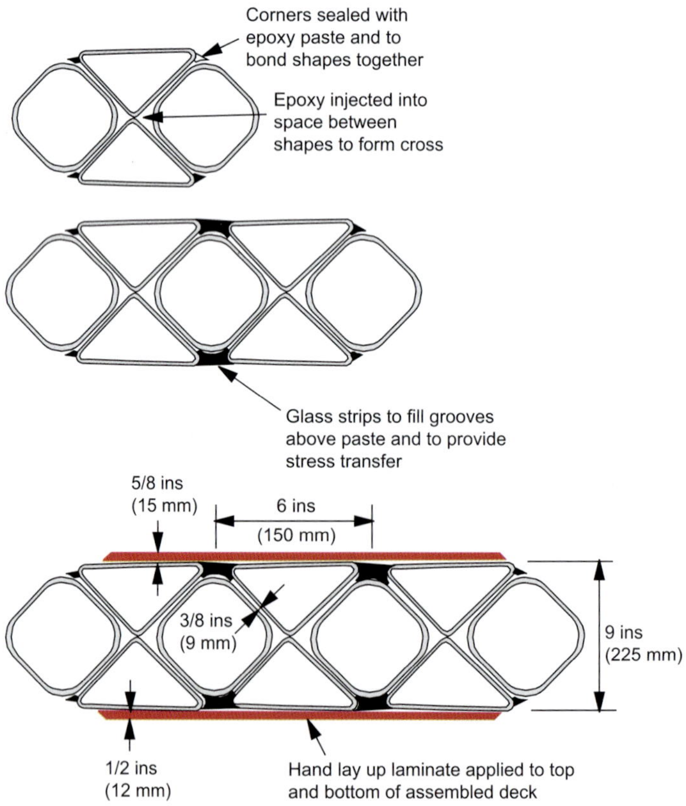

After exhaustive analytical work non-destructive load tests were carried out to study the effects of fatigue. It was found that only minor damage was observed after several million cycles, and it was confirmed that *stiffness and fatigue* (rather than stress and strength) control FRP deck design. Ways of connecting the deck to longitudinal steel girders have been developed and a suitable wearing surface to provide skid resistance, protection from chemical attack, abrasion, freeze-thaw cycles and weathering has been achieved.

The deck is currently undergoing field testing in Virginia. If successful, and economically viable, the FRP system could provide a satisfactory solution for the replacement of severely damaged timber or concrete decks and also be employed in new designs.

8.3.7 Modular FRP composite deck system for bridge deck replacement

Another FRP contender for replacing damaged timber or concrete decks is a *modular* system being developed in the USA under the auspices of the US Army Corps of Engineers (Lopez-Anido & Ganga Rao, 1997) and extensive research and development has been underway at West Virginia University to establish the most efficient shape and fibre architecture.

The chosen pultruded structural shapes and dimensions of the FRP deck modules are shown in Figure 8.14. The modules are bonded together and placed transversely to the direction of the traffic on longitudinal beams which can be spaced up to 2.74 m (9 ft) apart. To conform to the average thickness of concrete decks in the State, the Virginia Department of Transport have recommended a constant depth of 203 mm (8 ins) so that the FRP system can be used for concrete deck replacement. They have used the system on several bridges in Virginia and, so far, they have performed 'excellently' (Rao & Craigo, 1999).

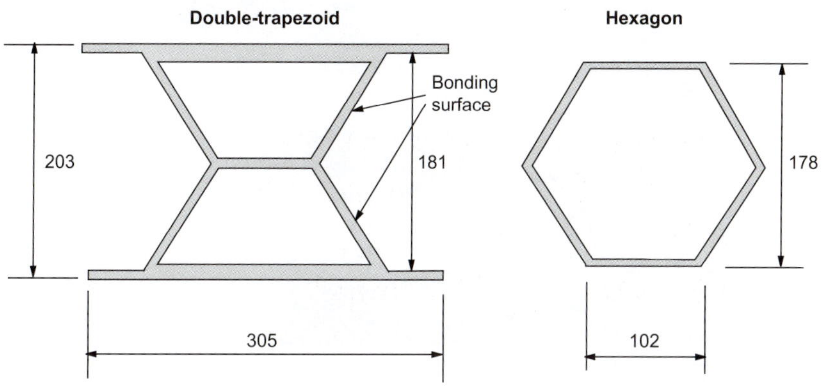

FIGURE 8.14
Modular FRP system.

The composite material comprises a group of untwisted parallel strands called *roving* and triaxial stitched fabrics with binderless chopped strand mats (CSM) in a matrix of vinyl ester 580 N resin. This material is pultruded to form the deck modules, which are later bonded together using an adhesive applied at the fabrication plant. Panels to the order of 10 m (32 ft) square can be fabricated in this way and easily transported to site (see Figure 8.15). Composite action between the FRP and the supporting beams is made possible by a combination of blind bolts (pioneered by Lopez-Anido and Ganga Rao) and a two-part polyurethane adhesive. Finally, the surfacing is applied in the form of a thin (1 cm) polymer concrete overlay. Although the initial costs are higher than for concrete decks, the authors quote the following significant advantages.

- Reduction in erection time and seasonal costs due to prefabrication.
- High load ratings due to high strength-to-weight ratio of the FRP deck which has 6 to 7 times the load capacity of a reinforced concrete deck, and is 80% lighter.
- Long service life and reduction in maintenance costs due to fatigue and corrosion-resistant properties.

The system has been proven theoretically, and is currently being tested in the field on two bridges to designs based on LRFD principles (AASHTO, 1994). Since the FRP deck is half as stiff as concrete, the serviceability limit states usually control. Both bridges are instrumented with sensors to monitor their long-term performance. If the tests are successful it is hoped that guides to good practice will be produced to allow the design, specification, fabrication and construction of this type of FRP deck.

FIGURE 8.15 *Installation of FRP modules (Rao and Craigo, 1999).*

8.3.8 Exodermic bridge deck panels for bridge deck replacement

A system that has been in use since the 1930s in the USA is the exodermic bridge panel (Bettigole, 2000) which is now installed on 40 bridges in the USA. It consists of a lightweight (245 kg/m^2) open steel grid upon which are laid galvanised steel sheets which act as shuttering for a reinforced concrete slab up to a maximum of 115 mm thickness, and is able to span up to 4.4 m (see Figure 8.16).

By virtue of its lightweight (usually lighter than the deck it is replacing) it allows for an increase in live load. The steel grid and the concrete act as a composite section, and it is claimed that the panels can resist both positive (sagging) and negative (hogging) bending. The panels themselves are made composite with the bridge by welding shear studs to the top of the steelwork through blockouts in the exodermic panels, after which rapid hardening concrete is poured. Once the whole deck is completed, the surface of the concrete is 'diamond ground for a smooth riding surface, and overlaid with a 9.5 mm epoxy concrete riding surface' (Bettigole, 2000).

Rapid replacement is possible, often working overnight to minimise traffic disruption, and the work progresses in bays so that, at the end of each working day, there is a transverse construction joint across the bridge (the new abutting the old). This has the advantage that *all* lanes can be opened in

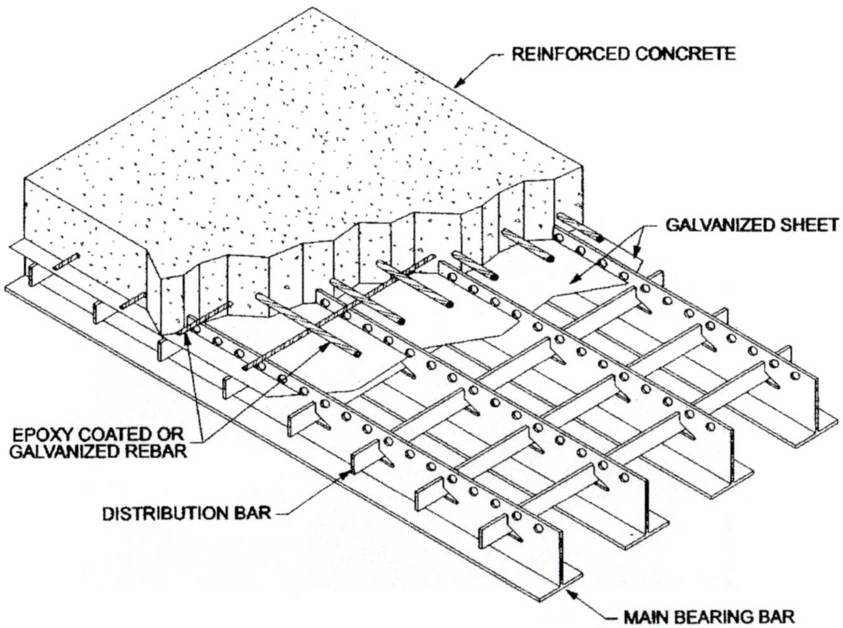

FIGURE 8.16
Exodermic bridge deck panels.

the early mornings, and the only discomfort for drivers is the slight 'bump' as vehicles travel over the joint.

The design of the panels was revised in the early 1990s to improve the means of shear connection and to simplify some of the details to reduce fabrication costs. The new design has been thoroughly tested under both static and dynamic loads and continues to perform well.

8.4 METHODS OF REPLACEMENT AND CONSTRUCTION

8.4.1 Lifting

Lifting is the most common way of erecting and is carried out by the use of inclined slings directly connected to the beam or *vertical* slings via a steel lifting beam. It can be carried side-on to the bridge or from the ends using a single or double crane arrangement depending upon the span, size of beam and available craneage. The compression stresses in the beam are calculated under all conditions and checks carried out to ensure that buckling does not occur. The lifting points are designed to ensure that local crushing or tension failure does not occur. Swinging is controlled by the use of tag lines attached to the beams (Figure 8.17 illustrates the use of slings). Stratford and Burgoyne (1999) have examined in detail the stability of precast concrete beams during transportation, erection and while simply supported. They have shown that lateral or lateral–torsional buckling under self weight conditions is the most critical case, and that, once incorporated into the structure, buckling is unlikely to occur.

The condition of the ground is important and must be assessed for strength and stability, and, if required, stillage or timber baulks are provided

FIGURE 8.17 *Lifting of beams for a replacement bridge.*

to ensure that the crane remains stable. For safety reasons the operational characteristics of the crane, that is, its lifting radius and weight restrictions, must be adhered to at all times when working out the erection sequence. If a double (tandem) crane arrangement is used (see Figure 8.17) then one crane should be capable of carrying 70% of the weight and precautions taken to ensure that swinging does not occur. In this respect erection should be carried out under favourable weather conditions, that is, *little or no wind*.

Steel beams are erected in pairs and, provided there is adequate cross-bracing, stability is assured. Concrete beams, however, are lifted singly and the first beam must be supported at its ends with inclined struts. Second and subsequent beams are braced across the tops at their ends.

Because of the relative lightness of steel beams it is often possible to construct the whole steel superstructure adjacent to the site and then lifted into place in one operation using a high capacity crane. With good management, a single span comprising either steel or concrete beams can be erected in an overnight possession with minimum disruption to traffic.

8.4.2 Side-by-side

As its name implies, this method of replacement involves constructing the new bridge alongside the old and re-routing the traffic and services on to it (Champion, 1994). The old bridge is then demolished. The clear advantage of this method is that traffic flow is uninterrupted, but it does mean that a small change in the alignment of the roadway at each end has to be accommodated.

8.4.3 Lateral sliding

This is similar to 'side-by-side' construction except that the new bridge is slid into position to occupy the same alignment as the original. Figure 8.18 illustrates the method. For small bridges, the new deck can be erected on temporary steel supports with transverse rails or rollers under, and man-oeuvred into place by the use of hydraulic jacks by either pulling or pushing.

An example of this on a grand scale is the Oberkasseler cable-stayed bridge over the River Rhine at Dusseldorf (Saul, 1996). The original steel truss girder bridge comprised four 94.5 m spans with 12.5 m carriageway. The new bridge is a 20 m wide, single plane cable-stayed bridge with a symmetrical four cable harp arrangement. The main span is 257.75 m clear and the other side consists of five spans 51.55 m each. The single tower has a height of 100 m above the box girder superstructure. The new bridge was built 47.5 m upstream of the original and the traffic re-routed to it. The old bridge (deck and piers) was then demolished and new piers constructed on the same

FIGURE 8.18
Lateral sliding.

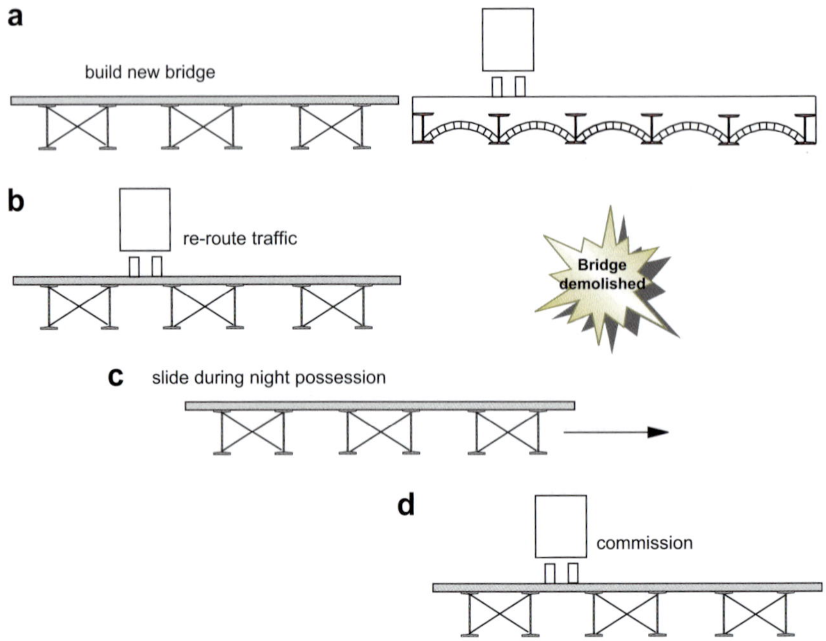

alignment. All of the supports were linked with sliding tracks and the whole bridge – weighing 12 700 t – was moved at 1 mm/sec using hydraulic jacks.

Another example is that of the replacement of the *temporary* Callender–Hamilton steel bridge carrying the A595 trunk road over Victoria Road near Whitehaven, Cumbria, UK, which was put in place in 1990 because of fears that the original bridge could not carry modern traffic loads (Leech, 1995). The bridge has a skew span of 23 m and a skew of 60°. The new composite steel and concrete bridge was built on temporary rigid steel supports alongside the existing carriageway without affecting the traffic flow of the trunk road. The slide path consisted of rolled steel channels (RSC) laid flat (web side down) and bolted to the supporting frame with the aim of using the sides of the RSC to guide the skates during sliding. The skates were covered with a Teflon derivative called 'Metoplast' which has a low friction property, and the channel was cleaned and coated with a 'high strength, water-resistant paste of molybdenum sulphide' (see Figure 8.18).

8.4.4 Slide and rotate

This special procedure was adopted to replace the three 30 m spans of the fish belly iron girders of the Schwarzbach Valley Bridge in Germany (Saul, 1996). An inspection in 1980 revealed severe corrosion and so it was decided to replace the 95-year-old bridge. Each section of the new prefabricated steel superstructure was transported *upside down* to the span it was replacing. The new and the old were then connected at their ends to giro-wheels and rotated in the *vertical plane* through 180° so that the old superstructure was upside down. The giro-wheels were removed and the new section was then lowered to its final position whilst the old superstructure was transported down the track to a breakers yard (see Figure 8.20).

Champion (1994) describes the rotation (in the horizontal plane) and sliding of the Green Street Bridge carrying the A282 over the M20 which was to be widened. The three span prestressed concrete voided slab bridge had a skew of 45° and, reducing this to 30° by sliding and rotating the bridge in plan onto a new set of abutments and piers, it was possible to gain the extra land required for the widened motorway.

8.4.5 Launching

This is a useful technique where a new over bridge is required to span across a widened motorway without interrupting the traffic flow. The new bridge is first constructed each side of the motorway with launching frames acting as temporary piers. Each section is then launched by the use of pulling jacks until in its final position when it is closed in an appropriate manner. A typical application is described by Champion (1994) where a new in situ post-tensioned concrete viaduct crosses over an existing widened motorway. The main stages are shown in Figure 8.20.

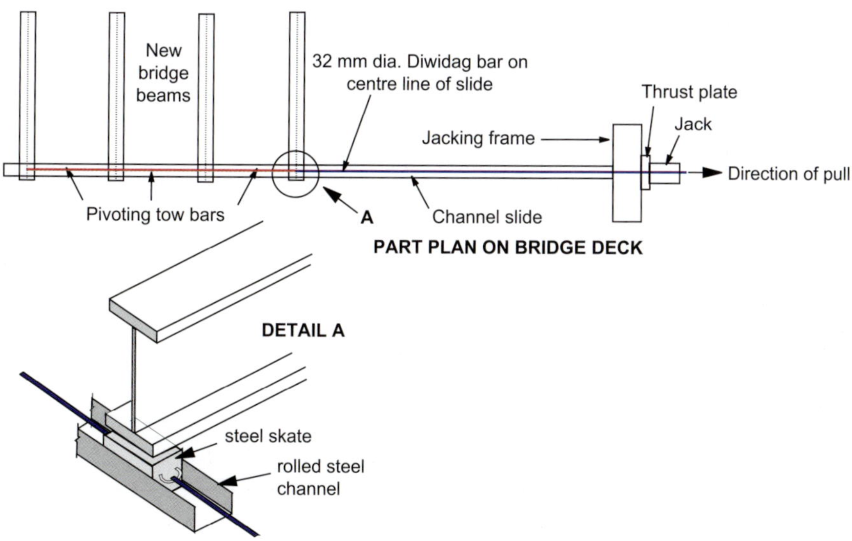

FIGURE 8.19
Sliding detail on Callender–Hamilton bridge replacement.

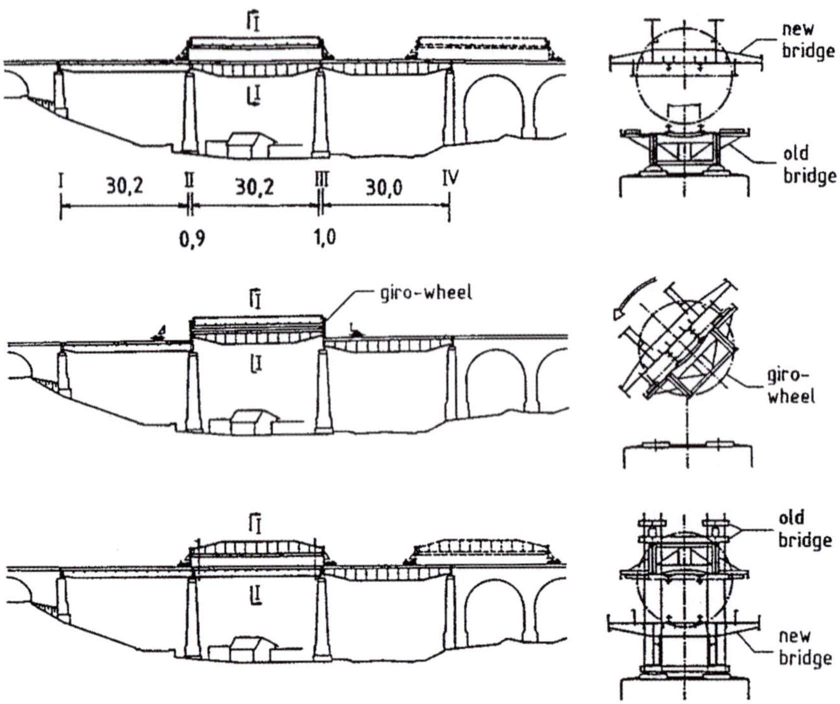

FIGURE 8.20
Construction procedure for the replacement of the Schwarzbach Valley Bridge at Wuppertal (Saul, 1996).

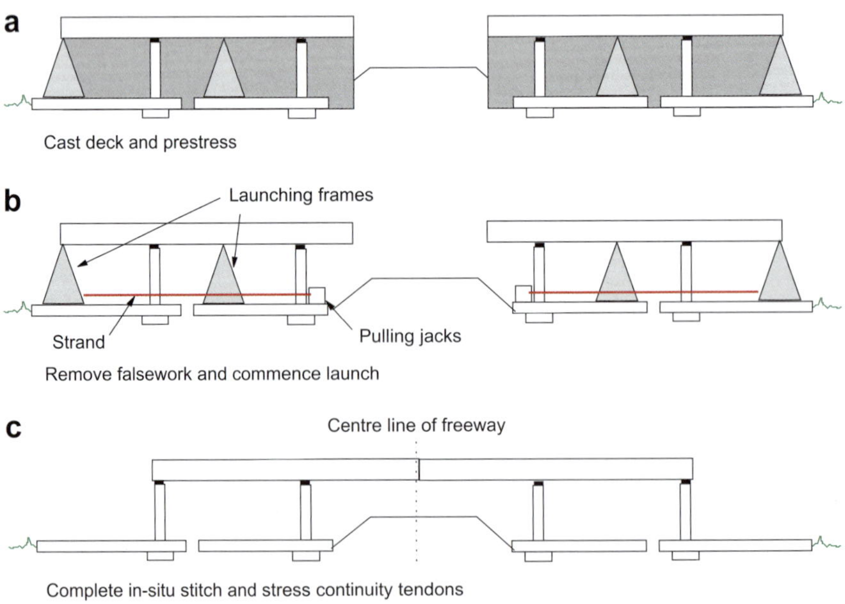

FIGURE 8.21
Launching method (after Champion, 1994).

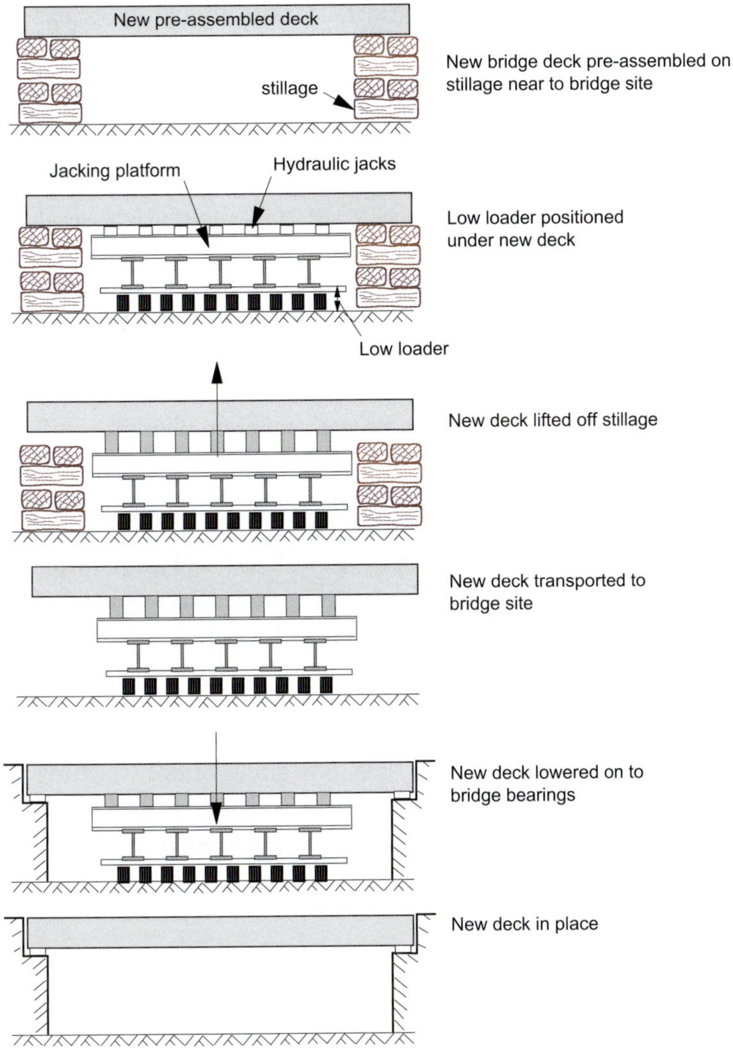

FIGURE 8.22
Low loader method.

Another example of an innovative use of launching for the removal and replacement of a bridge in South Africa is described by De Milander and Strydom (1996).

8.4.6 Low loader lifting

Lifting heavy sections using craneage in high winds of greater than 40 km/h is dangerous and time consuming. High winds can neither be 'predicted nor effectively managed' and more use is now made of low loaders, both to remove and replace unsafe bridge decks. Although more expensive, the low loader method can go ahead in practically any weather. The six steps in the procedure are shown in Figure 8.23. Two examples of this method are described by Hayward (1999).

The first was the removal and replacement of part of the badly corroded wrought iron deck of Mirfield station railway bridge, West Yorkshire, UK, built in circa 1900 (see Figure 8.18). The work was carried out in only 46 hours of a scheduled track possession period of 58 hours over the Christmas break of 1998 using a 22 axle, computer-controlled, low loader with a self-propelled bogie supplied by Econofreight. The damaged iron sections weighing more than 200 t were lifted out first and stored for cutting and removing. The new deck which was 25 m wide and 560 mm deep and comprising steel beams with precast concrete planking and an in situ deck was constructed on trestles 200 m from the site. The 850 t section was then lifted, transported to the site and installed on bearings atop the existing masonry abutments (see Figure 8.18).

The second example is the removal and replacement of a 100-year-old, 19 m single span bridge with a badly corroded 255 t steel deck. The bridge is situated at Gretna near the Scottish border and carries the west coast main line railway over a minor road. The contractor opted for the use of an 80 wheel loader supplied by Abnormal Load Engineering rather than a 1000 t mobile crane because of the exposed nature of the site and the consequent risk of high winds. The new steel box girder bridge weighing 430 t was pre-assembled on stillage in a nearby field, transported and lifted into position in 54 hours of a 60-hour scheduled track possession.

8.4.7 Pulling system

Another way of sliding is the *French Autofoncage or Autoripage pulling system* for the installation of bridges underneath roads and railways whereby a complete bridge superstructure (deck and supports) is built beside the existing bridge and then dragged into position over a reinforced concrete raft and through a pressurised film of Bentonite (Hughes, 1999). One really beneficial advantage of these methods is that traffic can be kept moving at all times – albeit, there may be a speed restriction in place. The old method of pre-tunnelling beneath embankments for the installation of the slide tracks is no longer necessary. (See Table 8.1.) The steps in the procedure for placing a bridge under an existing road or railway line are indicated in Figure 8.24.

A recent example was the construction of a rail track under bridge at Chalvey near Slough in the UK where the track was removed, the

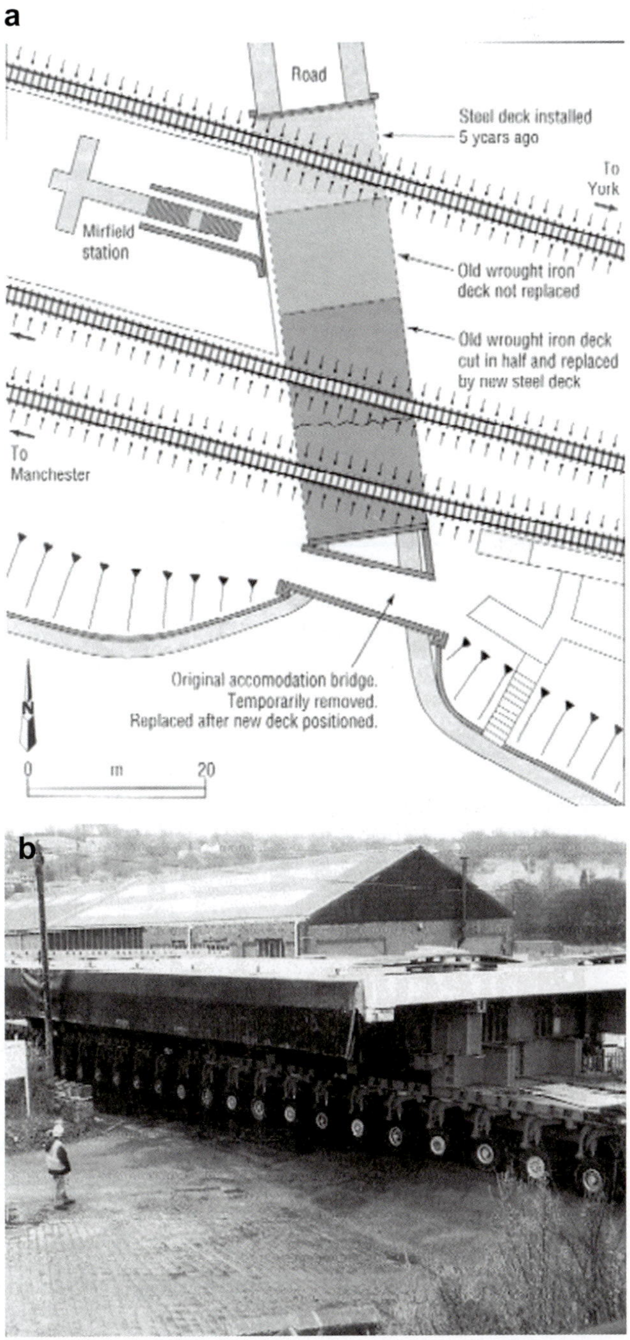

FIGURE 8.23 *(a) Plan of bridge and (b) Low loader moving 850 t deck (Hayward, 1999).*

Table 8.1 Construction Phases of the Autoripage System (after Hughes, 1999)

Phase 1	▪ Form the slope profile of the embankment with earth moving equipment. Prefabricate bottom slab, including two dead end anchorages to facilitate sliding of the structure
Phase 2	▪ Construct bridge structure on guide raft
Phase 3	▪ Fit pulling jacks (rams) in position and install pulling cables (strands)
Phase 4	▪ Remove existing railway tracks or roadway ▪ Excavate with earthmoving equipment ▪ Trial slide of the structure
Phase 5	▪ Begin backfilling sides and commence sliding of structure
Phase 6	▪ Complete side backfill operation position ▪ Lay new track or roadway ▪ Finishing for opening of structure

embankment excavated and the 4500 t bridge structure put into position during a 53-hour track possession. The contractor on the job maintained that the technique is faster and more accurate than existing systems (NCE, 1998). Another bridge – Ash Vale near Guildford in the UK – was a 2700 t reinforced concrete integral box-type bridge which was slid into position during a 52-hour railway possession and was only 5 mm out from its theoretical position (Hughes, 1999).

8.5 ENCLOSURES

An enclosure provides a sheltered environment from the weather, particularly rain. It is a means of bridge modification consisting of a durable skin secured to the sides and soffit of bridge superstructures, thereby keeping the weather out. The basic principle is very straightforward and is illustrated in Figure 8.25. Enclosures have almost exclusively been used around *composite* bridge superstructures, the steel beams of which, are particularly susceptible to *corrosion* without them. They protect the steelwork from waterborne contaminants in the atmosphere such as chlorides and sulphur dioxide and are in competition with 'high build' paint systems, weathering steel and galvanised steel.

Corrosion is not the only reason for having enclosures, they also provide a friendly, safe working environment for *inspection* and *maintenance*

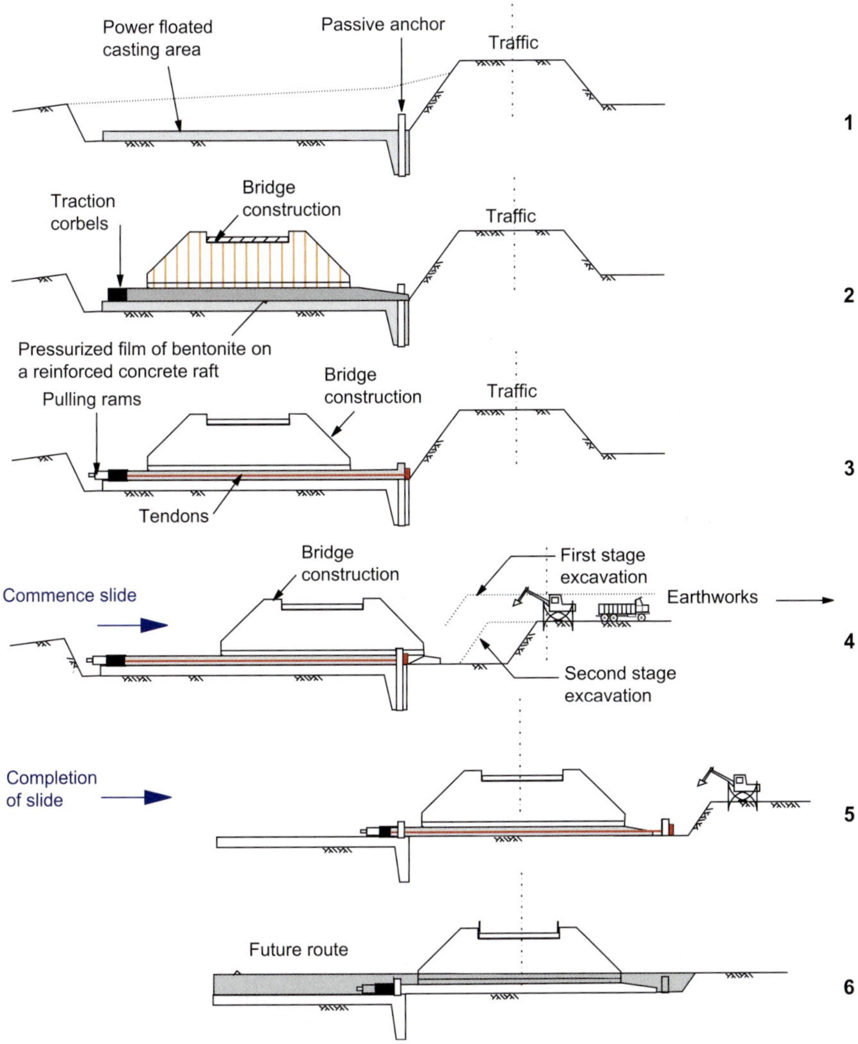

FIGURE 8.24
Autoripage sliding system (after Hughes, 1999).

personnel *without the need to disrupt traffic*, and can considerably *mitigate the costs* associated with road and rail closures for the purposes of inspection and maintenance. In some cases they may actually enhance the appearance of a bridge, and may even provide better *aerodynamic stability*. On the negative side, they can prove to be a trap for birds and small animals. They are now recognised as important elements to be considered as part of a strategy for newly constructed bridges or for rehabilitation of existing bridges.

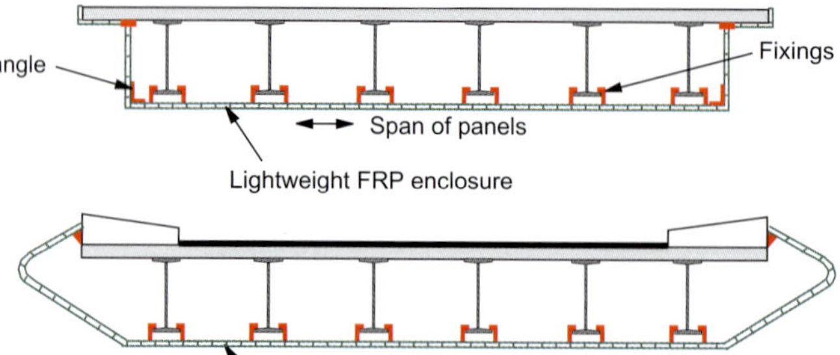

FIGURE 8.25
Enclosure principle.

8.5.1 Corrosion protection

The provision of an enclosure will reduce the corrosion rate of the steel (McKenzie, 1991) and will mean that not only can the periods between maintenance work be increased but that there will also be a concomitant reduction in the work necessary. The success of the enclosure in arresting corrosion can be seen in Table 8.2 from results measured on three of the bridges listed in Table 8.3.

8.5.2 Safe environment for inspection and maintenance personnel

An enclosure is designed as part of the *permanent* works and enables personnel to work in safety to inspect the steelwork, scantlings, bearings, expansion joints, service cables and pipes, and maintenance personnel can work in safety in a dry environment. This is more efficient than having to provide temporary works each time access is required for an inspection or maintenance work, involving lane closures, scaffolding, possibly some temporary enclosure using tarpaulins or tough polythene sheeting etc.

The space provided is normally regarded as a *confined space* during maintenance operations where risks to safety are increased, and is defined in

Table 8.2 Comparison of Corrosion Rates

			Interior Corrosion			Exterior Corrosion		
Bridge	**Type**	**Start**	**Year 1**	**Year 2**	**Year 3**	**Year 1**	**Year 2**	**Year 3**
Conon Bridge	Aluminium	1984	2	3	6	18	–	54
Tees Viaduct	Pultruded GRP	1988	1.5	2.5	10	–	49	150
Winterbrook Bridge	GRP	1993	4	6	–	–	–	–

Table 8.3 UK Bridges with Enclosures

Year	Bridge	Enclosure System	Reference
1979 & 1982	Iden, East Sussex	Timber	Bishop and Winnett (1980)
1980	Exceat, East Sussex	ABS panels on aluminium framework	Bishop (1986)
1982	Queenhill Bridge, Gloucestershire	Resin-bonded plywood-coated with epoxy resin	Bishop (1986)
1984	Conon Bridge, Scottish Highlands	3 mm aluminium profiled sheet and support brackets 'The Weaver System'	Bishop (1986) McKenzie (1991)
1988	Tees Viaduct, Middlesbrough, Cleveland, UK	GRP pultruded planks 'The Caretaker System' (ACCS)	Lee and Johnson (1994)
1991	Nevilles Cross Bridge, Durham, UK	GRP pultruded planks 'The Caretaker System' (ACCS)	Maunsell (1997) Irvine and Thorpe (1996)
1992a	Bromley South Bridge, Greater London, UK	GRP pultruded planks 'The Caretaker System' (ACCS)	Maunsell (1997) London Borough of Bromley (1996)
1992b	Spiceball Bridge, Oxon. Bluebird Bridge, Oxon.	GRP panels	McKenzie (1997)
1993	Winterbrook Bridge, Oxon.	GRP panels	McKenzie (1997)
1996	No.9 approach bridges to the second Severn crossing	GRP pultruded planks 'The Caretaker System' (ACCS)	Irvine and Thorpe (1996)

the Confined Spaces Regulations (1997) as 'any work area where there is a restriction of access, a lack of free breathable air, or a presence of dangerous gases, vapours or fumes', and the specified risks as 'injury due to fire or explosion, loss of consciousness due to an increased body temperature and asphyxiation'. It is essential, therefore, to provide suitable access/egress points via pier or abutment to the ground.

8.5.3 Traffic disruption

All motorists can testify to the frustration and anger sometimes felt upon seeing a bridge up ahead with coned-off areas in the distance, slow moving 'nose-to-tail' traffic and craneage or other plant and machinery in a coned-off lane or lanes, and no workers in sight! By its very nature, once in place, an enclosure (like the bridge it encloses) does not cause any traffic disruption. According to Irvine and Thorpe (1996) this is 'the pre-eminent benefit' of an enclosure. Work can be undertaken efficiently (without rushing), in safety and out of sight of the road users, or for that matter the road or rail users beneath.

8.5.4 Appearance

Providing that the criteria of the *context, form and detail* (Highways Agency, 1996b) of a newly, well-built composite bridge have been sympathetically applied, it can be a very pleasing artefact. Because all structural elements are exposed, their function and the way they work is easily understood. With time, however (and in spite of regular maintenance) exposure to wind, rain and pollution – of the steel beams in particular – will eventually result in corrosion. It then becomes unsightly and possibly unserviceable. A well-designed enclosure, on the other hand, can be aesthetically pleasing and not only will it not corrode but it will save on maintenance costs.

8.5.5 Brief history

The idea of protecting a steel bridge deck from atmospheric corrosion by means of a lightweight 'enclosure' was first mooted by the Transport and Road Research Laboratory (TRRL). The first experimental work took place in 1979 on a small composite bridge carrying the B2082 road over the River Rother at Iden in East Sussex (Bishop & Winnett, 1980). It had five weathering steel beams and a concrete slab, and part of it was enclosed in timber. The conditions on the inside and of the enclosure were studied using steel test coupons and it was found that there was no visible rusting after six months exposure.

This positive result encouraged the TRRL to carry out further trials using a more sophisticated enclosure material and construction. They developed a system using acrylonitrile butadiene styrene (ABS) lightweight sheeting which has a high tensile strength, low moisture absorption and good flame resistance, and this was attached to an aluminium framework between the bottom flanges of the steel beams of the Exceat Bridge in East Sussex (Bishop, 1986). The Iden Bridge was then revisited and a more comprehensive timber enclosure built, and the reaction of coupons internal and external to the enclosure were measured. Again, positive results were recorded.

Since then research has concentrated on the *choice of material*, which must be robust, lightweight and stable, form, which must be simple and easy to handle, and on *practical construction and fixing methods*. Development and take-up of the enclosure idea has been relatively slow (see Table 8.3). Prior (1998) recently contacted seven overseas highway authorities to inquire whether enclosures were used or being developed. Only four, namely, the US Federal Highway Authority (FHW), the Swiss Federal Institute of Technology (EPFL), the German Federal Highway Authority (BASt) and the New York State Department of Transportation replied, and they were all negative. In the UK there are now three systems currently marketed (Prior, 1998) and each one consists of a series of interlocking sheets connected together and hung from the superstructure.

8.5.5.1 Weaver system

Only one bridge is extant with this type of enclosure, namely the Conon Bridge in the Scottish Highlands. It uses ribbed anodised aluminium sheets 3 mm thick secured to the bridge with aluminium support brackets and it was designed for a uniformly distributed load of 3 kN/m^2.

8.5.5.2 The advanced composite construction system (ACCS)

Four stiffened glass-reinforced polymer (GRP) sheets and one cellular plank made by different manufacturing processes (see Figure 8.26) were tested as long ago as 1981 (Head, 1982). GRP has a high strength-to-weight ratio and very good corrosion resistance.

After many tests, Maunsell's research team opted for the pultruded plank which is described by Irvine (1998b) and the pultrusion process is described by Hollaway (1986). This basic standard unit (the plank) and other ancillary components (connector and trimmer units) were developed over a period of three years into the ACCS (see Figure 8.27). The design was based on a uniformly distributed load of 2.4 kN/m^2 and 2.5 kN point loads at 700 mm centres on a 150 mm square.

Non-standard components and adaptable fixings are also possible to enable designs to be produced for a range of different bridge types. It is now marketed and has written a manufacturing specification and a design guide for users. Each GRP components is made from E-glass (BS 3496 1989) embedded in isophthalic polyester resin, typically 60% glass by weight. The sections are pigmented and may be painted if required.

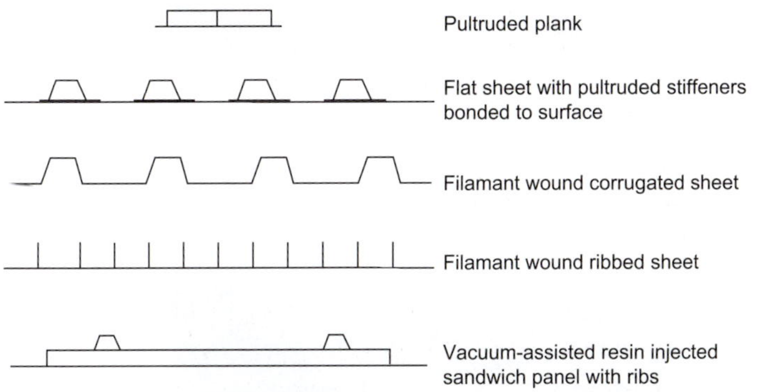

FIGURE 8.26 *Trial GRP profiles.*

FIGURE 8.27

(a) ACCS component data (courtesy of Maunsell Ltd.) and (b) GRP planks.

a

DATA SHEET

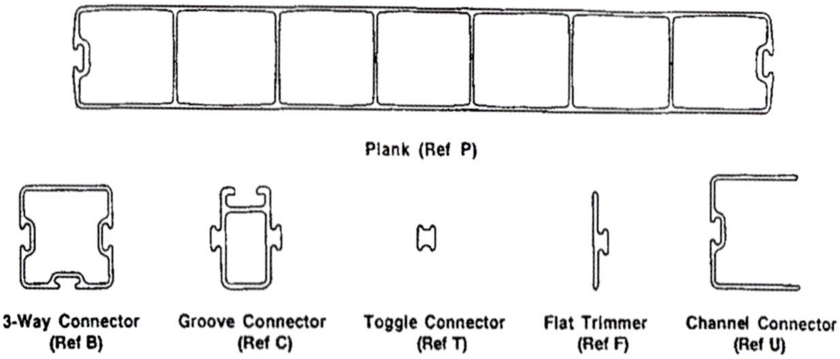

Geometric Properties

		Overall Dimensions (mm)	Typical Wall Thickness (mm)	Weight (Kg.m)
Plank	Ref P	600 x 80	3.5	11.0
3-Way Connector	Ref B	80 x 80	3.5	2.2
Groove Connector	Ref C	80 x 40	4.0	2.3
Toggle Connector	Ref T	Not Applicable	N/A	0.5
Flat Trimmer	Ref F	80 x 10	4.0	0.9
Channel Connector	Ref U	80 x 88	3.5	1.8

Material Properties

Axial Modulus	20–25000 MPa
Transverse Flexural Modulus	10–13000 MPa
Surface Spread of Flame (BS 476 : Part 7)	Class 0–2
Thermal Resistance, U-Value (when foam filled)	0.3–0.45 W/m^{2}K

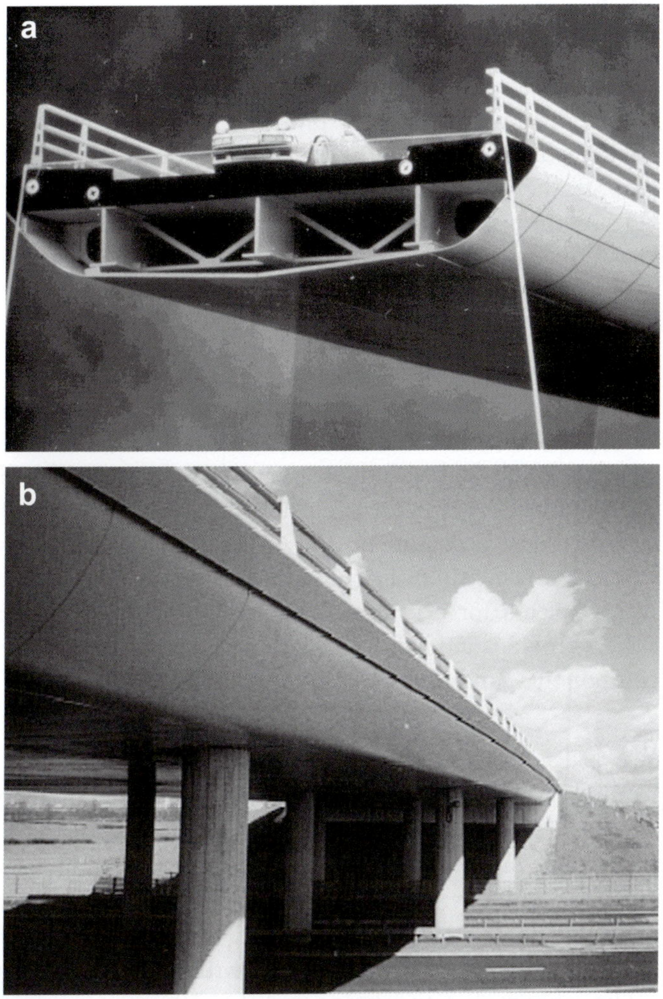

FIGURE 8.28 *(a) ACCS concept (courtesy of Maunsell Ltd.) (see colour plate section) and (b) A403 Pilning Bridge.*

Figure 8.28a shows the ACCS system in concept and applied to the A 403 Pilning Bridge on the approaches to the Second Severn Crossing, and illustrates how well the appearancc criteria are satisfied.

8.5.5.3 GRP panels

A less sophisticated system employing GRP technology was developed in the 1990s by Oxfordshire County Council and designed by consultants Mouchel & Partners in the UK. It comprises a basic ribbed GRP panel 1860 mm × 998 mm × 100 mm thick and various stainless steel fixings.

The design was first analysed by modelling the panel using finite elements and subjected to a range of loads to check torsion effects, local buckling, local stress concentrations and possible debonding of the laminate. The basic design load was 2.5 kN/m^2 plus a 1.7 kN patch load on a 75 mm square anywhere on the panel. Physical testing on five panels was then carried out at Strathclyde University to validate the numerical model results (ICCI, 1996). At the time of writing only three bridges have been enclosed using this system (see Table 8.3, years: 1992b, 1993; McKenzie, 1997).

8.5.6 Good practice

When the ACCS was first installed on the Tees Viaduct there were no design standards or codes of practice available. At the time of writing there are only three relevant codes of practice in the UK (BS 3496:1989; BS 3532:1995; BS 3691:1995) all of which deal with the constituents of the GRP itself. In order to help bridge engineers to evaluate enclosures for a new bridge the Highways Agency has issued design document (BD 67/96) (Highways Agency, 1996a). One important consideration is the comparison of the long-term cost benefits over 40 years for a bridge, *with and without an enclosure*. The document then provides guidance on performance criteria, choice of materials and components, and the design of the structural elements.

8.5.7 Cost

A cost-effectiveness analysis of enclosure system is required in the UK according to BD 67/96 (Highways Agency, 1996a) on a whole life cost basis, assuming a discount rate of 8% (see Chapter 9) and a design life of 120 years. This is unrealistic and normal practice, both in the UK and elsewhere, is to consider a design life of 30–40 years. Beyond that period very small sums accrue and can be ignored. The actual cost of the enclosure is variable but Prior (1998) has shown that an average rate of £150/m^2 is not unreasonable.

8.5.8 Design and durability

One of the stated objectives of the European Union's research project EU 468 EUROCOMP is that enclosures should use 'standardized sections which engineers can use with confidence' and profiles manufactured in the UK, and which meet this criterion are listed by Clarke (1996).

Design of the panels and fixings is carried out at the ultimate limit state subject to live loads (UDL and Point) given in both BD 67/96 (Highways Agency, 1996a) and BD 37/88 (Highways Agency, 1988) and checked for the serviceability limit states which often proves critical. In practice, enclosures will not be designed by the bridge engineer in the normal course of his work, rather the manufacturer of the system will design and test the system himself, leaving the engineer to consider the practical details and to include the estimated dead and live loads from the enclosure in the bridge design.

Although the bulk of the surface area of an enclosure is underneath a bridge, the polymer-based ones are subject to ultraviolet (UV) degradation. This is less of a problem now that resin formulations have been improved and there is the possibility that additives can be incorporated into the resins in the future to limit or prevent UV degradation (Hollaway, 1993). Surface protection treatments are, however, available to combat both moisture and UV degradation. Aluminium is protected by the anodising process. A summary of the usual treatments is given in Table 8.4 (Prior, 1998).

Other factors which affect the durability are the presence of pigeons which are able to roost and deposit waste on the inside of the enclosure panels, and water percolating from leaking drains or through expansion joints. There is also the possibility of 'bridge bashing' damage and the appearance of graffiti. The system should be designed, therefore, to be easily *demountable* so that affected panels can be removed and replaced.

In conclusion, the main *advantages* of enclosures are:

- Corrosion protection.
- Reduced user disruption.
- Improved access for inspection and maintenance.
- Safer working environment.
- Improved aesthetics.

The main *disadvantages* are:

- Few proprietary suppliers, hence difficulty in obtaining competitive prices.
- High initial cost.
- New FRP or metallic products will carry high development costs.

The jury is still out on whether enclosures will 'catch on'. Perhaps there is resistance in the bridge engineering fraternity to concealing the structural

Table 8.4 Surface Protection

Enclosure System	Protective Coating	Life to First Maintenance
Aluminium Weaver System	20 μm anodised layer	120 years
ACCS Pultruded GRP	Surface veil	30–40 years
Oxfordshire GRP panels	Pigmented gel coat	30–40 years

form of a bridge. This is probably the case in the 'new build' situation, but an already deteriorating bridge might benefit from concealment. There is also the high initial cost involved and the lack of a robust competitive market.

8.6 INTEGRAL CONVERSION OF EXISTING JOINTED BRIDGES

By designing multi-span bridges to be continuous over the intermediate supports, it is possible to reduce the deck thickness because of the better distribution of loads, to eliminate joints at intermediate supports, halve the number of bearings at each pier, use narrower crossheads or columns and to increase the degree of redundancy, thus making the bridge safer.

Such construction is now becoming the norm. In response to a mail survey carried out in 1990 to US transportation departments, it was discovered that 87% of respondents (26 out of 30 departments polled) routinely use continuous construction for the design of short to medium span bridges (Burke, 1990), the longest being the Long Island Bridge at Kingsport, Tennessee (nicknamed, The Champ), which has a total length of 2700 ft (823 m) made up of 29 continuous spans! But whilst continuous construction eliminates intermediate joints and bearings, those at the abutments tend to malfunction or fail outright. Many of the joints leak, thus allowing water containing de-icing salts and chemicals to percolate, causing unsightly staining and corrosion.

A remedy for this, which was developed in the USA in the 1930s, is the use of *integral bridges*, that is, *bridges with decks that have no joints between spans or at the abutments and form an integral structure with the approach embankments*. According to Hambly (1992) in spite of structural analysis which suggests that problems should occur from the induced high-restraint stresses, the general consensus in the USA is that they perform successfully, indeed none have so far failed.

The same concept can be applied to *existing* jointed bridges to arrest further corrosion problems. Guidance to retrofitting jointed bridge decks is available from the US Department of Transportation (1980) in the form of an advisory document, and some States in the USA, namely Texas, Utah, Wisconsin and Ohio, have developed their own ingenious solutions (Burke, 1990). Each solution appears to work satisfactorily, and they are shown in detail in Figures 5 and 6 of Burke (1990), but to illustrate the basic principles the Ohio and Utah solutions are reproduced here in Figure 8.29.

In the UK there is no published literature on the conversion of jointed decks to integral decks. Pritchard and Smith (1991) produced a contractor report detailing ways of achieving continuity in concrete composite bridge decks for use in new designs, but there is nothing specifically related to the *retrofitting* of existing bridges. Kumar (1998), whilst describing a novel way of providing continuity in new bridges, refers in passing to a continuity detail for existing simply supported bridges. It is very similar to the Ohio retrofit in Figure 8.29 except that the in situ part of the slab is made to bear on *compressible* material rather than polyethylene.

Any type of retrofit like this will alter the structural system, in particular at the intermediate piers where adjacent spans will rotate under live load thus causing negative bending in the slab. To prevent (or at least limit) transverse cracking in the *top* surface of the deck due to live load moments, and in the bottom surface due to creep moments, steel reinforcement is required.

8.7 SHOCK TRANSMISSION UNITS

Shock transmission units (STUs) are a means of providing load sharing in an otherwise articulated bridge structure. Pritchard (1992, 1994) has written extensively on their development and use in bridge engineering. *Basically it is a dashpot type device joining separate parts of a bridge which acts in two ways:*

1. It permits slow, long-term temperature, settlement or creep movements, and
2. It is capable of acting as a rigid link between the separate parts and sharing short-term earthquake shock, braking or traction forces.

The device was developed by R. Mander and J. Chaffe of the then Ministry of Transport who patented it in 1970 (Pritchard, 1992) and its construction is shown diagrammatically in Figure 8.30. It consists of a perforated piston which moves in a cylinder full of *silicone putty,* the common name for a chemical compound of boron-filled dimethyl syloxane developed in the USA in the 1960s as part of the space programme. Its special thixotropic properties made possible the invention of the modern STU because *it readily deforms under slowly applied pressure but acts as a rigid body under sudden impact*.

The device has been used in several ways in the field of bridge engineering. Pritchard (1992) cites three main areas:

- New or existing multi-span simply supported bridges.
- New multi-span continuous bridges.
- New continuous bridges in earthquake zones.

8.7.1 New or existing multi-span simply supported bridges

Multi-span simply supported bridges are built either for short-term economy or in areas where differential settlement is expected, such as, areas of mining subsidence. Typically, each span has a fixed bearing at one end and is free at the other. Thus, any sudden traction or braking load applied in any one span at deck level will be transferred to the fixed end and then to the pier. If an STU is secured to the free end of each span and the pier (or abutment), then under a sudden braking load, the induced horizontal forces will be shared between the piers as indicated in Figure 8.31. Exactly where the STUs are fixed in any particular bridge depends upon its articulation and type. In beam and slab decks they are usually bolted between bearings to the concrete diaphragm on one side (the free bearing side) and to the pier on the other side (the fixed side). Slab decks present no problems.

8.7.2 New multi-span continuous bridges

In this very popular form of construction, STUs can be used to great effect at abutments to relieve braking or traction loading on intermediate piers where the deck is monolithic with the piers. They do this by attracting the transient load away from the piers.

8.7.3 New continuous bridges in earthquake zones

This application is exemplified by the construction of the Saint André Viaduct on the A43 motorway in the Savoy region of France (Datry, 1999). It is a unicellular concrete box in cross-section over 18 varying spans (39.4–95 m). The area is in a region of earthquake activity and to attract horizontal forces due to earthquake loads, six damping devices were placed on each abutment. The actual device used was the 'Jarret Seismic Damper' (Datry, 1999) which, although used specifically to dissipate seismic forces, can be used in the same way as STUs, Figure 8.30(b).

If modified with STUs, shock-loading from an earthquake will be distributed in the same way; without STUs it is likely that each span will collapse leaving the fixed end perched on the top of the pier whilst the free end drops to the ground. This is well illustrated in Figure 8.32 which shows the collapsed condition of a viaduct after the earthquake in Istanbul, Turkey on 17 August 1999 (Thompson, 1999).

8.8 BEARING REPLACEMENT

Providing that bearings are in the same environment assumed in the initial design, then their maintenance requirements are generally little or none, and they should last for the life of the bridge. Sadly this is not the case, and many bearings exhibit a life only 20–30 years. This is because the bearing material

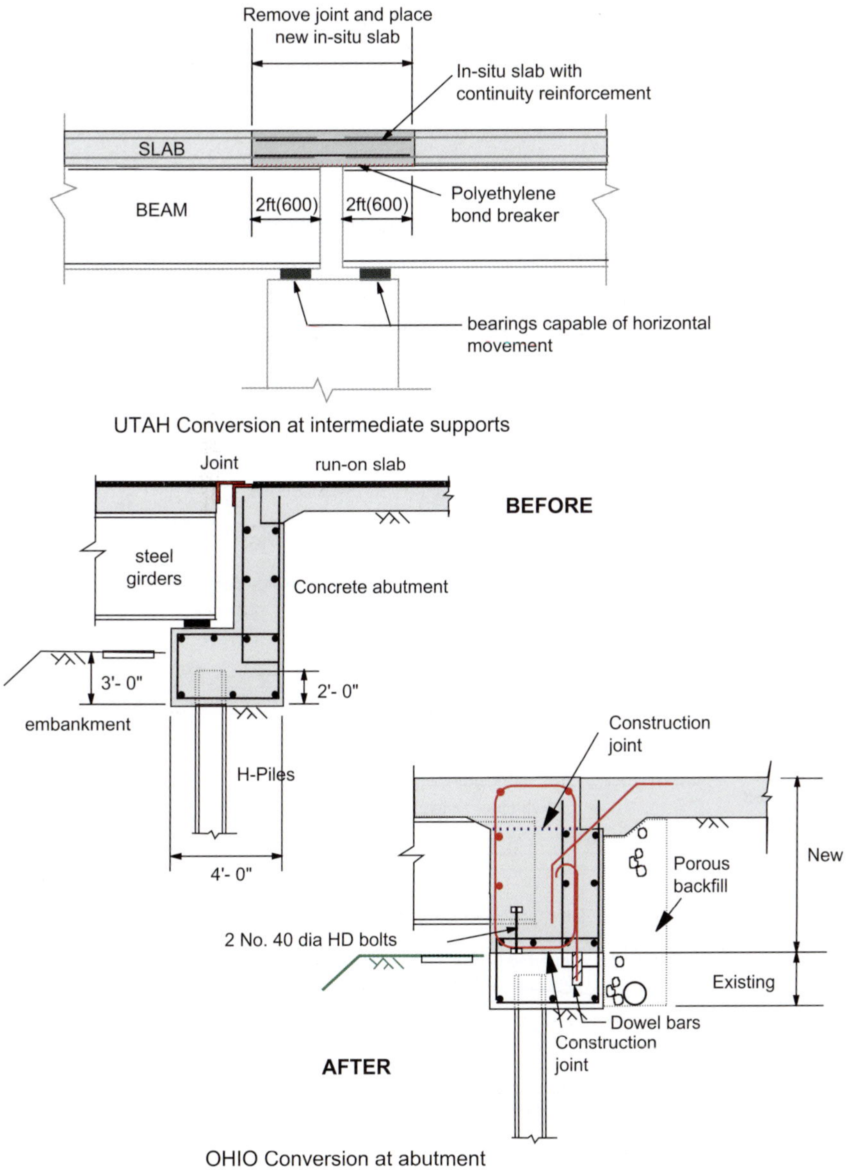

FIGURE 8.29 *Conversion of jointed bridge decks to integral bridge decks.*

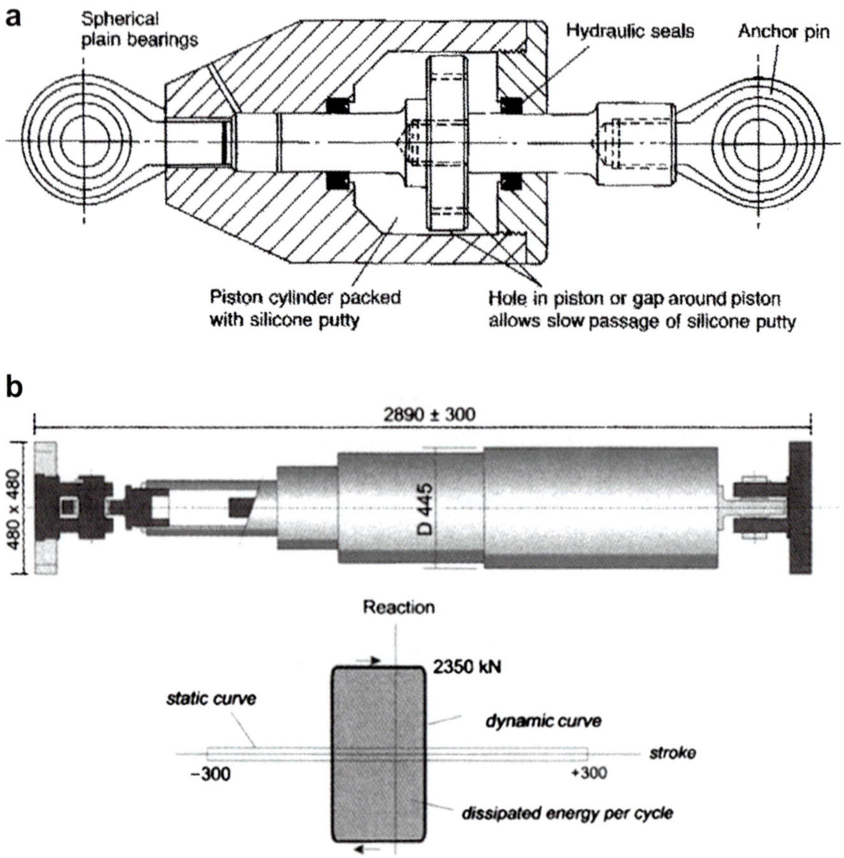

FIGURE 8.30
(a) Modern shock transmission unit (Pritchard, 1992) and (b) Jarret seismic damper (Datry, 1999).

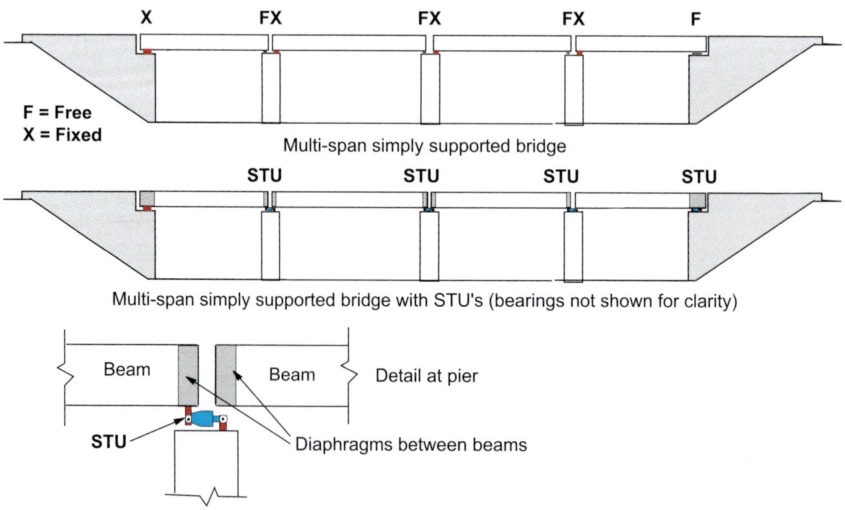

FIGURE 8.31
Multi-span bridge with and without STUs.

FIGURE 8.32 *Collapse of a Viaduct due to an earthquake in Istanbul, Turkey.*

has broken down due to unexpected chemical action, or has structurally failed due to overload and increased fatigue action.

With time, elastomeric bearings fail due to breakdown of the elastomer by attack from a hostile environment which may consist of oil, petrol, oxygen, leakage of salt water through joints and extreme temperatures where they literally melt.

Metal bearings fail due to cracking from overload, binding due to maintenance neglect or corrosion. The latter two problems can be dealt with by a suitable greasing regime, and may require cleaning and painting. Structural failure requires bearing replacement, and this can be a long and expensive process as demonstrated by the replacement of bearings on the 1.3 km long Orwell Bridge in Ipswich, UK (Mylius, 2000). The bridge comprises 21 spans of twin post-tensioned concrete boxes on twin piers, with a pair of bearings to each pier, and each pair carrying a load of 3250 t. Hairline fractures in the bridge's roller bearings resulted in a decision by the Highways Agency to replace all of the roller bearings with flat, sliding bearings. This involved a delicate jacking procedure illustrated in Figure 8.33.

Each box was modified in turn and the traffic was kept moving during the whole operation, albeit limited to two 3 m lanes during jacking. The bridge had to be strengthened above the bearing points of the jacks, which was different for each box due to the presence of a water main in the southbound

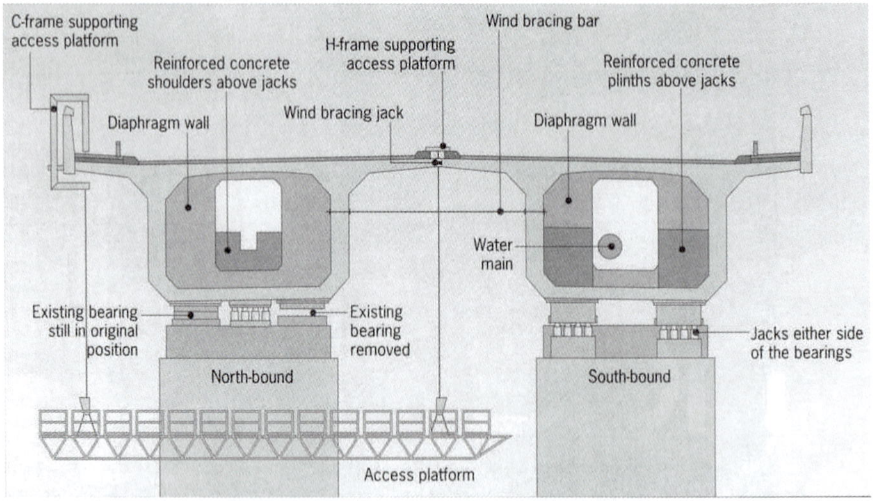

FIGURE 8.33
Orwell bridge bearing replacement (Mylius, 2000).

box (Figure 8.33). On the northbound box there is access on three sides of the bearings, but on the southbound box access is restricted to one side only. Human access was provided by a hanging platform which encircled each pier. The project is one of the Highways Agency's major refurbishment schemes under its new asset management system (see Chapter 1). It cost £1.2 million over a 15-week schedule.

8.9 CREATIVE THINKING

With a traffic flow in excess of 250,000 vehicles per day and rising, the existing expressway in the centre of Buenos Aries was becoming congested, causing delays and high levels of pollution. There was no way of providing diversion routes and so the local authorities decided to increase the number of lanes from four to nine. This meant the removal and retrofitting of many of the structures along the route. Two overbridges however, in close proximity to one another, presented a major problem. In order to accommodate the nine expressway lanes the three central piers of the two sixspan overbridges (namely the Superi Street and Sapiola Street bridges) would have to be removed!

It required some very *creative thinking* by bridge engineers to arrive at a solution which maintained traffic flow at *all* times, and was not only within strict budgetary constraints, but also produced a robust and notable-looking landmark structure in an otherwise drab and congested urban landscape.

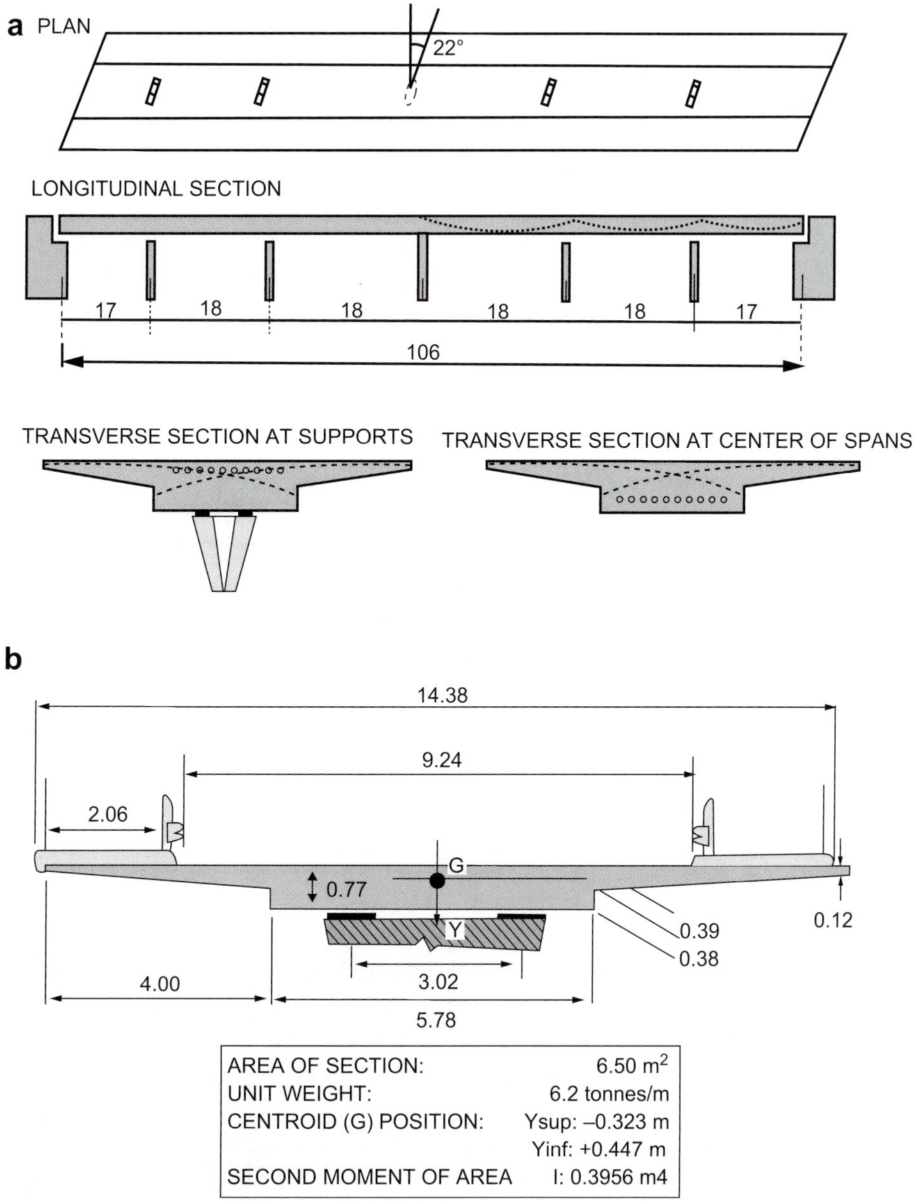

FIGURE 8.34 *General arrangement and typical cross-section.*

It was clear from the very beginning that the new section of overbridge would have to be supported from above to allow the increase in traffic lanes. Those investigated were: Cable-stayed; Suspension; Arch and finally a Truss all shown in Figure 8.39.

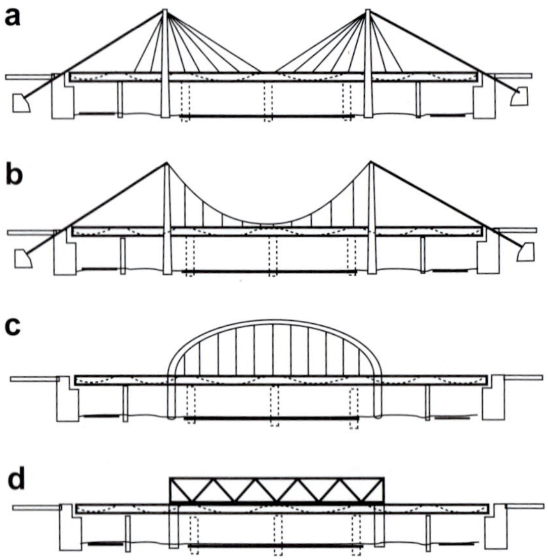

FIGURE 8.35 *Feasible alternative schemes.*

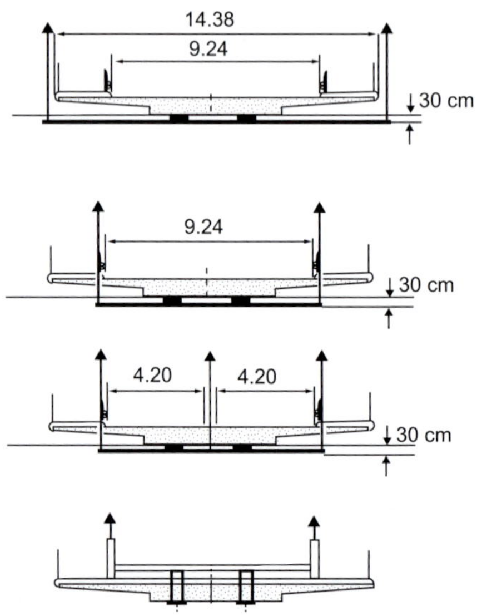

FIGURE 8.36 *Final configuration of whole structural system.*

FIGURE 8.37

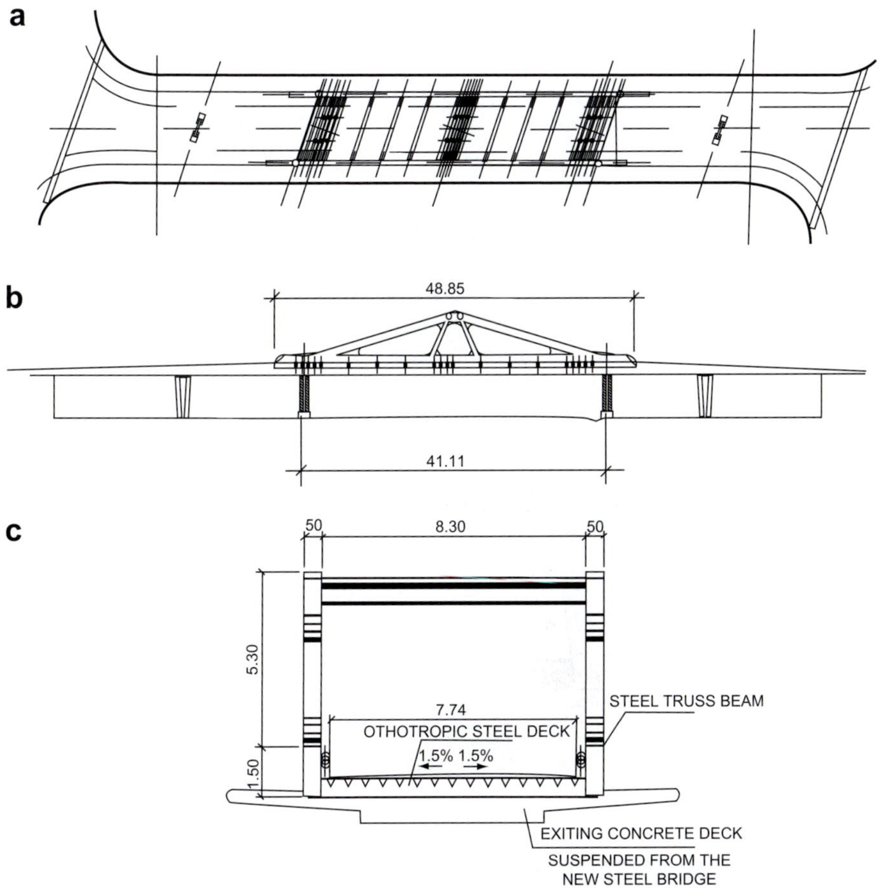

The ideal choice of bridge type would have to meet the following constraints:

- Be the lowest 'first cost'.
- At no time during construction or in-service were the present stresses in the post tensioned slab to be exceeded
- The new supporting structures (SS) should be designed to be stiff enough such that deflections in the existing deck supported by the new (SS) did not exceed the existing.

FIGURE 8.38

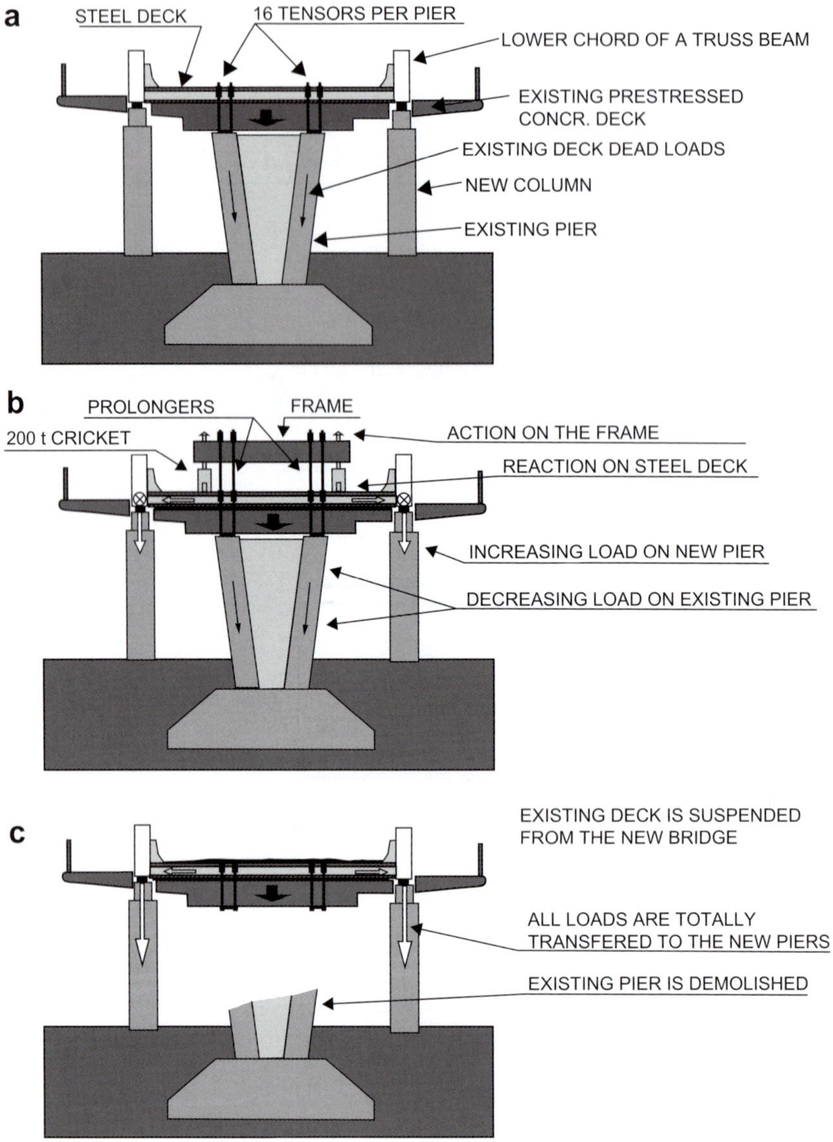

- The (SS) should be aesthetically pleasing and not be such that its presence was either dominant over, or subservient to the urban environment.
- Traffic must be kept moving at all times.

After detailed consideration of the four alternatives, a Trussed Beam was selected, which came to be known as 'the hanger' because of its appearance. See Figure 8.41.

The 'hanger' elements were constructed out of hollow steel sections (fy = 240 N/mm^2) which not only provided the stiffness required but also added a more pleasing aspect to the eye, and enabled support and connection details to be 'durable friendly'.

The central section of the existing bridge deck was a post-tensioned concrete slab (see Figure 8.41) with a monolithic post- tensioned concrete cantilever each side. A detailed analysis of the stresses in the deck due to dead and superimposed dead loads was carried out, these acted as a 'platform' to which were added the stresses due to the process of demolition and construction. The major design restriction here was that at no time should the existing stresses increase.

Finally, but not least, a traffic management scheme was put in place which limited the nuisance factor to the absolute minimum.

8.9.1 Construction

The new bridge had first to be constructed and supported on new circular concrete piers, two at each end.

At each panel point of the existing bridge, holes were drilled through the posttensioned concrete slab, through which the steel suspenders were placed to transfer the loads from the slab to the new transverse beams. Drilling the holes in the right place, and avoiding cutting through the existing prestressing cables was very slow and time-consuming process.

A jacking frame was then positioned over the existing piers, and the existing deck loads were incrementally transferred to the cross beams. When separation of the existing slab from the bearings was observed, the loads were deemed to have been transferred. Since the central pier was monolithic with the deck it was first cut and wedged, and when the wedges became free, the load was assumed to have been transferred.

The load transfer was a delicate operation, and had to be very carefully co-ordinated and monitored to ensure that all of the jacks were accepting load incrementally. As the loads on the existing piers diminished to zero, the loads on the new piers increased to the maximum. The jacks and jacking frame were removed and the existing piers were demolished.

The construction sequence for the load transfer of the existing concrete deck to the new trussed beams is indicated in Figure 8.42.

The project was a wonderful illustration of the need for *creative activity*. No two bridges are exactly alike. The analysis tools, materials and construction techniques vary from country to country.

The authors of this paper state: 'The main lesson is the importance of creativity in engineering or, as it is more usually called, *the art of engineering*' which, they state 'is an extremely difficult concept to teach other people'. The technological aspects of engineering, compared with the art of engineering, 'can be disseminated and taught without great difficulty'. To which most of us would say 'amen', but I suspect that it has more to do with subjectivity than raw technical understanding. A numerical problem (given a set of parameters) has only one answer; whilst a conceptual problem has many. At another time, in a different place, the problem may have been handled differently, but in any event there is a great feeling of satisfaction in exercising creative thought to arrive at a job well done.

8.10 CONCLUDING REMARKS

It is impossible in such a short space to enumerate all of the possible modifications that can be made to bridges in order to restore their satisfaction of all limit states or to alter them to cater for changed circumstances. In this chapter, apart from strengthening which is covered in Chapter 7, an attempt has been made to highlight the more common major modifications of bridge widening, deck replacement, bridge replacement, enclosure and load sharing. Hopefully, this will provide a useful overview of some of the options available.

REFERENCES

American Association of State Highway and Transportation Officials. (1994). *LFRD bridge design specifications*. Washington, DC: AASHTO.

Bettigole, R. A. (2000). Bridge deck replacement with lightweight exodermic bridge deck panels. *Fourth international conference on bridge management*, Guildford, Surrey, UK (pp. 608–616). London: Thomas Telford.

Bishop, R. R. (1986). *Enclosure, an alternative to bridge painting, TRRL Supplementary Report 621*. UK: Crowthorne.

Bishop, R. R., & Winnett, M. A. (1980). *An alternative to bridge painting, TRRL Supplementary Report 621*. UK: Crowthorne.

British Standards Institute. (1989). *BS 3496 E-glass fibre chopped strand mat for the reinforcement of polyester resins*. London: BSI.

British Standards Institute. (1995). *BS 3532 unsaturated polyester resin systems for low pressure fibre reinforced plastics*. London: BSI.

British Standards Institute. (1995). *BS 3691 glass fibre rovings for the reinforcement of polyester and epoxy resin systems*. London: BSI.

Burke, M. P. (1990). The integrated construction and corrosion of single and multiple span bridges. In: *First international conference on bridge management* (pp. 677–693). Guildford, UK: Elsevier.

Champion, M. A. (1994). *Some examples of bridge modification for motorway widening. Bridge assessment, management and design* (pp. 385–390). Amsterdam: Elsevier.

Clarke, J. L. (1996). *Structural design of polymer composites, EUROCOMP Design Code and Handbook*. London: E&F Spon.

Cole, M. (1996). Taking an interest. *New Civil Engineer, Roads Supplement*, 14–18.

Datry, J. (1999). The Saint Andrè Viaduct. *Concrete Engineering International, 3*(4), 16–22.

Del Carril, T. (2002). A creative solution for two bridges. *The Structural Engineer*, 21 May 2002, pp. 25–30.

De Milander, C. N., & Strydom, C. P. (1996). Innovative solutions for unusual bridge maintenance problems in South Africa. In: *Third international conference on bridge management* (pp. 618–625). Guildford, Surrey, UK, London: E & F N Spon.

US Department of Transportation. (1980). *Bridge deck joint rehabilitation (retrofit)*, Federal Highways Administration Technical Advisory T5140.16.

Hambly, E. C. (1992). *Bridge deck behaviour*. London: E & F N Spon.

Hayward, D. (1999). Decking out for winter. *New Civil Engineer*, (21 January), 30–32.

Head, P. R. (1982). GRP Walkway membranes for bridge access and protection. *British Plastics Federation, 13th Reinforced Plastics Congress*. London: BPF.

Henriquez, O. E., Plecnik, J. M., Munley, E. P., & Cooper, J. D. (1997). Development of an FRP system for bridge deck replacement, *Recent Advances in Bridge Engineering*. In: Zurich, (Ed.) U. Meier & R. Betti. In: *Proceedings of the US-Canada-Europe Workshop on Bridge Engineering* (pp. 197–206).

Highways Agency. (1988). BD 37/88 – Loads on bridges. *Design Manual for Roads and Bridges* (Vol. 3). London: HMSO.

Highways Agency. (1996a). BD 67/96 – Enclosure of bridges, *Design Manual for Roads and Bridges* (Vol. 3). London: HMSO.

Highways Agency. (1996b). *The appearance of bridges and other structures*. London: HMSO.

Hollaway, L. (1986). Pultrusion. In A. Whelan, & J. L. Croft (Eds.), *Developments in Plastics Technology – 3, Chapter 1*. Elsevier Applied Science Publishers.

Hollaway, L. (1993). *Polymer composites for civil and structural engineering*. London: Blackie Academic and Professional.

Hughes, D. (1999). Off-line bridge installation. *Concrete Engineering International, 3*(1), 73–77.

ICCI. (1996). Fibre composites in infrastructure. In: *Proceedings of the first international conference on composites in infrastructure* (pp. 929–943). Tuscon, Arizona, USA.

Irvine, R. A. (1998b). Verbal report on the manufacture of advanced composite materials.

Irvine, R. A., & Thorpe, J. A. (1996). *Bridge Enclosures to Meet the New Highways Agency Standard BD 67/96*. Beckenham, Kent, UK: Maunsell Structural Plastics.

Lee, D. J., & Johnson, P. F. (1994). A19 Tees Viaduct – strengthening and refurbishment. *Proc. Institution of Civil Engineers, Structures and Buildings, 104*, 145–166.

Leech, I. (1995). A595 bridge replacement. *Construction Repair, 9*, 26–28.

London Borough of Bromley. (1996). *Principle Inspection Report, Bromley South Bridge*, Report 56A, London Borough of Bromley.

Lopez-Anido, R., & Ganga Rao, H. V. S. (1997). Design and construction of composite material bridges, *Recent advances in Bridge Engineering*. In: Zurich, Meier U., & Betti R. (Eds.), *Proceedings of the US-Canada-Europe workshop on bridge engineering* (pp. 265–272).

Mabey and Johnson. (1992). *The quick bridge system. Reading*. UK: Mabey and Johnson.

Maunsell. (1997). *Nevilles cross bridge project data sheet*. Maunsell & Partners.

McKenzie, M. (1991). *Corrosion protection – the environment created by bridge enclosure*. Crowthorne, UK: TRRL Research Report 293.

McKenzie, M. (1997). *The corrosivity of the environment inside Oxfordshire county council bridge enclosures, Project Report PR/CE/67/97*. Crowthorne, UK: TRL.

Mylius, A. (2000). Ice cool operation. *New Civil Engineer,* (23 March), 30–31.

NCE. (1998). Slough slide, *New Civil Engineer*, News, April 6.

Paul, A. M., & Coke, R. S. (1994). Widening bridges for the M1 and M6 motorways. In B. Pritchard (Ed.), *Bridge modification* (pp. 56–68). London: Thomas Telford.

Pritchard, B. (1992). *Bridge design for economy and durability.* London: Thomas Telford.

Pritchard, B. (1994). Bridge strengthening with minimum traffic disruption. In: Pritchard B. (Ed.), *Proceedings of the conference on bridge modification*, Institution of Civil Engineers (pp. 185–203). London: Thomas Telford.

Pritchard, B., & Smith, A. (1991). *Investigation of methods of achieving continuity in concrete composite bridge decks*, TRRL/DTp Contractor Report CR 247, Transport and Road Research Laboratory, Crowthorne.

Prior, R. G. (1998). The enclosure of bridges. MSc Thesis.

Rao, H. G., & Craigo, C. A., II. (1999). Fiber-reinforced composite bridge decks in the USA. *Structural Engineering International, 9*(4), 286–288.

Road Vehicles (Construction and Use) Regulations. (1998). London: HMSO.

Saul, R. (1996). Replacement of steel and composite bridges without traffic interruption. In: *Third international conference on bridge management*, Guildford, Surrey, UK (pp. 295–303). London: E & FN Spon.

Stratford, T. J., & Burgoyne, C. J. (1999). Lateral stability of long concrete precast beams. *Proc. Institution of Civil Engineers, Structures and Buildings, 134*(2), 169–180.

Svensson, L., & Pettersson, L. (1990). Aluminium extrusion bridge rehabilitation system. In *First international conference on bridge management* (pp. 777–783). Guildford, Surrey, UK: Elsevier.

Taylor, H. P. J. (1991). *Precast concrete bridge beams and the new Y Beam, Bridge Replacement Symposium*. Leamington Spa: Construction Marketing Ltd.

Taylor, H. P. J. (1998). The precast concrete bridge beam: the first 50 years. *The Structural Engineer, 76*(21), 407–414.

Taylor, H. P. J. (1999). The precast concrete bridge beam: the first 50 years: discussion. *The Structural Engineer, 77*(15), 28–30.

Thompson, R. (1999). Panic in Istanbul as quakes head east. *New Civil Engineer,* (2 September), 8–9.

Trinidad, A. A. (1993). Aluminum bridges in the USA. In: *Second international conference on bridge management*, Guildford, Surrey, UK (pp. 190–199). London: Thomas Telford.

Upstone, T. J. (1996). Replacement of motorway overbridges maintaining continuous motorway traffic. In: *Second international conference on bridge management*, Guildford, Surrey, UK (pp. 385–392). London: Thomas Telford.

FURTHER READING

Boyes, D. S. (1994). Policy traffic management and available options. In B. Pritchard (Ed.), *Bridge modification* (pp. 1–11). London: Thomas Telford.

Irvine, R. A. (1998a). Verbal report on the development of curved enclosure panels.

CASE STUDIES

First international conference on bridge management. (1990). Guildford, Surrey, UK, Elsevier.

Second international conference on bridge management. (1993). Guildford, Surrey, UK, Thomas Telford, London.

Third international conference on bridge management. (1996). Guildford, Surrey, UK, E&FN, Spon, London.

Fourth international conference on bridge management. (2000). Guildford, Surrey, UK, Thomas Telford, London.

Report. (1995). Widening of the Rodenkirchen (suspension) Bridge. *Construction Repair, 9*(1), 10–15.

CHAPTER 9

Whole-Life Costing, Maintenance Strategies and Asset Management

CONTENTS

9.1 INTRODUCTION

Whole-life costing (WLC) – sometimes referred to as life cycle costing (LCC) – is a way of determining the total cost of a bridge structure from its initial conception to the end of its service life.

It attempts to quantify, in present monetary terms, the costs arising from all work undertaken on a certain structure. This is referred to as a *net present value* (NPV) or a single current cost. In other words, it tries to answer the question: '*How much will this bridge cost in today's money over its entire life?*' The PVA technique involves the calculation of the cost of alternative schemes in present day monetary terms, i.e. the amounts that are required in today's value to obtain goods and services at any future date. This enables the whole-life cost of a bridge to be determined and is based on a simple investment principle. The costs to be included would be those arising from design, construction, repair, maintenance and upgrading, traffic management and delays and possibly demolition.

Whole-Life Costing provides the client with a more realistic estimate of how much a bridge structure is going to cost in the long term, and enables the exercise of sound financial management in respect of initial design and construction costs, and ongoing maintenance costs. The choice is very often reduced to that between lower initial cost structures with higher maintenance costs, and higher initial costs structures with lower maintenance costs. Additionally attempts are made to consider the future cost to the community if there are delays, accidents etc.

WLC addresses the problem of the future maintenance of bridge structures (both new and existing) and behoves designers to consider the future consequences of their present actions in order to get the best value for money. In the UK procedures for WLC are provided in Departmental Standard BD36 (1992) and Departmental Advice Note BA28 (1992).

9.2 THE BASIC THEORY

The need for a present value analysis (PVA) method in bridge management is the fact that options need to be considered on an equal basis. Funds are not in inexhaustible supply and must therefore be carefully managed to achieve maximum efficiency in their use. The PVA assumes that there will be a consistent supply of regular funding, however, little that may be. Unfortunately this can vary due to political uncertainties and the vagaries of national life as experienced by whatever political party is in power at the time budget allocations are set.

The present value of a structure is based on a simple investment principle. A capital or principal P invested for n years at an interest rate r, compounds to a sum C:

$$C = P(1 + r)^n \tag{9.1}$$

This can be expressed another way to mean that the NPV P of an expenditure C in year n at a discount (or tests discount) rate r is given by:

$$P = C/(1 + r)^n \tag{9.2}$$

In reality there are a number of expenditures in a bridge, such as abutments, piers, deck, bearings, expansion joints etc., and so the cumulative present value would be:

$$\Sigma P = \Sigma C/(1 + r)^n \tag{9.3}$$

This then allows for the comparison of different schemes on an equitable basis, and generally the scheme with the least NPV is the preferred one. This process of calculating the NPV is known as *discounting* and the terms *interest rate* and *discount rate* are interchangeable.

The whole-life cost of the project is then the sum of the initial (capital) costs and long-term negative discounted costs of subsequent inspection, maintenance, repair and replacement costs P, as and when they occur based on best estimates. Bridge Engineers are very practised in estimating initial construction costs. The long-term intermittent costs, however, are difficult to quantify. They rely upon the past experience of costs on other projects (database information) and expert opinion where no previous information is in hand. Some Bridge Management Systems throughout the world have been accumulating data for a number of years, and as these bases expand, cost predications (barring unexpected events) should improve. Comparative cost estimates, therefore, are essential.

Table 9.1 Comparative Whole-Life Costs

Year	Operation	Cost	Discount Factor	Present Value
Option A				
0.000	Substantial strengthening	450,000	0.000	450,000
30.000	Concrete repairs	10,000	0.412	4120
60.000	New concrete bridge	600,000	0.170	102,000
90.000	Concrete repairs	10,000	0.067	667
120.000	Re-waterproof	21,000	0.033	693
120.000	Concrete repairs	20,000	0.033	660
			Total	**558,140**
Option B				
0.000	New concrete bridge	600,000	0.000	600,000
30.000	Concrete repairs	10,000	0.412	4120
60.000	Re-waterproof	21,000	0.170	3570
60.000	Concrete repairs	20,000	0.170	3400
90.000	Concrete repairs	30,000	0.067	2010
120.000	nil	0.000	0.000	0.000
			Total	**613,100**

The example in Table 9.1 illustrates the variation in cost that can result from such a comparative analysis. The discount rate has been chosen as 3%, and the 'Discount Factor' in Column 4 is the value of $1/(1 + r)^n$, thus at 30 years the factor $= 1/(1 + 0.03)^{30} = 0.412$.

This simplified example included to illustrate the method, shows with the assumptions made, that it is cheaper to strengthen than to provide a new bridge. *What this example does not indicate are the notional costs due to traffic disruption* during the maintenance operations. Such costs have to be estimated separately and added to the construction/repair costs and may produce a different answer.

9.3 DISCOUNT RATE

The discount rate is the 'opportunity value of money' and varies depending on the investment strategy adopted or the risk involved. It takes into account that money available now can be invested for growth in the future. In the UK, the discount figure is currently taken as 8% and can be considered to be the return which a potential investor might expect after allowing for inflation. This is generally considered to be too high and a lower rate of 2–3% would be more appropriate. There is even an argument put forward for taking a zero discount rate as Governments will doubtless continue to fund maintenance requirements directly out of taxes rather than putting the money aside. In the USA the rate is 6%; in Germany 3% and in Switzerland only 2%.

Any discount rate chosen cannot be considered as absolute and must be based on tax rates, the method of investment, market conditions and inflation etc. For any given project a number of rates can be chosen to test the sensitivity of various options, and this will then give the bridge engineer an indication of what items in the project either significantly change with discount rates, or quantitatively affect the overall cost. It might be possible to accept almost any figure for items which represent a small percentage (5% say) of the overall cost such as parapets, but for a significant item such as a concrete bridge deck which might represent as much as 45% of the cost, the discount rate must be chosen with care.

The actual variation of the discount factor with time is shown in Figure 9.1 and illustrates not only the highly sensitive nature of the discount rate, but why some clients prefer the higher rates since the apparent monetary outlay is bound to be less.

For example, the money required to replace an item after 40 years at 8% is only 1/9th of that at 2%. Alternatives are therefore very sensitive to the discount rate, and it is advisable to carry out a sensitivity analysis. The example in Table 9.2 shows how the total cost of replacing bearings in a bridge every 25 years varies with the discount rate. The life of the bridge is assumed to be 120 years and so the bearings are replaced four times during its life.

FIGURE 9.1
Variation of discount factor with time.

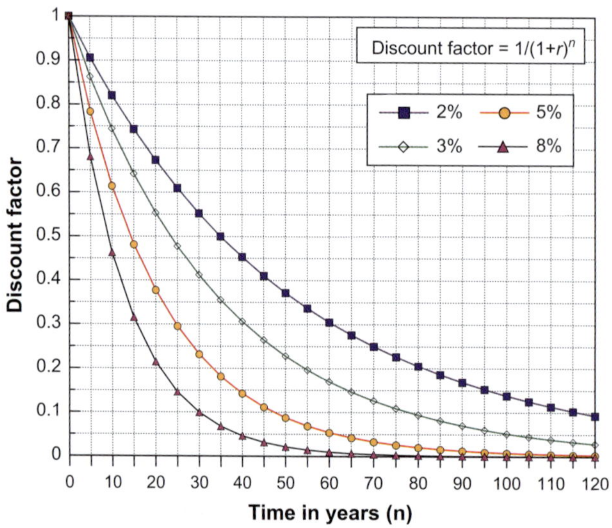

Table 9.2 Effect of Different Discount Rates on Replacement Costs

					Whole Life Cost (£)		
Element	**Number**	**Capital Cost (£)**	**Replace (£)**	**Life (Years)**	**2%**	**5%**	**8%**
Bearings	30.000	12,000	4000	25	17,382	13,363	12,683

$$\text{The whole-life cost} = \text{capital cost} + S(\text{replacement cost} \times \text{discount factor}).$$

Thus for the 2% Discount Rate:

$$\text{WLC} = 12{,}000 + 4000 \times [1/(1+0.02)^{25} + 1/(1+0.02)^{50} + 1/(1+0.02)^{75} + 1/(1+0.02)^{100}] = £17{,}382.00$$

The cost differences can be clearly seen. For just this single item there is a difference of nearly £5000 between the maximum and minimum rates. For a complete bridge the difference could be of the order of hundreds of thousands of pounds. An overestimate of the rate would mean a shortfall in funds and conversely an underestimate would mean a surplus of funds.

Another factor to consider is the *'replace or repair'* argument. It may be more economical to replace a bridge entirely after, say 60 years than to have to continually repair at regular intervals up to a design life of 120 years. Finely tuned social, engineering and financial judgement is required in order to arrive at the most favourable solution. A continuously repaired bridge may result in many relatively short traffic delays, but result in a bridge which not only looks 'patched-up' and run-down but with time begins to look out of harmony with its surroundings and possibly downright ugly. In some cases where a bridge becomes too narrow to cope with the volume of traffic or where it is under strength due to increased traffic loads, it is functionally obsolete and must be replaced. A new bridge may cause a longer disruption of traffic initially, but it will have a beneficial impact on the local economy and hopefully result in a structure that the local community can be proud of.

9.4 TRAFFIC DISRUPTION

In addition to maintenance and replacement costs, it is necessary to appraise the costs of traffic disruption which normally equates to traffic *delay* (or increase in travelling time), either due to a traffic-light controlled system or to a diversion during the construction work. In the UK these costs are

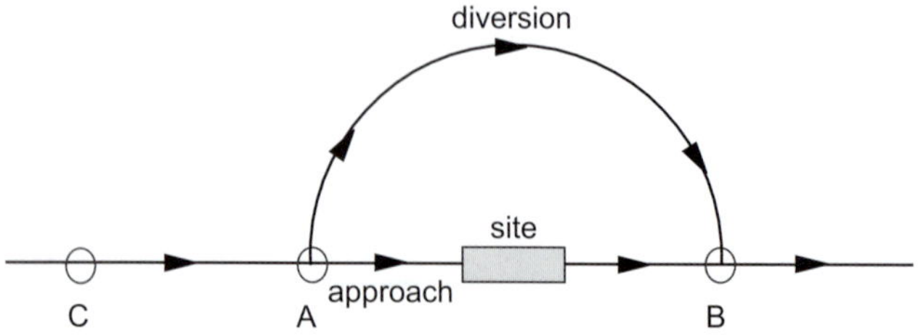

FIGURE 9.2
QUADRO network.

estimated by the use of a computer program QUADRO (**QU**eues **A**nd **D**elays at **RO**adworks), (Quadro, 1996) which breaks the problem down into elements of *time*, *accidents* and *vehicle operating costs*.

The Quadro system is generally for use in a *rural* environment and is based on a network which defines the main route and the diversion as shown in Figure 9.2 (Jones, 1995). The site lies on the *main route* from A to B. Point C is the first upstream junction.

For the general public, bridgeworks on a main highway represent a frustration; annoyance, and a potentially increased hazard, but for commercial travellers real costs are incurred such as fuel, increase in employees working day, vehicle maintenance etc.

The aim of programmes such as QUADRO is not only to minimise delay costs but to determine the best combination of traffic management and to some extent when best to carry out the works.

The actual calculated delay costs do not result in any tangible monetary benefit to the client (usually a national highway authority) but their calculation is an attempt to respect the effects of delays on working and non-working road users alike.

9.5 NEW BUILD

The principle of WLC can also be applied to new bridge projects where every conceivable cost is taken into consideration, which would include costs due to:

- Construction (C)
- Maintenance (M)
- Traffic disruption (TD)
- Traffic betterment (TB)
- Capital loss (CL)
- Scrap value (SV)

Then

$$\text{NPV} = C + \sum_{n=1}^{n=N} \frac{M_n + \text{TD}_n + \text{TB}_n}{(1+r)^n} - \text{CL}_N + \text{SV}_N \qquad (9.4)$$

where

N = services life
n = year of service life considered
r = discount rate

TB would be costs equivalent to time saved by traffic using the new bridge which previously had travelled by another slower route, for example in the case of a by-pass which normally speeds up journey times.

9.6 PERFORMANCE PROFILES

It is probably impossible to be sure that all of the factors that dictate the final cost of a bridge will be foreseen and consequently the final cost will not be known with a high degree of accuracy. Lee (1990) says that in his view '(...maintenance costs for bridges are similar to estimating the length of a piece of string....)'. Awareness of the fact that bridges cannot last forever, and that problems with bridge deterioration are now better understood, means at least that reasoned attempts can be made to assess the maintenance costs and hence the final cost of a bridge. Great emphasis is rightly placed on ensuring that a bridge is *strong* enough to carry the loads imposed upon it, but this should not be to the detriment of the *serviceability* of the bridge. It may never reach its ultimate limit state, but it surely will become unserviceable unless proper precautions are taken and every item of the bridge is considered.

Figure 9.3 is a purely *hypothetical* representation of the performance of a bridge (or a bridge element) which is expected to deteriorate from the day it is opened (point A), to the point where it cannot function at all (point B). It is in essence a 'do-nothing option'. The curve from A to B represents a *trend*, it is not a curve in the conventional sense that for any value of (x) there is a corresponding coordinate (y).

A modified form of Figure 9.3 can be used as a visual aid to build up a profile for a bridge (or element) to develop a maintenance strategy, and thereby determine the total cost of the bridge from conception to the end of its nominal life.

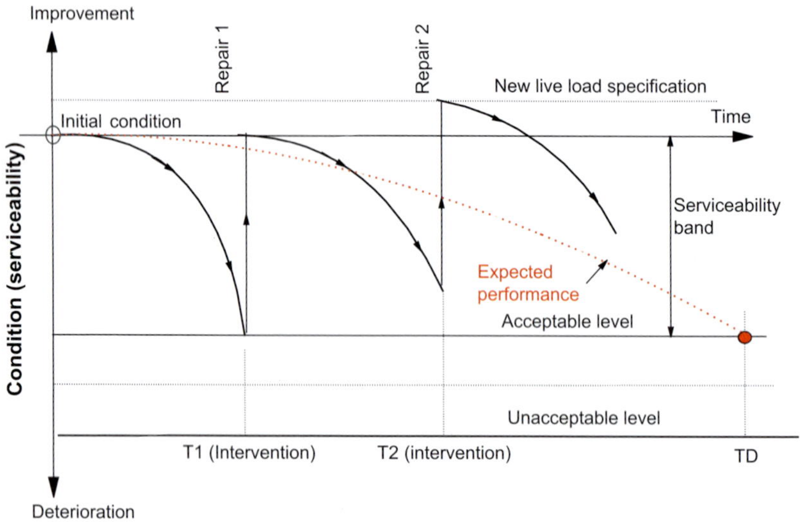

FIGURE 9.3
Deterioration model.

For all bridges the basic information to hand at the time of an assessment would be:

- Its location
- Its type
- When it was built
- Its present condition
- Its design load carrying capacity, W_D
- A history of works (if any)

For bridges already in service, the build-up of the profile for the *load carrying capacity* begins with the *datum or base line* design value of W_D at time T_0. Since all bridges when first built have a reserve of strength, their actual initial load carrying capacity W_I is greater than W_D as shown on Figure 9.3.

The *first* assessment indicates that the bridge has weakened a little, and a decision is made to *strengthen* the bridge up to the W_I level.

The *second* assessment indicates that the bridge has weakened further, and a decision is made to *strengthen* once more but only up to the design level W_D.

The *third* assessment indicates a further, but much smaller, reduction in strength and a decision is made to strengthen once more.

The fourth and fifth assessments indicate that the bridge is holding up well and no strengthening works are necessary. Then just when the bridge engineer thinks that he has the situation under control the bridge *owner*

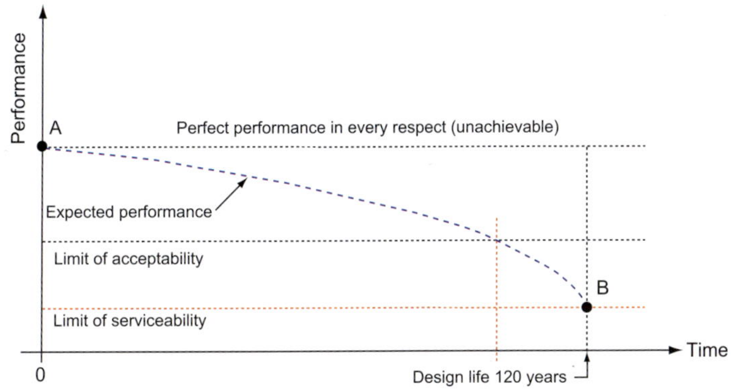

FIGURE 9.4
A typical bridge performance profile for load carrying capacity.

decides that the traffic specification has to increase and so further strengthening works are put in hand, and so the process continues....

Whenever W_I falls below the base line, or the applied load is increased then intervention is required, and appropriate action taken to ensure that the bridge remains in service. After a number of years a typical performance profile would look something like that in Figure 9.4.

Each intervention means that decisions have to be made and money has to be spent in order that the bridge remains 'fit for the purpose for which it was designed'. To optimise the cost, several maintenance strategies and maintenance options need to be considered (see section 9.6.4).

9.7 MAINTENANCE STRATEGIES

Here a *strategy* is defined as 'a method for making, doing or accomplishing something'. It considers plans in the long term. The clear maintenance objective in the case of a bridge stock is to ensure that each bridge continues to function properly up to the end of its design life in respect of serviceability and appearance and is able to carry the loads imposed upon it.

Maintenance work does not have the glamour of new bridge design work. It is often seen as tedious and it requires a different mind set to that for design. The last 10 years or so have shown that this is no longer the case, for not only is bridge management in itself immensely challenging but it requires great depth of thought to get the best service out of what is already on the ground and not simply on paper. In the past (pre-1980) maintenance issues were rarely considered at the design stage and were almost an after-thought. This was because it was considered that concrete was a durable material. Nobody knew of the terrible consequences to concrete in the presence of steel reinforcement and that paint systems on steel would not break down (or if they did, it was not broadcast). In addition the growth in traffic has far outstripped original predictions and as a consequence bridges are suffering stresses that they were not designed for. Alas! We now have to pay for our short sightedness and ignorance.

A sound *maintenance strategy* is essential to the good health of a bridge structure (see Chapter 1) and the best use of scant resources. It is tantamount to ensuring that the bridge has an acceptable level of safety and is serviceable. If no maintenance action is taken then bridges will become progressively less safe with time. The huge dilemma facing bridge engineers is the fact that there are large numbers of bridges to maintain of varying ages, type, condition, loading and importance, all in different environments. It's a bridge manager's nightmare. An old bridge may be of historical significance but require attention; one type of bridge may deteriorate quicker than another; the condition of a stock of bridges will vary widely; a bridge of otherwise medium importance may now be subject to loads greatly in excess of its design loads and an unsafe bridge may be on a trunk route of critical importance to industry; the choice of where to allocate funds is critical.

The 'whole-life cycle' cost for *new* bridges from conception to replacement or demolition enables precautions to be taken at the design stage to minimise damage from whatever source. For *existing* bridges, however, maintenance issues can only be considered in service. Precisely when to *intervene* to cure an ill can be made based on data supplied from the BMS tempered with experience and engineering judgement.

9.7.1 Maintenance planning

Maintenance history is a very helpful (nay some would say *essential*) aid to future planning. If there is a general trend of increasing cost of repair then a decision must be made on whether to continue with annual repairs or to invest heavily so as to restore the value of the bridge.

Bergg (1990) classifies maintenance work into five categories:

Cyclic – this is *preventative* work to prevent the malfunction of bridge equipment and general deterioration such as repainting, joint sealing, drain cleaning, flushing down of concrete faces (to remove chlorides), application of sealants to concrete etc.

Equipment replacement – *restorative* work to re-establish the proper functioning of the bridge and includes bearings, expansion joints, parapets, pinions etc.

Minor structural – to make good superficial damage such as local spalling, surface staining, minor cracking, water leaks, local rotting of timber etc. The repairs should be as good, if not better than, the original.

Major structural – carried out when the integrity of the structure is threatened, and aims to restore the composite action of the structure. Major concrete repairs, cold stitching of cast iron, epoxy resin repairs to timber, repairing tears and buckles in steel etc. (see also Chapter 6).

Bridge strengthening and replacement – carried out when the structure is too weak to carry the existing and projected traffic loads by strengthening, for example, steel and concrete bridges in bending and shear by the use of bonded CFRP (Carbon Fibre Reinforced Polymer) strips. Repairs depend on the nature of the bridge, its general condition (see Chapters 7 and 8), and its impact on traffic, and constitute a major capital cost.

The cost of repairs usually follows the pattern shown in Figure 9.4. If, however, funds are out of step with need then a large backlog of work will build up with possible catastrophic consequences and so priorities and programmes of work must be determined to use the money in the most cost-effective way.

Figure 9.5 also shows clearly that preventative maintenance is probably the best policy where it can be successfully applied, that is on a bridge which has no more than minor structural damage.

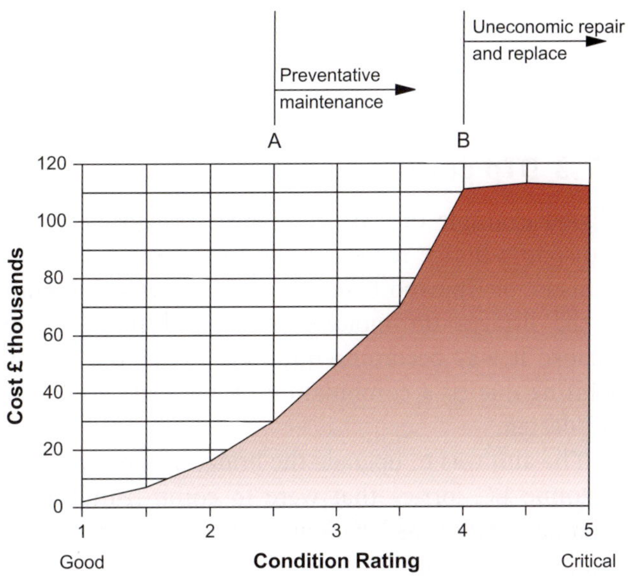

FIGURE 9.5
Cost of repairs-versus-*condition rating.*

9.7.2 Costing of maintenance work

The cost of maintenance work is notoriously difficult to estimate exactly, and very often has to be revised once work has begun – this is certainly the case for the major structural, strengthening and widening options. Quotations, therefore, include lump sum and daywork items and it is up to the bridge manager to ensure, to the best of his ability that a bidding contractor has good access to the structure to inspect the damage for himself and satisfy himself on the most efficient method of working. Maintenance works cannot be priced on paper and sometimes the repair method has to be revised. A recent classic example of this was the *repainting* of the Forth Railway Bridge with an epoxy resin-based paint which is forecast to last for 20 years before another repaint is necessary (see Chapter 6, sections 6.4.1 and 6.4.2.1) thus ending the ritual annual repainting using conventional materials. The work commenced in March 1998 but halfway through the £40 m four-year contract work was stopped and postponed for at least a year. The existing contract was cancelled due to exceptionally bad weather, and difficulties in applying the three-coat paint system (Haywood, 2000). No doubt a different method will be adopted and the price will escalate.

A good BMS will offer a number of maintenance options generated by whole-life costing from which the lowest two or three options can be put out to tender. It may be that a specialist contractor can undertake an apparently more expensive option for less than the cheapest.

Once all the works are completed, feedback on their effectiveness and performance should be fed back to the BMS to increase its information base and to improve the confidence limits of the output.

9.7.3 DTp programme

A good example of a maintenance strategy is that of the Department of Transport's bridge rehabilitation programme. In 1985 in the UK the then Minister of Transport announced a 15-year programme of rehabilitation of the bridges on motorways and all purpose trunk roads. This was because it was becoming quite clear that the nation's bridges were deteriorating due to a combination of a hostile environment, overloading and bad design.

The aim was to upgrade the bridge stocks by 1999 to be able to carry the 40-tonne EC lorries that were to come trundling into the country from 1 January 1999, as *efficiently*, *effectively* and *economically* to meet EC requirements whilst maintaining the nation's heritage.

Table 9.3 Typical Maintenance Strategy

Domain	Work Items
(1)	Core of the programme. Repair of reinforced concrete. Painting of steel beams. Replacement of bearings.
(2)	Concerned with bridges designed for loadings much less than they are today. Mainly short-span arch bridges built in the 1880s before national standards introduced in 1922 but includes some modern short-to-medium-span bridges.
(3)	Rectification of deficiencies where current standards are not being met involving safety and durability and problems peculiar to pre-stressed concrete bridges. Includes: Waterproofing of decks, Rehabilitation of post-tensioned PSC bridges, repairs to pre-stressed precast beams with deflected tendons, replacement of substandard parapets. Structural alterations to deal with effects of 'bridge bashing', strengthening of piers and columns and finally, health and safety aspects of access to structures.

The rehabilitation programme was commenced in 1985 and the *maintenance strategy* chosen was to divide the work into three main domains (Holland and Dawe, 1990): (see Table 9.3).

- Steady state maintenance
- Assessment and strengthening
- Upgrading substandard features

All of this work was planned for a 15-year period to even out the resources available and to ensure that there was not a widespread disruption of traffic. It is the function of the BMS to process all of the bridge condition and budgetary data so that a priority table can be formulated and the work undertaken.

9.7.4 Maintenance options

Once the *overall strategy* has been established, the detail of the process can be worked out by the consideration of various *maintenance options.*

In the past since maintenance was largely ignored it could be classed as the 'do-nothing strategy'. Lee (1990) suggests that this is just one strategy which may be set out in the form of an *array* where the maintenance option can be selected from one axis and the time and date from the other.

A hierarchy of maintenance options could be as follows:

- do-nothing
- monitor
- repair
- strengthen
- replace

Clearly the *do-nothing* option could last for many years until a defect is noticed and, depending on the size and importance of the defect, it would be sufficient to *monitor* it for a further period. Monitoring the situation might be sufficient, for example where a staining of concrete is observed. If during the next inspection delamination is observed, monitoring may still be in order. However, if small lumps of concrete are beginning to fall off onto the road below, then *intervention* is necessary, but it need not be to plan repair works, it might simply be sufficient (if it is practically possible) to provide netting or something similar slung around the offending area to contain spalling concrete. This was in fact done on the arch bridge which carries the A31 over the A3 near Guildford. The bridge was built in 1930. The main load carrying ribs were composed of curved steel beams surrounded with concrete. In about the year 1985 staining was noticed which later developed into cracking and eventually spalling of the concrete which dropped to the carriageway below. In order not to risk a fatality from a piece of concrete smashing into a car windscreen, each of the ribs was surrounded by strong, medium mesh netting. This solved the problem until the time came for maintenance work to be carried out.

On the other hand if, after a sudden event such as a flood or an earthquake, damage is sustained in the form of a sunken pier or a collapsed span, *intervention* must be immediate, albeit in the form of temporary measures to ensure the safety of the users. At such times budget restraints are overcome by events and it usual for the government to 'find' the money to restore the integrity of the bridge.

Lee (1990) provides a good example of *maintenance options* for reinforced concrete portal frame, where he lists five potential classifications of damage with one item subdivided further to represent worsening degradation (see also Roberts & Atkins, 1997). He then considers the maintenance options available and applies a cost to each. A cost comparison is carried out and then a maintenance option is chosen. This method was applied to various elements of the Midland Links Viaducts and the Tees Viaduct, and it was concluded that in some cases the remedial costs can be similar or exceed the original construction costs.

The decision of what to do is based on the Markov Chain of probability as in Chapter 2, section 2.9.6. Figure 9.6 illustrates its use. Suppose the Markov Chain on the left represents the probability tree for a particular concrete element and each number represents a condition classification according to the table on the right.

At point A if there is a greater probability that the element will remain in condition 2 after one year ($P_{2,2} > P_{2,3}$) then appropriate action is taken (apply surface coating). If after one year (point B) the probability of remaining the same is greater than further deterioration then again a surface coating is applied. At point C after another year the chances of remaining the same and deteriorating further is exactly the same ($P_{2,3} = P_{2,4}$) and so an engineering judgement is made to act or do-nothing (the latter is cheaper). At D if the probability of progressive deterioration to classification 4 is greater than remaining at 3 then action is taken (application of cathodic protection). The 'do-nothing' route A, F, G, H, J in this extreme example would result in condition 5 after only four years and be very costly. In fact this would be disallowed and at the target level (say level 3) intervention would take place and 'doing nothing' would not be an option. Clearly there are many possible pathways to consider resulting in a great amount of calculation which can be handled by suitable techniques within the BMS.

Performance profiles can also be used to define the various elements of a bridge in terms of their susceptibility to degradation and failure. The following list is a sample of how elements may be defined as *weak or strong*.

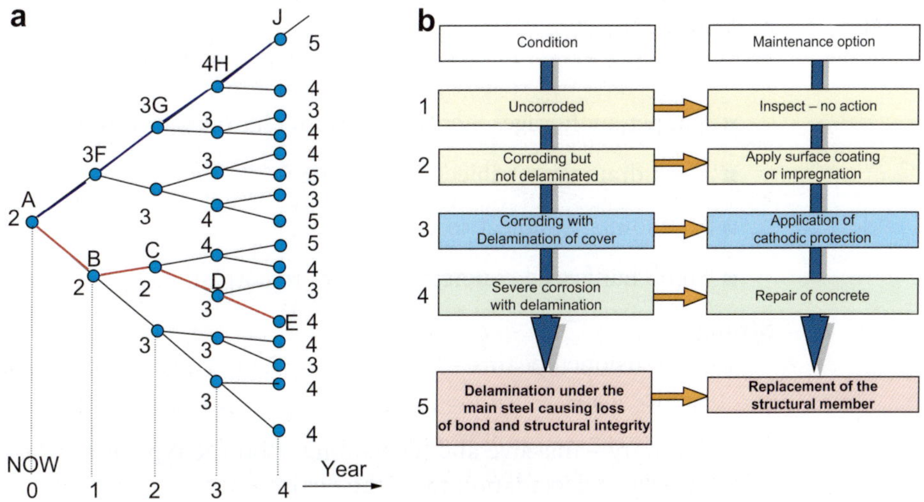

FIGURE 9.6 *Relationship between the Markov Chain and maintenance options.*

9.7.5 Weak and strong elements

In many countries of the world certain elements of a bridge are simply defined as *weak or strong* depending upon their inherent ability (or otherwise) to resist the factors which cause degradation. Some of the more typical ones are (see Table 9.4):

Table 9.4 Improving of Weak Materials

Country	Improvement
Canada	Recognise Micro-climate Zone defence policy *For example they provide epoxy-coated bars in sensitive areas*
Denmark	Policy of Sacrificial layers *e.g. Substructures: 100 mm cover and stainless steel mesh or 75 mm cover plus epoxy-coated bars*
Japan	Reduce bearings Reduce joints New bridges to be continuous
USA	New works to eliminate materials or details with low performance profiles (the definition of low is not given)

Weak

- Expansion joints – because the anchorage is an inherent weakness – no-one has yet invented the perfect expansion joint.
- Depending upon how they perform Metallic bearings – subject to corrosion.
- Parapet anchorages – corrosion from chlorides in water.
- Deck drainage – subject to blocking.
- Half joints – subject to corrosion.
- Some reinforced concrete – subject to corrosion.

Strong

- Pre-tensioned beams – because there is no cracking under transfer or service loads.
- Masonry – massive and (depending upon the type of rock used) not subject to degradation from the environment.
- Reinforced soil – generally made from corrosion resistant polymers.

Weak elements can be improved in a variety of ways as shown in Table 9.7.

9.7.6 The cost problem

In a particular County in the UK, the results of a survey of the state of the bridges in the County revealed that:

- 220 were substandard
- 35% under strength
- five had width restrictions
- 12 had weight restrictions
- 22 had been strengthened
- 11 had been re-decked
- 25 had been rebuilt

Of this list the 35% understrength bridges cause the most alarm and have to be prioritised. The first reaction is to panic! but there is no need to if the results are managed in a responsible way. The bridges will most probably have been assessed using simple techniques and so the next course of action would be to follow a logical sequence to see in fact whether their actual strength is greater by:

- Carrying out a more refined analysis
- Load testing to establish proof load

If this still indicates that the bridge is unsafe then either:

- the bridge will have to be strengthened, or
- a width and weight restriction will have to be imposed on the bridge.

If the bridge is deemed to be marginally under strength then it can be monitored over a period of time to see whether degradation is continuing or whether a steady state has been reached. This requires fine engineering judgement which can only come from an experienced senior bridge engineer in consultation with his colleagues.

At the very worst the bridge will have to be demolished and rebuilt.

Whatever decision is taken it has to be costed and the most efficient way of doing this to ensure optimum use of funds is to carry out a cost analysis.

9.8 HIGHLY DURABLE BRIDGES

We all know the frustration of traffic delays due to bridge works. These cannot always be quantified in economic terms, but they certainly can in terms of levels of stress and risk to health due to moving at a snails pace (or not moving at all) for hours in a traffic queue inhaling poisonous exhaust fumes.

There are considerable advantages to be had in spending more money at the start of a contract to produce a bridge which is both more durable and can be easily inspected and maintained. Maintenance can be reduced to a minimum by providing easy access to all parts of the bridge (especially expansion joints and bearings); providing surface treatments for the protection of both steel and concrete; ensuring that water is effectively drained, and paying attention to details. Although initially such measures add expense, in the long run the whole-life cost will be cheaper.

The old Britannia tubular railway bridge across the Menai Straits is a good example of this principle. It comprised wrought iron tubes supported on masonry abutments and piers and was constructed between 1846 and 1850. When dismantled in the early 1970s because of severe damage due to fire there was no evidence of deterioration from the environment. Considerable time and money had been spent in ensuring that all parts were durable. When the end sections were lifted, the bearings composed of timber packings impregnated with a mixture of putty and red lead, and the gun metal roller bearings were in pristine condition. No areas of the wrought iron had corroded, and the timber sleepers and iron rails were in sound condition. The masonry towers and piers were unaffected by time, and indeed have been used to support the new two level pedestrian/rail/road bridge. The whole wrought iron structure had been painted with red lead paint, and the inside was covered with a mixture of coal tar, slaked lime and turpentine. Later in the twentieth century tar soaked wooden planks were placed across the tops of the tubes to provide waterproofing, and the outside coated in tar to protect it against the ravages of the environment, and ironically it was this that did for it in 1970 when two boys were chasing bats from inside the tube with lighted tapers. They were disturbed by a local and the two boys dropped their flaming torches onto the pitched iron with disastrous consequences (North Wales Chronicle, 1970).

At the other end of the scale is the Forth Railway Bridge which was the first all steel bridge ever built. It was opened in 1890 and since that time has been continually painted to ensure its durability. Some angle plates and bracings corroded by steam and smoke have also been replaced, as also have some timber sleepers (Beckett, 1990). This has meant that the maintenance costs have far outstripped the original cost, but for such a landmark bridge both in its location and its historical and engineering importance it would be considered a price worth paying. It has now been decided that the original hot-dipped linseed oil protection will be grit blasted off down to bare steel and be replaced by a zinc rich priming coat followed by two further coats of two-part epoxy paint (Jones, 1997) to provide a 20-year protection. It should be very durable and require only occasional touching up.

De Brito and Branco (1996) has pointed out that too often bridge designers do not consider the long term, and the decision criteria has been to produce a bridge that is both serviceable and cheap. The result, very often, is a bridge that is *'difficult to inspect; expensive to maintain and quick to deteriorate'*.

9.8.1 Bridge enclosures

The concept of a bridge enclosure to provide both protection and a means of maintenance access was first mooted in TRL (1980) (see Chapter 8, section 8.4). In effect it consists of a durable material – usually glass fibre reinforced polymer – manufactured in sections in such a way as to provide an envelope around the bridge below deck level. Not only does it provide clean lines but it prevents corrosion (especially of steel bridges); provides a safe environment during maintenance operations at any time and prevents traffic disruption. To date, the few that have been provided have proven their worth in reducing considerably the whole-life cost (Irvine & Thorpe, 1996).

9.8.2 Whole life assessment

Whole life assessment (WLA) is a way of obtaining data for developing the *whole life performance* (WLP) profile of a new or existing bridge thus enabling a more accurate minimum WLC to be calculated.

In very broad terms a WLA focuses on specific structural elements or components as the outcome of a Particular Assessment (see Chapter 3) to establish their current and future performance levels in terms of *performance indicators* $I_{sub.}$ defined in Departmental Standard BA81 (2000). The indicators from the WLA (see Chapter 3) are then used to determine what sort of maintenance action is required – if any – and the associated (minimum) costs calculated. (see Das, 2000) for a fuller description of this process.

9.8.3 Asset management

Until a few years ago the term *asset management* did not appear in the plethora of published information on bridge management by way of Conference Proceedings; Research programmes, government initiatives, Codes of Practice or News items. Until, that is, the UK Bridges Board was set up in 2002 to bring together national and local government from across the UK to consider bridges and highway structures and engineering matters, in other words a *talk-shop*. Now, under the umbrella of the UK Bridges Board it is going to be a requirement for those managing highway networks to participate in the '*service wide application of asset management*' which, it says, is '*a new concept*'.

So what is Asset Management? In the business world, it is basically the 'art of looking after what we own', and a typical financial institution, which one may find in the City of London, seeks to increase the financial assets of its customers by investing in various funds throughout the world with a view to maximising the total value, or by purchasing a run-down business and then asset stripping, or to put it another way and relating it to transport infrastructure: 'A *strategic approach that identifies the optimal allocation of resources for the management, operation, preservation and enhancement of the highway infrastructure to meet the needs of current and future customers*' (County Surveyors Society, 2004). To help us understand what that means, a companion publication drafted by W. S. Atkins entitled: *Guidance Document for Highway Infrastructure Asset Valuation*, was published in 2005, and is available from The Stationery Office.

What little published data there is on this topic seems to suggest it will be incumbent upon all local authorities to produce a Transport Asset Management Plan (TAMP) which is to include all transport matters within their remit, and, according to Cole (2005) '*all local authorities will be required to prepare accounts which explicitly value all transport assets and evaluate the level of depreciation*' (Cole, 2005).

So far Surrey County Council is the only authority in the UK to have produced a TAMP, which is the first step towards a fully integrated Structures Life Cycle Plan (SLCP).

Cole (2005) in his conclusion to a recently published paper on the subject states:

> *'Asset management does not replace existing good practice, instead it provides the framework within which this practice may be more effectively implemented, managed and complemented by other processes and actions'.*

To help in the preparation of the information required, CIPFA (2005) have produced a Code explicitly aimed at Local Authorities. The information to be provided is set out under the following headings, and provides a panoramic view of what is intended for the framework for Highway Asset Management (HAM).

- **Starting Point**
  Goals, Objectives and Policies, Valuation and Inventory

- **Levels of Service**
  Condition Assessment and Demand Aspirations

- **Option Identification**
  Performance Gaps and Lifecycle Planning

- **Decision Making**
  Optimisation and Budget Considerations, Risk Assessment

- **Service Delivery**
  Forward Work Programme and Physical Works and Services

- **Reporting and Monitoring**
  Performance Measurement and Improvement Actions

A quick scan of this list is enough to see that the required engineering skills are considerably different from the traditional ones of 'Analysis, Design; Construction and Maintenance'. This will mean that suitably qualified staff will need to be recruited to undertake the tasks involved. It is money well spent, and is certainly better than sending some of the existing engineering staff on short courses etc. who will have to cope with new learning curves for work that they had not initially trained for, nor were they being paid for.

9.8.4 Design, Build, Finance and Operate (DBFO)

It has been suggested by Tilly (1995) that the introduction of design, build, finance and operate contracts will provide scope for changes in balance between construction and maintenance costs. The preferred option might be to reduce construction costs and accept a more expensive maintenance regime thus reducing the initial costs which might have to be financed by expensive loans. Tilly (1995) rightly indicates that although apparently financially attractive in the short term this will result in increased risks of unexpected maintenance requirements and lower levels of safety, and high costs associated with traffic delays.

DBFO contracts are priced on the profit motivated estimate of the true cost of a bridge, whereas the budget constraints of a public agency are not necessarily required to be nearly so exacting. The Government Agency appoints a concession holder on behalf of the Client (normally the Government) who would normally solicit the services of a Contractor and a Consultant. It is then up to all parties to agree an appropriate balance of construction and maintenance costs. The Consultant will want to ensure that the bridge is one that he can be proud of and which will enhance his reputation; the Contractor will want to build and depart with the minimum of fuss, and the concession holder will want to make as much profit as possible. The condition for the bridge at handover of the bridge to the Client is normally 30 years, and so the Contractor will want to know the latest date at which he can carry out maintenance to ensure that the bridge is in an agreed condition. A DBFO contract thus forces all interested parties to find accurate information pertaining to the quality of the bridge and the probable level of tolls because they are basing investment decisions on it and have shareholders to satisfy. (The tolls may be a hard collection at a toll gate or, what is more common, a shadow toll payment based on either the number of vehicles or vehicle journey times.) Somehow it has to be to the benefit of the DBFO contractor to maintain a bridge in good order. This could perhaps be achieved by ensuring that an agreed third party inspects the bridge at specified intervals who then reports directly to the Client and the Contractor, and that if the maintenance requirements are not met on a continuous basis by the contractor, rather than at 30 years only, then the Client has powers to activate a restrictive penalty on the amount of toll receipts.

9.9 CONCLUSIONS

The method of whole-life costing is considered to be a rational means of comparing the costs of alternative bridge designs and maintenance regimes, and to perform long-term economic analysis at the design stage. It is not an exact science and for that reason sound judgement is necessary based on many factors. After first cost, the main parameters to quantify are the test discount rate, the service life of the structure and the maintenance strategy. Individual items on a bridge to be considered would be the main construction materials, deck waterproofing, surfacing, parapets, expansion joints, bearings, drainage and service ducts, substructures and decorative panels. TD costs would also need to be estimated. Many countries of the world are now adopting some form of performance profiling to classify deterioration susceptibility, but unforeseen deterioration due to vandalism, accidental damage, environmental disasters, increase in volume and density of traffic cannot be allowed for. For the future it is anticipated that performance indicators using reliability methods (see Chapters 3 and 12) will be adopted and that as a result, an acceptable level of reliability can be achieved in tandem with a more accurate minimum WLC.

REFERENCES

Beckett, D. (1990). The Centenary of the Forth Rail Bridge, 1st International Conference on Bridge Management, Inspection, Maintenance. *Assessment and Repair,* pp. 1–16. London: Elsevier, University of Surrey, Guildford, UK.

County Surveyors Society. (2004). *Framework for Highways Asset Management*.

Departmental Advice Note BA28. (1992). *Evaluation of Maintenance costs in comparing alternative designs for Highway Structures*. London: SO.

Departmental Standard BD36. (1992). *Evaluation of Maintenance costs in comparing alternative designs for Highway Structures*. London: SO.

Departmental Standard BA81. (2000). *Whole life performance based assessment of Highway Structures and Structural Components, Design Manual for Roads and Bridges*. London: HMSO.

Das, P. C. (2000). *Reliability based bridge management procedures*, Proc. Fourth Bridge Management Conference, Guildford, UK. Thomas Telford, London.

De Brito, J., & Branco, F. A. (1996). *Whole Life Costing in road bridges applied to service life prediction*, 3rd International Conference on Bridge Management, Inspection, Maintenance, Assessment and Repair. University of Surrey, Guildford, UK, pp. 603–612. E and F N Spon.

Haywood, D. (2000). *Repainting of Forth Bridge abandoned*. New Civil Engineer. 9th March issue.

Holland, D. A., & Dawe, P. H. (1990). *Bridge Rehabilitation: Department of Transport's Fifteen-Year Strategy,* 1st International Conference on Bridge Management, University of Surrey, Guildford, UK, pp. 145–153, Elsevier, London.

Irvine, R. A., & Thorpe, J. E. (1996). *Bridge Enclosure: Facilitating construction, inspection, maintenance, upgrading and operation,* 3rd International Conference on Bridge Management, Inspection, Maintenance, Assessment and Repair. University of Surrey, Guildford, UK, pp. 429–436, E and F N Spon.

Jones, C. J. F. P. (1995). Performance Profiles and the concept of Zone Defence, One Day Seminar on Whole Life Costing – Concrete Bridges. *Concrete Bridge Development Group* 62–76.

Lee, D. (1990). *Comparative Maintenance Costs of different Bridges*, 1st International Conference on Bridge Management, University of Surrey, Guildford, UK, pp. 145–153, Elsevier, London.

North Wales Chronicle, *Fire at the Britannia Bridge*, August 1970.

Quadro 3. (1996). *DMRB Vol. 14, Economic Appraisal of Road Maintenance, (plus amendments May 1997 & Nov. 1997)* Department of Transport (Now Highways Agency), UK.

Roberts, M., & Atkins, C. (1997). Deterioration Modelling. *Concrete Engineering International, Vol. 1*(No. 1), 38–40.

Tilly, G. P. (1995). Financial Implications of Whole Life Costing, One Day Seminar on Whole Life Costing – Concrete Bridges. *Concrete Bridge Development Group* 1–14.

TRL. (1980). *Enclosure, an alternative to bridge painting*.

FURTHER READING

Browne, R. D. (1986). *Practical considerations in producing durable concrete*, Seminar on the Improvement of Concrete Durability, ICE, pp. 97–116.

Das, P. C. (1997). *Whole life performance-based assessment of highway structures: proposed procedure*. In P. C. Das (Ed.), *Symposium papers on The Safety of Bridges*. London: Thomas Telford.

Das, P. C., & Micic, T. V. (1999). *Maintaining Highway Structures for Safety, Economy and Sustainability, Ninth Conference on Structural Faults and Repair*. Edinburgh: Engineering Technics Press.

Ferry D. J. O., & Flanagan, R. (1991). *Life Cycle Costing (1991) – A Radical Approach*, CIRIA Report 122.

Frangopol, D. M., & Estes, A. C. (1998). *Optimum design of bridge inspection/repair programs based on reliability and cost, Conference on The Management of Highway Structures*. London: ICE.

Jones, D. R. B. (1995). The Appraisal Of Traffic Delay Costs, One Day Seminar on Whole Life Costing – Concrete Bridges. *Concrete Bridge Development Group* 27–36.

Jones, M. (1997). *At last, an end to the never-ending job*, NCE, 11/25 December, p. 4.

Roberts, M. B., Atkins, C., Hogg, V., & Middleton, C. (2000). *A proposed empirical corrosion model for reinforced concrete*. *Proc. Instn Civ. Engrs Structs & Bldgs, Vol. 140* (No. 1), 1–11.

The Institution of Asset Management and BSi. (2004). PAS 55–1 Asset Management: Part 1: Specification for the Optimised Management of physical Infrastructure assets.

The Institution of Asset Management and BSi. (2004). PAS – 2 Asset Management: Part 2: PAS 55-2 Guide Lines for the application of PAS 55–1.

International Infrastructure Management Manual, UK Edition, 2003.

CHAPTER 10

Durability and Protection

CONTENTS

10.1 INTRODUCTION

Proper management of a bridge includes making it as durable as possible and should be considered in detail at the design stage. Wood (1996), however, has observed that of the time allocated for a complete bridge design and specification, bridge engineers spend typically 1% of the time on durability design, yet durability faults contribute to over 99% of the deterioration defects in concrete bridges. In Chapter 1 reference was made to the appalling condition of many of the worlds bridges simply by neglecting this fundamental requirement of durability. Hopefully we have all learned from our mistakes and there is now a culture of *prevention* rather than *cure* relating to our bridge defects. At the heart of this culture is the desire to produce bridges which are *durable*.

Adequate *protection* is an obvious solution to some of the causes of deterioration, and *monitoring* is a way of warning us of an impending critical situation.

10.2 DURABILITY

The first explicit and codified recognition of *durability* can be found in Clause 210 of CP114 for the Design of Reinforced Concrete in Buildings, 1969 (see Appendix 10A). In the same Code there is reference to chemical attack (see Appendix 10B) and it clearly states that *'The rate of attack may be reduced by good compaction and the use of a hard impermeable aggregate'*.

In the first publication of a limit states code relating to reinforced concrete buildings (BS 8110) the durability requirements of *adequate cover to steel*

CONTENTS

reinforcement and sound concrete are reiterated, and the cover limits are given as a maximum of not less than 50 mm, and a minimum of 20 mm.

What both of these Codes imply is that no matter what cover is stated on the drawing, if the concrete is not properly *designed, mixed, poured, compacted and cured* then the durability of the structure is compromised. Beeby (1993) has pointed out with great force that it is futile to rely only on the Code requirements for cover if the manufacture and placing of the concrete is poor.

The obverse of this is also true, and can be illustrated by a very important event which took place during the second world war. On D-Day 6th June 1943, allied forces landed on the north coast of France via three multispan pontoon or floating bridges, which were part of a temporary harbour (Code name Mulberry) constructed to enable men and machines to cross the English Channel and commence the liberation of Europe (see Appendix 10A for general layout of the harbour). Each bridge consisted of steel battle deck spans supported on concrete 'floats'. Each 'float' measured 84 ft long × 22 ft 6 ins wide × 10 ft 6 ins high consisting of pre-cast concrete panels to the bottom, sides and top, between concrete stiffeners at 1.22 m c/c. (see Appendix 10B for mid-ship section). A typical panel was only 2 ins. thick with reinforcement consisting of a mesh of 10 mm bars @ 150 c/c vertically and 10 mm bars @ 75 mm c/c horizontally, leaving a cover to the reinforcement of just under 20 mm, and yet CP 114 specifies a minimum of 40 mm to 'all external work', and BS 8110 specifies 40 mm for a 'severe' condition of exposure.

Seven years after the event Mr K. Wigmore of the then Mouchel and Partners who designed the 'floats', quite by chance came across one of the panels on the sea shore at Bognor Regis. He found the panel was *'in excellent condition and the bars uncorroded'* (Hartcup, 1977). How does one explain the excellent condition of the panels, in a marine environment, suffering wetting and drying and the incessant pounding of the waves, and without any maintenance whatsoever. I doubt that we shall ever know, but the concrete mix might shed some light on it (see Appendix 10A).

From the foregoing it is clear that durability is difficult to define exactly, but a good definition is offered by Holland (1993) as: *'the ability of materials or structures to resist, for a certain period of time with routine maintenance, all the effects to which they are subjected, so that no significant change occurs in their ability to carry the loads or to remain serviceable'*, or to put it another way, the ability of the components or the materials that make up the bridge to resist unacceptable breakdown or deterioration.

Durability affects both serviceability and safety in that deterioration of materials can lead to loss of 'function' and also mar the 'appearance' of a bridge, and corrosion can lead to *loss of strength*. Durable bridges are ones which change little with time, that is, *they last.* They are economical bridges having a long service life and low maintenance requirement leading to low life-cycle costs.

By dint of our limited technology, imperfect construction practices, and inability to predict all of the causes and extent of deterioration and breakdown of protective systems, it is not possible to produce 100% durability. What can be done is to consider *on paper* the effect of the design, detailing, construction and inspection (from conception to eventual decommissioning) on the overall durability of the bridge, and to ensure that 'best practice' is followed through every phase of the life of the bridge. The whole process is one of iterative synthesis and is far removed from routine structural calculation and drafting.

By definition a structure is in a state of limiting equilibrium if a small increase in load will cause collapse. It is said to be at the *ultimate limit state* (ULS). Without question, this is the most important state to consider at the design stage, and ensures (all other things being equal) that the bridge has an acceptable factor of safety against collapse.

There is a danger of concentrating so hard on the ULS, however, that the serviceability limit states (SLS) such as *durability*, general *appearance* and *function* are considered as secondary matters (if they are considered at all!). They are, however, of vital importance if the bridge is to stay in service for its design life. A bridge, apart from possessing overall aesthetic appeal, should be of clean *appearance* with absence of staining, cracking, graffiti, flaking paintwork, rotting timber, and the like. Every part of the bridge should *function* properly so as to ensure a safe and comfortable ride. This includes sound operation of the bearings and expansion joints, proper drainage, smooth surfacing finish, adequate lighting, and acceptable limits of vibration and displacement.

The result is, that in addition to 'number crunching' to prove strength, it is necessary that the bridge should be:

- aesthetically pleasing
- of a favourable appearance
- well detailed
- able to be constructed with relative ease using appropriate construction techniques
- easily accessible for inspection
- cheap to maintain

- safe for users, with adequate containment; separation of pedestrian and motor traffic; drainage etc.
- comfortable for users by minimising deflections, vibrations, uneven running surface etc.
- free of unsightly cracking, staining, deflections etc. to ensure user confidence

It should then be possible to draw up an inspection programme for the entire life of the bridge.

The Department of Transport in the UK issued an advice note in 1995 (BA 57/95) entitled 'Design for Durability' the clear aim being to encourage prevention of deterioration through the application of sound, practical design and detailing. BA 57/95 and its companion BD57/95 are divided into four sections: the *Conceptual Stage; Problem areas; Detailed requirements and Steel bridges.* Although very prescriptive, they do encourage thought and hopefully will result in attractive, longer lasting bridges.

10.3 WATER

Water is enemy number 1 and is the cause of many of the ills that beset our bridges. Not only does it precipitate corrosion directly if in the presence of oxygen, but it carries harmful de-icing salts in solution to all parts of bridge decks in the form of running water, and spray from vehicular traffic. It flows over (and sometimes through) expansion joints, and it can trickle through to bridge bearings. Once on the bearing shelf it continues to flow down the vertical faces of piers and abutments. If it percolates into the fabric of the bridge then there is the danger of freeze-thaw damage.

Piers and abutments are splashed with salt laden water from vehicles passing under a bridge, and foundations are prey to ground water bearing harmful sulphates. Bridges built over water may have their foundations undermined by the effects of scour, and flood conditions can cause collapse. Bridges over salt water are subject to chloride damage in the splash and tidal zones, and microbiological and chemical attack at the mud line.

Long span 'lattice-type' steel bridges are subject to corrosion where water can pond within the structural members and seep into bolt fixings and gusset interfaces. The main cables and towers of suspension and cable-stayed bridges are vulnerable to wind-blown water which may have salt in solution if situated in coastal regions.

Water needs careful management. Either the water has to be directed away from vulnerable areas or the bridge must be protected in some way. Thus there are two major ways of dealing with water problems which are:

1. to ensure that the water can be collected and carried away quickly
2. to protect the bridge by surface treatments, electrochemical apparatus or structural intervention.

10.4 DRAINAGE

10.4.1 Surface drainage

Surface drainage is vital and it is essential to ensure that water is collected and transported away from the deck quickly and efficiently so that no ponding occurs on the deck and water is diverted from expansion joints. Conventionally this is achieved by laying the surfacing to a fall of not less than 1 in 40 to allow the water to flow to collection gullies placed at intervals along the kerb line. The run-off is then transported to an off-bridge drainage system via pipes hung below or cast into the deck.

Hydraulic design of the *drainage pipes* is essential to ensure that there is no backup of water in a downpour resulting in ponding on the deck surface. If possible it is better *not* to cast the pipes into concrete, but rather to hang them from the deck (either in a void or between beams) or placed in a service trench at the edge of the deck. This facilitates replacement in the event of pipe failure. The drainage system should be robust and made from materials that are tough and resistant to the spillage of normal chemicals.

Off and on the bridge the connections to the main drainage system should incorporate practical details such as rodding eyes to facilitate cleaning, and junctions should be formed within manholes. If there is the possibility of leaking, then an alternative escape path should be provided to prevent build-up of water in the cells of box girders and cellular or voided slabs. This can be in the form of small diameter *scupper pipes* in the soffit with their ends clear of the soffit line as in Figure 10.1.

10.4.2 Sub-surface drainage

10.4.2.1 At expansion joints

Because of their permeability, sorptivity (the rate of absorption into a dry material measured as the 'slope of the line obtained from a plot of the volume of water absorbed per unit area of surface and the square root of the absorption time') and defects, the surfacing layers (especially porous asphalt)

FIGURE 10.1
Scupper pipe surface drainage and sub-surface drainage at expansion joints.

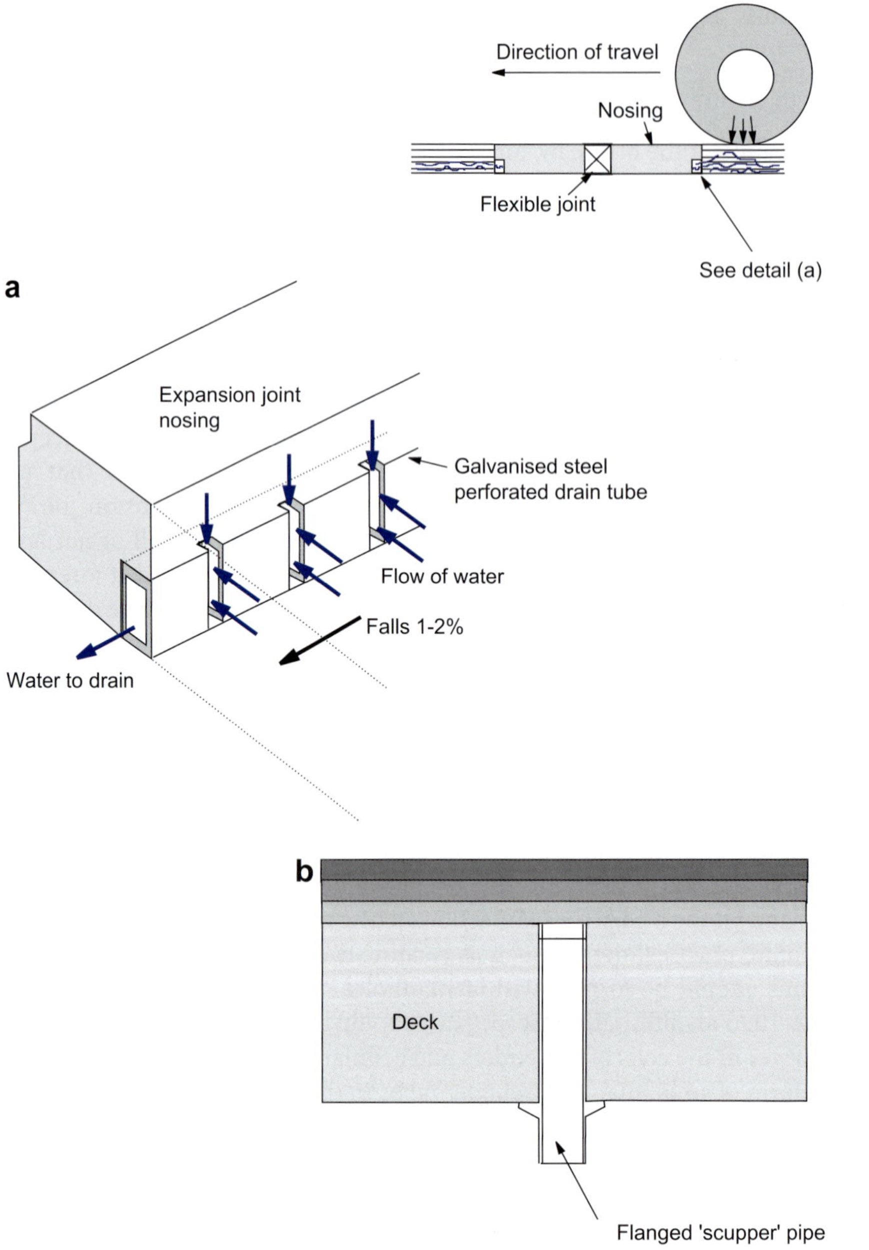

on bridge decks allow water to migrate through and along the layers and to penetrate down to the waterproofing layer (see later). At this level it is free to flow in all directions, but when it reaches the point where the surfacing abuts up to the edge of an *expansion joint nosing or kerb stone*, it stops. Of itself this is not harmful (unless it is salt-laden or there is so much water that it lies above surface level and ponds), but the presence of moving traffic means that when the wheels reach the surfacing/expansion joint/kerb intersection, there is a build up of water pressure which can damage the surfacing locally.

In *expansion joints*, therefore, it is prudent to provide a small continuous perforated drainage channel at that location which will not only dissipate the water pressure but collect the water and carry it to a drainage system within the deck (see Figure 10.1).

Cuninghame (1998) carried out a limited study on the transport of water through some typical samples of bridge deck surfacing materials. Permeability, capillary and sorptivity rates were measured and recorded, and he concludes that a generous number of deck drains [of the scupper type] should be installed, of sufficient size to enable rodding and avoid blockage by crystallising salts (see Figure 10.1).

10.4.2.2 At the kerb line

There is a similar problem in areas adjacent to the kerb line and until recently the problems of *sub-surface* water had not been fully appreciated. Although water has been removed from the surface of the deck, contaminated water still remains trapped within the sub-surface layers. Heavy wheel loads travelling adjacent to the kerb units result in the sudden and repeated build-up of hydrostatic pressures in the sub-surface water which is pushed away from the kerb in one direction, but is pushed towards the kerb in the opposite direction. The water also travels upwards carrying fine particles from the lower layers causing erosion. Eventually the carriageway breaks up and repairs are necessary.

In 1996 an innovative *dual purpose drainage system* was launched that dealt with these problems (Cooper Clark,1999). Instead of individual gullies placed at intervals along the deck, this system comprises a run of kerb units with built-in drainage inlets which allow a continuous drainage aperture for the surface water and the sub-surface water together with an integrated drainage system. In effect *it replaces the traditional combination of separate concrete kerb units, drainage gullies and drainage pipes by one multipurpose unit* (see Figure 10.2). Each unit comprises a base unit of cast or ductile iron which interlocks with its neighbour so that vehicle impact forces are distributed, and a top unit made from a high strength polyester composite.

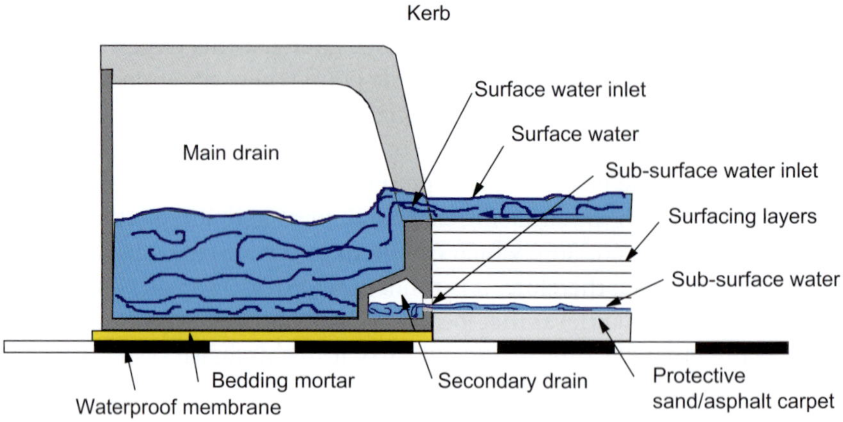

FIGURE 10.2
Kerb drainage units.

They are 500 mm long with variable widths from 300–450 mm and variable heights from 75–125 mm. One very interesting feature is that the surface and sub-surface water each flow through separate apertures and into separate drainage channels. This ensures that back pressure from the surface drainage channel does not prevent the flow of the sub-surface water. The units can also be adapted to be continuous at expansion joint locations. Along any given length there are a number of special inspection units.

Behind and parallel to the kerb line, High Density Polyethylene (HDPE) pipes (typically 100 mm diameter) direct the water in to the abutment drainage channel where it is transported to the main drainage system at ground level.

A similar idea has been used to produce bridge parapets with built-in drainage channel support made form Glassfibre Reinforced Cement (GRC) – see section 10.3.12.

10.5 EXPANSION JOINTS

Most expansion joints in the UK designed to operate in the 0–80 mm range fail between 5–10 years after installation (Cuninghame, 1992). They have a poor durability record and seem unable to cope with the shock loadings and vibrations from the daily passage of traffic on our roads, neither do they prevent the passage of shock through them to other vulnerable parts of the structure (Raggett, 1992). Depending upon the type of bearing, they are subject to rutting, cracking, debonding, binder flow, binder stiffness, anchorage failure and general deterioration from environmental effects such as temperature and humidity (Johnson, et al., 1992).

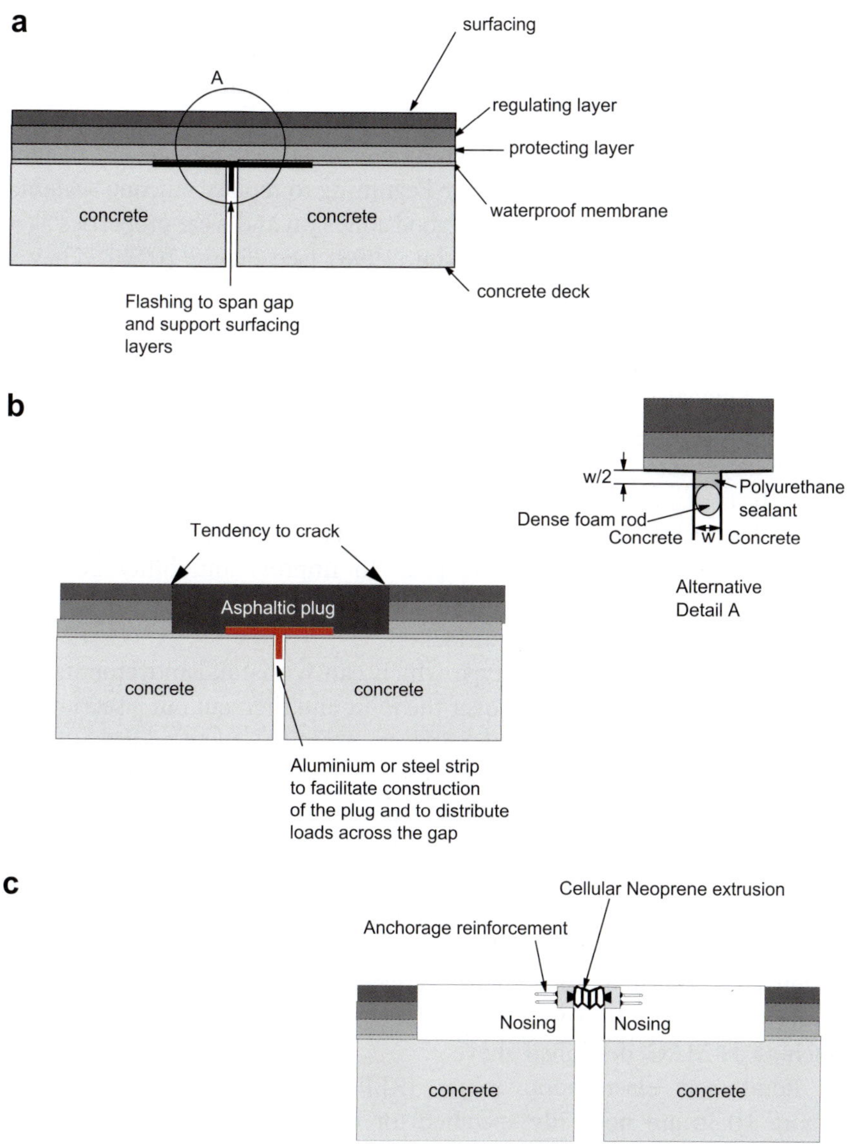

FIGURE 10.3

(a) Surface;
(b) Asphaltic plug and
(c) Preformed extrusion between nosings.

On short span bridges (up to about 50 m) the annual movement expected due to temperature is of the order of ±9 mm. Up to 5 mm movement can be accommodated by utilising the flexibility properties of the surfacing which is simply carried over the top of the joint without a break as in Figure 10.3(a). Unfortunately the surfacing on this type of joint often *cracks* because it *lacks* the necessary flexibility. To deal with this and improve durability, Woodside et al. (1990) have recommended the application of a strip of 'H-MAC' which uses 'Icosit Membrane H – a hand-applied two-pack solvent-free fast-curing polyurethane resin combination...' as a *binder* which results in a stronger and more flexible surfacing material than traditional bituminous-based materials. Instead of the metal flashing it is possible to provide one of a range of polyurethane sealants (which are beginning to replace silicone sealants in the building industry) as they have good adhesion and wear properties as well as other desirable attributes (Wilson, 1999) (see Figure 10.3a). They are, though, strictly limited to movements of ± 5 mm.

For movements of between 5–10 mm the use of a *buried joint* which is similar to the surface joint, but with an additional elastomeric pad sandwiched between a local thickening of the waterproof membrane and set in a recess in the top of the concrete deck is recommended. There are many proprietary joints of this type on the market (Lee, 1994), the common factor being the continuation of the surfacing over the joint. These are very prone to failure because the local wheel loads cannot be spread over a wide area due to the softness of the elastomeric pad. To improve durability Lee (1994) recommends the use of expanded metal reinforcement in the joint and a rubberised polymer asphalt surfacing.

A slight variation of this type which can withstand movements in the range 15–45 mm is to surface over the joint and then cut out a section about 500 mm wide and replace it with an *asphaltic plug joint* (APJ) as in Figure 10.3b. They are very sensitive to the quality of installation, but providing that there is a good bond between the contact surfaces with the bridge and the surfacing, then they have proven to be very reliable (Johnson et al., 1992). Good bond can be obtained if after the saw-cut has been made, the vertical sides of the surfacing could be roughened, and also the APJ joint installed shortly after the deck surfacing has been laid to minimise any differential thermal movement between the plug and the surfacing (Cuninghame,1994). Another solution could be to use an alternative binder such as 'H-MAC' described above.

Reinforced elastomeric joints (REJ) such as the one shown in Figure 10.3c are normally specified for movements up to 80 mm. They have two major weaknesses: failure of the nosing and/or breakdown of the bond between the nosing and the bridge deck. Nosings made from epoxy resin should be avoided because the resin shrinks, and when set, it has different thermal properties to the neighbouring concrete. This results in unfavourable shear stresses on the contact surfaces and a breakdown of the bond. A more effective nosing material is that of fibre reinforced cement (FRC) which, probably because it has similar properties to the surrounding concrete, has proven to be very durable.

Elastomeric in Metal Runner joints (EMR) catering for movements up to 480 mm generally perform well but it has been found that they are prone to leakage and break-up of the surfacing due to water pressure.

The issue of break-up due to water pressures was dealt with in section 10.3.2.

10.6 DECK WATERPROOFING

Although sound, well-mixed, well-placed and properly cured concrete is practically impervious, waterproofing *is normally* provided to the top of the concrete deck slab to protect it from water ingress; freeze-thaw action and de-icing salts. It can be applied in the form of either a *cold* sprayed membrane or *hot* hand laid sheets, and protected by a layer of sand-asphalt approximately 20 mm thick (see Figure 2.17).

The hand laid sheets (or membrane) are about 5 mm thick and may consist of a modified bitumen coating on a spun-bonded polyester carrier or a polymer modified bitumen membrane reinforced with polyester and glass. They are supplied in rolls typically 1 m × 8 m long. Both types are finished with a 'torch-on' polythene film on the underside and talc on the topside.

A typical hand laid procedure (Flexideck, 1994) consists of ensuring first of all that the bridge deck is *'clean, dry and free from all loose material, oil, grease or laitance'*, applying a primer directly on to the concrete surface, leaving it to dry, after which the membrane is laid using what is called the 'torch-on' method which means that the sheet is applied directly from a roll with torch heads that heat the underside of the membrane just before it sticks to the primer. Laps are generally not less than 150 mm.

The sprayed (or liquid) form such as the 'Eliminator' system (Stirling-Lloyd, 1998) requires that *first of all the deck surface (steel or concrete) is thoroughly cleaned*. A *primer* is then applied to penetrate and seal the substrate. This is followed by spraying on the methylmethacrylate resin membrane which is the main waterproofing barrier with a finished dry film thickness (DFT) of 2 mm. Finally a tack coat is added to enhance the bond between the membrane and the subsequent surfacing.

The spray system is a *cold* one and by its very nature can follow the contours of the deck closely to produce a seamless (or jointless) membrane.

10.7 ABUTMENTS AND WING WALLS

Abutments and wing walls are subject to attack from water seeping through the fill material behind, underneath or in front of them which is potentially

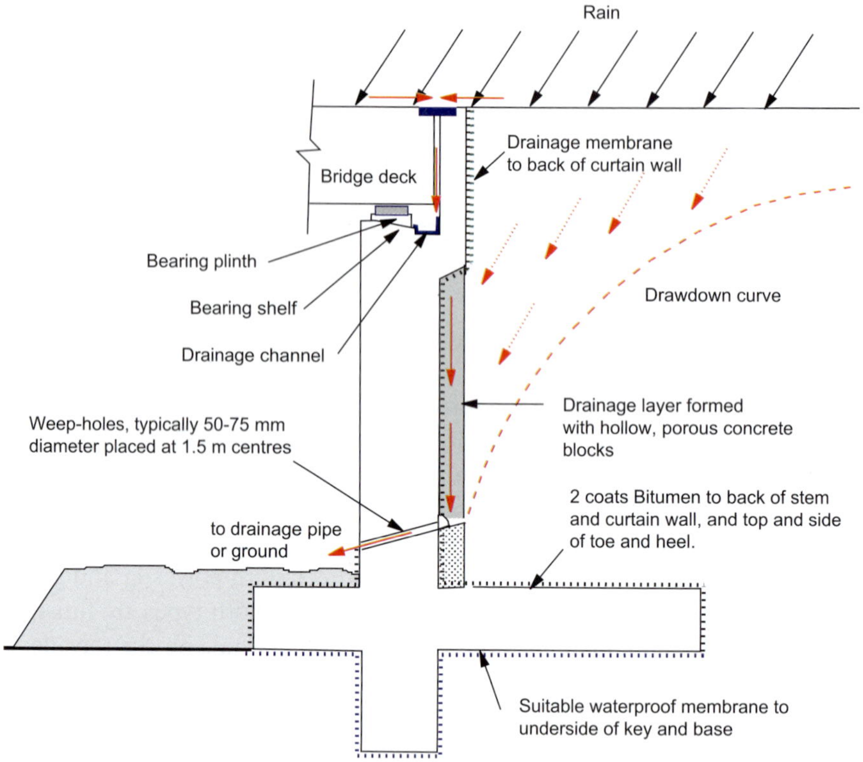

FIGURE 10.4 *Drainage details to abutment.*

harmful to the reinforcement (see Figure 10.4). To prevent this, the underside of the base can be protected by the provision of a heavy duty waterproof membrane, and the first line of defence elsewhere is a coating of *bitumen* applied to all of the concrete surfaces in contact with the fill material.

A drainage layer (such as 'Filtram') which is less than 25 mm thick a well graded aggregate 300–400 mm thick, or a 450 mm wall of porous concrete blocks) can be placed at the back of the abutments and wing walls which tends to draw down the water so that pore water pressures are dissipated, and it then drains away through weep-holes made from clay, cast iron or polymer based pipes provided at the base of the abutment stem (or wing walls) at regular intervals. These may drain into an underground drainpipe in the front of the abutment or wall at base level, or simply drain onto the ground.

Any water which finds its way to the bearing shelf can be collected by means of a small channel at the rear of the shelf, and drained away at one or both ends into a drainage system. This is preferable to providing small diameter pipes cast in just below the curtain wall to drain the water to the back of the abutment, because the channel can be easily cleaned, cast-in pipes cannot.

The provision of *drip checks or drip noses* at all edge beams, deck ends and abutments, and at copings to wing walls and retaining walls, acts as a collection point for the water, which, instead of spreading, drips away from the structure.

10.8 MATERIALS

10.8.1 Concrete

The *durability* of concrete depends upon its ability to withstand attack from internal agencies such as Alkali-Silica Reaction (ASR) and external agencies such as sulphates and chlorides. Its vulnerability to such attacks depends primarily upon its *permeability* (Neville, 1981). Concrete with a low permeability is able to inhibit the flow of water and salts (sulphates) in solution, which reduces its vulnerability to the effects of frost, and the danger of carbonation, and corrosion of steel reinforcement. An important factor is the water/cement (w/c) ratio. ACI Standard 301–72 infers that *structural concrete is watertight* if the w/c ratio is not greater than 0.48 when exposed to fresh water, and not greater than 0.44 when exposed to sea water. Generally the lower the w/c ratio the greater the resistance of concrete to attack. A low w/c ratio can also be indirectly achieved by specifying cement contents for durability and not strength, and that *rigorous attention be given to the cement content, compaction, cover and curing (the 4C's)* Bowman et al. (1996).

10.8.2 Air entrained concrete

Another way of improving durability is by the use of *air entrained concrete.* This is concrete which has small air bubbles (about 0.05 mm diameter) formed by the addition of suitable agents such as animal fats and wood resins. According to Neville (1981) air entrainment has the following effects on the hardened concrete:

- it reduces permeability
- increases the workability
- easier to place
- reduces bleeding
- reduces segregation
- lowers the density of the concrete

- increases frost resistance
- increases resistance to de-icing salts

Menn (1990) maintains that entrained air pores equal to 5% of the total concrete volume can significantly reduce the damage done by the combined action of frost and de-icing salts. The short term extra costs of providing air-entraining agents can be offset from the savings in the amounts of cement and sand required in the mix and, what is more important, since maintenance costs will be reduced then *in the long term the whole life cost of the bridge will be reduced* – this is manna from heaven for beleaguered bridge engineers who know how difficult it is to obtain funds for maintenance. Air entrained concrete should be seriously considered as an option to ordinary concrete.

Browne (1986) has investigated the relationship between Chlorides, Carbonation, Cover and Concrete strength in an attempt to provide guidance at the design stage and these are reproduced in Chapter 9.

10.8.3 Steel reinforcement

Steel reinforcement or pre-stressing ducts are liable to corrode when the chloride content in the concrete is greater than 0.4% by weight of the cement. Since this small amount is easily obtainable in many of the country's bridges, methods are required to mitigate or eliminate corrosion by protecting the steel in some way or by using other materials.

10.8.3.1 Fusion-bonded epoxy-coated reinforcement (FBECR)

As long ago as 1982 a report issued by the then TRRL in the UK concluded that durability of steel reinforcement could be considerably enhanced by the protection provided by the use of *fusion-bonded epoxy-coated reinforcement*, (Willis, 1982). Read (1990) gives an example to show that although the cost of coated rebar is almost twice that for uncoated rebar, as a percentage of the whole cost of a replacement reinforced cross beam (for example), it is only a fraction of the total cost of the works. He cites numerous bridges and structures built between 1980 and 1989 which used FBECR and which to date have shown no signs of corrosion. Treadaway et al. (1989) carried out corrosion tests on slabs reinforced with bars which were untreated, galvanised and epoxy coated (FBECR), and found that after five years exposure to high chloride contamination the epoxy-coated bars performed *significantly better* than either the untreated or galvanised bars. He did, however, find some 'under-film corrosion', which was localised at the cut ends of the bars and near defects in the coating. Figure 10.5 shows a typically epoxy-coated bar. There is still an inherent weakness in that the epoxy coating, if damaged

FIGURE 10.5
(a & b) Epoxy-coated reinforcement (Courtesy of TRL,UK).

in any way, renders the steel beneath liable to the corrosion process just as if it were uncoated.

10.8.3.2 Electrostatic epoxy-coated reinforcement (EECR)

Another form of epoxy coating is the electrostatic, epoxy-coated reinforcement (EECR) which was developed in the USA on a National Experimental

Evaluation Programme (NEEP) 16 under the auspices of the Federal Highway Administration (FHWA). Several systems had been tested but the FHWA finally came out in favour of the EECR, and it has since been used extensively in the USA and elsewhere since 1973 (Safier, 1989). Further very severe tests carried out by the FHWA in their own laboratory in Virginia revealed that even poor quality EECR is 41 times more resistant than uncoated steel (Safier, 1989). Because of the coating the bond strength is only slightly less than an uncoated bar and in the USA 85% of the bond stress for uncoated steel is allowable in the ASTM standard (A775-81) published in 1981.

10.8.3.3 PVC vinyl coated reinforcement

Unlike epoxy coated steel, PVC vinyl is more flexible than epoxy coatings and therefore in situations where the steel is bent (in links, cranked bars and anchorage points) the PVC is able to accommodate both the stretching on the outside and crumpling on the inside of bends. Priming and finishing coats are required, which adds slightly to the total cost.

10.8.3.4 Stainless steel reinforcement

Reinforced concrete bridge structures designed and built using *stainless steel* instead of conventional mild or high yield reinforcement will have a considerably enhanced durability in hostile environments (Walker, 1998). The main reason for this is that high yield reinforcement will corrode when the chloride content exceeds 0.4% by weight of cement, whilst stainless steel will not corrode in conditions where the chloride content is as high as 8% by weight of cement at 20°C – *that is it can resist the effects of up to 20 times more chloride than ordinary steel* (Abbott, 1999). On the basis of several hundred structures tested Hendriksen (1996) has showed that the critical chloride level on the surface of the concrete which initiates damage is only 0.05% by weight of dry concrete and assuming a linear relationship between deterioration with time, then it would take only 15 years before serious damage is incurred if carbon steel reinforcement is used. Indeed a recent survey in the USA indicated that the corrosion damage to the nations reinforced concrete bridges built in the immediate post-war period was put at US$200 billion! (Haynes, 1999).

Stainless steel clearly has a lot to offer, and for bridges in severe environments it is suggested that material grade 316S33(1.4436) to BSS 6744:1986 is used (equivalent to grade 1.4436 to BS/EN 10088-1:1995 or grade 316 to AISI). The properties of this steel is compared with carbon steel in Table 10.1.

Table 10.1 Comparison of Properties for Carbon Steel Bars and Stainless Steel Bars (After Abbott, 1999)

Steel Type			Youngs Modulus kN/mm^2	Density kg/m^3	Coefficient of Expansion $10^{-6}/^{o}C$	Proof Stress N/mm^2				Max. Chloride Resistance (%)
						Plain bar	Ribbed Bar Diameter			
BS EN 10088	BS 6744	BS 4449					3–5 mm	6–16 mm	20–40 mm	
1.4436	316S33		200	8000	16	200	650	500	500	8
		460	200	7850	12	250	460	460	460	0.4

It is clear that stainless steel is not only more durable but is also stronger.

Stainless steel does not have to be placed throughout an entire structure but can be targeted to those places or zones (see section 10.8.3.11) which are vulnerable to chloride attack. Intelligent use of combinations of stainless and high yield steel should dramatically reduce, if not eliminate, corrosion. McDonald (1995) has shown conclusively that there is no danger of bimetallic corrosion where the two metals are in close proximity in concrete, and at least one major bridge (the A48 Higham Bridge in Gloucestershire) has successfully incorporated the two types of steel (Abbott, 1999). One of the reinforcement cages constructed is shown in Figure 10.6.

In the USA stainless steel has been used as a direct replacement of carbon steel on a bridge in Trenton, New Jersey (Haynes, 1999), and a recent examination of a core sample showed no corrosion at all.

FIGURE 10.6

Stainless steel reinforcement cage for a bridge foundation.

There are some drawbacks in using stainless steel:

- At the time of going to press, stainless steel bars were only available in 6 m lengths, but this is due to change in the year 2000 when (like carbon steel bars) 12 m lengths will be manufactured.
- The material cost of stainless steel is about 7 times that of high yield carbon steel. However, in whole life cost terms, it may well prove that over the long term the stainless steel solution is the cheapest because it is probable that very little (if any) maintenance costs will be incurred and also plain Portland cement can be used without any costly additives. A survey carried out by Queens University, Belfast (Sha et al., 1999) showed that both consultants and contractors consider the 'high initial cost as the main barrier to using stainless steel'. In the USA significant direct cost savings of up to 18% have been made by the process of *weld depositing ribs and threads on to plain round bars* (Haynes, 1999) and the product is marketed as 'Rotherbar'. Some research from the TRL (Cuninghame et al., 1996) suggests that if stainless steel is used as a replacement for carbon steel, then if the initial capital cost of the carbon steel is 1.00 its whole life cost (WLC) on a bridge carrying 60,000 vehicles per day (VFD) is 3.61, whilst for the stainless steel its capital cost of 2.00 is unchanged. In Canada, stainless steel has been used in the construction and repair of bridges for over 20 years, and their experience has indicated that 'the total project costs will increase by 1–10%, depending upon the size and complexity of the bridge' (Smith, 2007).

  The Mullet Creek Bridge in Canada (Gedge, 2000) was built with grade 316LN stainless steel reinforcement in the deck and the increase in the initial cost over epoxy-coated bars was 20%. In hindsight this figure could have been reduced to 5% by the use of grade 304, although it does not have the same corrosion resistance as 316. In the Schaffhausen Bridge in Switzerland (Gedge, 2000) the targeting approach was used and stainless steel was incorporated into the parapet edge beams (which are vulnerable to corrosion from chlorides) and this increased the construction costs by 0.5%. In both cases the long term benefit is *reduced maintenance costs*.
- Because of the cold forming of the product it is more difficult to bend. This can be overcome in the bending process, but can cause problems on site when the steel-fixers are trying to put links in place.

In spite of these objections, stainless steel has a future, and certainly if it is as durable as claimed it could greatly enhance the design life of our bridges giving us better value for money.

10.8.3.5 Galvanised steel reinforcement

Galvanised reinforcement has been tested on a number of occasions but has been found wanting (Treadaway et al., 1989). When tested in a slab subjected to a hostile chloride environment (5.4% NaCl) over a period of five years it corroded severely resulting in serious cracking at the top, bottom and sides of the slab compared to an epoxy-coated steel reinforced slab which showed no cracking at all. In some states in the USA, however, it is still regularly specified as a second line of defence together with an epoxy coating.

10.8.3.6 Glass fibre reinforced polymer (GFRP) reinforcement

GFRP is close to becoming a viable alternative to carbon steel and epoxy-coated reinforcement. Many years of research have borne fruit and now it is possible to obtain GFRP from a number of manufacturers world-wide (Gremel, 1998). Whilst GFRP is stronger than high yield steel – typically 655 N/mm^2, with a maximum of 900 N/mm^2 for small diameter bars) it does possess a lower Youngs Modulus. It also exhibits a linear stress/strain relationship up to brittle failure. It is generally held that GFRP bars in concrete are highly durable, but nevertheless more research is needed to prove that view (Cuninghame et al., 1996). As passive reinforcement it will probably be used in small footings, small lintels, minor structural applications and in areas where electro-magnetic interference has to be minimised such as hospitals, airports and structures near high voltage transporters.

In 1986 the first ever bridge *prestressed* with GFRP was the Ulensberg-strasse Bridge in Dusseldorf (Wolff et al., 1990). It is a 15 m wide slab bridge with two spans: one of 21.3 m and the other of 25.6 m. Since then GFRP has been used as prestressing strands in some short span (20–25 m) bridges throughout the world in Germany, Austria, Japan, Spain and at Chalgrove in Oxfordshire (Cuninghame et al., 1996) and all of these are being monitored.

10.8.3.7 Polypropylene fibres

Another recent breakthrough in the search for durability and strength is the advent of *synthetic collated fibrillated polypropylene fibres* as a direct replacement for carbon steel reinforcement (CFPF). This material is very much in the developmental stage, but at least one bridge has been constructed with a *steel-free* deck, namely the Salmon River Bridge in Kemptown, Nova Scotia, opened in December 1995 (Pitts, 1999) which was constructed with elements containing 0.5% fibre by volume of concrete, chosen to meet the requirements of 'fatigue strength, toughness and long term durability'. The bridge is heavily instrumented with fibre optic sensors which are monitored remotely and so far the results have been very impressive.

Polypropylene is used as the fibre because it is strong and can resist the highly alkaline environment of concrete. It is also *hydrophobic* and is suitable for the manufacturing of fibrillated fibres (Pitts, 1999). Typically the fibrillated network fibre lengths are 19 mm to 38 mm, but longer lengths up to 60 mm perform better than the shorter ones but do require more sophisticated equipment to ensure a thorough three-dimensional mix of fibres.

This is an exciting development, and if its *reliability* can be proven then it heralds well for the future of short span bridges. What may be more difficult is to convince users that it is a safe alternative for traditional steel reinforcement.

10.8.3.8 Carbon fibre reinforced polymer (CFRP)

One notable use of CFRP reinforcement (in the form of Carbon Fibre Composite Cables CFCC) in concrete, is the deck of the Herning Footbridge in Denmark (Christofferson et al., 1999). The bridge has two 40 m spans supported by sixteen CFRP cable stays from a steel pylon and six unbonded post-tensioned CFRP tendons in the concrete deck. Half of the bridge is reinforced with CFRP bars and stirrups and the other half has a combination of stainless steel and mild steel reinforcement.

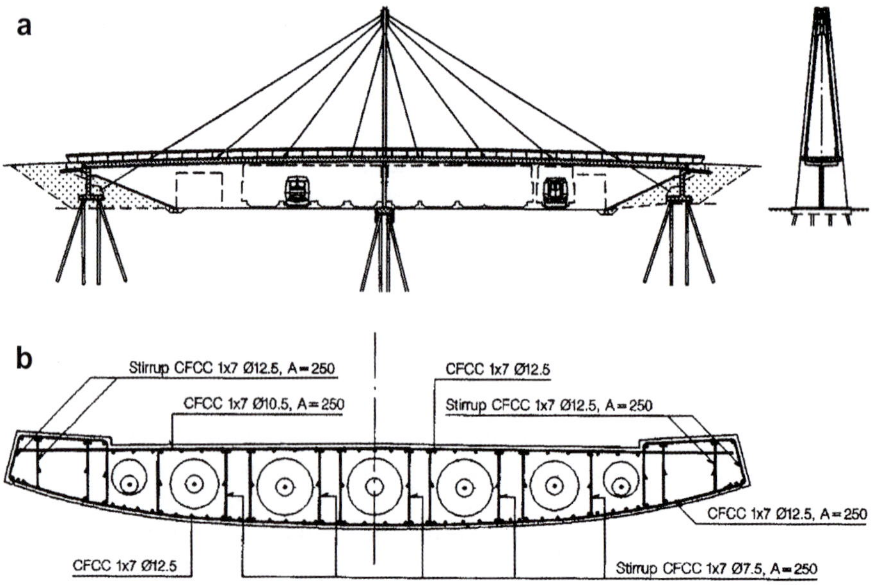

FIGURE 10.7(a, b) *Herning Footbridge, Denmark. (a) General Arrangement (b) Cross section through deck.*

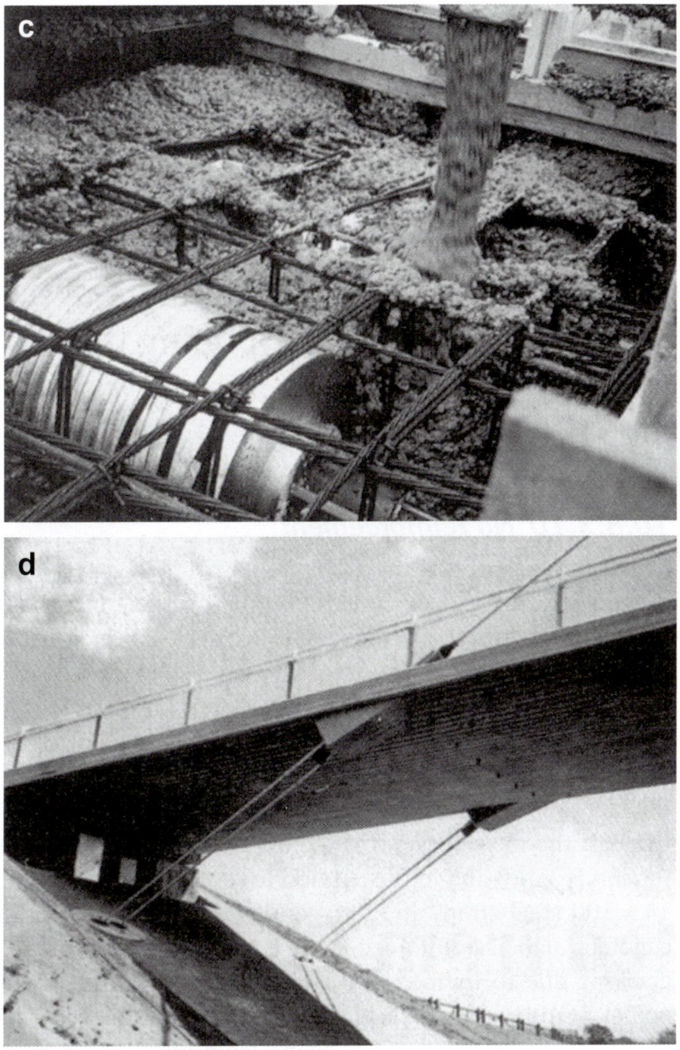

FIGURE 10.7(c, d)
(c) Pouring concrete to deck
(d) Stainless steel backstays.

The bridge was opened in September 1999 and its behaviour is being monitored for corrosion, bond, strains in the CFRP reinforcement and load cells to measure the loads in the cable stays. The long term data on this bridge should prove invaluable for bridge designers in the future.

10.8.3.9 Reinforcement identification

Since there can be several different types of steel on site (due to zoning, see section 10.8.3.11) it is useful to be able to identify them according to their finish 'ex-factory'. This, compared with their tag identification will lessen the risk of reinforcement ending up in the wrong place.

Each of the reinforcement types listed can be easily identified by its distinctive colour:

Carbon steel (mild or high yield)	brown
Stainless steel	silver/grey
Galvanised	mottled silvery grey
CFRP	black
Fusion bonded epoxy coated	light green
FBECR	green
GFRP	yellowy/green

10.8.3.10 No reinforcement

One way of preventing the problems associated with steel reinforcement is to leave it out of the deck construction altogether. This has been proposed as a workable solution for concrete bridge decks by mobilising the strength due to *internal arching action* (see Chapter 5, Figure 5.3) of the deck between steel girders (similar to the behaviour of jack arch bridges) and has been put into practice in Canada (Cole, 1998; Mufti et al., 1997) where 5 so called 'steel-free' bridges have been constructed since 1995. The principle relies on the fact that the girders are restrained laterally thus enhancing the strength through the development of membrane or 'arching' action (see Chapter 5, section 5.2.4.3) and in the Salmon River Bridge this was achieved by welding 14 × 100 steel straps at 1.2 m centres to the top flanges of the girders. The concrete slab is reinforced with chopped polypropylene fibres to prevent cracking due to temperature and shrinkage changes. All of the bridges are being monitored and the data to hand so far suggests that they are functioning adequately. What has been accomplished is that the structural action of the deck has been changed from bending/shear to compression due to the introduction of the 14 mm straps which themselves are susceptible to corrosion! It is difficult to see how the provision of steel straps and on-site welding can be cheaper than conventionally reinforced decks, but maybe the price will reduce as Contractors become familiar with the system.

Another system without conventional reinforcement was employed for Merseyside's Millennium Bridge (NCE, 2000). It is a 34 × m long cable-stayed footbridge made using ferrocement, and supported via a single 25 × m pylon. The deck is 50% lighter than conventional reinforced concrete and is claimed to be more durable.

10.8.3.11 Zoning

In every part of a bridge there are areas where the reinforcement is well protected and there is no need to provide any one of the *enhanced* reinforcement types as described above. But there are exposed areas where attack is more likely to occur and *selective use (or zoning)* of stainless steel reinforcement, for example is clearly beneficial. These can be identified as:

- Edge beams and parapet plinths of cantilevered walk-ways
- Main longitudinal edge beams
- Adjacent to expansion joints
- Crossheads to reinforced concrete piers
- Lower areas of columns
- Concrete in the splash zone
- Underwater structures such as bridge piers or cutwaters.
- Foundations

Put into proper perspective, the additional costs of adopting such a philosophy would probably amount to no more than 0.1% of the total contract price.

10.8.3.12 Splices and terminators

The traditional way of transferring load from one bar to another in tension or compression is by *lap splicing* which relies on the bond between the steel and the concrete. Also, at lapped positions the effective area of the reinforcement is doubled, and more often than not results in areas of congestion, which inhibits the flow of the concrete during construction and makes compaction difficult. It is also possible in tension laps for transverse stresses to be induced which have to be contained by the use of links (akin to the bursting stresses in prestressed concrete anchorages, though, of course, at a much lower level). An alternative, and more efficient way is to provide a mechanical splice which is independent of the surrounding concrete, by *linking bars end to end* connected by a mechanical coupler, which in fact is *stronger* than lapping. One particular system has *tapered* threads, which, according to Gruson, P. (2001) 'are self-aligning and self-centering, and because of the small bar point relative to the mouth opening of the coupler, the end can be easily slammed into the coupler, after which the system requires less than five turns to fully engage'.

Traditional end anchorages are in the form of bends or hooks and rely upon the *compression* resistance of the projected plane of the reinforcing bars, and the *bond* resistance of the embedded bars. The hooked and bent ends if not properly stored can prove to be a health hazard on site.

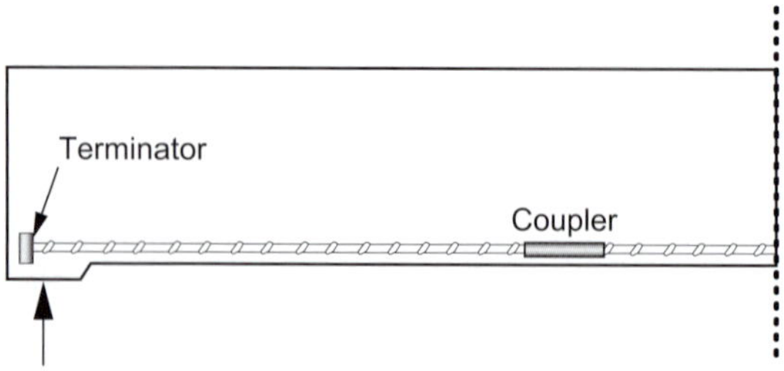

FIGURE 10.8 *Typical terminators and splices to steel reinforcing bars.*

Terminators (or end anchorages) are a modern alternative to the familiar hooked and bent bars. They are more compact and have a well defined failure mechanism, very similar to the cone of concrete in the pull-out test for anchorage bolts. The absence of bent ends also leads to less congestion at anchorage points and easier concrete placement. They are more efficient in that the load transfer from tension in the reinforcement to compression on the end plate is direct.

10.8.4 Protective coatings on concrete

10.8.4.1 Water repellent impregnation

One way of protecting reinforced or prestressed concrete from the ingress of chlorides and water and thereby arresting corrosion of the embedded reinforcement is to apply a coating of suitable liquid compound which is colourless and allows the concrete to 'breathe' whilst keeping the water out. This is known as *water repellent impregnation*.

Montgomery et al. (1990) listed three such protective systems currently favoured in the USA which were:

- monomeric alkylalkoxy *silane*
- *silane-siloxane* overcoated with a methacrylic polymer
- *epoxy* coating

Although research has been carried out in the USA since 1980 on the efficacy of these treatments there is little data on the *longevity* of each system.

In the UK (BD 43, 1996 and BA 33, 1990) impregnation is recommended for new structures and existing structures not yet subject to corrosion. The system specified is a hydrophobising *monomeric alkyl (isobutyl)-trialkoxy-silane (silane)* (Pearson-Kirk et al., 1993) which is applied directly on to the surface by spraying. The liquid then penetrates the surface and initiates a reaction with the silicates and moisture in the

Table 10.2 Comparison of Sealers from Laboratory test performance (after Alam et al., 1997) [HMWM = High Molecular Weight Methacrylate]

Type	Solvent	Material Cost $/m^2	Overall Ranking
>90% Silane	NA	2.80	1
40% Silane	Water	1.29	2
100% Silane	None	4.41	3–4
HMWM over 100% Silane	None	8.50	3–4
40% Silane	Alcohol	1.51	5
20% Siloxane	Min. Spirits	1.51	6
50% Linseed oil	Min. Spirits	0.97	7
HMWM	None	3.98	8
10% Siloxane	Min. Spirits	1.08	9
20% Siloxane	Water	NA	10
Sodium Silicate	Not Reported	0.97	11

concrete to produce a 'water-repellant but vapour-permeable layer' that keeps both water and chlorides out.

Pearson-Kirk et al. (1996) carried out tests on two bridges in the UK to test the effectiveness of the silane and found that by means of a Karsten Tube the water absorption of concrete was reduced by 55 to 80% with the first application and by up to a further 15% with the second impregnation.

Alam et al. (1997) carried out extensive tests in the USA to compare the ability of eleven commercial sealers to protect reinforced concrete structures from corrosion due to the ingress of chlorides and the effects of freeze-thaw cycles. An indication of performance from their laboratory tests is shown in Table 10.2. They are ranked from 1 (best)–11 (worst).

From this limited study it can be seen that the best value for money would be the 40% silane with a water solvent. The general conclusions made by Alam et al. (1997) were:

- *new construction* – a combination of good quality concrete and silane should provide adequate corrosion protection. Silane with poor concrete cannot be expected to perform as well.
- *existing structures* –
  1. if the concrete has surface cracks then a coating of HMWM to seal the cracks followed by an application of silane should provide an effective solution.
  2. if the concrete has already suffered chloride ingress then a coating of silane will help to prevent further deterioration.

Used wisely silane appears to offer the most effective and economic solution to chloride and carbonation invasion through concrete.

However tests by Hassan et al. (1997) on a range of commercial protective treatments (*penetrant, pore liner or surface coating*) were applied to sand/cement mortar patch repairs and left in place for 90 days. The performance of each treatment on mortar test prisms 200 × 100 × 100 mm was assessed on the basis of:

- Degree of saturation versus depth
- Median pore diameter profile
- Apparent average pore diameter profile
- Oxygen permeability profile

An efficiency rating (Ei) was defined in order to rank the efficacy of the individual treatments as:

$$Ei = 100[(Pc - Px)Pc]$$

where Ei = efficiency index %

Pc = intrinsic permeability of the control specimen measured using oxygen, m^2

Px = intrinsic permeability of the surface treated specimen measured using oxygen, m^2

The results are shown in Table 10.3.

These results apply only to mortar repairs of a very limited age (90 days) and cannot be compared to results on mature concrete structures.

10.8.4.2 Migratory corrosion inhibitors (MCI)

The use of sodium monofluorophosphate (MFP) as an inorganic corrosion inhibitor was first considered in 1984 by the Domtar Research Centre in Montreal, Canada (Haynes et al., 1999) who were looking for an additive to

Table 10.3 Efficiency Rating of Different Treatments

Code	Description	Mechanism of Protection	Rating
Epoxy	Primer + top coat of solvent-free epoxy	Sealer/coating	1
P + A.WB	Primer + Acrylic water-based top coat	Sealer/coating	4
P + A.SB	Primer + solvent-based pigmented methacrylate	Sealer/coating	2
A.SB	Pigmented acrylic solvent-based top coat	Coating	6
SS	Sodium silicate	Penetrant (Pore blocker)	3
Silane	100% monomeric alkyl(isobutyl)tri-alkoxy silane	Penetrant (Pore liner)	5

road salts to protect or limit corrosion damage in reinforced concrete structures. It has now developed to the point where it can be used curatively by applying MFP in solution to the surface of existing structures.

The chemistry is quite complicated but in solution most of the product is in the form:

$$Na_2PO_3F = Na_2 + PO_3F_2$$

The monofluorophosphate (PO_3F_2) ion is the active ingredient which has a corrosion inhibiting action on steel reinforcement by forming around the steel and, by acting as a catalyst, actually making the natural film of oxides more resistant to aggressive elements.

It has been shown (Haynes et al., 1999) that:

- MFP can significantly slow down (and even stop after about 75 days) the corrosion process even when reinforcement is already covered with corrosion products
- The MFP treatment can reduce the rate of carbonation by up to 500%
- The application of MFP consolidates the surface zone of the concrete and makes it more resistant to freeze-thaw cycles even when the concrete is already contaminated with chlorides
- MFP can also be applied to uncontaminated (neutral) concrete to prevent or stop corrosion due to carbonation

The MFP treatment is simple in application but *takes longer to complete than silane impregnation*. A typical time period is 10 days where MFP in solution is brushed, rolled or sprayed on to the surface in up to 10 passes with short drying periods in between, after which the surface is sprayed with water to promote the impregnation of the MFP through the concrete to the reinforcement by a process of capillary absorption and diffusion. Once carried out the treatment does not have to be repeated. Used in conjunction with a surface coating, it produces a synergistic effect, that is, the total effect is greater than the sum of the two effects.

Its greatest advantage seems to be that contaminated or carbonated concrete need not be removed. This can result in considerable savings of money and time and the means to rehabilitate an otherwise ailing bridge.

Another proven inhibitor is *calcium nitrite* which, according to Gianetti (1998) has the properties of:

- delaying the initiation time to corrosion
- lowering the corrosion rate after the onset of corrosion

- not having any detrimental effect on the mechanical properties of the concrete
- being compatible with other corrosion protection measures

It is recommended that calcium nitrite be used in conjunction with *good quality* concrete when it will improve long term durability.

10.8.5 Cathodic protection (CP)

The principle of *cathodic protection* has been known for many years. In reinforced concrete an anode (−) is installed in the concrete and connected to a DC power supply which in turn is connected to the reinforcement (+). This forms an electric circuit making the reinforcement act as a cathode. Electrons start to flow from the steel thus reversing the corrosion reaction and re-establishing passivity (see Figure 10.9a).

The surface anode can be formed in several ways:

- coating of arc sprayed zinc
- graphite in a resin binder applied in thicknesses of between 250–400 microns
- metallic mesh (titanium) overlaid with sprayed concrete or super plasticised mortar

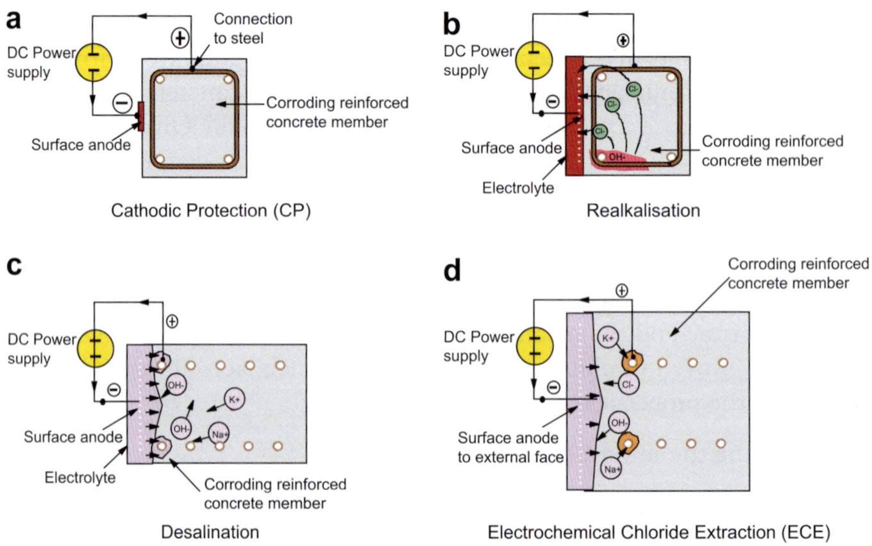

FIGURE 10.9
Principle of Cathodic Protection, desalination and realkalisation.

- discrete anodes fixed at close centres and secured with a cementitious mortar, such as cast iron rods overlain with a conductive asphalt which in turn is covered with an asphaltic concrete wearing course. This method is ideal for use on bridge decks.
- sacrificial anodes
- internal anodes

CP has been used successfully on the Midland Links highway viaducts where they were installed in the crossbeams in 1986 (Hall et al., 1996) and four years later they were checked and it was found that corrosion had halted. Overall the performance has been good and it is planned to continue monitoring the existing installations every 6 months, and proceed with CP installation on the remaining crossheads well into the year 2000. As a chosen maintenance technique it has proven to be *low cost* and is constantly being improved as new technology emerges.

The same technology can also been applied to arrest the corrosion of *steel piles* by shifting the potential of the steel to the immune region, and this has been done successfully on the White Cart Viaduct in Aberdeen, Scotland on piles up to 50 m long where CP was installed in 1968 and upgraded in 1985 (Wallace, 1996).

Another way of artificially shifting the potential of a metal so that it becomes either immune or passive is by the use of a *sacrificial anode*. A galvanic cell is set up by connecting the steel to a more reactive metal such as zinc which corrodes (thus becoming an anode) and the steel becomes cathodic and ceases to corrode. The current densities used in CP are relatively low and so the small amount of acid produced can normally be taken up by the alkaline concrete. The Renderoc Galvashield XP system has been specifically designed to provide protection of steel reinforcement in localised areas where there has been a patch repair. It utilises a zinc sacrificial anode embedded within a capsule of porous cementitious mortar. It is fitted with ties for attachment to reinforcement in any orientation. A typical installation is shown in Figure 10.10.

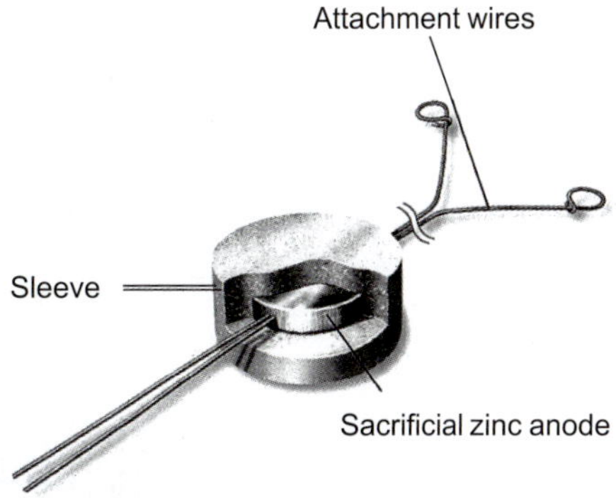

FIGURE 10.10 *Sacrificial anode system for use in patch repairs.*

A very successful form of internal anode has been in use in Australia for many years. The anode consists of a 400 mm long ceramic rod 3 mm in diameter made from magneli suboxides of titanium (Solomon et al., 1998). The anodes are

FIGURE 10.11
Access gallery between deck and abutment. (a) After C. Menn (1999) and (b) BA 57/95.

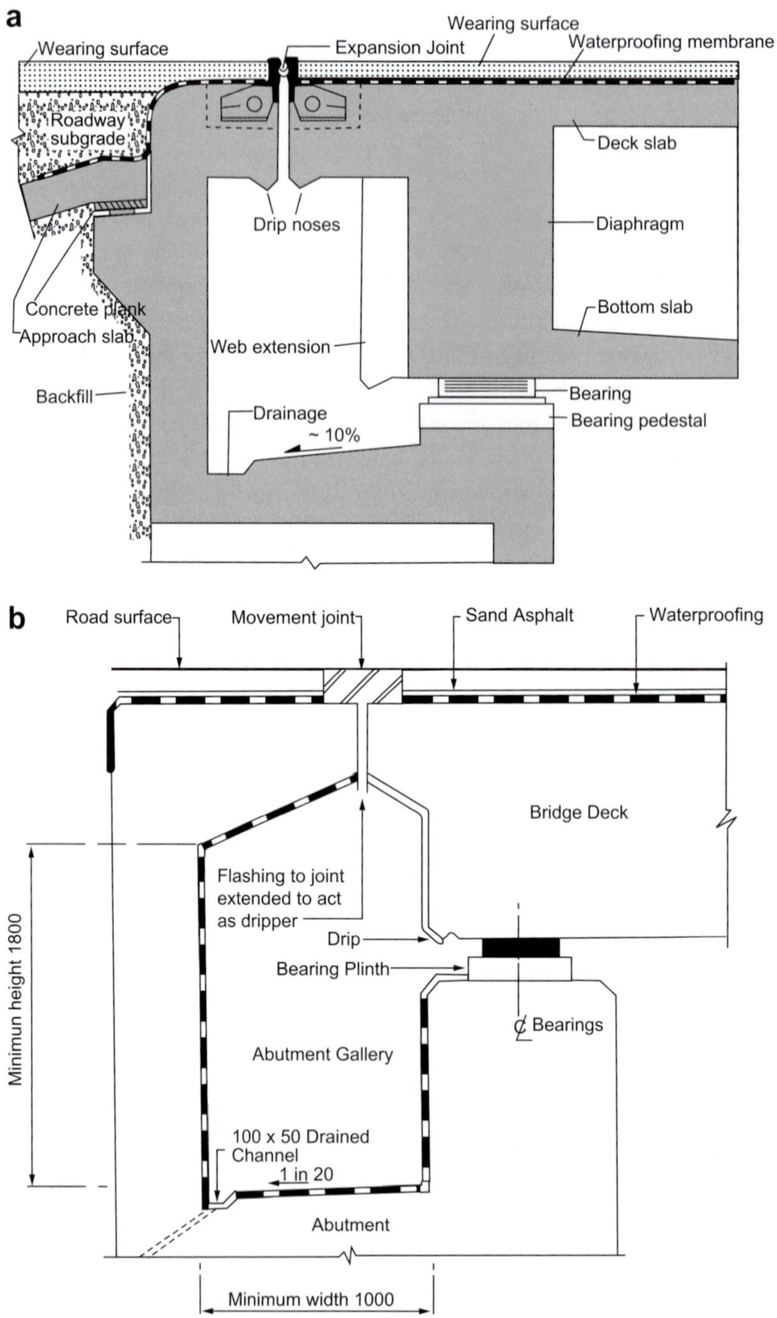

grouted into pre-drilled holes adjacent to the reinforcement to be protected with low shrinkage cementitious grout at spacings of about 300 mm. They are unobtrusive and a number of important claims are made for them, namely:

- gas generated at the anode/concrete interface is able to be vented away to the atmosphere
- ease of installation
- design flexibility in that they can be placed at different depths in the concrete
- can be cast into new construction projects
- isolation of the anode from the reinforcement can be ascertained to ensure electrical isolation of the anodes from the steel
- current control

The system can also be remotely monitored after installation and minimal maintenance is required. The system has been installed in the Kelvedon and Newmans Creek Bridges in Tasmania which are located in a marine tidal environment exposed to wind-borne sea water spray (Soloman et al., 1998), and 'so far the monitoring has not indicated any deficiencies or output variation'.

10.8.6 Realkalisation and desalination

These are electrochemical processes originating from Norway (Hollinshead et al., 1997). *Realkalisation* (Fig 10.9b) is a process whereby an anode is placed in an electrolyte on the surface of the concrete and connected to the steel reinforcement. Current then flows from the anode to the cathode and the hydoxyl ions drift away from the steel thus restoring its alkaline environment. *Desalination* (Fig 10.9c) is a process designed to remove the chloride ions from the vicinity of the steel by the use of a cell mechanism thus stabilising its environment.

The Building Research Establishment (BRE) in the UK are at present investigating the use of six coating systems to be used with these processes with five different electrolytes. It is hoped that a report will be available by the year 2000.

10.8.7 Electrochemical chloride extraction (ECE)

This method is similar to realkalisation and desalination and cathodic protection, and was first used on a bridge deck in the USA in 1975 (Lankard et al., 1975) where is now an accepted alternative to repair. The principle is shown in Figure 10.9d. It is an ideal method of arresting the corrosion process in areas where conventional repairs are difficult if not impossible because of access or because of the particular structural detail such as the half-joints at the ends of suspended spans. This was the precise problem with the Bann River Bridge over the M1 in Northern Ireland (Cromie et al., 1999) a three span beam and slab bridge - the centre span being suspended. Leaking joints had resulted in salt contamination to the parts of the bridge soffit and substructure.

According to Cromie et al. (1999) the differences between ECE and desalination are:

- Current densities for ECE are less than for desalintion
- ECE is carried out over a longer period of time
- In ECE the electrolyte is changed and replenished regularly to remove the acidic by-products of the anodic reaction

The steel becomes cathodic producing hydoxide ions which migrate towards the anode at a slower rate than the chloride ions resulting in a layer of the highly alkaline hydroxide ions in the vicinity of the steel thus effectively arresting the corrosion process.

Very thorough tests were carried out before and after the treatment over a period of a year, viz:

- visual survey
- covermeter survey, carbonation depths
- acid soluble chloride, electron microprobe and petrographic analysis
- porewater analysis (using a new method developed by the University of Ulster)

The test programme was successful and Cromie et al. (1999) claim that 'reinforcement corrosion has ceased'. After a lengthy explanation of the *kinetics* of electrochemical corrosion reaction the authors have effectively introduced a caveat on the application of the method to areas where concrete repairs have been first carried out, and they further state that '*It is imperative that a suitable barrier coating is applied and maintained on concrete surfaces after treatment and that, in future, an effective drainage system is designed to ensure removal of the chloride-contaminated run-off water from the deck above*'. In other words supplementary works are required in order for the treatment to remain effective.

10.8.8 Moisture control

One of the main factors affecting both the degree and extent of corrosion in RC bridges is the level of the Relative Humidity (RH). As long as it remains below about 70% corrosion will not occur at a measurable rate. If the ingress of moisture, can be controlled, therefore, corrosion can be minimised. Instead of applying a surface coating it is possible by the process of *electro-osmosis* to control the movement of moisture through the concrete (Lambert, 1997).

The method works by the application of low voltage DC pulses through the concrete and has successfully reduced (and maintained) excess free moisture levels at below 60% RH (Lambert, 1997). This environment also results in reduction in dissolved salts (chlorides) and reduction in chloride ions to below the critical levels associated with chloride-induced corrosion. A degree of cathodic protection is also provided as the system negatively polarises the reinforcement.

10.8.9 Protective shuttering

The soffit of concrete slabs cast onto the top of steel beams (new build) can be protected by the use of permanent shuttering formed either with Glassfibre Reinforced Cement (GRC) for spans up to 1.25 m and prestressed concrete planks such as 'Topdeck' (Tarmac, 1999) for greater spans. GRC has been manufactured since the 1960's (Ferry, 1998) and is now made with a sand:cement ratio of 1:1; less than 900 kg/m^3 cement and undergoes only 0.13% shrinkage. It is lightweight material (approx. 1/6th that of concrete) and has a high resistance to carbonation and chloride penetration.

10.8.10 Selection

Which remedial method to use for protecting concrete can be decided after a close inspection of the affected structure. The location, nature and degree of the damage will in most cases be enough to focus on the best method, thereafter budgetary restraints will decide what is possible. Some useful guidance is provided by Vassie et al. (2000) on which preventative or arresting methods to use, from which a short list can be drawn up and the final choice made.

10.8.11 Formwork

The role of formwork when considering the durability of concrete is often overlooked simply because it is primarily seen only as containing the concrete when fluid, and providing a pleasing appearance when dry. But there is more to it than that.

Table 10.4 Process Control

Problem	Cause
Surface blemishes such as:	
Blowholes and pinholes	Trapped air and water at the
Scouring	formwork interface
Crazing	ditto
Plastic cracking	ditto
Dusting	
Retardation	
Oil discoloration	Residues from oil/release agents
Surface contamination	ditto
Reduction in surface properties of the finished concrete	Improper placing and compaction methods
Excessive porosity at the concrete/formwork interface	Improper and ineffective curing

According to Wilson, D. (2001) '*the formwork face contact is critical in determining the quality of the outer 20 mm of the concrete surface*'. The concrete/formwork boundary is extremely sensitive to the movement on one side and the nature of the formwork surface on the other. The traditional practices of *pouring, compacting, curing and formwork removal* (process) require good quality control in order to avoid the problems listed in Table 10.4. But by the simple expedient of introducing a *textile drainage liner* sandwiched between the finished concrete surface and the formwork surface, most, if not all of the problems can be solved.

The generic name for the system is *Controlled Permeability Formwork (CPF)*. It consists of a 'combined filter and drain supported by a rigid backing'. The usual arrangement is to stretch the liner over the backing (formwork) and to secure it with staples. To retain the majority of cement fines and to control the movement of the air and water, the pore size of the filter fabric should be less than 0.05 mm. The total thickness of the system is generally less than 2.5 mm. The liners are re-usable, depending upon the primary use, that is: cosmetic or durability, but in no event should release agents be applied to the CPF.

In summary, if CPF is properly applied then the benefits are:

- elimination of surface blemishes (hence no need for preparatory work for coating application)
- no need for release agents (hence no surface contaminants)

- increase in surface cement contents (vastly improved durability)
- low porosity surface (no rapid loss of moisture when formwork struck)

10.9 STEEL BRIDGES

The traditional way of protecting steel is by the use of a paint system which involves first cleaning of the steel surfaces by shot or grit-blasting; the application of a primer such as Micaceous Iron Oxide (MIO) and two finishing coats of a high-build paint to the clients colour preference. The thickness of each coat is dependent on the degree of exposure to the environment. Very good guidance on paint systems is provided by SCI (1995).

10.9.1 Weathering steel

The advent of *weathering steel* in the 1960's offered another option which has proved to be one of the most effective ways of containing corrosion of steel beams and truss girders. The steel is similar to mild steel except that it has additional alloying elements of copper, chromium and phosphorus. The steel is left unprotected and allowed to form its own barrier in the form of rust as it corrodes leaving the members with a light to dark brown coloration. This is a particularly attractive solution where either access is difficult, or the disruption of traffic due to lane closures would result in costly delays. Guidance on its use in the UK is found in BD7 (1981). A recent survey of 141 such bridges in the UK (McKenzie, 2000) revealed that 85% were classed as having no defects and only 5% had defects affecting the steel corrosion generally. It was no surprise to find that the remaining 10% were affected by local corrosion due to leaking expansion joints and bad detailing and a few of the bridges had suffered some corrosion in a marine environment. The worse case was that of a steel box girder footbridge where salt water had leaked through an access hatch in the top flange and caused corrosion that was so advanced that the bridge had to be demolished and rebuilt in concrete. All of the bridges were over 20 years old and so the use of weathering steel can be judged to have been a success. Although adding to the initial cost of a bridge some thought should be given to providing an additional thickness of steel (in effect providing a sacrificial layer) to compensate for the loss of strength over the life of the bridge due to corrosion.

10.9.2 Hot dip galvanising

The galvanising of steel is the process whereby a layer of zinc is alloyed to its surface by 'dipping' it into a hot bath (450°C) of molten zinc thus forming a 'tough, hard wearing' layer which does not react to water. This prevents corrosion of the steel in three ways (Smith, 2000) namely:

1. Zinc weathers at a slow, predictable rate giving a long life
2. Zinc corrodes preferentially to provide cathodic (sacrificial) protection to small areas of exposed steel formed either by drilling or cutting
3. In larger damaged areas, sacrificial protection prevents the sideways creep of rust which can undermine paint coatings

The corrosion rate of zinc is directly related to the level of sulphur dioxide (SO_2) in the atmosphere and with reducing levels its longevity increases. Smith (2000) demonstrates this with data from Stockholm where the corrosion rate dropped from 15 g/m^2/year for an SO_2 level of 90 μg/m^3 of air in 1978 to only 4 g/m^2/year for an SO_2 level of 9 μg/m^3 in 1992.

The case studies of two steel and concrete composite bridges built in 1974 are presented by Smith (2000) which involved very close site examination to determine the amount of salts that had been built-up on the surface, and the remaining thickness of zinc.

The first bridge crosses the A174 Parkway interchange near the A19. Contrary to a Department of Transport report in 1990 which stated that there was 'significant loss of zinc and voluminous zinc corrosion products', the independent survey carried out in 1998 showed that 'over 75% of the surface area still exhibited the original (distinctive) zinc spangle without any significant zinc salts'. When built, the zinc thickness had been a nominal 150 microns in accordance with BS 729 and the thickness of remaining zinc varied from 72 to 159 microns. After 24 years without maintenance this is quite remarkable and it was considered that protection would continue for at least another 20–25 years when some maintenance, probably in the form of painting, will need to be carried out. This means that *the life to first maintenance painting* will have been more than 50 years.

The second bridge was the Burn Rew Access Bridge over the Paddington to Penzance railway line. Over 95% of the surface of the beams showed the original zinc spangle with zinc coatings of between 120–140 microns. The remaining worst affected areas (with a zinc layer of 70 microns) were due to leakage of salt water through the fibreglass permanent shuttering at the top flange/fibreglass interface.

CEB Bulletin No. 211 (1992) reports that of the many bridges built in the USA and Canada, then later inspected, the ones with black reinforcement showed signs of corrosion even in the presence of low levels of chloride, whilst those with galvanised reinforcement showed no signs of corrosion or concrete distress. It also reports that of 237 bridge deck evaluations carried

Table 10.5 Comparison of Different Protective Systems for Steel

System Number	Environment	Protective System	Rank
	Air exposure		
1	– ditto –	Phenol varnish finishes with aluminium–graphite powder fillers	1
2	– ditto –	Paints with vinyl resin binders pigmented with rutile titanium oxide and barium sulphate	2
3	– ditto –	Paint finishes with phenol resin binders pigmented with 36% chromium oxide	3
4	– ditto –	Zinc alone	All sites excellent
	Immersed exposure		
5	– ditto –	First layer of zinc + phenol varnish based paint with aluminium–graphite powder fillers	Very good but with blistering and flaking of paint
6	– ditto –	Paint alone on sand blasted steel	Poor
7	– ditto –	As system 2	Poor
8	– ditto –	Micro talc	Good
9	– ditto –	Barium sulphate bound by epoxy polymide resins	Good
10	– ditto –	Coal system	Good

out by the Pennsylvania Department of Transportation only one of the 237 showed any signs of corrosion.

Some very useful tests were carried out in Belgium from 1980–1993 on steel plate and bolt specimens to discover the efficacy of zinc coating in preventing corrosion (Degueldre et al., 1993). Eight test sites were chosen around the country each with different environments, including industrial, marine, industrial urban, rural and immersed in water. It was found that *the use of zinc as the first coating of a traditional paint system (whether galvanised or metallised) proved to be the most effective barrier to corrosion.* In full they found that for the steel sheets the protection systems were ranked as shown in Table 10.5 and that their results were confirmed by BD7 (1981).

The range of protective coatings given to the *bolts* was 20–30 microns of sheradisation, hot galvanising, mechanical galvanising and combiflon (hardened resin), of which all performed satisfactorily.

Clearly galvanising is a sound option, and providing that regular maintenance (every five years) is undertaken in the form of rubbing off the accumulating salts, there is no reason why the beams should not last 100 years or more. At that time an epoxy based paint system could be used to further prolong its life.

10.10 DESIGNING FOR DURABILITY

The best way to maximise durability and to minimise degradation of a bridge is to act in a pre-emptive way and deal with all of the conceptual and detailing matters that have a direct bearing on durability. In the UK this has been addressed by the Highways Agency through the introduction of a Departmental Standard (BD 57/95) and an Advice Note (BA 42/96).

10.10.1 Structural continuity

A general summary of the recommendations in the Department's documents is included here after Holland (1993). They should be considered as being of a *guidance* rather than a prescriptive nature.

- *Continuous* (as opposed to articulated) decks have proven to be more durable than simply supported ones because deck joints leak and allow water to leak through to bearings, piers and abutments thus causing staining and corrosion. Continuity can be limited to the deck slab alone or extended to the whole deck structure.
- Simply supported spans should be limited to situations where differential settlement might occur, but even then they should only be used where possible displacements, bending moments and shears cannot be accommodated by a continuous structure. The use of slender decks reduces the effect of settlements but this should not be to the detriment of user comfort resulting from 'liveliness' of the deck.
- Continuity can also be achieved by the use of 'portal' single span construction in that the deck and abutments are connected, i.e. *monolithic* construction. Portal ends such as this have been used extensively in the USA without any problems from the global effects of temperature. In many cases the bridges were completely jointless with no discontinuities at either abutments or piers often referred to as *integral bridges*. In the UK some bridges of this type have been built but with spans less than 50 m. In the UK *the use of integral bridges* is recommended for bridges with an overall length of 60 m and skews less than 30°.
- *Half joints are to be avoided* if at all possible as they are very vulnerable to problems caused by corrosion. They are virtually impossible to inspect, and maintenance and repair is exceptionally difficult.

- *Buried structures* should be considered if circumstances allow. These are usually limited to single spans of the portal type or the box type where the top slab and/or the bottom slab are integral with the abutments.

10.10.2 Materials

The two most common building materials for bridges are concrete and steel combined in the well known forms of reinforced concrete, prestressed concrete and composite construction. Provided *that there is sufficient cover and that water (especially salt-laden) can be kept out*, then corrosion should not be a problem. The concrete will retain its alkalinity and continue to provide an exceptionally effective environment for providing corrosion protection to the steel reinforcement. The steel beams (or boxes) in composite structures can be protected either by shot-blasting and painting to a stringent specification, or by the use of weathering steel.

Other requirements include:

- Where circumstances allow, the use of glass fibre or steel fibre reinforcement should be used to cope with tensile stresses due to shrinkage and temperature.
- Generous cover to reinforcement, not merely limited to the Code requirements.
- Generous amounts of secondary reinforcement to deal with stresses resulting from *early thermal shrinkage as well as normal shrinkage and ambient temperature changes.*
- Generous cover to anchorages, ducts and passive reinforcement in prestressed concrete bridges.
- Close monitoring of grouting of the ducts.
- Generous amounts of bursting reinforcement in the anchorage zones and that it is well detailed and uncongested.
- Sound detailing to facilitate the replacement and re-stressing of the individual tendons in externally prestressed structures. Tendons should also be easily accessible.

10.10.3 Inspection and maintenance

Adequate access for inspection is an important factor at every stage of the life of a bridge. The aim should be to provide a reasonably comfortable and pleasant working environment for the bridge inspector. This should include good access for (Holland, 1993):

- Cleaning and painting
- Replacement of components such as bearings, expansion joints, cross-bracings, parapets etc.
- Maintaining proper functioning of drainage, lighting, lubrication of machinery (moveable bridges)
- Replacement of bearings by the provision of jacking pockets at traditional free-standing abutments
- Internal inspection of closed structures (box girders) – via access galleries properly ventilated and lit. These should be large enough to provide inspection and maintenance personnel with unhindered access, and should be well ventilated and illuminated (see Figure 10.12).

10.10.4 Post-tensioned concrete structures

Apart from access problems, post-tensioned concrete bridges have been the cause of some concern in the UK, especially after the sudden, unexpected collapse of the Ynys-y-Gwas Bridge in Wales in the early hours of the morning of 4th December 1985 when the single span Ynys-y-Gwas Bridge over the River Afan in South Wales collapsed without warning (Woodward et al., 1988). Fortunately there was no traffic on it at the time and so there were no casualties. The bridge was built in 1953 and had a cross section similar to that shown in Figure 5.6 (Chapter 5). In the longitudinal direction, the beams were not continuous but of *segmental* construction. The reasons for the collapse were primarily due to a *combination of poor detailing, unusual design incorporating segmental post-tensioned segments and the absence of a waterproof membrane on the deck.* The ingress of water (both clear and salt-laden) to both the longitudinal and transverse prestressing tendons resulted in severe and widespread corrosion of the prestressing steel. As the steel corroded, the cross section diminished and the stress in the wires increased to the point of sudden fracture. *Since the contribution of the steel wires to the overall stiffness of the section was small, the loss was not detected during the routine monitoring of the bridge deck deflections and so it was assumed that all was well and that the bridge could continue in service.*

Some modern post-tensioned bridges have suffered an *excessive loss of prestress*, which may lead to reduced factors of safety and deterioration due to cracking. Others have suffered from *corrosion of the prestressing*, largely associated with incomplete grouting of the cable ducts; and yet others have

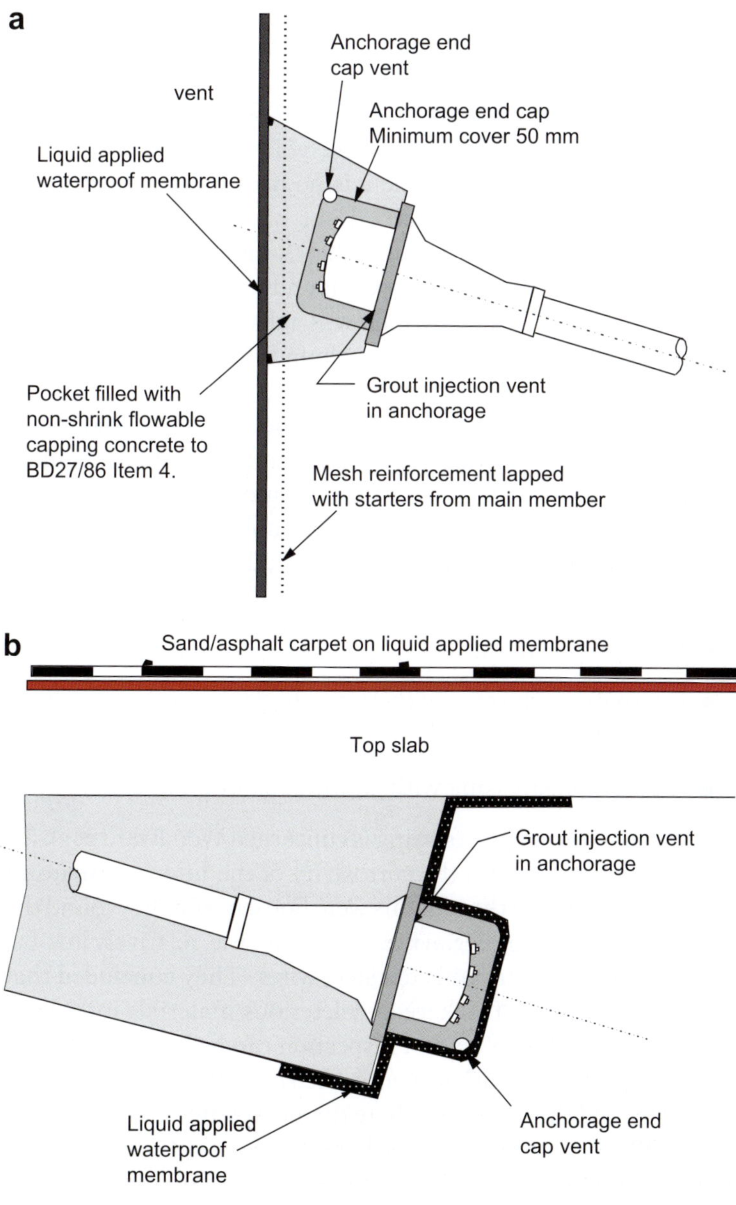

FIGURE 10.12

Durability details at anchorages.

suffered *failure of the anchorage block*, resulting, in one case, in the complete failure of a tendon as it slipped from its anchorage.

Grouted post-tensioned tendons are thought to be well protected by virtue of their being surrounded by a cement-based grout within a steel or polyethylene duct. However if there are voids where water can collect then a corrosion risk is imposed. There is no hard evidence in the UK to suggest that grouting has not been carried out with due caution and good quality assurance. Indeed, Buchnor et al. (1994) found that in over twenty structures which they had inspected most were found to have been well grouted. Even in the worst bridge examined (which was subsequently demolished because of some severely corroded cables, they found that 'over 90% of the cables were found to be in excellent condition and fully grouted'.

Between 1992 and 1996 the Department of Transport imposed a moratorium on the construction of precast segmental construction during which the industry undertook a close examination of all aspects of design and construction. This culminated in the publication of TR47 (1996) which contains a series of recommendations including three design options, mainly applicable to box girder construction, namely:

- External prestressing
- Internal prestressing with non-corrodable tendons such as aramid or carbon fibre
- Internal prestressing with continuous internal ducts

As far as *external* prestressing is concerned Woodward et al. (2000) carried out an investigation into the corrosivity of the internal environment in two concrete boxes where the tendons were housed, and they found that in the five years since monitoring began the 'corrosivity was relatively low' but there were spots of rust on the unprotected steel wires. They concluded that 'as long as there is no evidence of leakage of deleterious materials into the box then the frequency and detail of tendon inspection can be kept to a minimum'.

They infer that *strands in grease-filled polypropylene sleeves* give sufficient protection, and that systems where the sleeves are themselves housed in wax-filled HDPE ducts is excessive as there is little risk of corrosion in the normally dry, well ventilated and clean environment encountered inside a concrete box.

The second option is one which is being taken ever more seriously, and one where the prospect of a complete corrosion-free solution can be a reality. Iyer (1994), Nakai (1994), Maissen et al. (1997) and Shehata et al. (1997) and many others have shown that the use of GFRP, CFRP and CFCC (Carbon Fibre Composite Cables) for prestressing is viable and competitive, the three main problems being the lack of ductility, low Modulus of Elasticity (GFRP) and difficulty in designing a suitable anchorage. Corrosion is not a problem as the composites are virtually inert in the presence of water.

The third option infers the use of segmental in-situ concrete construction, but with an *improved grouting regime* and *dual protection to tendons* in the form of traditional grouting in a steel duct followed by a second high density polyethylene (HDPE) duct in which the annulus is filled with a petroleum wax grout.

Another vulnerable area is evident at the *anchorages,* and the problems here were addressed by a Working Party (Raiss, 1994) set up by the Concrete Society whose recommendations resulted in the publication of TR47 (1996). To ensure that water does not get to either end anchorages or intermediate 'blister' anchorages, a detailing and waterproofing regime is proposed as shown in Figure 10.13 (see also Raiss, 1995).

Another weak spot in glued *segmental* post-tensioned bridges is the joint between segments. No reinforcement crosses the joint but prestressing cables do, and to ensure a tight joint it has been recommended that a minimum compressive stress of 2–3 N/mm^2 across the joints should be provided – in effect making such bridges Class 1 structures and limiting deflections. Collings (2000) quotes the results of some laboratory research carried out in the USA on the durability of the joints in match cast segments (West et al., 1998) which verbatim are:

- The use of gaskets in joints did not appear to be beneficial
- Dry joints allow significant chloride ingress
- Plastic ducts provide additional corrosion protection

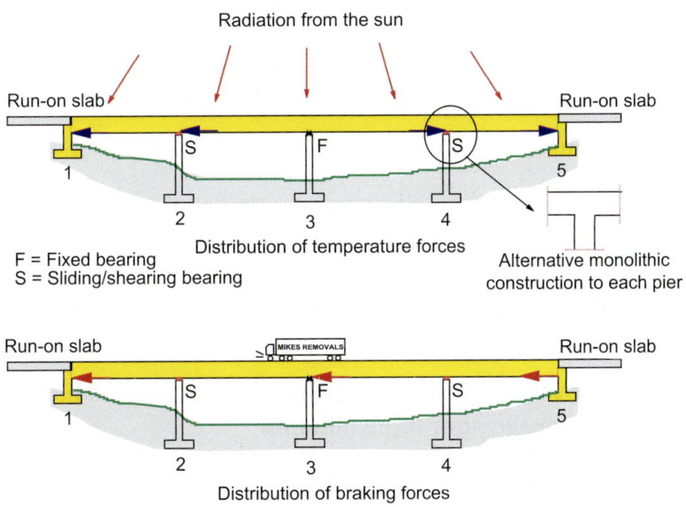

FIGURE 10.13
Principle of Integral Bridges (Idea after Hambly, 1992).

- The effect of pre compression of the joint is not significant in reducing corrosion
- Epoxy joints appear to provide excellent corrosion protection

VSL Management (Ganz, 1996) have developed a plastic duct system PT-PLUS which connects to a plastic-protected anchorage for complete encapsulation. It not only protects but reduces frictional losses in the tendons.

The upshot is that the use of *plastic ducts and epoxy joints* in *match cast* segmental construction provide a satisfactory durable product.

10.11 EXPANSION JOINTS AND BEARINGS

Water will eventually seep through *expansion joints* (even though they should be watertight) and so the structure must be detailed accordingly to provide drainage under the joints. All surfaces at risk should be treated with a protective coating, and a drainage channel provided beneath the joint to carry the water away from the inspection gallery and into a drainage system (see section 10.3.3).

10.12 CONSTRUCTION

A bridge may be well conceived, analysed, designed and detailed, but if it is not built to high standards then all of the pre-contract work is in vain. Quality control is vital. Some of the points to consider are:

- Shuttering to be clean and free from debris
- Use of CPF (Controlled Permeability Formwork – see section 10.3.15)
- Reinforcement to be accurately placed and clean
- Concrete to be well mixed, poured and compacted
- Concrete to be properly cured
- All drainage facilities to be tested, and deck waterproofing to be laid in dry weather
- All bearings, expansion joints, parapets etc. to be thoroughly checked for damage prior to installation
- All steel beams and scantlings to be checked for damage to surface protection (paint, galvanising etc.)

10.13 INTEGRAL BRIDGES

The benefits of continuity were listed in section 10.5.1 and one way to achieve this is to employ the concept of the *Integral Bridge* which may be defined as:

> *'a bridge which has no movement joints between spans or between spans and abutments, and so they form an integral structure with the ground and approach embankments'*

They were developed in the USA and Canada in the 1930's as a way of avoiding expensive maintenance problems due to seepage of salt laden water from the decks through the joints and onto the bearings and substructures (Burke, 1996). They are very popular and have proven to be successful, and overall lengths of 200 m for concrete and 100 m for steel have been achieved (Burke, 1990). Chakrabati (1996) states 'that in the State of Tennessee alone over 300 steel and 700 concrete bridges have been built with integral abutments, the maximum lengths for steel and concrete being 121 m and 243 m respectively'.

In the UK they are less popular, and the Advice Note BA 42 (1996) which deals with the design of integral bridges limits the overall length to 60 m and horizontal movement to ±20 mm.

10.13.1 Design

Articulated bridges *accommodate* the effects of cyclic temperature changes by the provision of movement joints and (apart from the horizontal forces induced due to bearing friction) very little stress is induced in the deck. Integral bridges, on the other hand, *resist* deck movements and considerable stresses result.

If a concrete bridge was 200 m overall and with its stagnant point (or thermal centre) at the *middle*, then it would move by βTL. Thus for example if $E = 30\ kN/mm^2$; $\beta = 12 \times 10^{-6}$ and $T = 30°$ then the movement $\Delta = 12 \times 10^{-6} \times 30 \times 100 \times 10^3 = 36$ mm equivalent to ±18 mm about the mean position. A steel/concrete composite bridge would be expected to undergo a greater movement.

Conversely, if the bridge was fully restrained between abutments, the induced stress is given by $E\beta T$ then the induced compressive stress $= 30 \times 10^3 \times 12 \times 10^{-6} \times 30 = 10.8\ N/mm^2$. This is very high and would, of course, have to be added in to the dead and live load stresses. For a steel bridge with $E = 200\ kN/mm^2$ then the induced stress $= 72\ N/mm^2$, again very large, and it could cause problems of a buckling nature. This sort of restrained movement could be expected to cause considerable damage.

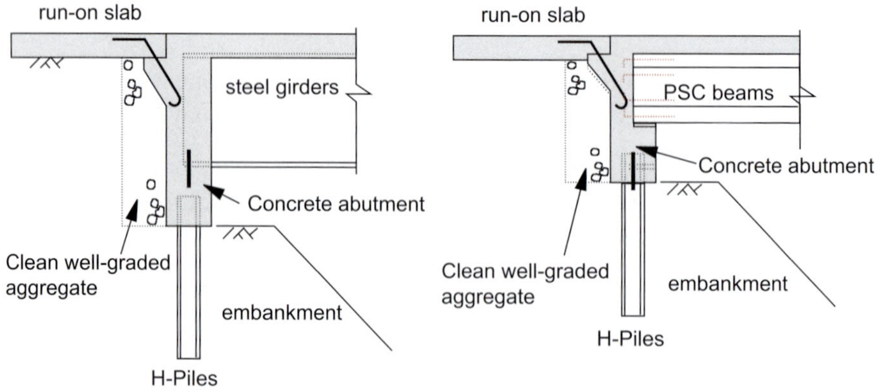

FIGURE 10.14
Two types of 'flexible' integral abutments.

Chubb (1996) shows that the effect of a fully restrained deck in concrete or composite construction in central England would be to induce restraint forces of 70 MN and 120 MN respectively which could not be resisted by conventional sliding or piled resistance.

However, in reality the abutments offer only a *partial restraint* and so the stresses are much lower. In practice, in the USA, it has been found that concrete integral bridges of 200 m perform satisfactory and very little damage has been observed in the abutment regions. The bridge deck, piers, abutments and ground all act as a single system and the *soil/structure interaction* becomes much more important. The abutments generally have enough flexibility to partially absorb thermal deck movements and braking forces. Figure 10.13 illustrates the general principle under the action of both temperature loading and traffic braking loads. The actual distribution of the forces will depend on the *bearing type and the stiffness* at the abutments.

10.13.2 Construction

Clearly the sensitive areas are at the abutments and this is where most research effort has been directed. Some common abutment 'flexible'designs are shown in Figure 10.15 where temperature movements and stresses can be absorbed by the steel piles.

The general features of integral abutments are:

- They are generally kept small to minimise the amount of structure and backfill that must move with the deck and to limit the passive earth pressures
- The bridge deck is cast into the abutment wall

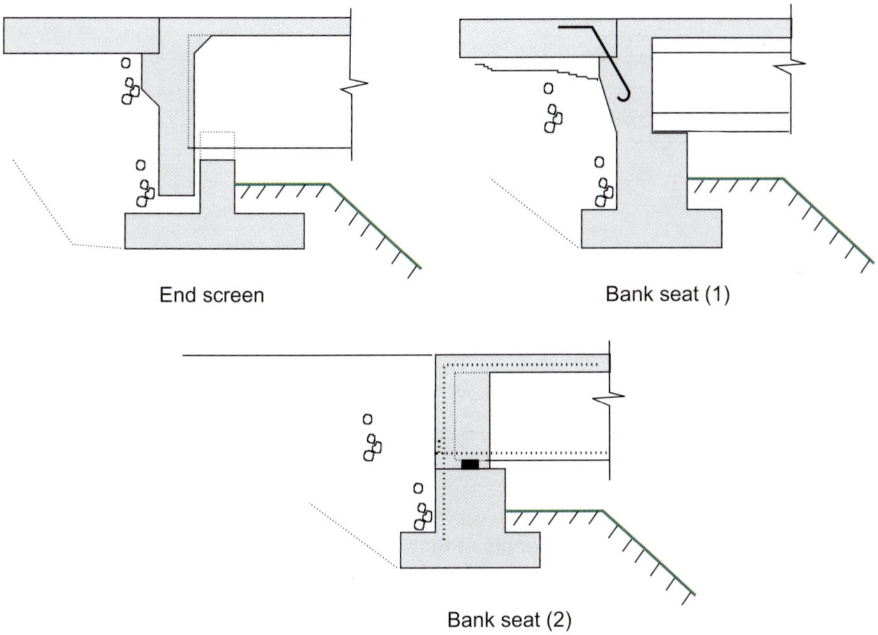

FIGURE 10.15
Other types of 'sliding' integral abutments.

- The backfill provides a deformable lateral restraint whilst also accommodating thermal movements
- An approach slab (or run-on slab) is provided to prevent traffic from compacting the backfill, and it also prevents rutting of the roadway. In the USA the minimum overall length is specified as 4.25 m with at least 1.2 m at the end of the slab bearing on the ground. In the UK they are not favoured because of problems encountered in the 1970's where high approach embankments were founded on *soft compressible clay soils* (Thornburn, 1996). In the UK now, the fill is well compacted up to the top of the curtain wall
- It is recommended that the slab be tied to the abutment wall to prevent a gap opening up when the deck shrinks (and then filling with deleterious material prior to its next expansion)
- When an integral abutment adjoins a *rigid* pavement a movement joint is located between the end of the approach slab and the pavement

Most abutments are connected to piles as they are flexible thus offering minimum resistance in bending to horizontal thermal movement but maximum vertical resistance. Other shallow types without piles are shown in Figure 10.16 which rely on bearing pressures to support vertical loads and

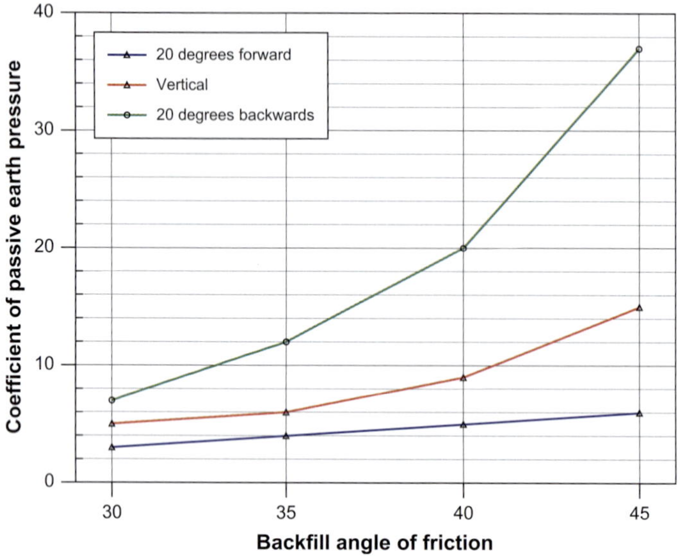

FIGURE 10.16
Variation of coefficient of passive earth pressure.

granular fill (under the base) to resist thermal movements and where thermal effects are absorbed by sliding.

O'Brian et al. (1996) have studied the effects of creep and shrinkage on a model of a concrete portal frame and found that from data over a period of 28 days a horizontal displacement of the bridge support of 3 mm will result in zero strain due to creep and shrinkage at time infinity.

Springman et al. (1996) have carried out laboratory tests on the effects of slow *cyclic* temperature induced expansion and contraction of the bridge deck. They tested 9 configurations over a period of one year, and found that most of the damage occurred after the first 10–20 cycles from particle orientation in the backfill causing densification of the fill material, and increased forces on the abutment. The precise level and extent of the damage was not specified, but presumably it was related to displacement of the fill which would cause a dip in the road surface, and it is this which the authors feel has been the greatest benefit of their analysis, confirming as it did that a 'typical "compaction" type profile is created by passive wall rotations at the serviceability limit state'.

The Advice Note BA42 gives advice on the pressure coefficient K to use in earth pressure calculations, viz:

$$K = \left(\frac{d}{0.05H}\right)^{0.4} K_p$$

where K_p – coefficient of passive pressure

ϕ' – backfill angle of friction
δ – wall friction $= \phi'/2$
H – abutment height
d – distance moved towards the backfill

and K should not be taken as less than the at-rest earth pressure K_0, nor less than $K_p/3$. Also to take account of the inclination of the back face of the abutment on the magnitude of the horizontal force a table is provided from which values of K_{ph} can be interpolated for values of ϕ' between 30° and 60° and for vertical, and 20° forward and 20° backward (see Figure 10.15). Under live load only BA42 recommends the at-rest value of:

$$K_0 = (1 - \sin\phi)$$

where ϕ is the effective angle of shearing resistance. Bush (1996) points out that the backfill needs to be sufficiently resilient to accommodate the imposed expansion and contraction of the deck, and that the designer has to assess the effects of the disturbed and repacked fill during typical heating and cooling cycles and then to make a judgement as to whether a run-on slab is required.

Low (1996) has studied the effects of the interaction between the bridge and its approaches using a finite element analysis and found that 'a retained abutment with a stiff embedded wall could take repeated horizontal abutment movements of ±18 mm without any internal shearing of the fill'. He also points out that the *fill may be considered as loading and/or structure*. In BA42 it is considered as loading, but in a conventional soil-structure interaction analysis it acts as structure. He raises a number of searching questions, shortcomings and proposals in relation to BA42 but on the question of run-on slabs his prediction for UK designers is that the normal practice will be the use of compacted fill and no run-on slab.

10.14 PROVISION OF CONTINUITY IN PRECAST BEAM CONSTRUCTION

A less radical way of improving durability is to provide continuity over internal piers whilst leaving joints at each end. There are many bridges built using precast concrete beams, either of the *contiguous type or the spaced beam-and-slab type* which are usually erected span by span as it was perceived that this was easier and less costly than dealing with the complexities of continuity. However simply supported spans have to be deeper, the substructure has to be wide to accommodate a double line of bearings, and the maintenance costs are high because of the leaking movement joints.

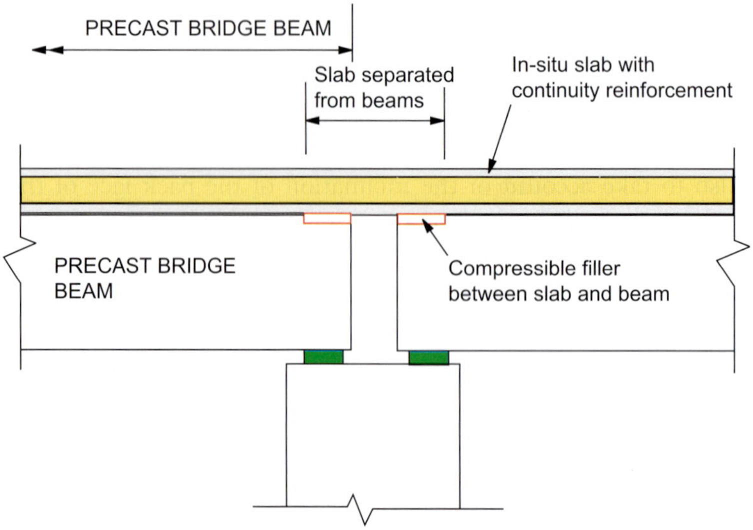

FIGURE 10.17
Type 4 – Continuous separated deck slab.

The provision of continuity counters all of these objections and reduces the number of movement joints to one if the bridge is fixed at one end, or two if it is fixed at an internal pier, but they are required to accommodate larger movements, and piers or abutments with fixed bearings have to resist larger horizontal forces.

A study carried out in the UK by (Pritchard et al., 1991) revealed *five* possible methods of providing continuity at internal supports and these are illustrated in BA57. Types 1 and 2 are the most common and are reproduced here along with Type 4 which confines continuity to the deck slab only.

- **Type 1 – Wide in-situ integral crosshead**
  This type uses precast beams which are shorter than the spans between supports and is the most common one in the UK.

  The beams are generally inverted Tee, U or M beams which are supported on temporary supports built off the pier foundations. The wide in-situ integral crosshead is then cast between and around the beams to provide about 1m embedment. Longitudinal hogging bending continuity is achieved by reinforcement within the continuous deck slab and passive reinforcement that extends from the ends of the beams. Transverse strength is achieved by pretensioning or reinforcement. The crosshead is then supported on a single row of bearings set centrally on the pier. A typical arrangement is shown in

Figure 10.18 (after Figure 3.1 in BA57).

This method is preferred in the UK for the following reasons:

1. The wide crosshead means that spans larger than those dictated by the maximum manufactured lengths are achievable.
2. A bridge curved in plan can be easily accommodated by varying the width of the crosshead from one side to the other, and varying the width of the facia concrete.
3. Drainage pipes can be concealed within the thickness of the crosshead.
4. Only a single row of bearings is required, which also implies that piers can be thinner.
5. There are savings in foundation materials since there are no moments applied at bearing level.
6. The crossheads can be cantilevered out each side of narrow piers and reinforced with either passive reinforcement or prestressing. The cantilevering also means less ground intrusion, and a saving in the number of bearings.

■ **Type 2 – Narrow in-situ integral crosshead**

This is a very popular form of construction in America and is shown in Figure 10.18. In this situation the precast beams can be seated on a twin row of *permanent* bearings on the piers, or initially placed on

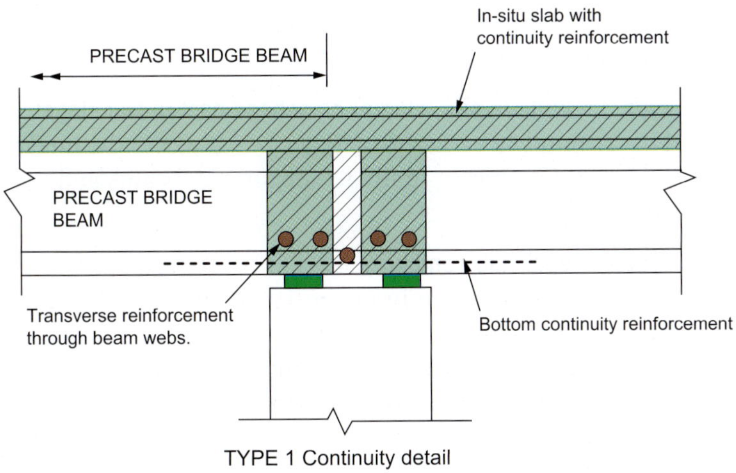

FIGURE 10.18
Type 1 – wide in-situ integral crosshead.

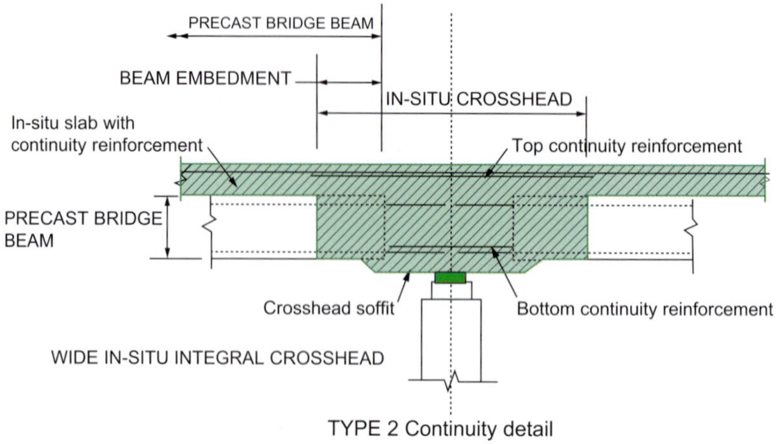

FIGURE 10.19
Type 2 – narrow in-situ integral crosshead.

a twin row of *temporary* bearings, which, after the crosshead has been cast (in and around the beams as for Type 1), can be removed and replaced by a single line of *permanent* bearings. Alternatively, depending on the magnitude of the reactions, a single permanent strip rubber bearing can be provided which acts as a seating for both beams.

The crosshead is narrower than Type 1 and it can prove difficult to place transverse reinforcement between the beam ends. Longitudinal bending is catered for by reinforcement in the top slab. Reinforcement in the crosshead and through holes in the beams provides transverse strength. Bottom continuity may also be provided to cope with sagging moments but is not mandatory. Without reinforcement any cracking that may occur has no durability implications.

- **Type 3 – Wide in-situ integral crosshead cast in two stages**
  This is simply a variation of Types 1 & 2 where the integral crosshead is cast in two stages, firstly to support the beams and then to provide continuity.

- **Type 4 – Deckslab continuity**
  This method is shown in Figure 10.20 and uses the deck slab only to provide local longitudinal continuity.

  The deck slab flexes to accommodate the rotations of the simply supported deck beams by means of a layer of compressible material between the tops of the beams and the underside of the slab. Reinforced diaphragms are provided transversely as for the other types.

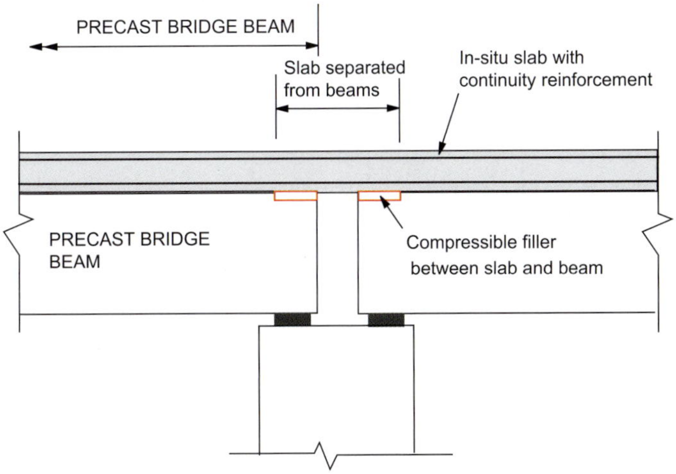

FIGURE 10.20
Type 4 – continuous separated deck slab.

10.14.1 Choice of type

All of the types provide a degree of continuity. The precast beam self weight acts over a simply supported span, and the continuity provided resists extra moments and shears from the live load, creep and shrinkage, and temperature variations.

The embedment of the precast beams into the in-situ crossheads in Types 1 & 3 restrains the ends of the beams thus giving rise to *sagging* moments at the internal supports due to creep. When the in-situ slab is poured and starts to set, the effect of differential shrinkage is to cause the beams ends to rotate in the opposite direction giving rise to *hogging* moments at the internal supports. This is illustrated in Figure 10.21.

Very often the two moments cancel each other out, but in other circumstances either transverse cracking can ensue at the bottom of the crossheads or the design serviceability hogging moment is exceeded.

There is usually a delay of one or two months between beam embedment and pouring of the deck slab, and the moments due to restrained creep during that period must be considered as a separate load case.

10.15 PIER PROTECTION

Piers can fail to the effects of scour (see Chapter 11, section 11.6), and they can suffer damage due to collision from river traffic. Whitbread et al. (1996) give a very enlightening description of what scour actually is and propose

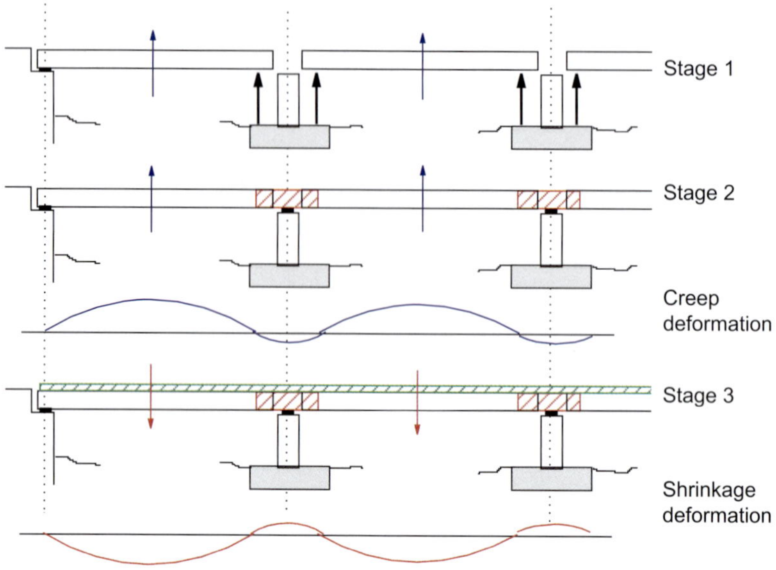

FIGURE 10.21
Construction stages and effects of creep and shrinkage.

ways of reducing or even preventing it. One method is to provide a mattress made from used lorry tyres placed upstream which produces a reduction in scour whatever the shape of the pier. Another method (Paice et al., 1993) proposes groups of piles installed upstream of the pier which interrupts the flow of water and causes scour around the base of the (sacrificial) piles and not at the pier. Circular piles were chosen since they are not affected by the angle of attack; they are easy to construct and they require only one (central) setting out point. Two configurations using such piles have been applied to two bridges in the UK which have different pier shapes, namely: Tavy Viaduct in Plymouth and Over Viaduct, River Severn. Both have been monitored since 1993. The general layouts are shown in Figure 10.22.

Piers also have to be protected to withstand vehicle impact, whether from highway traffic or water traffic, and an example of the former is shown in Figure 10.22.

Protection is best afforded by thickening the lower portion of the bridge piers or columns with mass concrete fitted with a steel impact barrier at low level.

There are also unusual loads such as *ice forces* for which the Confederation Bridge in Canada was designed (Brown, 2000), and which resulted in each pier of this 13 km long bridge being fitted with a conical *ice-breaking collar* located at the waterline. Each concrete collar is 6.6 m high with a 52° slope, and topped by a 3 m cone with a 76° slope. The diameter of the collar at mean waterline is 14.15 m.

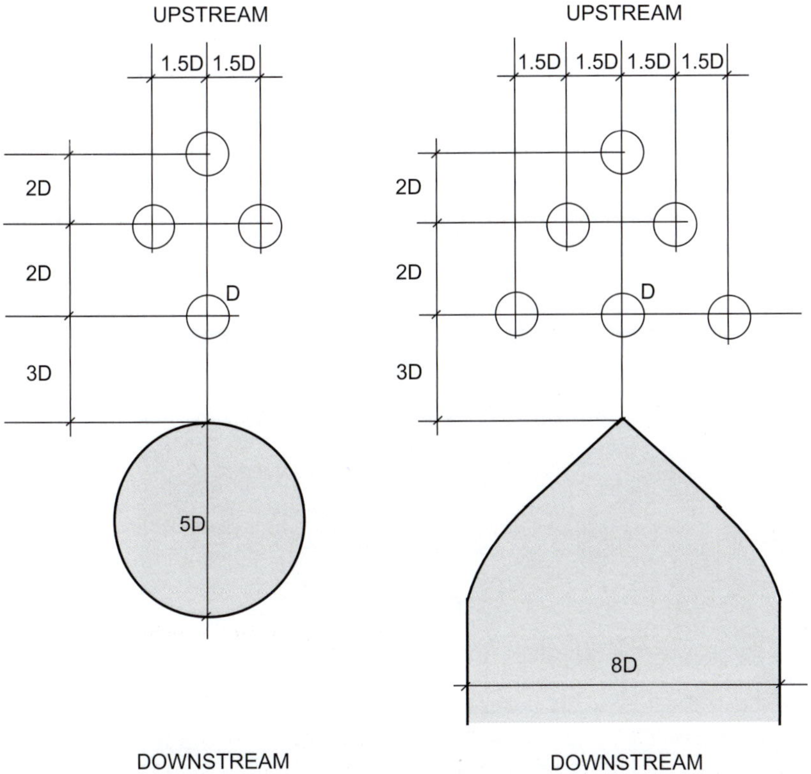

FIGURE 10.22
Scour protection using circular piles.

10.16 ENGINEERED CEMENTITIOUS COMPOSITES (ECCs)

Steel and concrete are by far and away the most common building materials used in the bridge engineering industry. They are used separately or in combination. Concrete is strong in compression but weak in tension and is prone to cracking; steel is strong in tension but prone to buckling and susceptible to corrosion.

In reinforced concrete construction, concrete is employed in an 'active' sense by way of its compressive strength; and in a 'passive' sense in that it provides protection to steel reinforcement by way of cover. Embedded steel bars provide both tensile and compressive strength in areas of bending or direct compression.

Concrete is 'brittle' in that it fractures without warning at very low tensile strains (typically 0.003). Attempts have been made in the past by many researchers to render concrete less brittle by the addition of metal or plastic fibres to the traditional concrete mix of cement, fine aggregate, coarse aggregate and water to form Fibre Reinforced Concrete (FRC) but it has not met with any great success from a structural point of view. The fibre volume is usually about 5% or more, and very often it has to be aligned to the main tensile stress trajectory, making it very difficult to place on site.

Research workers at the University of Michigan, USA claim to have developed a truly non-brittle material which they call Engineered

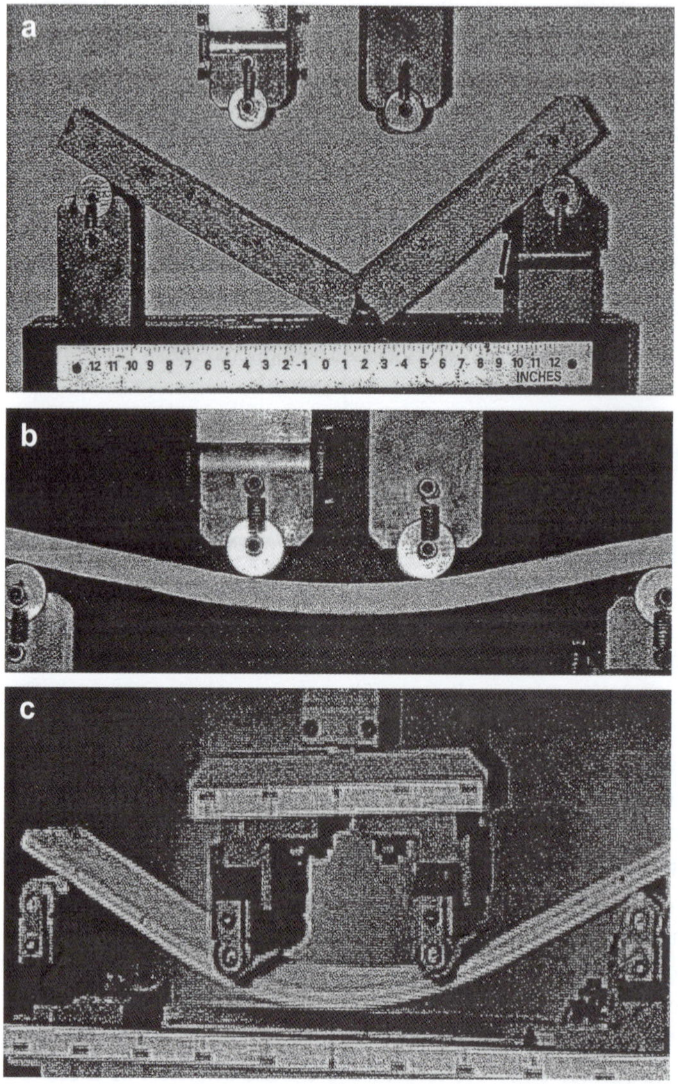

FIGURE 10.23
Highway bridge column damaged by vehicle impact.

Cementitious Composites (ECC) (Fisher et al., 2007). They comprise a mix of *'Portland cement, fly ash, sand, water and superplasticiser in controlled portions, combined with surface-coated fibre-reinforcement'*. They have no coarse aggregate, and the fine aggregate (usually sand) together with the fibres are 'systematically tailored'. The resulting material is strong in tension, has a 'tight' crack formation and is ductile in behaviour.

Li et al. (1998) state that 'the most fundamental mechanical property difference between ECC and FRC is that ECC *strain-hardens* rather than *tension-softens* after first cracking'. This unique ductile property is achieved by 'composite constituent tailoring guided by micromechanical analyses' (Li et al., 1992), (Li, 1993), (Kanda et al., 1998).

Figure 10.24 demonstrates the high deformability of ECC under a four point bending test. The failure configuration lies somewhere between the fracture of plain concrete and the yield hinge of steel, illustrates just how 'bendable' ECC is. It is 'springy' rather like a cricket bat behaves when the batsman is settling into his crease. The material has also been tested for restrained shrinkage, fatigue and bonding tests, freeze/thaw exposure, wearing/abrasion tests and accelerated environmental tests. It has also been subject to chloride immersion and de-icing salt exposure.

In many cases where a structural element is subject to high shear forces, it has been found that, since ECC remains non-brittle in shear then smaller amounts of shear reinforcement are needed, if any at all (Li.V 2005).

ECC is also very beneficial as a repair material as it eliminates surface cracking and internal interface delamination. A typical mix comparison is shown in Table 10.6 and it can be seen that in simple terms the coarse aggregate of a conventional OPC concrete is replaced by Fly ash in the ECC.

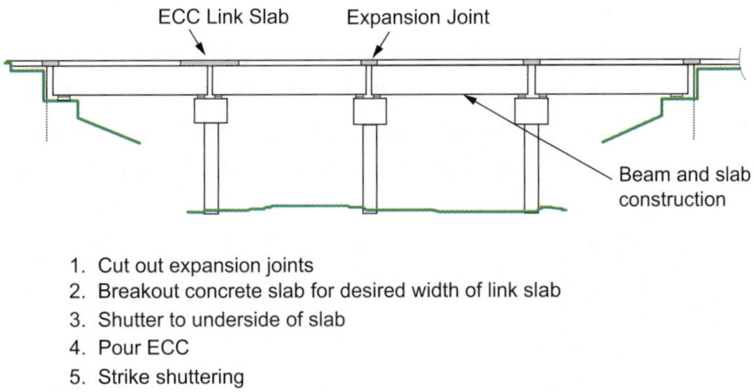

FIGURE 10.24

ECC in bending (reproduction of Figure 1: ECC in bending – flexible concrete.

Table 10.6 Typical Mix Comparison (kg/m^3)

	Cement	Gravel	Sand	Fly Ash	Water	SP	PVA Fibre
Concrete	432	864	864	0	192	4.3	0
ECC	561	0	449	673	327	14	25.99

(Taken from The Bridge, LTAP, April/June, 2005).

10.16.1 Applications

In the USA and around the world there are trials in progress in order to assess the efficacy of ECC's. One application is the rehabilitation of simply-supported multi-span viaducts with an expansion joint at each support location. The bridge deck section immediately adjacent to each support line can be replaced by a *link slab* comprising an ECC, thus rendering the viaduct *semi-continuous*, that is: simply supported under dead loads and continuous under live loads (see Figure 10.25). More importantly it improves the durability of the bridge by eliminating expansion joints and all of the attendant problems related to salt-laden water seeping through damaged joints.

The Grove Street Bridge over Interstate 1-94 (Figure 10.25) was rehabilitated by removing the expansion joints, cutting out a 5.5 m wide strip of deteriorated slab concrete and replacing it with ECC using conventional ready-mixed concrete equipment (Fischer et al., 2005). After two winters subject to appalling Michigan State weather conditions the link slabs have shown no sign of deterioration. It remains to be seen how it will fare over the next 30 years or so.

10.16.1.1 Ultra-high performance fibre reinforced concretes (UHPFRC)

These are a family of concretes tailor-made for a specific purpose. Rather like ECC's they do not have fixed codified mix proportions but are considered on a case by case basis. As with ECC's they have 'fibre contents of between 2–6% by volume; a low water/cement ratio typically 0.15–0.2, and no coarse aggregate content'. (Jones et al., 2005). Typical properties are shown in Table 10.7

Other properties which make UHPFRC desirable are:

- If the materials are heat treated at 60–90°C then the *creep coefficient* can be as low as 0.2 (one tenth of the normal concrete).

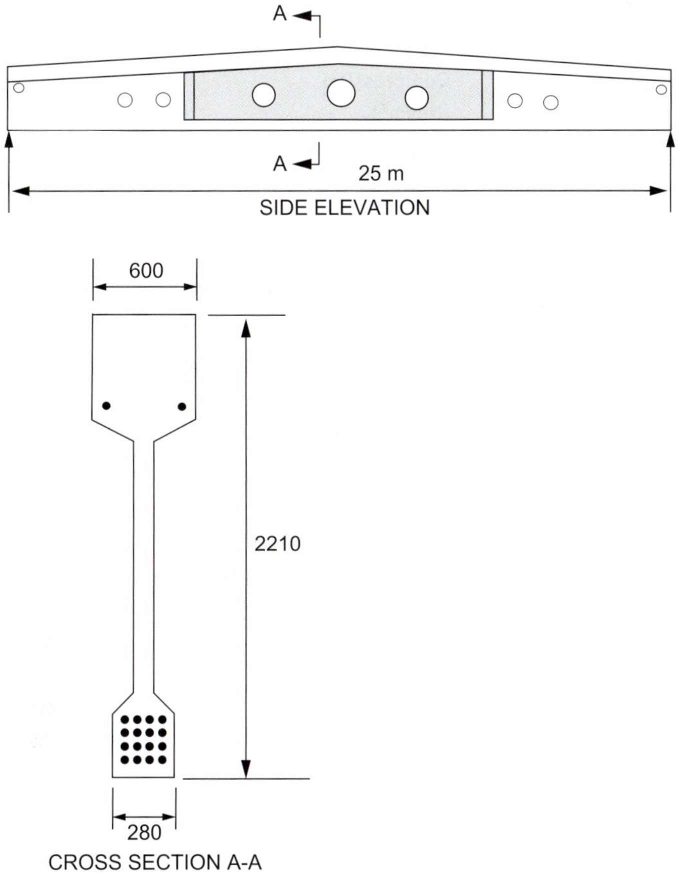

FIGURE 10.25
Link slab application.

- Shrinkage can be quite high (500 micro-strain), but again, if the materials are heat treated then this can be reduced to almost zero.
- Because of its dense nature, the permeability of the concrete is low, which means enhanced durability and low cover concrete (as low as 10 mm).
- It is possible in areas of low shear and tension to dispense with conventional bar reinforcement altogether, in which case the provision of 'cover' is redundant.
- Sections can be slender and may enhance the architectural appeal of a bridge, but not to the point where it appears 'flimsy'. Proportionality must remain the predominant dictactor at the serviceability limit state.

Table 10.7 Typical Properties of UHPFRC

Bending (N/mm^2)	Compression (N/mm^2)	Tension (N/mm^2)	Youngs Modulus (kN/mm^2)
10–50	50–300	10–15	45–60

(After Jones et al., CEI, Spring 2005).

10.16.1.2 Self compacting concrete (SCC)

Self compacting concrete has been the dream of building contractors for many years. Fixed and mobile vibrators are noisy, intrusive, and are often unable to get the concrete where it should be such as re-entrant corners, in and around heavily reinforced sections, and in narrow sections, and this has lead to inconsistent concrete, honeycombing and other blemishes once the shuttering has been removed. Now, at last, a well developed and proven concrete that '*compacts itself*' is on the market. The front runners in Europe for this type of concrete are Sika products (Sika product information sheet, 2008) who make the following claims:

- The extremely fluid and soft consistency of SCC allows fast placing of concrete
- No vibration is required therefore there is no noise pollution
- Can be placed speedily to the highest quality
- 'White finger syndrome' injury brought about by handling vibration equipment is obviated
- It can 'flow' into the tightest of corners or through heavily congested reinforcement and produces a more homogeneous surface layer thus reducing permeability and increasing resistance against chloride ingress, carbonation and chemical attack.

10.16.2 The German experience

In 2006 and 2007 the German company of Bauuntemehmung Gmbh & Co.KG. undertook the designing, detailing, construction and testing of four pre-stressed steel-fibre-reinforced beams made of self compacting concrete (SCC) *without* conventional steel reinforcement. Four such beams, each with different types of steel fibre, were constructed and tested. The elevation and section are shown in Figure 10.26.

As there is no National Standard for SCC at the present time then 'Technical Approval' had to be sought on a case by case basis.

The beams were loaded in increments up to failure. A typical load/deflection profile at mid-span under the action of a point load is shown in Figure 10.26.

10.17 THE WAY AHEAD

We have to live with the deterioration problems resulting from various causes in the past and try to rehabilitate where we can, but for the future we should try and *prevent* where possible the same problems arising again. Bamforth (1999) argues that some way should be found of *defining performance requirements and ways of testing for them at the design stage*. He rightly states that researchers should work together and not compete, and that in Europe there still persists the 'not-invented-here-syndrome. Why do we find it so difficult to accept someone else's ideas'?

We are confronted with a wide (often bewildering) range of options at the design stage (see also 10.3.14) and in order to choose the best option for a particular problem in reinforced concrete, a guidance document is available from the Concrete Society (Bamforth, 1999). The approach will supercede the current 'deemed to satisfy' philosophy for durability design 'by modelling the processes leading to the onset of corrosion and subsequent cracking'. Other parameters are input to model the particular circumstances and proposed protective measures based on extensive research data. The method can be approached in either a deterministic or probabilistic manner, and there is no reason why a similar approach cannot be adopted for other materials.

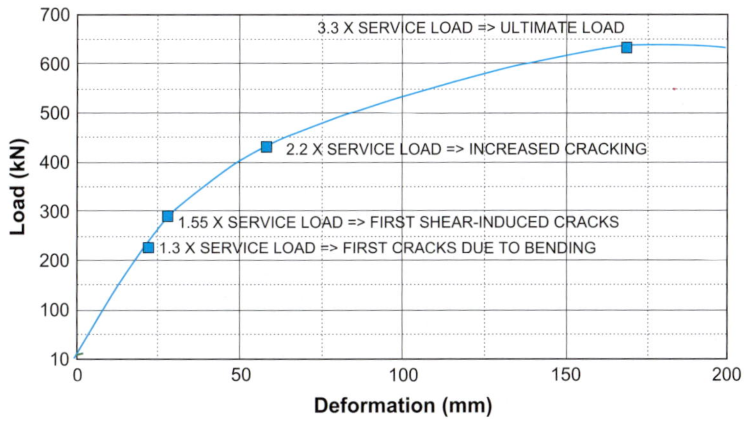

FIGURE 10.26
Typical Load -Deformation curve for a prestressed, steel-fibre-reinforced beam made from self compacting concrete.

There is also an exciting development of new materials for both building and bridge structures which do not only have improved strength and mechanical properties but which also exhibit *enhanced durability characteristics*. Some fully composite structures are already built such as:

- the cable-stayed Abefeldy Bridge in Scotland, UK (Burgoyne, 1999) built with a GFRP deck, supported by Parafil aramid ropes from GFRP pylons.
- the Bonds Mill Bascule Bridge near Gloucester, UK (Head, 1994) with a lightweight deck made from GFRP pultrusions.
- the cable-stayed Herning Footbridge, Herning, Denmark with 16 CFRP stays and a concrete deck incorporating 6 non-bonded post-tensioned tendons, CFRP bars and links and a mixture of carbon and stainless steel reinforcement (Christofferson et al., 1999).
- the cable-stayed Strandhuse pedestrian footbridge at Kolding, Denmark (Braestrup, 1999) constructed entirely of GFRP.

Some proposed bridges are:

- the two span beam and slab Kings Stormwater Channel Bridge in Salton Sea, California, USA which utilises an E-Glass FRP deck system on Carbon/Epoxy shell girders supported on Carbon/Epoxy shell encased PC/PS concrete piles (Seible et al., 1999).
- the cable-stayed I-5/Gilman Bridge at the University of California, San Diego, USA, which utilises three different materials for the cables, a concrete filled carbon/epoxy shell pylon and a concrete filled carbon/epoxy shells transverse hybrid FRP beams and polypropylene fibre-reinforced arch-action concrete deck! (Seible et al., 1999).

The use of new materials for bridge structures is in its infancy but there is a whole range of possibilities and already we are seeing fibre reinforced concrete (FRC) replacing traditional reinforced concrete (RC) construction in areas where, for example, links were necessary to resist shear forces. Add to this the advent of self compacting concrete (SCC), then it is possible to foresee a revolution taking place in concrete bridge construction.

The major behaviourable difference is that in traditional (RC), steel bars are designed and positioned to attract and resist 'lumps' of load for example tension, compression, shear and torsion – in the form of bars and links, and the load paths are well defined. In (FRC), however, fibres are distributed evenly throughout the structure and resist the loads in a complex *tension/compression* field of discrete elements.

It is possible that corrosion due to water or sodium chloride in solution may be eradicated but there may be other hidden sources of deterioration such as ultraviolet (UV) degradation for which we have only limited data. In the final analysis no method of either prevention or cure will ever eradicate corrosion problems, and neither will they be a substitute for sound specification, testing, design, detailing, construction and maintenance.

BIBLIOGRAPHY

Hartcup, G. (1977). *Code Name Mulberry.* London: David & Charles.

REFERENCES

Abbott, C. J. (1999). Steel Corrosion Solution. *Concrete Engineering International, Vol. 3*(No.4), 39–43.

Alam, H., Fowler, D. W., & Carrasquillo, R. L. (1997). *Protective coatings for bridge decks*, Proceedings of the Seventh International Conference on Structural faults and Repair, Edinburgh, July, pp 239–247.

ASTM: A775-81. (1981). *Standard specification for epoxy coated reinforcing bars*, American Society for Testing Materials.

Bamforth, P. (1999). An enduring problem. *Concrete Engineering International, Vol. 3*(No.8), 28–31.

Beeby, A. W. (1993). *Design for Life*. *Proc. Concrete, 2000*, 37–49.

Bowman, S. A. W., Lau, C. K., & Wong, K. W. (1996). *Influence of maintenance on the design of highway structures*, Third International Conference on Bridge Management, Guildford, Surrey; E. & F.N. Spon, London, April, pp. 378–384.

Braestrup, M. W. (1999). Footbridge constructed from Glass-Fibre-Reinforced Profiles, Denmark. *Structural Engineering International, Vol. 9*(No.4), 256–258.

British Standards Institution. (1969). *BS 449: The Use of Structural steel in Building*. London: BSI.

British Standards Institution. BS 5493 BSI, London.

British Standards Institution. (1986). *BS 6744:1986, Austenitic stainless steel bars for the reinforcement of concrete*. London: BSI.

British Standards Institution. (1995). *BS EN 10088-1:1995, Stainless steels. Part 1, list of stainless steels*. London: BSI.

Brown, T. G. (2000). Ice loads on the piers of the Confederation Bridge, Canada. *The Structural Engineer, Vol. 78*(No.5), 18–23.

Buchnor, S. H., & Lindsell, P. (1994). *The condition of existing prestressed concrete bridges*, Bridge Assessment and Design, Proc. of the Centenary Year Conference, Cardiff, Ed. Barr & Harding, Elsevier, pp. 443–448.

Burgoyne, C. J. (1999). Advanced Composites in Civil Engineering in Europe. *Structural Engineering International, Vol.* 9(No.4), 267–273.

Burke, M. P. (1990). *The integrated construction and conversion of single and multiple span bridges*, First International Conference on Bridge Management, Guildford, Surrey; Elsevier, London, pp. 677–693.

Bush, D. (1996). *Integral Bridges – Soil/Structure interaction effects*, One day Seminar on the Design of Integral Bridges at The Institution of Structural Engineers, London.

CEB. (1992). *Bulletin D'Information No.211*. Lausanne, Switzerland: Protection systems for reinforcement.

Christoffersen, J., Hauge, L., & Bjerrum, J. (1999). Footbridge with Carbon-Fibre-Reinforced polymers, Denmark. *Structural Engineering International, Vol.* 9(No.4), 250–253.

Christofferson, J., Hauge, L., & Bjerrum, J. (1999). Footbridge with Carbon Fibre reinforced Polymers, Denmark. *Structural Engineering International, Vol.* 9(No. 4).

Chubb, M. (1996). *Integral Bridges, design of piers and abutments*, One day Seminar on the Design of Integral Bridges at The Institution of Structural Engineers, London.

Cole, M. (1998). Arching action. *New Civil Engineer.* May, 32–33.

Collings, D. (2000). *The durability of Precast Prestressed Segmental bridges*, Fourth International Conference on Bridge Management, Guildford, Surrey; Thomas Telford, London.

Cooper Clarke. (1999). *Private communication*.

Concrete Society. (1999). *Guidance on the use of Stainless Steel Reinforcement*, Technical Report No. 51.

Cromie, J. A., Abu-Tair, A. I., Lyness, J. F., McFarland, B., & McCullough, W. P. (1999). The independent assessment of electrochemical chloride extraction treatment on a 30-year-old motorway bridge. *The Structural Engineer, Vol.* 77(Nos 23 & 24), 38–41.

Cuninghame, J. R. (1992). *Expansion joints and continuity, a two day seminar on Bridge Design for Durability.* Crowthorne, UK: TRL. pp 1–16.

Cuninghame, J. R. (Ed.). (1994). *Bridge Deck expansion joints: Interim report*, (Unpublished) TRL Project Report PR/BR/4/94, E548B/BC.

Cuninghame, J. R., Chakrabarti, S., & Clarke, J. L. (1990). *Non-ferrous prestressing and reinforcement for concrete highway bridges*, Third International Conference on Bridge Management, Guildford, Surrey; Spon, London, pp. 503–515.

Cuninghame, J. R. (1998). *The management of water on bridge decks, Sixth Annual Surveyor Bridges Conference and Exhibition - "The Essential Link"*. London: TRL.

Degueldre, J. P., Hougardy, M., & Hubert, M. (1993). *Ten years of operating paint-testing stations*, Second International Conference on Bridge Management, Guildford, Surrey; Thomas Telford, London, pp. 931–939.

Department of Transport. (1981). *Departmental Standard BD 7/81, Weathering steel for Highway bridges*. London: SO.

Department of Transport. (1995). *Departmental Standard BD 57/95. Design for Durability*. London: DOT.

Department of Transport. (1996). *Departmental Standard BD 43/96. Criteria and Material for the Impregnation of Highway Concrete Structures*. London: DOT.

Department of Transport. (1996). *Departmental Advice Note BA 42/96. Design for Durability*. London: DOT.

Ferry, R. (1998). Glassfibre Reinforced Cement. *Concrete Engineering International, Vol. 2*(No.3), 60–63.

Fischer, G., Lepech, M., & Li, V. (1998). *Engineered Cementitious Composites. Concrete Engineering International (CEI)*. Winter 2007.

Flexideck. (1994). Bridge deck waterproofing system. *Technical literature*.

Ganz, H. R. (1996). *Plastic ducts for enhanced performance of post tensioning*. London: FIP Symposium. pp. 37–44.

Gedge, G. (2000). *Stainless steel reinforcement for concrete - a solution to chloride induced corrosion of reinforced concrete structures*, Fourth International Conference on Bridge Management, Guildford, Surrey; Thomas Telford, London.

Gianetti, F. (1998). Corrosion Inhibitor (part 2). *Concrete Engineering International, Vol. 2*(No.3), 47–51.

Gremel, D. (1998). GFRP Rebars. *Concrete Engineering International, Vol. 2*(No.3), 52–54.

Gruson, P. (2001). Mechanical versus lap splicing. *Concrete Engineering*. Autumn 2001.

Hall, R. J., & Jones, G. H. (1996). *Application of Cathodic Protection to Highway Viaducts*, Third International Conference on Bridge Management, Guildford, Surrey; Spon, London, April, pp. 180–187.

Haynes, M., & Malric, B. (1997). Use of migratory corrosion inhibitors. *Construction Repair, Vol. 11*(No.4), 10–15.

Haynes, J. M. (1999). A Stainless reputation. *Concrete Engineering International, Vol. 3*(No.4), 44–45.

Head, P. R. (1994). *The World's first advanced composite road bridge, Symposium on Short and Medium Span bridges*. Canada: Calgary.

Holland, D. A. (1993). *Design of bridges for durability and maintenance*, Second International Conference on Bridge Management, Guildford, Surrey; Thomas Telford London, April, pp. 889–896.

Hollinshead, K., & Bassi, R. (1997). Assessing Coatings Systems. *Concrete Engineering International, Vol. 1*(No.1), 36–37.

Iyer, S. L. (1994). *Prestressing concrete bridge decks with Advanced Composite cables using innovative anchorage systems*, Bridge Assessment and Design, Proc. of the Centenary Year Conference, Cardiff, Ed. Barr & Harding, Elsevier, pp. 353–360.

Johnson, I. D., & McAndrew, S. P. (1992). *Research into the condition and performance of Bridge Deck Expansion Joints, a two day seminar on Bridge Design for Durability*. Crowthorne, UK: TRL, pp. 17–33.

Jones, T., & Cather, B. (2005). Ultra-high performance fibre-reinforced concrete. *CEI*. Spring 2005.

Lankard, D. R., Slater, J. E., Hedden, W. A., & Niesz, D. E. (1975). *Neutralisation of chloride in concrete*, Federal Highways Administration Report No. FHWA-RD-76–60.

Li, V. C., & Leung, C. K. Y. (1992). Steady state and multiple cracking of short random fiber composites. *ASCE J. of Engineering Mechanics*.

Li, V. C. (1993). From Micromechanics to Structural Engineering. *ASCE J. of Structural Mechanics and Earthquake Engineering*.

Li, V. C. (2005). *Non-brittle concrete*. *Concrete Engineering International*. Winter 2005.

Kanda, T., & Li, V. C. (1998). *A new micromechanics design theory for pseudo strain-hardening cementitious composites*. *ASCE J. of Engineering Mechanics*.

Lee, D. J. (1994). *Bridge Bearings and Expansion Joints* (Second edition). London: E&FN Spon.

Low, A. (1996). *Integral View - Designers Viewpoint*, One day Seminar on the Design of Integral Bridges at The Institution of Structural Engineers, London.

Maissen, A., & de Smet, C. (1997). Prestressed concrete using carbon fibre reinforced plastic (CFRP) strands statically indeterminate systems. In Meier Betti. (Ed.), *Recent Advances in Bridge Engineering* (pp. 255–264), EMPA.

McDonald, D. B., & Krauss, P. D. (1995). Stainless steel reinforcing as corrosion protection. *Concrete International, Vol. 17*(No.5), 65–70.

McKenzie, M. (2000). *The performance of in-situ weathering steel in bridges*, Fourth International Conference on Bridge Management, Guildford, Surrey; Thomas Telford, London.

Menn, C. (1990). *Prestressed concrete bridges*. Germany: Birkhauser.

Montgomery, F. R., & McC. Murray, A. (1990). *Reinforced concrete bridge protection in Northern Ireland*, First International Conference on Bridge Management, Guildford, Surrey; Elsevier, London, April, pp. 259–265.

Mufti, A. A., Jaeger, L. G., & Bahkt, B. (1997). Field performance of steel-free deck slabs of girder bridges. In Meier., & Betti. (Eds.), *Recent Advances in Bridge Engineering* (pp. 239–246). EMPA.

Nakai, H., Asai, H., & Noritake, K. (1994). *Studies on characteristics of beams using Aramid Tendons*, Bridge Assessment and Design, Proc. of the Centenary Year Conference, Cardiff, Ed. Barr & Harding, Elsevier, pp. 489–494.

New Civil Engineer. (2000). *Steeling the Show*, Editorial, NCE 20.1.2000, p. 9.

O'Brien, E. J., & Flanagan, J. W. (1996). *Relief of creep and shrinkage stresses in integrally constructed bridges*, Third International Conference on Bridge Management, Guildford, Surrey; E & F N Spon, London, April, pp. 194–172.

Paice, C., Hey, R. D., & Whitbread, J. (1993). *Protection of Bridge Piers from scour*, Second International Conference on Bridge Management, Guildford, Surrey; Thomas Telford, London, April, pp. 543–552.

Pearson-Kirk, D., & Jayasundara, P. (1993). *Impregnation of concrete highway structures*, Second International Conference on Bridge Management, Guildford, Surrey; Thomas Telford, London, April, pp. 39–48.

Raggett, S. (Ed.). (1992). *Bridge Design for Durability: Expansion joints & Continuity*. Crowthorne, Bucks: TRL Seminar. PA2138/92.

Raiss, M. E. (1994). Durable Post-tensioned Concrete Bridges, Bridge Assessment. *Management and Design*. Cardiff: Elsevier.

Raiss, M. E. (1995). Lasting Effect. *NCE Supplement* 46–48.

Read, J. A. (1990). *Rebar Corrosion - FBECR: The fight to Cure the Problem*, First International Conference on Bridge Management, Guildford, Surrey; Elsevier, London, April, pp. 267–284.

Safier, A. S. (1989). Development and use of electrostatic, epoxy-powder coated reinforcement. *The Structural Engineer, Vol. 67*(No.6), 95–98.

Seible, F., Karbhari, V. M., & Burguenio, R. (1999). Kings Stormwater Channel and I-5/Gilman Bridges, USA. *Structural Engineering International, Vol. 9*(No.4), 250–253.

Sha, W., & Kee, R. V. E. (1999). Stainless Steel: A users survey. *Concrete Engineering International, Vol. 3*(No.8), 49–52.

Shehata, E., Abdelrahman, A., Tadros, G., & Rizkalla, S. (1997). FRP for large span highway bridge in Canada. In Meier & Betti. (Ed.), *Recent Advances in Bridge Engineering* (pp. 247–254). EMPA.

Smith, F. N. (2007). The use of stainless steel reinforcement. *Concrete Engineering International, (CEI)*. Summer 2007.

Smith, W. (2000). *Hot Dip Galvanising of Bridge Beams for Long term Corrosion protection*, Fourth International Conference on Bridge Management, Guildford, Surrey, Thomas Telford.

Soloman, I., Vimpani, P., & Argyriou, S. (1998). Ceramic anode technology. *Concrete Engineering International, Vol 2*(No. 2), 23–26.

Springman, S., & Norrish, A. (1996). *Integral Bridges - Researchers' Viewpoint*, One day Seminar on the Design of Integral Bridges at The Institution of Structural Engineers, London.

Steel Construction Institute. (1995). *Design of steel bridges for durability.* UK: Ascot. Technical Report No. 154.

Stirling-Lloyd. (1998). Bridge deck waterproofing. *Technical literature*.

Strobach, Claus-P. & Bogl, Max. (2007). Practical use of steel-fibre reinforced precast beam. *CEI*. Spring 2007.

Tarmac,. (1999). *Building and Civil Engineering products*, Stamford, UK, p. 31.

Thornburn, S. (1996). *The use of run-on slabs - a solution or part of the problem*, One day Seminar on the Design of Integral Bridges at The Institution of Structural Engineers, London.

Transport Research Laboratory (TRL). (1988). *Overseas Road Note 7: A guide to bridge inspection and data systems for district engineers*. Crowthorne: TRL.

Treadaway, K. W. J., & Davies, H. (1989). Performance of fusion-bonded epoxy coated steel reinforcement. *The Structural Engineer, Vol. 67*(No.6), 99–108, March.

Vassie, P. R., & Arya, C. (2000). *Selecting appropriate remedial measures for corroding concrete structures*, Fourth International Conference on Bridge Management, Guildford, Surrey, Thomas Telford.

Wallace, A. A. C., & Colford, B. R. (1996). *White Cart Viaduct Inspection, Maintenance and design*, Third International Conference on Bridge Management, Guildford, Surrey; Spon, London, April, pp. 626–634.

Wallbank, E. J. (1989). The performance of concrete bridges: A survey of 200 highway bridges. *Report prepared for the DoT.* London: HMSO. April, pp. 1–96.

West, J., Vignos, R., & Breen, T. (1998). *Laboratory Durability study of internal tendons in precast segmental bridges*, ASBI Convention, USA.

Whitbread, J. E., & Riddell, J. (1996). Bridge Scour, Informal discussion, Proc. Instn. *Civ. Engrs. Transp., Vol 117*, 67–69.

Willis, J. (1982). *Epoxy coated reinforcement in bridge decks, Transport and Road Research Laboratory Supplementary Report 667*. Department of Transport.

Wilson, D. (2001). *Durability problems: is formwork the cause? Concrete Engineering, Vol. 5*(No.2), 35–38.

Wilson, I. (1999). PUR and simple. *Concrete Engineering International, Vol. 3*(No.4), 54–56.

Wollf, R., & Miesseler, H. (1990). *Prestressing with Fibre Composite Materials and Monitoring of Bridges with Sensors*, First International Conference on Bridge Management, Guildford, Surrey; Elsevier, Telford, London, pp. 395–402.

Wood, J. G. M. (1996). *Durability design: applying data from materials research and deteriorated structures, Third International Conference on Bridge Management*. Guildford, UK: Spon. pp. 723-731.

Woodside, A. R., & Woodward, W. D. H. (1990). *The design of a flexible surface mix for use at bridge expansion joints, First International Conference on Bridge Management*. Guildford, UK: Elsevier. pp. 737–745.

Woodward, R. J., & Williams, F. W. (1988). The collapse of Ynys-y-Gwas Bridge, West Glamorgan. *Proc. Instn. Civ. Engrs., Part 1, Vol. 84*, 634–669, Aug.

Woodward, R. J., & Milne, D. (2000). *Monitoring the environment in the box section of externally post-tensioned concrete bridges*, Proc. Fourth International Conference on Bridge Management, Thomas Telford, Guildford, UK.

FURTHER READING

Broomfield, J. P. (1997). *Corrosion of steel in concrete*. London: Spon.

Derucher, K. N., & Heins, C. P. (1998). *Bridge and Pier protective systems and devices*.

Wallbank, E. J. (1989). *The Performance of Concrete Bridges: A Survey of 200 Highway Bridges*, pp. 1–96, HMSO, Report prepared for the DoT, London.

APPENDIX 10A

210. Durability. In order to ensure adequate protection to the reinforcement with the cover specified in Clause 307, it is essential that it should be *dense, impermeable and of a quality suitable for the conditions of exposure involved.* The greater the severity of the exposure, the higher the quality of concrete required, and thus although a weaker concrete may be adequate from structural considerations this weaker mix may not be suitable from the durability viewpoint. For example, a concrete having a specified works cube strength of 25 N/mm^2 at 28 days may be suitable structurally and when used in exposed conditions in a non-industrial area; however, if the structures were sited in an industrial area, a higher grade of concrete with a greater cube strength would probably be required.

APPENDIX 10B

Concrete mix for panels.

11 cwt cement, 13 cu ft foamed slag and 27 cu ft of normal shingle. Cured for three days with water.

CP114 Cl. 352. Effects of chemical reagents. Concrete structures are slowly attacked by some chemical reagents such as vegetable oils and fats, sugar solutions and sulphates, prolonged exposure to which will bring about gradual deterioration. Increased resistance to some forms of chemical attack may be obtained by the use of high alumina cement or sulphate resisting cement, or by the use of protective coatings. Where destructive agents are liable to come into contact with floors or ceilings, the finish should be designed so as to prevent the entry of these agents into the structural concrete.

CHAPTER 11

Stress Measurement and Monitoring

11.1 INTRODUCTION

CONTENTS

The assessment of bridge structures as part of ongoing bridge management programmes often requires a knowledge of the *in situ* stresses in the deck and superstructure. These consist of the serviceability stresses due to dead loads and, in some cases, prestressing loads. They are also important if damage to the structure has been identified which might indicate an *overtsress* such as the shearing of a bolt or tearing of a weld in metal structures or cracking in a concrete structure. It is also necessary to measure the *in situ* stress levels prior to a supplementary load test (Chapter 5).

In the case of prestressed concrete decks, which have always been designed to maintain stresses at or below codified limits, unacceptable losses of prestress may be contributing to the general deterioration of the bridge as evidenced by cracking and spalling and large deflections under dead loads. Very often, as in the case of grouted tendons, it is not possible to determine the stresses in the prestressing tendons themselves, but it is possible to measure the stresses in the surrounding concrete (Lindsell & Buchner, 1987). If the measured stresses indicate an unacceptable loss of prestress (*an understress situation*), then bridge engineers can be alerted to take appropriate action.

11.2 *IN SITU* OR RESIDUAL STRESSES

In situ stresses in concrete are due to a combination of shrinkage, creep, temperature, dead (self weight) and superimposed dead loads and loads from prestressing. These *in situ* stresses can be obtained using the *stress relief* principal, either *directly or indirectly*, which results in a value of a *total stress*, which then has to be broken down into its component parts.

Tyler (1968) reports that until 1968 there were only a few clumsy methods of determining these stresses which were neither practical nor accurate.

11.3 STRESS RELIEF PRINCIPLE

If any structural element subject to a stable stress field is disturbed by drilling or cutting in any way, then there is an immediate redistribution of stress in the region of the disturbance. The redistribution can be measured as a strain change, either in the material adjacent to the disturbance (a level 1 change), or it can be detected in a measuring device which is installed into the discontinuity caused by the first disturbance (a level 2 change). Both methods are *indirect* in that they first measure a strain in the region of the disturbance which is then converted into a strain in the body of the host material.

Figure 11.1 shows the basic principal. The plate becomes de-stressed and the elastic change at the cut (Δ) can be used to determine the original stress in the plate namely:

$$\sigma = E \times \Delta/L \tag{11.1}$$

where L is the length of the plate.

A more complex situation shown in Figure 11.1(iii) is where the stress pattern is disturbed by drilling a hole, for example, and the maximum stress adjacent to the hole ($k\sigma$) can be related to the original stress by a simple formula:

$$[p] = [k][\sigma] \tag{11.2}$$

which takes account of stress concentration constants and the elastic modulus of the material.

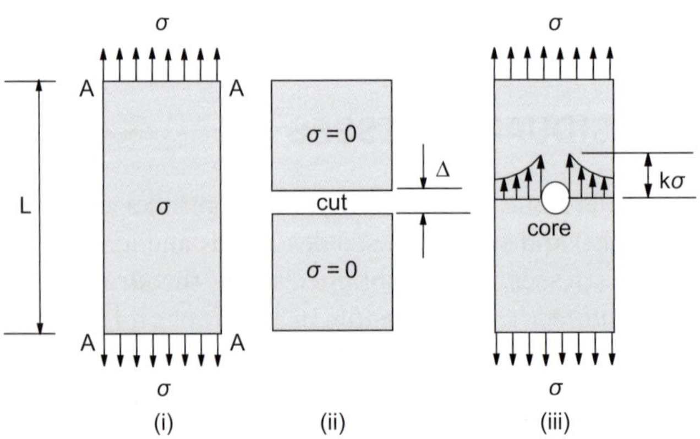

FIGURE 11.1
Stress relief principal.

11.4 INDIRECT STRESS MEASUREMENT

In situ stresses in the concrete can be measured *indirectly* by stress relief methods so-called because the measure of the disturbance caused by coring is translated into the final stress in the concrete. There are currently four ways of doing this namely: the standard overcoring technique (Mehrkar-Asl, 1996); modified overcoring (Owens et al., 1994); inclusion method (Ryall, 1993) and the saw cut method (Forder, 1992).

11.4.1 Standard overcoring

This method is essentially an extension of the *centre hole technique* used to assess the residual stresses in steel components due to rolling and welding (Beaney & Procter, 1973; Beaney, 1976) which involves attaching a special strain gauge rosette to the surface on which the residual stresses are to be measured. A hole is then drilled at its centre. The drilling relaxes the radial stress at the edge of the hole and causes a local redistribution of stress; the resulting strain change is detected by the gauges and is related to the residual stress in the material by equations developed by Beaney and Procter (1973), (see also section 11.4.8).

In concrete, because of its tendency to microcracking due to shrinkage and temperature and the coarse nature of its constituents (cement, sand and aggregate), it is not feasible to use the above method which uses very small rosette type strain gauges and a drill size of about 2 mm diameter. In what might be termed the *macro* version of the centre hole method, Mehrkar-Asl (1988) measured strains on the surface of the concrete using vibrating wire (vw) gauges, and used an overcoring drill of 150 mm diameter to produce a centre hole. It relies upon the measurement of surface strains around the periphery of a large diameter core using both vw gauges around and remote from the core circumference, and micrometer measurements across pop marks at the centre of 6 mm diameter metal discs (pips) glued to the surface of the concrete within and just outside the core perimeter. The two ways of measuring strain are used to cross verify the strain readings. The core is then cut incrementally; that is the drilling proceeds in steps of 25 mm and readings taken. This provides a diagnostic record of strain readings and any unexpected change of strain would be immediately apparent, and would indicate the breakdown of a vw gauge, the debonding of a pip or some unknown factor which would need to be investigated (see Figure 11.2).

An accurate knowledge of the physical properties of the concrete is required, namely Poisson's ratio and the modulus of elasticity, to translate the strain readings into stress. These are established from the core of

FIGURE 11.2
Standard overcoring method.

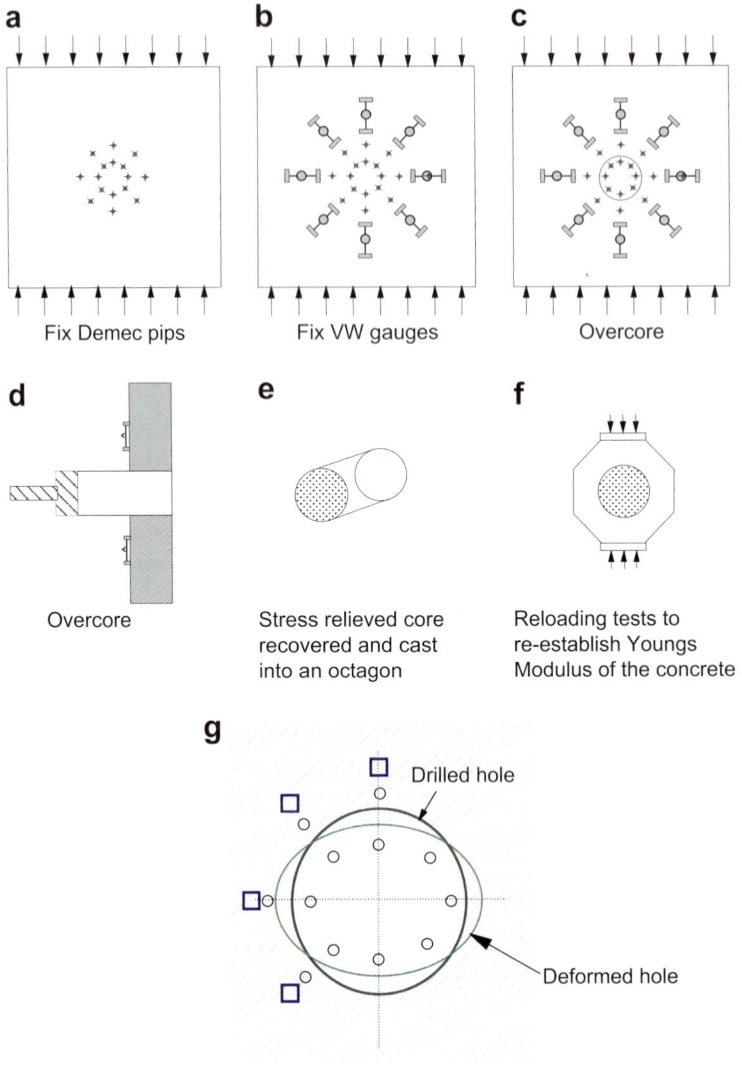

concrete extracted during the overcoring process which is cast into an octagon shaped specimen of concrete and allowed to cure for 28 days. An array of gauge points is secured to the opposite faces of the specimen and the initial readings are taken.

Readings are then taken in steps of about 20 kN up to 100 kN and then unloaded. This procedure is repeated for the vertical, horizontal and transverse directions. It is then possible to obtain relationships between stress and strain by re-establishing the original measured strain release in the gauges during the overcoring process and by applying the principal of superposition it is possible to establish the modulus of elasticity and Poisson's ratio of the concrete in the release direction (Mehrkar-Asl, 1988). A fuller version of the method has been published with governing equations in *FIP Symposium Papers* (Mehrkar-Asl, 1996).

As a back up to the octagon test, a modified version of the Goodman Jack (Goodman et al., 1970) can be used which was originally used for determining the properties of rock.

11.4.2 Modified overcoring

Owens et al. (1994) have developed a technique for determining the stresses in concrete bridges based on the standard coring method and similar to the trepanning process used in rock engineering (Leeman, 1964a, 1964b) and is shown in Figure 11.3.

The first innovation is that the strain measurement can be made on both the surface and at small depths below the surface. vw gauges of 50 mm active length are attached radially around the outside of the nominally 75 mm diameter overcore and one vw gauge is attached to the surface of the core in the direction of the principal stress which is usually the same as the main prestressing tendons (see section 11.5). The core is then drilled to a depth of 50 mm in 10 mm increments and strain readings taken at each increment and recorded against depth. Below 50 mm the surface gauges are not, according to Owen, sensitive to stress relief, and so the 50-mm core is removed, the bottom of the hole is ground flat, and the same procedure is

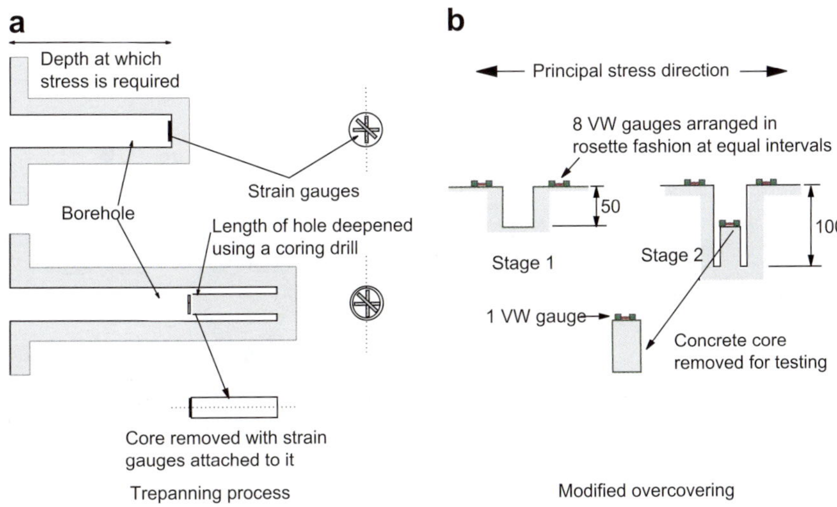

FIGURE 11.3
Comparison between trepanning process and modified overcore.

continued to a depth of 100 mm and readings recorded as before. The core is recovered and, because of practical difficulties, drilling is discontinued. The relieved strains are then used to determine the mean *in situ* stress acting over the hole depth. Poisson's ratio is assumed to be 0.2 and the elastic modulus is determined from independent core samples. Owen quotes a non-percentage accuracy of ± 0.2 N/mm^2 and states that *the non-percentage accuracy is directly proportional to the modulus*.

Owens states that the concrete *manufacturing stresses* (by which he probably means shrinkage stresses) are balanced between different signs in the matrix of the aggregate and therefore, if the gauge length and hole diameter are greater than or equal to twice the aggregate size, these stresses are localised and do not affect the strain readings. This statement is not substantiated, but if true, means that measured surface stresses do not have to be adjusted to take account of shrinkage stresses, a conclusion which is in direct contradiction to the conclusions of Mehrkar-Asl (1988). For a normal aggregate size of 20 mm this would mean that the smallest vw gauge could be used in conjunction with holes of about 40 mm diameter.

11.4.2.1 Multiple small hole coring

The second, more radical innovation, uses a progression of smaller 36–52 mm diameter holes with a radial configuration of vw gauges attached to the concrete using a quick setting adhesive (Owens et al., 1994). The gauges have a 50-mm active length and are recoverable for further use. Typical gauge layouts are shown in Figure 11.4.

Owens claims that the use of a series of smaller holes provides a more general distribution of stress than the localised stresses determined from the

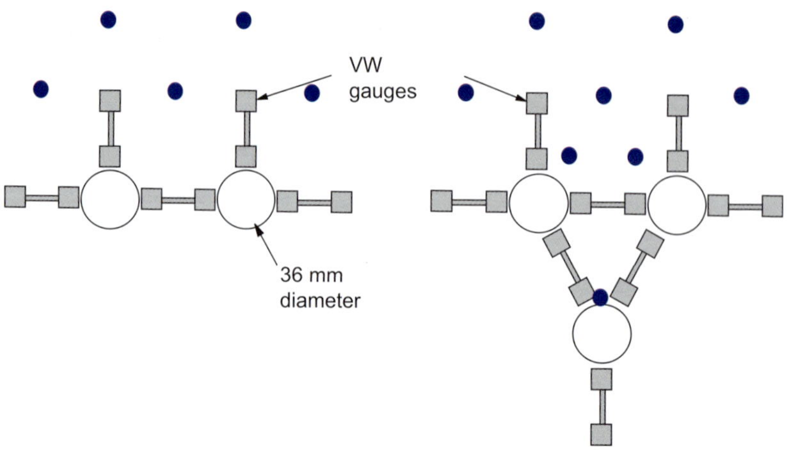

FIGURE 11.4
Typical gauge layout for multiple holes.

standard technique and does less damage to the structure under test. It is also useful where the distance between reinforcing bars is small and access difficult. A disadvantage is that shrinkage stresses become more significant as the hole size is reduced and have to be taken into account. The method was fully verified by experimental work and finite element simulation, although no comments were made about the effect of locked-in shrinkage stresses.

11.4.3 Saw cutting

Saw cutting has been developed by both Forder (1992) using an *indirect* approach and by Abdunur (1985) using a *direct* approach (see section 11.3.1); both employ stress relief principals.

In Forder's case a slot is cut through a concrete member over a defined length and the strain changes due to the cut are measured. A typical concrete sample used by Forder in this research measured 1 m × 1 m × 100 mm thick which was subjected to a uniaxial vertical compressive stress (see Figure 11.5).

The Modulus of Elasticity was first determined by incrementally loading and then unloading the slab. Subsequently the slabs were cut under load from both sides to represent site conditions and finally the cut slabs were unloaded and loaded incrementally. The idealised strain distribution along the horizontal datum line between the vw gauges would be as shown in Figure 11.6a and the actual profiles obtained are as shown in Figure 11.6b. The actual *in situ* stress is calculated by the product of the measured (release) strain × E called the *apparent stress* and a coefficient k,

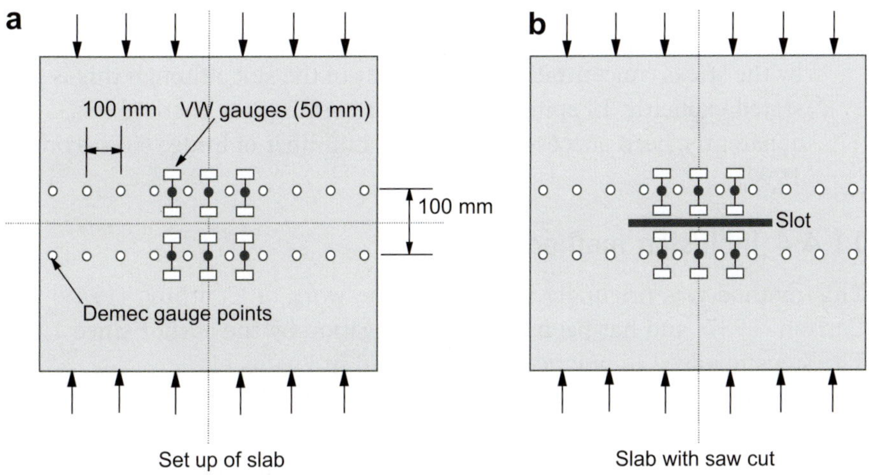

FIGURE 11.5
Saw cutting method.

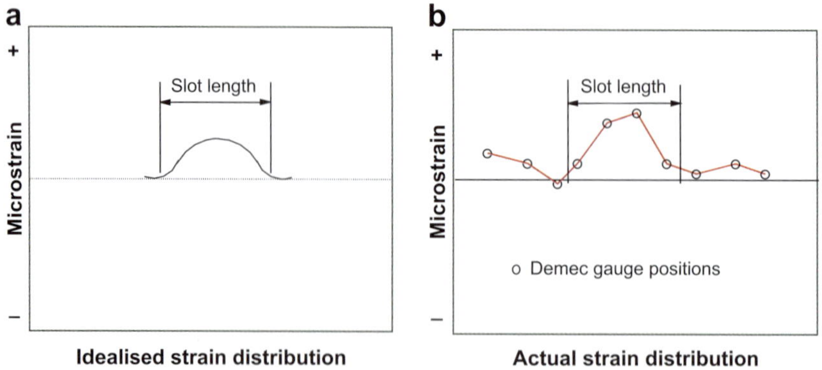

FIGURE 11.6
Comparison of ideal and actual strain distribution across a saw cut.

called the *release coefficient*. Shrinkage strains were taken into account by cutting the slabs under no load.

The difference between the absolute (measured) strain and the shrinkage strain is the actual strain.

Forder's research is incomplete for the following reasons:

- The results from the different types of gauges, namely, vw and micrometer, gave different values of the release coefficients.
- No independent check was carried out to calculate the stress on a slab from saw cut results.
- No work was carried out to ascertain the effect of a blind saw cut (which would be the case in practice) compared with the full penetration saw cut used in the tests.
- The absolute stress is based on the maximum strain measured across the slot (usually at about mid-length) which presumably is unaffected by the stress concentrations at the ends of the slot although this is not stated explicitly. In spite of these shortcomings the method has, apparently, been successfully used at a number of bridge sites (Forder, 1992).

11.4.4 Inclusion method

This method was originally inspired by the work of Couthino (1949) and Carlson (1937) and has been under investigation by the author since 1990. The steps involved in practice are illustrated in Figure 11.7.

A small 42 mm pilot hole is drilled into the concrete at the point of interest using a diamond-tipped drill bit. An instrumented cylindrical steel inclusion – the *Stressmeter* – is then inserted into the pilot hole and bonded

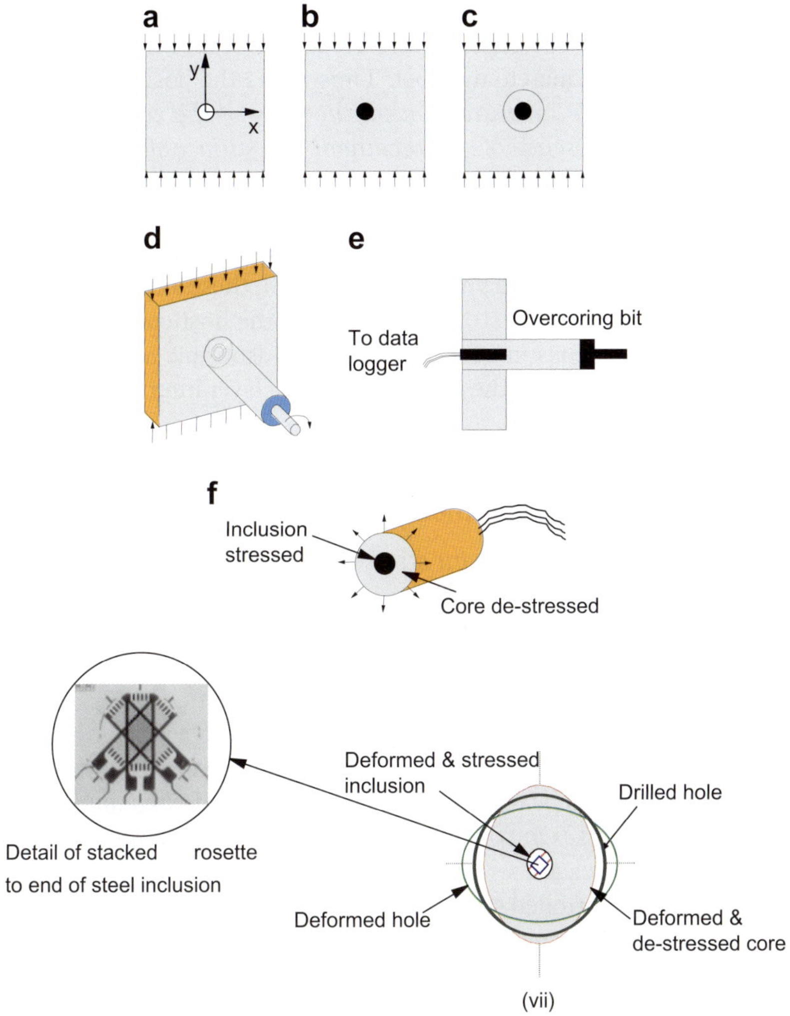

FIGURE 11.7
Standard inclusion method.

into place. The stacked rosette strain gauges mounted on the diametric ends of the stressmeter measure the strain when it is overcored incrementally using a 150-mm diamond-tipped drill bit. The action of overcoring relieves the core from the surrounding stresses and the concrete annulus expands or contracts to an unstressed state, but stresses up the inclusion itself. The electrical resistance strain (ERS) gauges attached to the inclusion record the resulting strain changes. The inclusion used is made up of 75 mm long steel sections locked together to form a single unit. Each section has a strain rosette bonded to one end to measure the strain when overcoring takes place.

There are five units and so this enables readings to be taken at five stations within the body of the concrete and the results form a stress profile through the depth of the concrete member. Thus it has the positive advantages that *the stresses can be measured within the body of the concrete in areas less affected by the presence of reinforcement and surface shrinkage effects*. The inclusion *is itself* shown in Figure 11.8 along with its use in a concrete box girder bridge.

Using a theory based on the work of Muskhelishivili (1953), Savin (1961), Duncan Fama (1979) and Spathis et al. (1988), these strain changes can then be directly related to the strain and stress in the host concrete (Ryall, 1993). The materials are assumed to be elastic and isotropic and a perfect bond is assumed to exist at both the concrete/bonding layer interface and the bonding layer/inclusion interface.

11.4.5 Direct measurement

Direct measurement of *in situ* stresses was originally proposed by Abdunur (1982) based on the slot cutting method used in rock mechanics (Rocha et al., 1966). After an extensive research and testing programme, they eventually formalised the method and reported to the Ministere de l'Equipment, du Logement, des Transports et de la Mer (Abdunur, 1985). Its continued development and use were restricted for a time to France until after 1990 when the method was reported in the UK (Abdunur, 1990) which led to its use elsewhere. *It appears to be the first recorded method applied directly to determining stresses in concrete bridges*. Abdunur (1985) has described it as a partial stress release which consists in the local elimination of stresses, followed by a controlled stress compensation.

Figure 11.9 illustrates the steps involved. In practice, the location of the required stress is identified and a displacement (d_0) reference field marked out on the concrete surface.

A thin, semicircular slot is then cut at right angles to the required stress field to a depth of 80 mm and (d_0) changes to (d). Finally, a semicircular flat jack is inserted into the slot and pressurised until the initial displacement field (d_0) has been restored. The final pressure in the jack is then a measure of the *average in situ* compressive stress over the cut segment.

The method is used reasonably successfully in rock mechanics but is not easily transferable to concrete. This is due to two main reasons:

The complex nature of concrete as a material.

The absolute need for miniaturisation to minimise damage to bridges.

Tests were carried out to analyse the effects of shrinkage and to enable its effects to be taken into account when interpreting the measured *absolute*

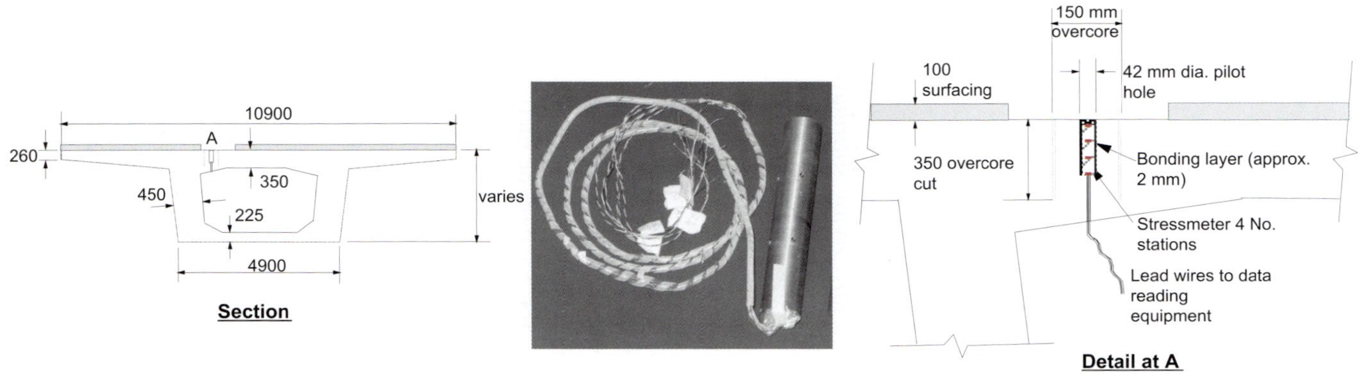

FIGURE 11.8 *Application of the instrumented inclusion (stressmeter).*

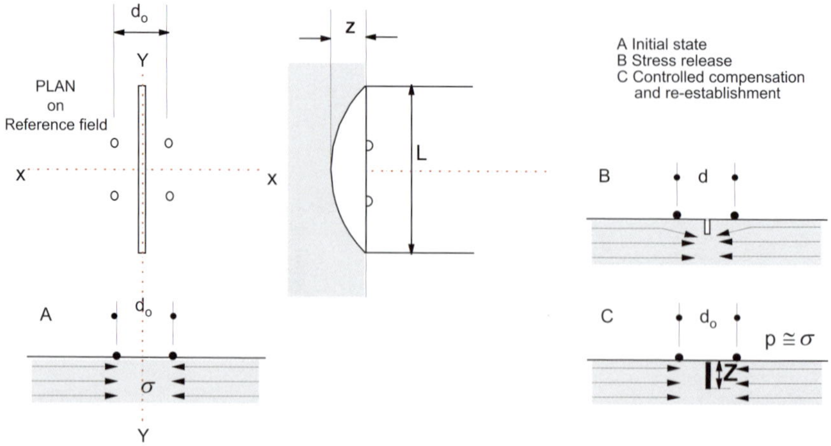

FIGURE 11.9
Slot cutting method (after Abdunur, 1985).

stress σT and breaking it down into its two main components: *applied (in situ stresses σA) and internal (mainly drying shrinkage σS)*. The superposition principal is valid, and enabled the two stress components to be separated out, thus:

$$[\sigma_A] = [\sigma_T] - [\sigma_S] \tag{11.3}$$

Abdunur designed prototype jacks to deal with the problem of size and these were successfully tested in the laboratory. They are only 4 mm thick and have a maximum depth of 80 mm.

Each of the three methods of 'stress measurement' has certain merits and demerits as shown in Table 11.1. Each of the methods is highly technical and requires suitably trained operators in the field for drilling, closely supervised by professional engineers. The collection and interpretation of data also needs to be carried out by suitably qualified staff. In many cases testing agencies can provide a fully comprehensive service, or alternatively a structural consulting engineer can bring together a team of tried and trusted staff.

11.4.5.1 Embedment gauge

Another less sophisticated but nevertheless accurate way of measuring the stresses in concrete is by the use of a recent innovation from America called the *embedment gauge* (Measurements Group, 1995) which consists of a rosette gauge mounted on steel plate (approximately 150 × 20 × 5 mm) with a pre-attached electrical cable. The whole assembly is specially encapsulated to resist both chemical attack and damage during the placement of wet concrete, and also to enhance strain transfer from the surrounding concrete matrix.

Table 11.1 Comparison of Stress Relief Methods

Question	Coring	Inclusion	Direct
Measurement of stress direct?			✓
Measurement of stress indirect?	✓	✓	
Concrete properties required?	✓		
Supplementary tests needed for concrete properties?	✓		
Concrete properties can be determined from test?			✓
Only surface (or near surface) stresses calculated?	✓		✓
Full stress profile possible with depth?		✓	
Extensive pre-test instrumentation required?	✓		✓

11.4.6 Blind hole technique applied to reinforcing bars

One way of indirectly determining the *in situ* stress in reinforced concrete structures is to find out the stress in the reinforcement and thus, assuming strain compatibility and knowing the geometry of the section, to deduce the stresses throughout the depth of the member.

Owens (1988) carried out an investigation using the centre hole or blind hole method (Beaney, 1976) to determine the load in a reinforcing bar. A small area of the bar was ground down and a target three-element rosette gauge as shown in Figure 11.10 was bonded to the prepared surface with an impact (instant setting) adhesive. A small hole was then drilled at its centre target. The effect of the drilling is to completely relieve the stresses at the edge of the hole resulting in a perturbation of the stress field at that location. It is then possible to determine the residual stresses in the bar local to the hole (Beaney, 1976).

This centre hole method is extensively used in the field of mechanical engineering for determining the residual (manufacturing) stress state in metal machine parts, large pipes and structural elements with effectively flat surfaces. In a reinforcing bar (placed in a bridge deck for example) there are additional stresses due to the applied loads. Owens (1988) refers to these as *locked-in* stresses. In order, therefore, to separate the mechanical stress from the locked-in stress, it is necessary to take measurements on an unloaded reinforcing bar and by a known relationship between them they can be separately determined.

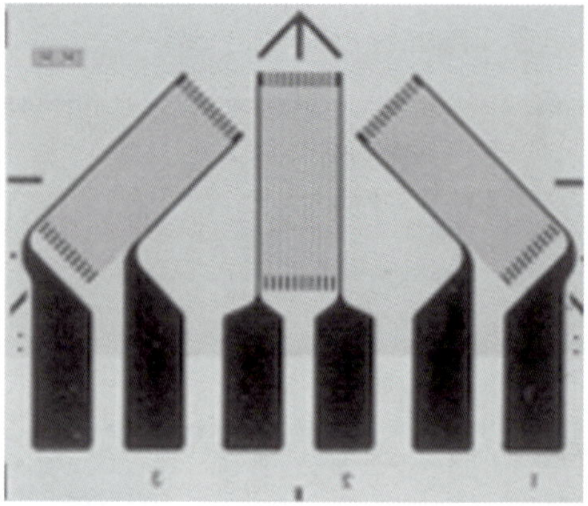

FIGURE 11.10 *Strain gauge geometry* (d = *1.57 mm).*

The method was shown to be suitable for bars of 20 mm diameter or greater and accuracy of ±10 kN was claimed for determining the load in the bar. Owens also claimed that it is relatively non-destructive in that only a small amount of concrete with a depth equal to the cover need be removed. The method is still in current use, but is heavily dependent upon an accurate assessment of the manufacturing stresses.

11.4.7 Cut bar technique

A method of measuring *dead load stresses in reinforcing bars* embedded in concrete bridges has recently been developed by LoBEG (London Bridges Engineering Group). Chalkley (1994) claims that the method is simple and extremely accurate and allows a full reinstatement of the rebar stress and strength on completion of the procedure. It is claimed to overcome the problem of locked-in stresses which bedevils stress assessment in steel elements.

The procedure is shown in Figure 11.11 and has been used successfully in conjunction with the supplementary load testing of bridges and means that a strain gauge can be secured to the bar to monitor actual stresses in the bar during the load test. The stress is only measured locally, but providing that the bridge deck has good load distribution properties it can give a reasonable assessment of how all of the bars are behaving and provide a way of ensuring that the test load is not exceeded. It also, of course, allows some assessment to be made of the *in situ* stresses within the structure.

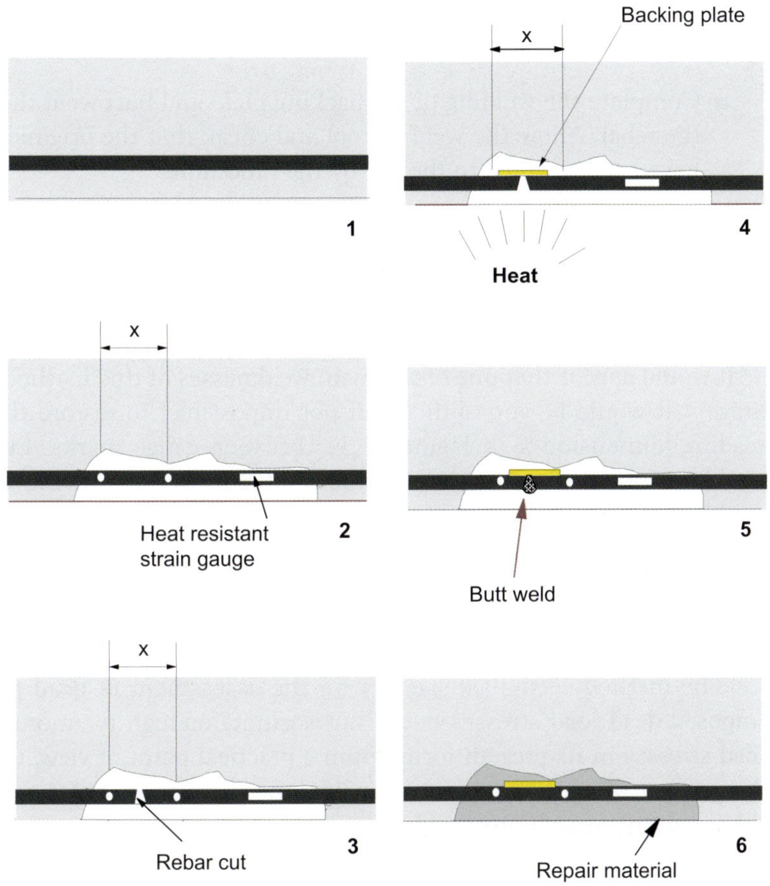

FIGURE 11.11
Cut bar procedure.

The stages of the procedure (after Chalkley, 1994) are as follows:

- Using a covermeter, identify the position of the rebar to be tested.
- Expose at least 450 mm of the rebar by water jetting or similar procedure allowing sufficient clearance around the bar for a backing plate to be inserted. Punch gauge marks either side of where the bar is to be cut and attach a heat resistant strain gauge at the end of the exposed bar remote from the cut position. Measure the distance between the gauge marks and take the initial strain gauge readings.
- Cut the rebar between the gauge marks using a cutting disc and shape the ends for butt welding. Take the *zero stress* strain gauge reading. Calculate the original dead plus superimposed dead loading stress in the rebar.

- Weld the backing plate to the rebar on one side of the cut and heat the rebar until the gauge marks are the same as originally measured.
- Complete the welding of the backing plate and butt weld the ends of the rebar. Allow the weld to cool and check that the original stresses have been restored to the bar by the procedure.
- Repair the concrete.

Chalkley (1994) states that generally a stress of between 75% and 125% of the original stress can be restored to the bar which presumably is measured by the strain gauge although this is not clear.

It would appear that one of the main weaknesses of this method is that at Stage 4 it would be very difficult (if not impossible) to restore the original reading (dimension x in Figure 11.11) between gauge marks. Even if this could be done by heating the rebar, it is difficult to see how the dimension could be maintained when further heat is applied during welding of the backing plate to the other side of the rebar. In fact it is not clear why it is necessary to restore the original stresses to the bar. Providing the bar has been welded and a further gauge measurement taken, then any change of stress locally due to live load will be recorded.

The method seems satisfactory for the assessment of dead plus superimposed dead load stresses but is not accurate enough for monitoring live load stresses in its present form. From a practical point of view, there is the danger that when the concrete is repaired the electric resistance gauge will be damaged unless well protected.

11.4.8 Centre hole technique applied to prestressing tendons

This technique was investigated by Richardson (1993) and also relies on the principals and theory developed by Beaney (1974). The aim of the work was to assess the feasibility of determining directly the stress in a prestressing tendon. This had never been attempted before and consequently the main objectives were to assess whether the method was theoretically and practically sound.

By comparing stress concentration factors used in the Kirsch solution (Timoshenko et al., 1951) for stresses around a hole drilled in a flat plate, with the stress concentration factors used in the solution for stresses around a hole in a round bar (Jessop et al., 1959) it became clear that the Kirsch solution would not apply directly in converting measured strains to stress.

Another problem was that high stress levels in localised areas are likely to exceed the yield strength of the material where the uniform stress exceeds 60% of the material yield stress (Beaney & Procter, 1973), and this is particularly so in prestressing systems where the final stresses after losses range between 55% and 70% UTS. This means that any calibrated stress release measurement system used *in situ* would have to rely on empirical factors to account for plastic deformation.

A final problem was that, like manufactured machine parts, there are residual or locked-in stresses in prestressing wires due to the heating, cooling and cold-drawing processes involved. *A major part of the investigation was to determine the locked-in stresses and their distribution across the diameter.* If this is not known then the proportion of the stress attributable to the applied loads is impossible to determine. This was achieved by calibrating the stress release mechanism *versus* applied load using a loaded strand and a free strand with identical residual stress distributions. According to Richardson typical stress variations before and after drilling are shown in Figure 11.12, although the residual stresses are somewhat higher than typical levels quoted by Owens (1994), who quoted values of 10 N/mm (±2) with compression on the circumference and tension in the middle.

The wires used in the test were taken from the Ynys-y-Gwas Bridge (Woodward & William, 1988) and were of nominal 5 mm diameter. These were cut into lengths of 600 mm and tested in tension in an Amsler stressing machine under loads varying from 0 to 24 kN (0–65% UTS). Two studies were then made: one using a three-gauge rosette with a hole drilled at its centre and the other a two-gauge arrangement with gauges aligned along the centre line and a hole drilled mid-way between them (see Figure 11.13).

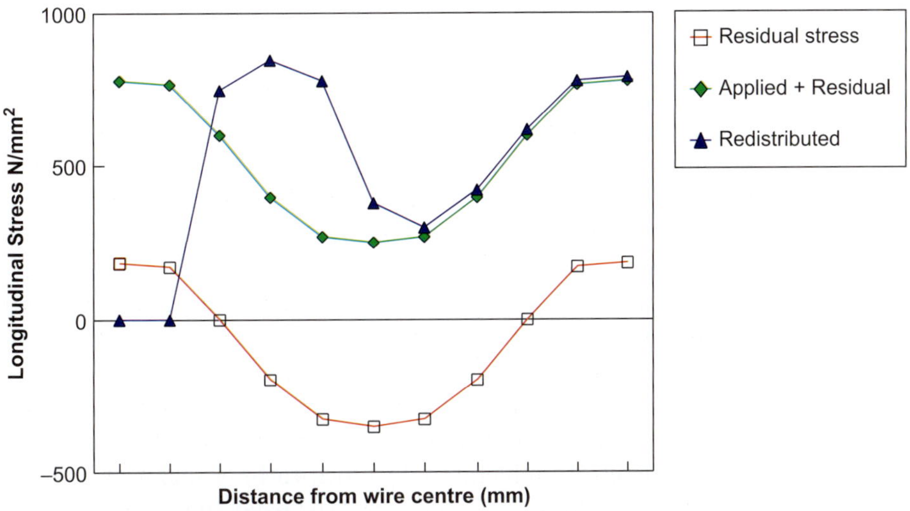

FIGURE 11.12 *Typical stress variations in a steel wire.*

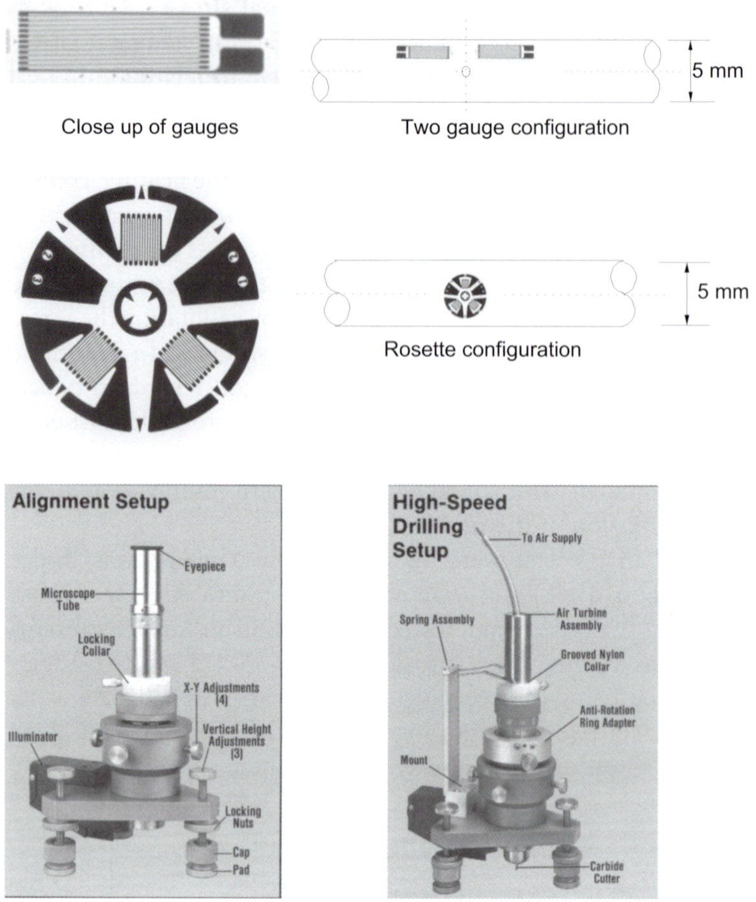

FIGURE 11.13
Gauge configurations and milling guide (courtesy of Measurements Group UK, Ltd).

The drilling of the centre hole was carried out using a high-speed air drill with a 0.8-mm tungsten carbide tipped bit mounted in an RS-200 milling guide. This was clamped to a steel plate secured to the Amsler machine. Aligning and drilling required very great precision and laboratory conditions ensured that it was possible (see Figure 11.13).

Some mistakes were made in the test procedure and so unfortunately the final results were of limited use. The tests did, however, show that the method was feasible provided that calibration factors could be reliably determined for the residual stress determination and that it is possible to mount the drilling machine accurately over the wire to be investigated in the field. This latter problem could prove to be a major stumbling block. In practice, most post tensioned bridges have their tendons embedded in concrete by means of grouted ducts, and so a considerable amount of work would be necessary to first expose the tendon, and then to isolate a single wire for testing. Assuming this was possible and that drilling was carried out precisely, the information would be limited to one wire and would not provide information about the global state of the prestressing. Externally tensioned bridges pose less of a problem as far as access is concerned but since the tendons run in groups and are unsupported for long lengths, a method has to be found to clamp and support a short length of tendon in space so that it is rigid enough to allow testing.

11.4.9 Stresses in prestressing tendons by the crossbow method

A technique pioneered by Godart (1987) called the 'Crossbow' method has been proposed for determining the *in situ* stresses in prestressing tendons. It consists essentially of breaking out a short length of tendon (500–800 mm) and anchoring it over a suitable base length (2L) as shown in Figure 11.14a. A carefully controlled increasing force (*F*) is then applied to the tendon until it has been displaced by a maximum (Δ) of 4 mm from its original alignment. Then since the effort required to deflect a taut wire is

a

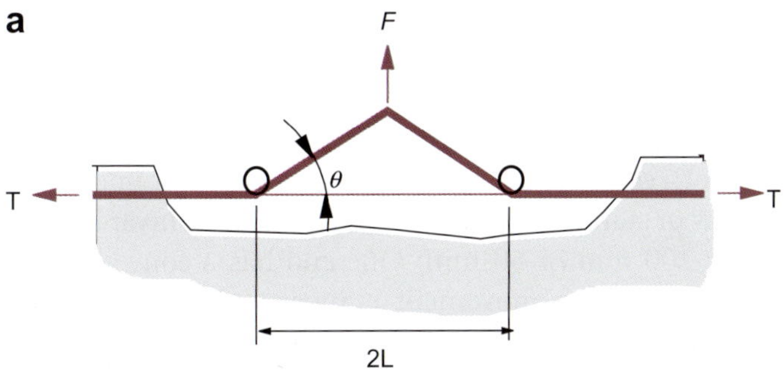

b

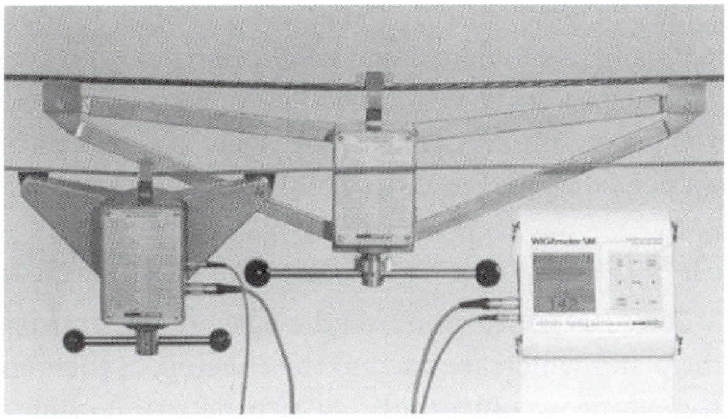

FIGURE 11.14

(a) Crossbow principal, and (b) Crossbow instrument.

proportional to the tension in the wire, the force F (Abdunur, 2000) can be calculated from:

$$F = 2(T + k)(\Delta/L) + K(\Delta/L)^3 \tag{11.4}$$

where k and K are constants which account for friction, flexural stiffness and overstretching.

There are now dedicated instruments like the one shown in Figure 11.14 which, although not specifically designed for measuring *in situ* stresses in post-tensioned bridges, can be used if good access is available. The force in the tendon is displayed directly on a digital display.

11.5 LIVE LOAD STRESSES

The magnitude of live load (or change in live load) stresses at the concrete surface in uncracked locations can be determined in a variety of ways by measuring the strain and then converting it to stress using the elastic and constitutive laws appropriate to concrete. In concrete bridge structures which are cracked as a result of tensile stresses, the measurement of residual stresses at such discontinuities is not possible but the change in strain due to live loads can be measured. When cracks appear they need to be monitored as described in Chapter 2, section 2.3.1.1.

11.5.1 Mechanical displacement gauge

This is a very useful instrument for making strain measurements at different parts of a concrete or steel bridge with one instrument. It consists of a digital display or standard dial gauge mounted on an invar bar with a gauge length of either 200 mm or 50 mm. One end has a conical point which is 'live' (or moveable) whose movement is measured by the dial gauge, the other has a conical point which is fixed. The measurement is made between two steel 'pips' with small depressions made with a punch, and which are pre-mounted on to the concrete with an epoxy resin compound (BSI 1984). The 'pips' are located using a standard Invar calibration bar with two fixed ends spaced at the nominal gauge length apart. The sensitivity varies from $2 \times 10\ \text{mm}^{-4} \times 10$ mm depending on the type of display and the gauge length. Figure 11.15 shows a 'Demec' (DEmountable MEChanical strain gauge) in use.

11.5.2 Tell-tale

This device is made from acrylic plastic and consists of two transparent sliding plates which are fixed to the concrete at their ends either with screws or an epoxy resin compound. The two plates then slide independently of each

FIGURE 11.15
Mechanical (Demec) gauge.

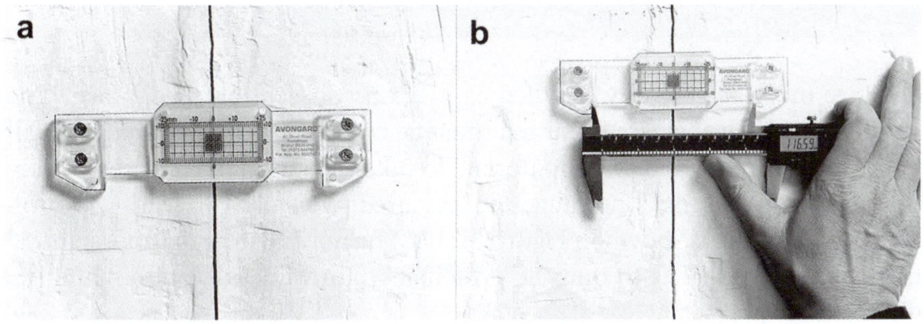

FIGURE 11.16
(a) Plastic tell-tale in use, and (b) Digital caliper (0.1 mm accuracy).

other to measure movement of up to 20 mm to an accuracy of 1 mm. It can be used to monitor cracks in conjunction with a digital caliper, and Figure 11.16b shows one in use.

11.5.3 Electrical devices

There are two such commonly used devices, namely:

the vw gauge
the ERS gauge

The vw gauge is described and illustrated in Chapter 5, section 5.2.7.1. It is a very popular, robust device which is accurate to 1 μm or better, and can be surface mounted or embedded in the concrete.

Of the strain measuring devices available, the ERS gauge is the most popular, and is available for unidirectional, bidirectional (usually in an orthogonal configuration) and tridirectional measurements, the last named often being used to determine principal strain values and directions. This type

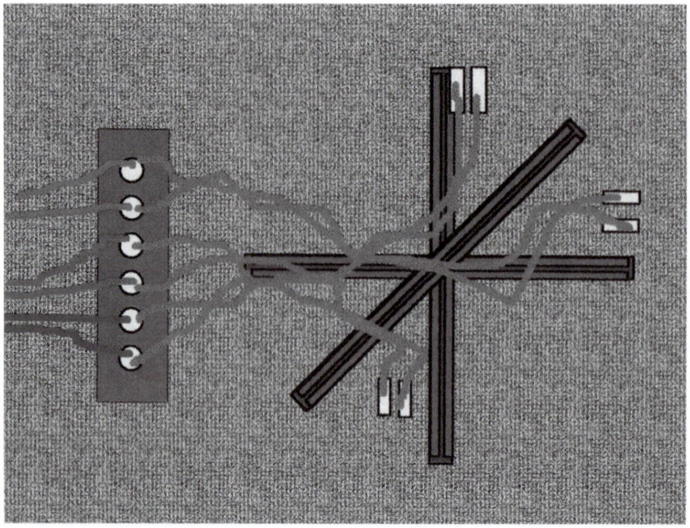

FIGURE 11.17
Long electrical resistance strain gauges.

of gauge must be carefully handled as it is very susceptible to damage. The surface on which they are mounted must be completely dry and free from all laitance, oil, grease and loose material. To take account of the microcracking that is always present, long gauges are required (typically 100 mm × 4 mm). A typical example is shown in Figure 11.17. The normal information required using the ERS gauge is to find the principal strain values (ε_1 and ε_2) and the direction of ε_1.

In general the strains at any point are measured by means of a tridirectional strain gauge rosette containing three gauges 1, 2 and 3 orientated relative to the two principal axes p and q as shown in Figure 11.18 for a Delta (0–60–120°) Rosette and a Tee (0–90°) Rosette. By resolving the strains it can be shown that the relationships between the strains ε_1, ε_2 and ε_3 measured by the Delta strain gauge and the state of strain at 0 are given by:

$$\varepsilon_{p,q} = \frac{\varepsilon_1 + \varepsilon_2 + \varepsilon_3}{3} \pm \frac{\sqrt{2}}{3}\sqrt{(\varepsilon_1 - \varepsilon_2)^2 + (\varepsilon_2 - \varepsilon_3)^2 + (\varepsilon_1 - \varepsilon_3)^2} \quad (11.5)$$

$$\sigma_{p,q} = \frac{E}{3}\left\{\frac{\varepsilon_1 + \varepsilon_2 + \varepsilon_3}{1 - \nu} \pm \frac{\sqrt{2}}{1 + \nu}\sqrt{(\varepsilon_1 - \varepsilon_2)^2 + (\varepsilon_2 - \varepsilon_3)^2 + (\varepsilon_1 - \varepsilon_3)^2}\right\} \quad (11.6)$$

$$\phi_{p,q} = \frac{1}{2}\tan^{-1}\left\{\frac{\sqrt{3}(\varepsilon_2 - \varepsilon_3)}{(\varepsilon_1 - \varepsilon_2) + (\varepsilon_1 - \varepsilon_3)}\right\} \quad (11.7)$$

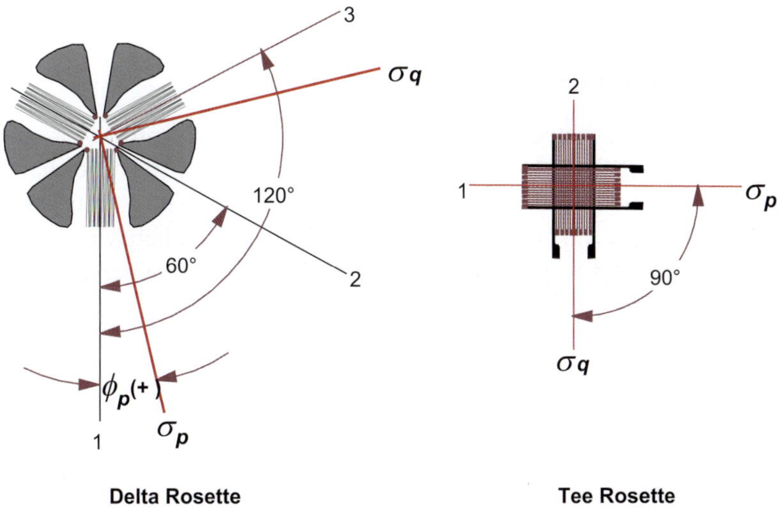

FIGURE 11.18
Layout of typical rosette gauges.

$$\text{if} \quad \varepsilon_1 > \frac{\varepsilon_2 + \varepsilon_3}{2} \quad \phi_{p,q} = \phi_p$$

$$\text{if} \quad \varepsilon_1 < \frac{\varepsilon_2 + \varepsilon_3}{2} \quad \phi_{p,q} = \phi_q$$

$$\text{if} \quad \varepsilon_1 = \frac{\varepsilon_2 + \varepsilon_3}{2} \quad \text{and} \quad \varepsilon_2 < \varepsilon_1 \quad \phi_{p,q} = \phi_p - 45^\circ$$

$$\text{if} \quad \varepsilon_1 = \frac{\varepsilon_2 + \varepsilon_3}{2} \quad \text{and} \quad \varepsilon_2 > \varepsilon_1 \quad \phi_{p,q} = \phi_p + 45^\circ$$

$$\text{if} \quad \varepsilon_1 = \varepsilon_2 = \varepsilon_3 \quad \phi_{p,q} = 0 \text{ (indeterminate)}$$

and for the Tee gauge are given by:

$$\varepsilon_p = \varepsilon_1$$

$$\varepsilon_q = \varepsilon_2$$

$$\sigma_p = \frac{E}{1 - \nu^2}(\varepsilon_1 + \nu\varepsilon_2)$$

$$\sigma_q = \frac{E}{1 - \nu^2}(\varepsilon_2 + \nu\varepsilon_1)$$

11.6 MONITORING

11.6.1 Introduction

Information gathered during bridge inspections will provide a snapshot of the condition of the bridge at one point in time. On the other hand continuous monitoring of bridges during service helps to gauge their performance, and to act as a warning device by recording unusual behaviour. Continuous monitoring during the construction stage and during service can provide invaluable information for the technical management, and, since generally the cost is minimal compared with a complete project, there is a strong argument for installing instrumentation as a matter of course. In short it is a way of assessing the health of a structure. The sort of information that can be collected is:

deck deflection and rotation
stress levels and changes
change of level
vibration characteristics temperature variation in the deck
wind speed and direction
corrosion rates and crack widths
water speed and rise of rivers
scour at abutments and piers
rise and fall of tides

It is expensive in terms of the cost of the technology and manpower required, and therefore it is usually confined to larger bridges where the cost benefits of the information gained, and the consequences of failure are highest. Prudent examination of how, where and when to monitor is therefore required.

11.6.2 Monitoring strain in steel

11.6.2.1 Reinforcing bars

The strain in reinforcing bars can be measured by the simple expedient of securing ERS gauges (section 11.5.3) to the bar itself. The normal preparations to the steel surface are required, and it is essential to protect the gauges with a soft or polystyrene wrap to prevent damage when concrete is being poured and vibrated. It is also advisable to install some 'back-up' gauges in the event of gauge malfunction. Some manufacturers of ERS gauges provide 'cabled strain gauges' specifically for civil engineering applications. Each gauge is supplied with a pre-attached 3–4 m cable ready for connection to the strain reading equipment. This saves time by eliminating the need to attach leaders to terminals or tabs.

Alternatively the gauges can be secured to a small mat of mesh of reinforcement which is remote from the main steel and connected to it. This lessens any interference with the contractors' programme as the mat can be installed just prior to concreting. Cairns (1990) reports on the successful outcome of such an installation in various parts of the concrete slab on the Milton Bridge on the M74 just 15 miles south of Glasgow, Scotland. It is a three span continuous composite bridge with spans of 47, 54 and 31 m, and has two plate girders (2.5 m) and an overall depth of approximately 3 m with a 14-m wide top slab. The purpose of the monitoring was to assess whether a clause on fatigue effects to the reinforcement was needed in BS 5400: Part 10 (1980). Data were obtained for the strains in the slab at 20 and 42 days after casting and for the dynamic response under the action of a standard vehicle. It is now being continuously monitored in service.

11.6.2.2 Other metal sections

ERS gauges are the natural choice for metal structures. There is a large range available for various specialised uses (Measurement Group, 1995) but generally the single and rosette types are the most popular in bridge engineering. The single for uniaxial situations and the latter where there are likely to be complex or high local stresses such as at post-tensioning anchors, truss members in the vicinity of bearings and joints etc.

As well as glue mounted gauges there are weldable strain gauges for use where conditions preclude the use of bonded gauges. Little surface preparation is needed and they are specially designed to be spot welded to steel and wrought iron structures.

11.6.3 Corrosion monitoring

Corrosion monitoring systems (CMS) are of vital importance for measuring the rate of deterioration in a structure. The data obtained are quantitative and play an important role in deciding the maintenance strategies to be adopted, and the prediction of future degradation.

For implanting in new structures, Broomfield et al. (1999) describe a sophisticated and compact assembly (utilising Linear Polarisation Principles) designed to fit into a rebar cage. It has electrodes to measure chloride and pH concentrations; probes to measure humidity, a thermocouple to measure temperature and stainless steel auxiliary electrodes to pass the

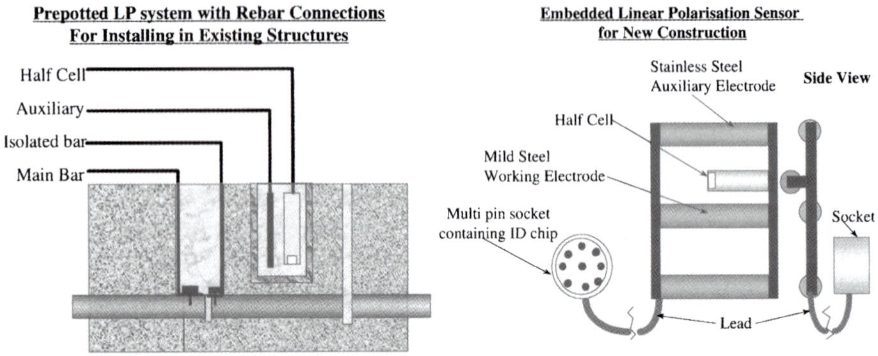

FIGURE 11.19 *Linear polarisation sensor in new and existing structures (Broomfield et al., 1999).*

current. The system can also be 'retro-fitted' into existing structures by breaking out concrete and repairing with mortar (see Figure 11.19).

There is some doubt about the stability of the sensors and the authors indicate that it 'is difficult to guarantee over a period of years, and calibration is almost impossible'. Their results are vague but they do report that changes have been recorded over a two-year period.

Another CMS for assessing the corrosion risk of steel reinforcement on a permanent basis is reported by Schießl and Raupach (1993) which was developed after an extensive two-year research programme and is presently installed in three bridge structures. The device consists of a number of anodes spaced at 50 mm intervals forming a series of macrocells so that the corrosion risk can be measured at various depths from the surface of the concrete. Type A device has four anodes and four cathodes, and Type B device has six anodes and one cathode. The anodes are made from carbon (or black) steel and the cathodes of a noble metal (stainless steel or platinum etc.). The general arrangement is shown in Figure 11.20. Electrical current begins to flow when a critical chloride content is reached or when the pH of the concrete decreases due to carbonation. The magnitude of the current provides a measure of the corrosion risk. The whole set-up is controlled by computer and readings are taken four times a year. The data are closely scrutinised as one by one the microcells begin to indicate a flow of electricity. By measuring the time between successive current flows a relationship can be formed between the depth of carbonation or the chloride content and the local concrete properties and environment. By extrapolation (presumably linear), the time to corrosion of the existing reinforcement can be forecast and preventative measures taken. Schießl and Raupach (1993) suggest that additional information about corrosion rates can be obtained from measuring the electrolytic resistances, potentials and temperatures.

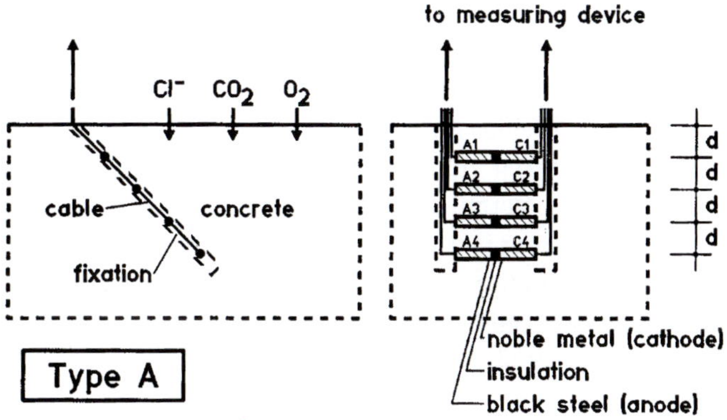

Arrangement Type A of the anodes and cathodes to monitor the corrosion risk for the reinforcement (schematically).

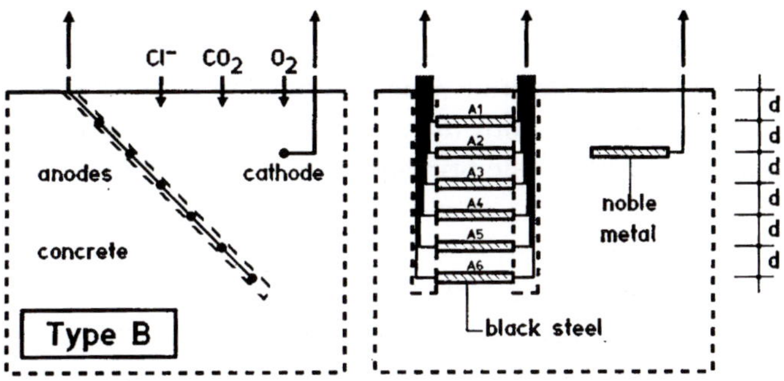

Arrangement Type B of the anodes and cathodes to monitor the corrosion risk for the reinforcement (schematically).

FIGURE 11.20

Type A and Type B arrangements (Schießl & Raupach, 1993).

11.6.4 Corrosion analysis

An expert system named 'CorroScope' which analyses monitoring data from a number of NDTs such as half cell potential, covermeter, resistivity logger, carbonation tests, chloride tests, visible defects etc. has recently been developed in the UK (Jordan et al., 1997). It purports to be the means whereby it is possible to overlay all of the data and assess the current and future corrosion parameters of a particular structure. Survey information from the NDTs (along with visual information) is stored in a logger and downloaded into a PC for analysis. Weightings are given to each data type, from the highest for visual data to the lowest for half cell potential, also allowing each data type to have a specified radius of influence. Limits are defined within the software for cover depths, resistivity etc. and according to Jordan et al. (1997) the CorroScope performs several types of analysis, including assessment and significance of corrosion, rate of corrosion and also the particular cause (e.g. chloride, carbonation etc.). The analysis area is either taken as an individual point, a chosen rectangle or as the whole object. CorroScope provides algorithms of data analysis on the basis of the available measured material. The system then calculates a total score for the indication of possible active corrosion using individual weighting factors.

The weighting factor is a subjective judgement by the authors, but with an expanding database built up over the years which includes the current state of affairs and a prognosis about future patterns of corrosion, the weighting factor should cease to have a significant effect on the results.

11.6.5 Optical fibre sensors

The use of optical fibre sensors in the field of bridge engineering is a relatively recent innovation which springs from their use in the evaluation of materials used in both the aerospace and hydrospace industries (Wolff and Meisseler, 1993). The principal is based on the measurement of the change of intensity of light as it travels down a small glass fibre (about a hair's breadth in diameter) and is reflected back by means of a tiny mirror to a sensor such as a Fibre Bragg Grating (FBG). The FBG sensor can be likened to an ERS gauge, but instead of measuring changes of electrical resistance, it measures changes of light intensity. If the optical fibre does not move, then there is no change of intensity. If microbending occurs due to a tensile load then there is a loss of light intensity, and if this is compared with the original light supplied, it can be translated into a measurement of crack formation, crack width variations, stress changes and temperature measurement. They are particularly compatible with advanced composite materials such as GFRP and could be incorporated into prestressing wires, for example, at the manufacturing stage. They are also very lightweight, corrosion resistant and immune from electrical or magnetic interference. Because of their relative fragility, however, care has to be taken to ensure that the sensor is not displaced or damaged in any way, as could well be the case in civil engineering applications. New ways of providing robust sensors are needed before they can be routinely used in bridge structures.

The first recorded use of fibre optics was in a footbridge in Berlin (New Civil Engineer, 1989) where sensors were incorporated into the Polystal glass fibre rods forming the external prestressing to the bridge. Additional fibres were laid in the webs of the twin Tee beams and the bridge floor to detect strain changes in the concrete.

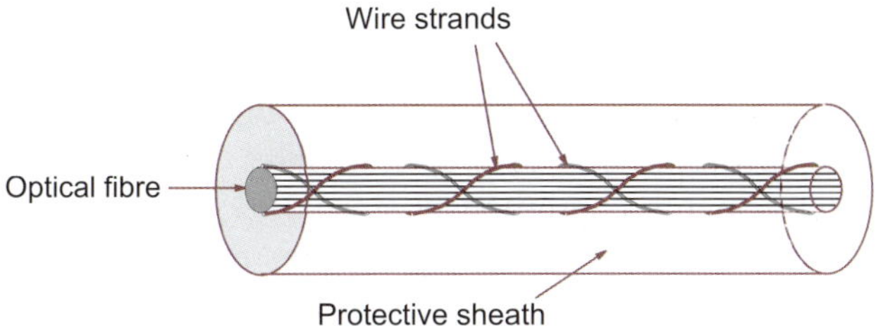

FIGURE 11.21
Optical fibre sensor (after Wolff and Miesseler, 1993).

A typical sensor (which is approximately 2 mm in diameter) is normally stranded with one or two steel wires as in Figure 11.21, and can be embedded into the centre of a glass fibre bar for protection. It is the pressure of the spiral of steel wires onto the fibre that causes microbending resulting in a loss of intensity or attenuation of the light. Meisseler and Wolff (1993) used this system to detect and monitor cracks and the behaviour of glass fibre prestressing tendons in two, three span, solid concrete bridge decks. Using a test load to traverse the bridges, they were able to detect crack positions and measure deflections of the order of 100 μm.

Another type of optical fibre sensor consists of a single fibre with numerous partial mirrors at intervals along the length. Pulses of light are sent down the fibre and the time taken for a pulse to be reflected back to the transmitter is measured. Any movement in the mirrors will result in a change of time, which can be converted back to a quantity of the variable being monitored.

The sensors can also be installed in existing buildings to monitor deformations after damage repair or if unusual behaviour has already been detected by other means. A modified system is suggested by Meisseler and Wolff (1993) for measuring crack widths in existing buildings whereby two loops are formed in the area between cracks and the remainder fixed to two load transfer rails connected to the structure. They quote 'a measuring sensitivity of 0.02 mm and a measurement range of 0.1–10 mm'. For some idea of the range of such sensing systems used in practice, refer to Hofer (1987).

One very important piece of research in this area was carried out by Idriss (1997) at New Mexico State University who installed an optical fibre monitoring system into a full scale 40 ft (12.2 m) span non-composite steel girder and concrete bridge in the laboratory. The network of sensors included the use of optical fibres side by side with ERS gauges so that direct comparisons of the strains measured by each could be made. He found that the difference in the measured strains on the undersides of the girders was less than 1% – testimony to the accuracy of the optical fibre sensors.

Sennhauser et al. (1997) report the use of optical fibres and ERS gauges for monitoring the behaviour of two 35 m long CFRP cables which were installed in an otherwise conventional cable-stayed bridge (the Stork Bridge, Winterthur, Switzerland). The optical fibres have been measuring strains and displacements continuously since 1996 and the results indicate that the bridge is performing well, and strains measured by the ERS gauges and the FBG sensors are almost identical. The authors state that 'Optical Fibre Bragg Grating sensors are well suited for strain monitoring of large civil structures like bridges as demonstrated by their multiplexing capability, high resolution order of 1 μm (0.001 mm), reliability and negligible creep'.

11.6.6 Lasers

Where it is not possible because of accessibility to use conventional ways of measuring deflections such as LVDTs, precise levelling or accelerometers, then the use of lasers may be an option. Sloan et al. (1990) developed a system using 'helium neon lasers fitted with beam expander and focussing optics' for monitoring the Foyle Bridge near Londonderry, Northern Ireland. The bridge is a twin, three span (approximately 144 m–235 m–144 m) variable depth steel box girder with 9 m deep haunches at the two pier positions reducing to 3 m at the midspan positions. The laser was set up inside the boxes at midspan and projects a small (3 mm) spot of light to targets positioned at the pier positions. Any movement of the bridge will cause the spots of light to move, and the movement is recorded using a solid-state camera with a light-sensitive computer chip which presents to the lens an array of 128 × 256 light-sensitive cells. The cells are scanned at regular intervals by a computer and the position of the spots stored in memory for later downloading to a main computer and subsequent processing. Each reading is the mean of 1024 camera scans. The arrangement is illustrated in Figure 11.22.

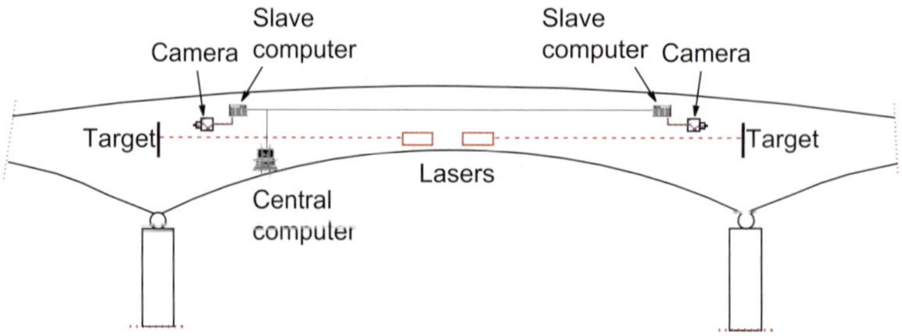

FIGURE 11.22 *Measurement of deflection at the centre of the main span of the Foyle Bridge using lasers (after Sloan* et al.*, 1990).*

The system is reported as operating satisfactorily, and other devices have been installed to measure steel strains, temperature changes, wind speed and direction and expansion joint movement. The whole integrated monitoring set-up is unattended and has been recording the behaviour of the bridge since 1990. For full details of all the hardware used, refer to Sloan et al. (1990).

The Dee Bridge in North Wales, UK also has an integrated monitoring system called Smart Bridge (though it is not smart in the sense that corrective action is automatically initiated when an unusual event occurs) but it is smart in that it has sensors which 'report' on the bridge's structural integrity by supplying a constant stream of information. The company responsible for manufacture and installation of the system use what they call 'isolated measurement pods' (IMPs) which will collect data from all of the sensors in place such as 'anemometers, thermocouples, strain gauges, corrosion probes and accelerometers' and transfer the data to a remote receiver for analysis *via* a PC located in one of the bridge's towers.

11.6.7 Telemetry

Instrumentation to assess the performance of three glued segmental concrete box girder bridges was commenced in 1985 by Waldren et al. (1990) by measuring variables such as wind velocity, solar intensity, strain, bridge temperature and ambient temperature. As with the Foyle Bridge (section 11.6.5) it became apparent that the volume of data collected on site was becoming unmanageable and a way was sought to deal with it. The problem was overcome by the installation of a telemetry system to efficiently collect and transfer data from the bridge sites to the office (see Figure 11.23).

A standard telephone line was used which enabled the remote computer on site to mimic the host computer in the office. The collection and frequency of data collection are now controlled remotely. The cost has been 'modest' and measures both long term effects such as temperature and creep, and short-term effects due to traffic – in total 150 sensors are read as required – and some very interesting results have been obtained.

11.6.8 Continuous acoustic monitoring (CAM)

This is a non-intrusive way of monitoring a bridge to detect 'events' caused by the fracturing of high tensile steel wires in post-tensioned concrete bridges. It is an acoustic method that has recently been assessed by TRL (1999) in the UK. The method is based on detection of small

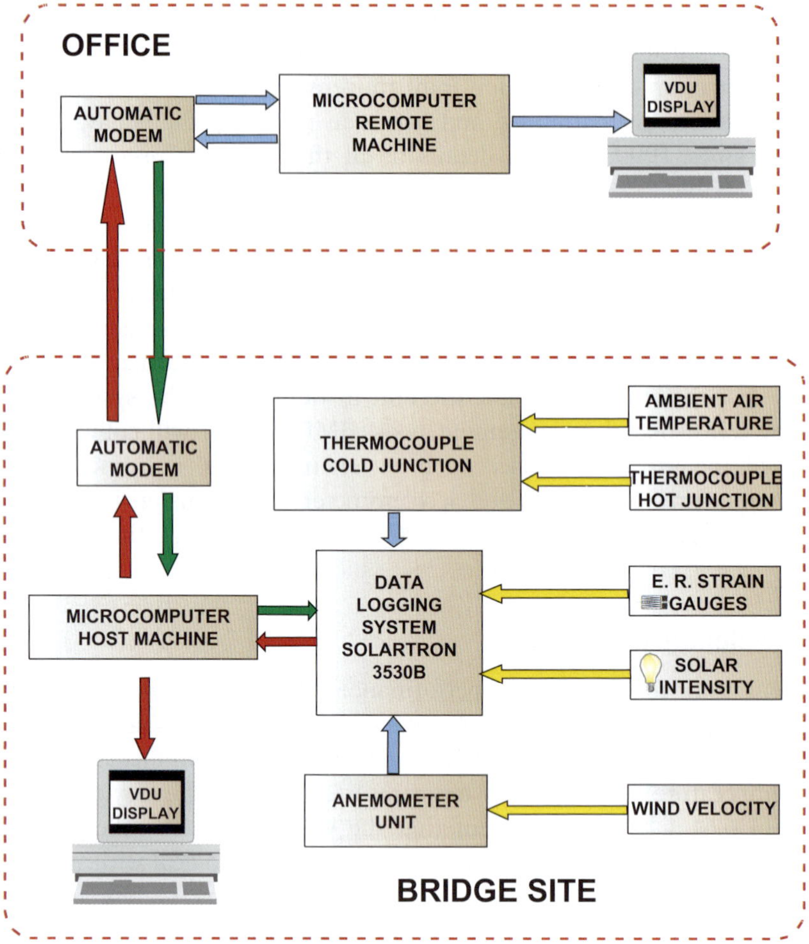

FIGURE 11.23
Data acquisition and telemetry system (after Sloan et al., *1990).*

amplitude stress waves which are emitted, for example, when a steel wire is strained beyond the elastic limit and it fractures. Bungey and Millard (1996) report that when used to detect fracturing in concrete, 'the emission rate and signal level both increase slowly and consistently until failure approaches, and then there is a rapid increase up to failure' the same will clearly happen when a steel wire breaks. TRL (1999) have found that it meets their criteria in that a single wire fracture can be detected above the ambient noise level, be distinguished from other acoustic events and located in position. The basic equipment requires a piezo-electric transducer which is secured to the concrete. This detects the signal and then it is amplified, filtered, processed, recorded and then displayed as a plot of emission rate (1000 s/sec) *versus* applied load on a computer or printer.

11.7 SCOUR SENSING

Scour is the single most common form of bridge failure (Smith, 1976; Watson, 1990). A few notable failures in the UK include, for example, the 127-year-old, five span, masonry arch, Ness (railway) Bridge in Scotland which collapsed during unprecedented flood conditions in February 1989 (Haywood, 1994). Water in the River Ness was flowing at a rate of 800 m/s and caused localised scouring beneath the bridge piers. Fortunately no one was injured or killed. The bridge was replaced in 1990 with a steel and composite superstructure on two reinforced concrete piers and the existing abutments.

On 19 October 1987, passengers travelling on the Swansea to Shrewsbury line were not so fortunate. Part of the five span, single track railway bridge at Glanrhyd had collapsed into the River Tywi in Wales which was in full flood, and the front car of the train plunged into the waters. Four people 'including the driver and a teenage boy died when the car was swept away' (Hayward, 1994) – this was the first collapse of a railway bridge with loss of life in the UK since 1915 (Whitbread & Riddell, 1996). The failure was due to scour around the central pier which subsided and initiated progressive collapse of the complete bridge. Another disastrous failure was that of the 140-year-old, 55 m long, five span bridge over the River May at Forteviot, south of Perth in Scotland (Hayward, 1993). The bridge carried two railway tracks supported on twin wrought iron girders on masonry piers and sank into the water when half of the short 2 m high piers 'settled and tipped' into the river due to local scour of the gravel river bed. The Scoharie and Hatchie Bridges in the US and the Kufstein Viaduct in Austria were also victims of scour in the late 1980s and early 1990s.

The Highways Agency have published an Advice Note (BA 59/94) which gives guidance on the hydraulic design of bridges. Meanwhile existing bridges have to be assessed and appropriate action taken. In Chapter 2, section 2.6 it was stated that 'prevention is better than cure' as far as scour is concerned. To aid the task of post-construction prevention, a monitoring device has been developed by Hydraulics Research Wallingford (Meadowcroft & Whitbread, 1993). It consists of 'robust but sensitive motion sensors mounted on flexible rubber tails'. Divers excavate the river bed at specified locations and specified depths and the sensors are

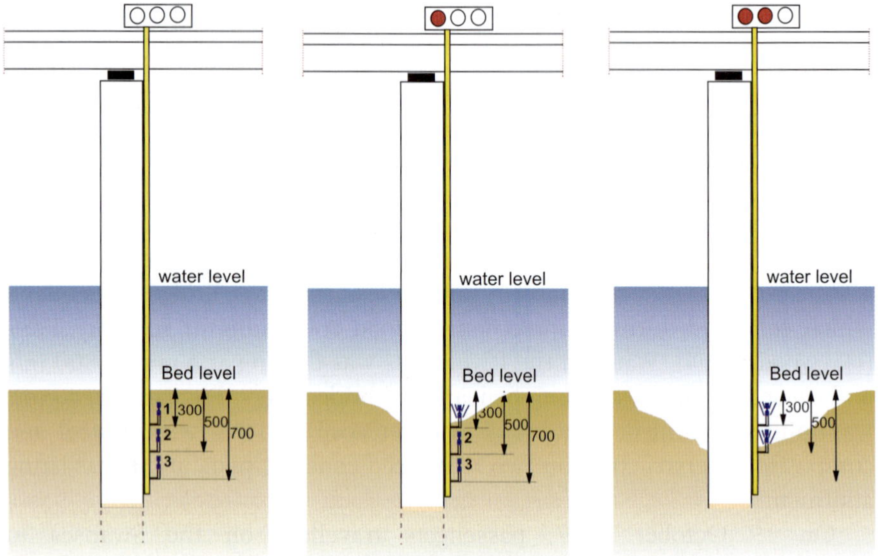

FIGURE 11.24
Buried motion sensors to monitor scour.

installed by mounting them at different levels onto brackets bolted to the pier. If scour occurs and the sensor is exposed, the tails oscillate and send an electrical signal to a data-logger or warning device (see Figure 11.24). This can consist of a plot of river discharge (m/s) and scour (pulses/min) *versus* days (dates) on the *x*-axis (Meadowcroft & Whitbread, 1993). As each sensor is activated, the pulse rate suddenly peaks from zero to 300–400 pulses/min or more. When the lowest sensor has been activated this would mean that the river bed had scoured to a level which placed the bridge at risk and action would be necessary. Meadowcroft and Whitbread (1993) report that in their tests once a flood begins to recede, the sensors get reburied and the signals cease. Prior to installing such devices at every bridge over water, a risk analysis is carried out, and in the UK the British Rail Handbook on assessment is used, known as Handbook 47, which though considered to be conservative, does at least provide an initial starting point for decision making.

11.8 LOAD CELLS AND BEARINGS

Load cells installed during construction enable the direct measurement of force at bridge supports, in cable stays, hangers, tendon anchors and foundations. Load measuring bearings can also be installed at the construction phase to monitor the change in load that often occurs at piers or abutments

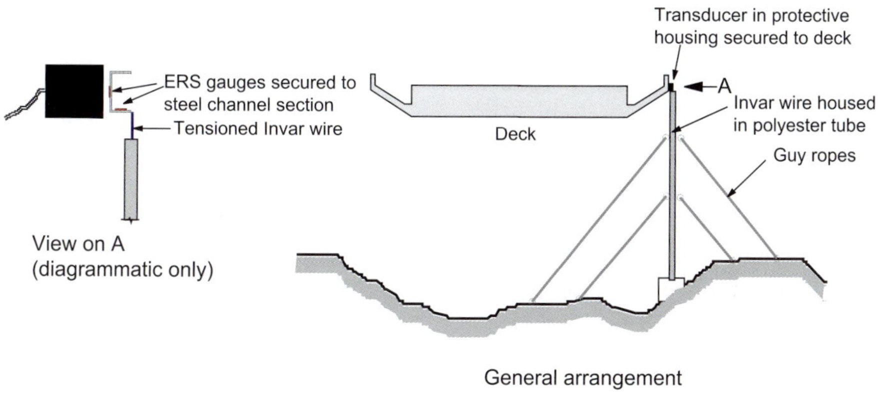

FIGURE 11.25
Transducer set-up.

which may settle (or subside) if founded on soft soils or in mining areas. Excessive changes can not only alter the assumptions on which the original design was based but also can cause damage at movement joints, and overstressing due to torsion. Direct digital readout is available and data are easily transferable to a PC for processing. Any slight changes are detectable which make it possible to start planning early for any structural alterations that may be necessary to stabilise the situation.

11.9 DISPLACEMENT TRANSDUCERS

Ways of measuring deck displacement using transducers were described in Chapter 5 (section 5.2.7.1) but an interesting variation of this was used on a bridge over the A38 near Bristol, UK (Cogswell et al., 1993). As the bridge was elevated, a dedicated system called the Deflected Cantilever Displacement Transducer (DCDT) was designed (see Figure 11.25). The transducer (consisting of a U-section of steel with ERS strain gauges mounted on it) was fixed to the cantilever section of the deck and anchored to it. A pre-tensioned Invar wire was connected to the tip of the U-section and secured to a standard concrete cube at ground level. The wire was protected with a polyester pipe and stabilised by supporting guy ropes. As the deck deflected downwards the tension in the wires was reduced, and the change measured by the transducer to an accuracy of ± 0.01 mm.

11.10 TRAFFIC MONITORING

The monitoring of heavy traffic is a way of protecting bridges from rogue drivers and transport companies who flout the law by overloading their lorries or trucks. Traditionally, monitoring was (and still is) carried out in a random fashion by police officers directing trucks into a weighbridge station and weighing the loads on their axles. If the loads were greater than the statutory values then a penalty was imposed and some of the pay-load had to be removed before the truck could resume its journey.

Early sensors dating from the late 1960s used bending plate sensors, and later piezo-electric strip sensors to capture wheel loads. Systems using optical fibres to detect both the speed and weight of vehicles (New Civil Engineer, 1989) have not been developed due mainly to the high costs involved (Betten et al., 1995).

Since the late 1980s there has been a move towards the installation of weigh-in-motion (WIM) sensors which weigh all vehicles travelling at normal speeds both beneath and on top of the carriageway using capacitive strip sensors. Such devices are now considered to be the way forward in weight enforcement. This information, in conjunction with axle counts using the familiar pneumatic tubes fixed to the road surface, and a vehicle detection system provide invaluable site-specific data for bridge assessments.

The one-sensor WIMs have proven to be unreliable and so the move is towards multiple sensors (Barbour, 1993) which, it is claimed, can estimate the weight of nearly 90% of vehicles (axles) to within 10% of their 'true' values.

A seven-month trial by the TRL at Abingdon in the UK (Barbour, 1993) using an equally spaced eight-sensor array demonstrated both the practicability and accuracy of such a system. Future research is underway to develop the prototype system to the point where it can become commercially available and would include 'additional software and an inductive loop (to detect the beginning and the end of the vehicles) which would provide WIM weights in real time'.

Portable WIMs using capacitive-strips have been used in Switzerland to gather data on roads rather than motorways (Bailey, 1996). They comprise the sensor housed in flat trapezoidal rubber casting which is clamped with metal plates screwed in to the road surface. They can be assembled in 4 hours with only minimal disruption to traffic and can gather data on two traffic lanes simultaneously. They have a capacity to store more than 12,000 records and can 'compile histograms based on any vehicle characteristic in real-time'.

11.11 CONCLUDING REMARKS

Monitoring a structure is one way of Prolonging its Active Life (PAL). Monitoring devices may be implanted during construction of new bridges or fixed onto the outside of existing structures. As well as providing invaluable information about the behaviour of a bridge, monitoring devices act as an early warning system to alert the bridge engineer to intervene to prevent at worst, a catastrophe, and at the very least further deterioration.

Mehrkar-Asl (2000) proposes a strategy for bridge structures for assessment leading to monitoring. It is probably not necessary to monitor every existing bridge and that is why a clear strategy is necessary. Mehrkar-Asl (2000) illustrates the value of his strategy with some interesting case studies, and his conclusions as to the benefits to be gained are:

facts not assumptions become the basis for decisions regarding safety
priorities can be based on performance criteria
funds can be distributed in a more effective way
expenditure can be limited

In addition he rightly states that the information gained from monitoring provides greater understanding of structural behaviour and allows some refinement of the analysis when estimating performance and the effects of strengthening options. It also helps to improve design methods and the data gained can help in the unhappy event of a dispute.

11.11.1 The use of remote sensing systems (RSR)

Just about anything on our planet can now be 'looked at' by deploying satellites in the stratosphere equipped with sensors of various kinds which can 'pick up' the information we require. We seem to be moving by degrees from land based methods of data collection to extraterrestrial based methods. The former often involves the braving of the weather and battling against hostile terrain to collect and store information we need either on paper or electronically *via* a data-logger. Once back at the office the data are off-loaded into the BMS, and used for whatever Bridge Management activity is being analysed. The latter is termed satellite remote sensing and information from the ground can be collected and relayed to a workstation where it is processed and needs only the press of a button or the scrolling around on a monitor, and 'hey-presto', what we want is blinking at us from a workstation monitor.

Satellites were originally designed to observe terrain, weather, vegetation, urbanisation, the oceans, cloud formation, surface temperatures and a host of other characteristics on a global scale. Observation was usually *via* optical sensors or through the barrel of powerful cameras which scanned vast areas and transmitted the data to earth where it was processed and converted to a 2-D coloured image on paper. The technology involved is extremely complex and has advanced to a stage where vast amounts of information can be gathered, managed and analysed.

11.11.2 Geographical information systems

The first major use of this idea was what is termed Geographical Information Systems (GIS) where geographical information was gathered and transformed into maps. The modern or GIS has at its heart raw geographic information which can be accessed and manipulated to solve a particular problem. It provides a spatial platform on which any object can be linked to for the purposes of analysis. In effect the two are merged and become one. For example, it is possible to pinpoint the location of a bridge by reference to a specified datum in terms of its latitude, longitude and altitude using a Global Positioning System (GPS). This information can then itself be manipulated upon and stored safely in the bosom or database of a Bridge Management System (BMS) and used, for example, for monitoring purposes.

Parker (1996) believes that GIS is a misnomer in that it 'gives the impression of computer software packages which analyse geographic information and produce results presented as maps'. In fact this is far from the truth. He believes that there is a wide range of remote sensing systems where GIS can play a substantive role in providing sophisticated, comprehensive and relatively inexpensive GIS operations. Some of these listed by Parker (1996) which he believes 'have potential use in civil engineering' are as follows:

Search
Location analysis
Terrain analysis
Flow analysis and network
Spatial/analysis and statistics
Measurements

In the field of bridge engineering there is very little in the way of research or practice relating to GIS. The following two examples indicate where GIS might be used.

11.11.3 Wilford suspension bridge

Using GIS, Furneaux (2004) studied both the vibration and deformation characteristics of the 65 m span Wilford suspension bridge in Nottingham, designed to carry water, pedestrians and cyclists across the Trent River. In both cases dual GPS sensors were strategically placed and secured to the deck to pick up and relay (*via* the satellite) the response signals to the GIS for processing. The results, depicted as graphs of displacement *versus* real time showed that the GIS was quite capable of sensing small displacements of a few millimetres in both static and dynamic modes, and also of monitoring displacements in real time. However, there was no clear-cut evidence that the results compared favourably with those from a conventional structural analysis.

11.11.4 Humber suspension bridge

In 1997, the Humber suspension bridge (which is situated in Kingston Upon Hull, UK and carries the A15 over the River Humber from Barton upon Humber to Hessle), was chosen as a vehicle for a real time monitoring experiment using kinematic Global Positioning System (GPS). Compatible GPS antennas were positioned at 'strategic points on the bridge deck and the support towers'. Longitudinal, vertical and lateral displacements were recorded against real time, at each location and exhibited traces typical of vibrating structures. Because of the wind conditions, and the antenna design, unwanted wind displacements (akin to noise in acoustic emissions) were present which could not be separated out from the total displacements. No defective conclusions could be drawn from the experiment. However, the authors make the statement that they 'are confident that kinematic GPS can be used successfully to monitor the three dimensional movements of large structures'.

Using GIS is expensive, and requires a considerable amount of expert input. It would seem that the cost-benefits are not substantial enough to warrant its use, and the extant information relating to bridges is practically non-existent. It is doubtful that in the future there will be any take-up relating to the monitoring of bridges, but, according to Mingins (1996) we should 'Keep an open mind'....

11.12 Continuous acoustic monitoring

The use of high strength steel wires in pre- and post-tensioned bridges, suspension bridges and cable-stayed bridges is well known. The most common failure is due to stress corrosion which is not easy to detect – witness the ill-fated Ynys-y-Gwas Bridge in the UK (see section 5.2.6). A post-mortem on the deck of the bridge showed that it had failed due to stress corrosion of the prestressing wires over a period of 32 years. There were no outward signs of incipient failure such as staining or excessive deformation – so how could the engineers have known at the time? What is required is a non-invasive health monitoring system which can plot the health of a bridge over its entire life – in other words a built-in system installed during construction. For extant bridges, a full scan of the bridge where wires are installed would first have to be carried out to identify any local failure of the wires, and after repair, to continuously monitor the situation.

Fuzier et al. (2000) have developed and tested what they term as a continuous acoustic health monitoring system. The system is unattended, remote and cost-effective, and has been in use since 1994. The working premise is that 'failures of prestressing wires generated an audible acoustic response'. This response would be in terms of a local change in frequency or energy, causing a local disturbance in the ambient situation.

The system requires the installation of suitable sensors glued to the structure at appropriate positions such that a wire break can be detected by at least three of them. Data acquisition and processing are *via* ground-based cable and the Internet; the hardware is both durable and robust, and all data are obtained in real time and it is possible to interrogate and obtain up-to-date reports.

So far the system has been successfully installed on the following structural types:

Bonded and unbonded post-tensioned structures and bridges
Concrete floor slabs
Suspension bridges
Cable-stayed brides

This clearly has great potential, and in due course will no doubt will become the industry standard. For a wider and deeper understanding of the system, refer to the paper by Fuzier et al. (2000).

REFERENCES

Abdunur, C. (1982). Direct measurement of stresses in concrete structures, IABSE Symposium on Maintenance, Repair and Rehabilitation of bridges, Washington DC.

Abdunur, C. (1985). Mesure de contraintes sur ouvrages d'art par une methode de liberation miniaturisee, Ref. 3010, Bull. liaison Labo. Ponts et Chaussees 138.

Abdunur, C. (1990). Direct access to stresses in concrete and masonry bridges. *Proc Second Int. Conf. on Bridge Management* pp. 217–226, Guildford, Surrey, Thomas Telford.

Abdunur, C. (2000). Structural evaluation methods for concrete bridges. *Proc Fourth Int. Conf. on Bridge Management* pp. 408–415, Guildford, Surrey, Thomas Telford.

Ashkenazi, V., & Roberts, G. W. (1997). Experimental monitoring of the Humber bridge using GPS. Proceedings of the Institution of Civil Engineers – Vol. 120, Issue 4. pp. 177–182. Thomas Telford, London.

Bailey, S. F. (1996). Basic principals and load models for the structural safety evaluation of existing road bridges, Thesis submitted to the Ecole Polytechnique Federal de Lausanne for the award of Doctor in Scientific Techniques.

Barbour, I. (1993). Multiple-Sensor Weigh-in-Motion Trials at Abingdon (A34), TRL Research Report 375.

Beaney, E. M. (1976). Accurate measurement of residual stress on any steel using the centre hole method. *Strain* 99–106.

Beaney, E. M., & Procter, E. (1973). A critical evaluation of the centre hole technique for the measurement of residual stresses. *Strain* 7–14.

Betten, E., Caussignac, J. M., Teral, S., & Siffert, M. (1995). Optical sensor system for WIM and vehicle classification by photoelastic effect and polymetric fringes, Proc. First European Conference on Weigh-In-Motion of Road Vehicles, *Zurich,* pp. 341–348.

British Standards Institution. (1984). BS 6319: Testing of Resin and Polymer/ Cement Composition for Use in Construction, BSI, London.

Broomfield, J. P., Davies, K., & Hladky, K. (1999). *Permanent corrosion monitoring for new and existing reinforced concrete structures*. London: Structural Faults and Repair Conference.

Bungey, J. H., & Millard, S. G. (1996). *Testing of Concrete in Structures*. Blackie.

Cairns, J. (1990). Monitoring of traffic induced strain in the steel reinforcement of a concrete bridge deck, Proc. First Int. Conf. on Bridge Management, Guildford, Surrey, pp. 619–629, Elsevier.

Carlson, R. W. (1937). Drying shrinkage of large concrete members. *J. Am Concrete Institute, Proc., 33*, 327–336.

Chalkly, C. (1994). *Modelling problems in short-span bridges and the basics of load testing. One Day Seminar in Hidden Strength*. London: The Sureyor.

Cogswell, G. & Healey, P. (1990). In service monitoring of A38 overbridge, Bristol, Second International Conference on Bridge Management, pp. 635–642, E. & F.N. Spon, London.

Cogswell, G. & Healey, P. (1993). In service monitoring of A38 motorway overbridge, Bristol, Proc. Third International Conference on Bridge Management, Surrey, UK, pp. 635–642, E. & F.N. Spon, London.

Couthino, A. (1949). *Theory of an experimental method for determining stresses, not requiring an accurate knowledge of the Modulus of Elasticity.* Paris: IABSE Congri. Vol. 9, pp. 83–103.

Department of Transport. (1994). Departmental Advice Note BA 59/94. *Design of Bridges for Hydraulic action*. London: HMSO.

Duncan Fama, M. E. (1979). Analysis of a solid inclusion *in-situ* stress measuring device, Proc. 4th. Congr. ISRM Montreaux., *Vol. 2,* pp. 113–120.

Forder, S. (1992). *Calibration of the Saw Cutting technique for in-situ stress determination. MEng Project Dissertation*. University of Surrey.

Furneaux, C. (2004). Structural integrity monitoring by global positioning systems. *The Structural Engineer* 25–27.

Fuzier, J. P., Domage, J. P., Elliott J. F., & Youdan, D. G. (2000). Continuous acoustic monitoring of bridges. Proc. Fourth International Conference on Bridge Management, Guildford, Surrey, UK. pp 120–128. Thomas Telford, London.

Godart, B. (1987). Apporche par l'auscultation et le calcul du fonctionnement de ponts en beton precontraint fissures, First Advanced Research US–European Workshop on Rehabilitation of Bridges, CEBTP, Saint-Remy-Chevreuse, France.

Goodman, R. E., Van, T. K., & Heuze, F. M. (1970). *The measurement of rock deformability in boreholes, Proc. 10th. Symp. Rock Mech., American Institute of Mechanical Engineers*. Austin: University of Texas, pp. 523–555.

Hofer, B. (1987). *Composites, 18*(4), 309–316.

Idriss, R. L. (1997). Nondestructive evaluation of a bridge using an optical fibre sensing system, Recent advances in Bridge Engineering, (Eds.) U. Meier and R. Betti, pp. 325–331, EMPA.

Leeman, E. R. (1964a). The measurement of stress in rock, Parts I, II and III, J. Soute African Inst. Min. Metall. 65, 45–114; 254–285.

Leeman, E. R. (1964b). Rock stress measurements using the trepanning, stress-relieving techniques, Mine Quarry Eng 30, 250–255.

Lindsell, P., & Buchner, S. H. (1987). *Prestressed Concrete Beams: Controlled Demolition and Prestress Loss Assessment, Technical Note 129*. London: CIRIA.

Meadowcroft, C., & Whitbread, J. E. (1993). *Assessment and Monitoring Bridges for Scour.* Oxon, UK: HR Wallingford Ltd., Paper No. 77.

Measurements Group. (1995). *Technical Bulletin 320*. USA and Basingstoke, UK: North Carolina.

Mehrkar-Asl, S. (1988). *Direct measurement of stresses in concrete structures. Ph.D. Thesis*. University of Surrey.

Mehrkar-Asl, S. (1996). Concrete stress-relief coring: theory and practice, FIP Symposium on Post-tensioned Concrete Structures, London, pp. 569–576.

Mehrkar-Asl, S. (2000). Monitoring to prolong service life, Proc. Fourth International Conference on Bridge Management, University of Surrey, Thomas Telford, London, pp. 137–144.

Meisseler, H. J., & Wolff, R. (1993). *Experience with permanent monitoring of the bridges Schiessbergstrasse and Notsch, Proc. Second International Conference on Bridge Management*. University of Surrey, pp. 236–243.

Mingins, P. (1996). The future of GIS. Proceedings of the Institution of Civil Engineers, Vol. 114 Special Issue, pp. 44–50., Thomas Telford, London.

Muskhelishivili, N. I. (1953). Some Basic Problems of the Mathematical Theory of Elasticity, Translated by J. R. M. Radok, Noordhoff, Netherlands.

NEWS. (1989a). Fibre prestressing monitors bridge state, New Civil Engineer April, 14.

NEWS. (1989b). Bedded optics to check lorry weights, New Civil Engineer November, 10.

Owens, A. (1988). Application of residual stress techniques in the determination of *in-situ* loads in reinforcing bars. Experimental Stress, pp. 23–27.

Owens, A. (1994). Measuring stresses in post-tensioned concrete. ICE Seminar (London Association) 24 February.

Owens, A., Begg, D. W., Gratton, D. N., & Devane, M. A. (1994). A new stress determination technique for concrete bridges, Bridge Assessment, Management and Design – Proc. of the Centenary Year Bridge Conference, Cardiff, UK, pp. 423–428.

Parker, D. (1996). An introduction to GIS and the impact on civil engineering. Proceedings of the Institution of Civil Engineering - Vol. 114, Special Issue, pp. 3–8., Thomas Telford, London.

Rocha, M., Lopes, J. J. B., & Da Silva, J. N. (1966). *A new technique for applying the method of the flat jack in the determination of stresses inside rock masses, Proc. 1st. Congr.* Lisboa: ISRM, pp. 57–65.

Ryall, M. J. (1993). Measurement of stresses in concrete structures using an instrumented hard-inclusion technique. *Journal of the Institution of Structural Engineers, 74*(15), 255–260.

Ryall, M. J., & Abdul-Rahman, A. (1990). *Assessment of* in-situ *stresses in prestressed concrete bridges, Proc. Second International Conference on Bridge Management*. University of Surrey, pp. 484–501.

Savin, G. N. (1961). *Stress Concentration Around Holes*. Oxford, UK: Pergamon Press.

Schießl, P., & Raupach, M. (1993). Permanent monitoring of the corrosion risk for the reinforcement of concrete bridges, Proc. Second International Conference on Bridge Management, University of Surrey, pp. 708–716, Thomas Telford, London.

Sennhauser, U., Anderegg, P., Bronnimann, R., & Nellen, M. (1997). Monitoring of Stork Bridge with fibre optic and electrical resistance sensors. In U. Meier, & R. Betti (Eds.), *Recent advances in Bridge Engineering* (pp. 366–373). EMPA.

Sloan, T. D., Kirkpatrick, J., & Thompson, A. (1990). *Remote computer-aided bridge performance monitoring, Proc. First Int. Conf. on Bridge Management*. Guildford, Surrey: Elsevier, pp. 361–371.

Smith, D. W. (1976). Bridge Failures. *Proc. Instn. Civ. Engrs., Part 1, 60*, 367–382.

Spathis, A. T. (1988). A bi-axial visco-elastic analysis of Hollow Inclusion Gauges with implications for stress monitoring. *Int. J. Rock Mech. Min. Sci. & Geomech. Abstr, 25*(6), 473–477.

Spathis, A. T., & Truong, D. (1990). Analysis of a bi-axial elastic inclusion Stressmeter. *Int. J. Rock Mech. Min. Sci. & Geomech. Abstr, 25*(4), 309–314.

Tyler, R. G. (1968). *Development in the Measurement of Stress and Strain in Concrete Bridge Structures, Road Research Laboratory, Ministry of Transport, Report LR 189*. UK: Crowthorne.

Watson, R. (1990). Hundreds of bridges to undergo scour tests, New Civil Engineer 9 August, 5.

Whitbread, J. E., & Riddell, J. (1996). Bridge scour, informal discussion, Proc. Instn. Civ. Engrs. *Transp, 117*, 67–69.

Woodward, R. J., & Williams, F. W. (1988). The collapse of Ynys-y-Gwas Bridge, West Glamorgan. *Proc. Instn. Civ. Engrs., Part 1, 84*(8), 634–669.

Woolf, R., & Meisseler, H. J. (1993). Glass fibre prestressing system. In J. L. Clarke (Ed.), *Alternative Materials for the Reinforcement and Prestressing of Concrete* (pp. 127–152). Blackie Academic and Professional.

FURTHER READING

Cullington, D. (1999). TRL evaluates tendon monitoring system, Research Focus, Issue No.36, ICE.

Matthews, S. L. (2000). Deployment of instrumentation for in-service monitoring. *The Structural Engineer, 78*(13), 28–32.

CHAPTER 12

Risk and Reliability

CONTENTS

12.1 INTRODUCTION

Risk and Reliability concepts are not generally understood by bridge engineers because they are not explicitly used in the everyday practice of bridge assessment or design. In the last decade or so, however, there seems to have been a growing appetite to understand both the concepts, and the practical application of them to make more confident decisions about the maintenance requirements and priorities of a bridge stock. Very good overviews of the meaning and application of reliability methods have been produced by Bailey (1996), Rosowsky (1997) and Melchers (1999) but essentially *reliability* is a probabilistic way of defining the limit states performance of a bridge system or bridge component. This is expressed as the probability of a certain event (limit state) occurring, the most important of which is that of *collapse*.

The loss of life or serious injury is the paramount consideration, and the users perception is encapsulated in the phrase *Is it safe to cross?* To answer the question it is necessary to define what is an acceptable risk.

12.2 RISK IN THEORY AND PRACTICE

A precise definition of risk is not possible as it is a composite idea. We take risks everyday of our lives, in fact life itself is 'risky'. Risk can only reasonably be measured in terms of past performance. For example in 1997, I did a bungee jump off the Victoria Falls bridge in Zimbabwe – a drop of 350 ft – and survived! I took a *calculated risk* based on the statistic that roughly 50 people a day did the jump over a six-month period every year and to the date of my jump there had been no fatalities. I was also reassured just before I jumped with the words 'Don't worry, its 100% safe'. The jump was a risk. In other words it had the *potential to do harm*. Although I had placed myself in jeopardy, I judged the risk (or danger) to be small. Had the information given to me been that on average one person a month dies because the bungee line breaks, then I probably would have declined. In other words I made an *informed* decision.

Risk implies that there is a *hazard* involved. If there is no hazard, there is no risk. If I want to cross a road there is a risk of being hit by the traffic – if there is no traffic, there is no risk from it. The road is not the hazard, but the traffic is, because it has the potential to do harm.

A new hazard is not so easily identified or assessed in terms of risk. For example, for many years in the early to middle part of the twentieth century there was a very low awareness of the risk attached to using asbestos in the building industry. It was considered to be safe until, that is, it was found that certain illnesses in people associated with the mining, manufacture and installation of asbestos products were directly linked to the asbestos itself. The risk had not been identified until severe illnesses and loss of life had occurred.

Another case in point is the terrible legacy that bridge engineers have inherited from the 1960s and 1970s because of the ignorance about the devastating corrosive effects from water laden with de-icing salts. The risk had not been identified until after the damage had been done. Now clients are having to pick up the tab to rectify the situation. In truth we should be grateful that we have not been sued for negligence.

A bridge is something that societies generally accept as being necessary, and users do not perceive crossing as a risky event, it is taken for granted that the structure is completely safe. The truth is, however, that every bridge that is built is itself 'at risk'. In other words something could happen to it to cause it to fail partially or outright. For example, in very broad terms:

- it might be blown up – *low* risk
- it might be hit by traffic – *medium* risk
- it could suffer material breakdown due to the environment – *medium to high* risk
- it could suffer structural damage due to the increasing intensity and weight of traffic – *high* risk

From risks to the bridge follow risks to people and the pollution, namely:

- injury or loss of life
- financial loss
- damage to the environment

All three can be mitigated if more money is spent on the initial product; or, alternatively, there is the option of *changing the concept of the bridge* to reduce the risk of degradation and possible failure.

Words like *low, medium and high* are not sufficiently precise to enable an informed judgement to be made on what is to be done at any given time, now or in the future. There are other factors to consider which complicate matters such as the class of highway on the bridge, the type of bridge, its environment, the quality of the components, the quality of construction, and how well it has been maintained.

A *new* bridge designed by a competent engineer and built to a high standard by a competent contractor would normally be considered very safe, and the risk of collapse when crossing it, minimal. An *existing* bridge which has performed well for many years and was not exhibiting any obvious signs of decay, would 'probably' be deemed as safe to the local population with little risk of collapse. An *older* bridge with many observable defects is weaker than when built and there is a greater risk of collapse.

Risk management is concerned with controlling risk by ensuring a certain level of safety. The two options listed previously imply that we can exercise *control*. A bridge which is already in place merits a different approach where risk assessment revolves around *curative* measures.

Thus risk analysis is a management tool which seeks to *balance risk against resources in some way so as to achieve the lowest overall risk for a given investment … and to ensure that no failure exceeds a certain acceptable risk level* (Elms, 1992). This is a complicated process and has to take into account many variables.

An acceptable risk level is not easy to determine, and when it has been evaluated it cannot be said to be the correct answer. It is an estimate. It may be satisfactory but *not* conclusive. It is therefore necessary to decide what event is the most critical in terms of the expected consequences. In bridge management it is unquestionably the *strength* of a bridge which is the primary concern and once a bridge has been assessed and found to *fail* then we seek to carry out a *risk assessment* (or risk analysis) in order to prioritise intervention. An accepted definition of risk often used in the structural world is:

$$\text{Risk} = \text{Probability of Failure } (P_f) \times \text{Consequences of Failure } (C_f) \tag{12.1}$$

Using the most accurate information available (either measured or simulated) probabilistic models have then to be constructed for the geometry, loads, analysis uncertainty, material strengths, deterioration and the consequences of failure. Some sort of scoring or ranking then has to be defined and applied to both P_f and C_f ending up with a number as a quantifiable measure of risk. Not an easy task!

Shetty, Chubb and Halden (1997) have suggested that the probability score be directly related to the probability of failure in an exponential form from $1E - 24$ to $1E + 00$ versus the score from 1 to 10 with four ranges defined thus:

$P_s = 0$–2 Negligible probability of failure $0 =$ completely safe
$P_s = 2$–5 Low probability of failure
$P_s = 5$–7 Medium probability of failure
$P_s = 7$–10 High probability of failure $10 =$ completely unsafe

These numbers are then modified to take into account the road type, impact effect, span, reserve strength and redundancy and finally the inspection regime.

12.2.1 Consequences of failure

A bridge failure is usually a spectacular and significant event, taking as it usually does a toll in human life and suffering. Table 12.1 lists some of the more tragic collapses that have occurred worldwide in the last 30 years or so.

In the UK there is about one bridge failure every one to two years (Menzies, 1997) and so the probability is well defined. On the other hand, however, the consequences of failure will vary considerably and are more difficult to quantify. Shetty et al. (1997) have suggested grouping the consequences of failure into four well-defined categories:

- *Human Factors* affecting loss of life and serious injury are the intensity of traffic and pedestrians passing under or over the bridge; the extent of the failure, i.e. partial or complete, the likelihood of a 'pile up', the nature of the failure, i.e. ductile or brittle.
- *Environmental* damage due to spillage of dangerous substances; the nature of the crossing (road, rail, river); and whether the locality is urban, industrial, rural or coastal; and whether or not it is subject to extreme events such as hurricanes, typhoons or earthquakes etc.
- *Traffic* delays and detours (which may not only be frustrating, but can result in noise pollution in narrow town roads and break up of the road surface from heavy vehicles).
- *Economic Factors* resulting from the whole cost of clearing away debris, rebuilding, compensation for vehicle damage, environmental clean up and legal costs.

Table 12.1 Bridge Failures in the Last 30 Years

Bridge	Cause of Failure
Silver Bridge, USA 1967	An undetected crack formed in an eye bar during manufacture had deepened due to corrosion and eventually the eye bar broke at the eye. Loss of 38 cars and 46 dead.
Milford Haven, UK 1972	Buckling of box girder diaphragm during construction
Yarra, Australia 1972	Ditto 35 dead.
Koblenz 1972	Ditto
Kempton, West Germany 1974	Road bridge centre span collapse as concrete being poured. Nine dead. Failure of A-Frame falsework supports.
Humber 1980	Deck gantry located on main suspension cables fell on to erected sections. Three sections damaged. Caused by failure of clamp securing gantry haul rope.
Almo, Sweden 1980	Ship hit unprotected twin tube pier. Arch deck fell into fjord. Led to tighter regulations for pier protection.
Ynys-y-Gwas Bridge, South Wales, UK 1985	Single span, 32-year-old segmental prestressed concrete bridge over the River Afan collapsed without warning. Cause: corrosion of the longitudinal and transverse prestressing wires.
Schoharie Creek, New York State, USA 1987	Foundation collapse leads to mains span falling into creek. Ten motorists dead. Scour undermined concrete spread footings.
Aschaffenburg, West Germany 1988	Cable stayed launch girder failed as main span over River Main approaches closing abutment.
Ness, Scotland 1989	Scour undermined central river pier of 127-year-old masonry rail bridge. Led to progressive collapse into the River Ness.
Changson Viaduct, Seoul 1992	Pier failure shortly after opening causing collapse of main span. One dead, seven injured.
Haeng Ju cable-stayed, Seoul, Korea 1992	Collapse of a temporary pier during construction. Only half of the required volume of concrete had been placed in the pier's foundation.
Songsu, Seuol, Korea 1994	Collapse of the midspan truss members due to fatigue exacerbated by traffic overloading. Thirty-two dead.
Palau, Micronesia 1996	Collapse of the world's longest post-tensioned balanced cantilever bridge. Likely cause: overstressing during recent strengthening work.
Jakarta, Indonesia 1996	Collapse of 30 m long, simply supported *in-situ* concrete bridge during construction. Three dead, 18 injured. Cause: early removal of the falsework.
Injaka Bridge, Mpumalanga Province, South Africa 1998	Failure during launching of a prestressed concrete box girder bridge. Cause: failure due to inadequate shear resistance at bottom flange/web junction. 14 dead.
Seo-Hae Grand Bridge, South Korea 1999	Partial collapse during construction possibly due to failure of the steel launching truss.
Footbridge over the River Yarkon, Israel 2000	Failure of the aluminium space frame bridge during a procession across the bridge to mark the start of the Maccabiah Games. Two dead. Bridge only three weeks old.

(*continued*)

Table 12.1 Bridge Failures in the Last 30 Years (*continued*)

Bridge	Cause of Failure
Prestressed concrete footbridge at Lode's Motor Speedway, Concorde, Carolina, USA 2000	One 25 m span of this four-span bridge collapsed suddenly probably due to grout failure and heavily corroded strands at the midspan section.
Millennium footbridge, London, UK 2000	'Wobbling' of the bridge at its opening on 10 June 2000. Attributed to sections of the crowd marching in step, but most probably a fundamental design error in not realising the sensitivity of the shallow suspension system to the effect of eccentric loading on the deck, akin to ancient jungle suspension bridges.
Ponto de Ferro bridge, Entre-os-Rios, Portugal 2001	Scour undermining one of the six masonry piers during flood conditions, exacerbated by illegal sand extraction around the piers, and hitting of piers by local ships. Seventy dead.

Work is underway in the UK to develop a consequence scoring system in terms of these four factors evaluated in terms of *cost* (Shetty et al., 1997). It is hoped that once the probability of (an assessed) failure is quantified the concomitant risk can then be classified as high, medium, low and negligible.

12.3 HEALTH AND SAFETY

The important and more general issues of health and safety are covered by the Highways Agency in the UK who have issued a Safety Memorandum based on the CDM – *Designing for Health and Safety* document. The idea is to arrive at a *Risk Rating* for the particular *activity* involved which then allows a maintenance programme to be planned based on a table of results of the risk rating for each bridge under consideration. The higher the risk index, the higher the priority.

The general hierarchy of from 'Intolerable' to 'Broadly acceptable' as defined by the Highways Agency is defined in Figure 12.1.

12.3.1 Determination of risk rating

A number of tables are provided by the Engineering Consultancy Division (ECD) of the Highways and Transportation Department to determine the Risk Rating. They are based on:

- Hazard Severity categories
- Occurrence Probability categories
- Risk rating matrix (RRM)

A number from 1 to 4 is taken from the RRM called the Rating Index which is then defined from the risk rating description table. All of these are shown in Table 12.2. The highlighted cells show typically, how a final index is arrived at. Inevitably this system is faulty because it relies to some extent on subjective judgement, but it does represent an attempt to minimise subjectivity and to provide a logical consistent approach to the problem of risk assessment. The process is further aided by a Hazard Inventory supplied by ECD and guidance notes on how to complete the inventory.

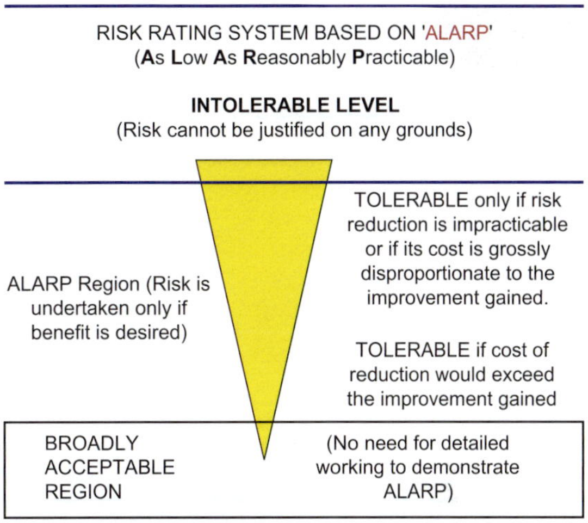

FIGURE 12.1 *Health and safety.*

12.4 STRUCTURAL SAFETY ANALYSIS

At this time there are three methods of determining how safe a bridge or a bridge component is, they are:

- Allowable stress method
- Limit states method
- Reliability method

Each one tries to calculate a 'number' which may be called the Factor of Safety in the cases of (i) and (ii) and the Probability of Failure in the case of (iii). The meaning of the number depends on the level of experience and understanding of the person making the judgement, but it is usually assessed according to a generally accepted codified value which has been derived by exhaustive research and testing and is deemed *acceptable* according to the collective wisdom of the engineering fraternity.

12.4.1 Allowable stress

This method says that provided the working stress level due to an expected maximum working load is lower than the failure stress level by an *acceptable* margin then it is safe, that is:

$$\text{Factor of safety against failure (FOS)} = \frac{f_L}{f_A} \qquad (12.2)$$

The FOS is normally within the range of 1.5–2.0 for bridge structures, but is 3.0 in the field of soil mechanics. It is higher in the latter case

Table 12.2 Risk Rating

Description	Definition			
	Hazard severity categories			
Catastrophic	Multiple fatality			
Critical	Single fatality or single/multiple major injury/illness			
Marginal	Minor injury/illness (resulting in lost time)			
Negligible	Very minor injury (no lost time)			
	Occurrence probability categories			
Probable	Occurrence not surprising. Will occur several times			
Possible	Likely to occur sometime			
Remote	Unlikely to occur, though conceivable			
Improbable	So unlikely that probability is close to zero			
	Risk rating matrix			
	Catastrophic	Critical	Marginal	Negligible
Probable	4	4	4	2
Possible	4	4	3	2
Remote	4	3	2	1
Improbable	3	2	1	1
Index	Risk rating descriptions			
1	The risk is tolerable; no further analysis of design alternatives is necessary			
2	The risk is tolerable; design alternatives only need to be applied if this can be done without extra cost or detriment to other design aspects			
3	The risk is on the border of tolerability, design alternatives should be sought			
4	The risk is intolerable; further design alternatives must be identified			

simply because there is more uncertainty about the consistency and behaviour of soil.

This method is understandable and has worked well for hundreds of years, in fact it is probably true to say that most structures in general and bridge structures in particular have been designed this way. Engineering experience allows some modification of the FOS up or down depending upon how well structures have performed throughout their design life. It is a simple – but not simplistic – deterministic approach to a probabilistic problem, and it works. Its main drawback is that it implies that all uncertainties have been taken into account, but it is one up on the process of trial and error, and can be justified by the fact that its use has resulted in structures which have not failed.

12.4.2 Limit states

This method of analysis employs the same concept of an FOS but is more detailed in that it employs the use of *partial safety factors* at a *defined limit state* so that if for example we are assessing the safety against collapse it can be stated as:

$$\begin{array}{r}\text{Resistance of a component}\\ \text{or the entire bridge}\end{array} \geq \begin{array}{l}\text{Effects of the applied dead}\\ \text{and traffic loads}\end{array}$$

that is

$$\begin{aligned}\text{Bending resistance of a beam} &\geq \text{Applied bending moment}\\ \text{Resistance} &\geq \text{Load effects}\\ R &\geq S\end{aligned} \qquad (12.3)$$

where representative values used for R and S are *random*.

Deterministic values of both R and S are impossible because they are the product of many uncertainties which are accounted for by the introduction of *partial safety factors* which allow for uncertainty and 'provide safety'. The load effects are maximised to γ_L,S and the resistance is minimised to:

$$\frac{R}{\gamma_m}$$

where both γ_L and $\gamma_m > 1.0$. In the UK this is expressed as:

$$\frac{R}{\gamma_m} > \gamma_L S \qquad (12.4)$$

This enables the calculation of an overall safety factor $\gamma = \gamma_m \cdot \gamma_L$ which ranges from 1.5 to 2.0 and is considered to be acceptable. It is by far the most universal approach in structural engineering and involves making judgements as to a suitable value of γ_m for the material strengths (e.g. 1.5 for concrete and 1.15 for steel) the lower value reflecting the fact that steel can be manufactured under much greater quality control than concrete, and likewise load factors γ_L (e.g. 1.4 for dead loads and 1.6 for live loads) the lower value reflecting the fact that the (nearly constant) dead loads can be estimated more accurately than fluctuating live loads. These values are based on engineering judgement to take into account of geometrical, load, strength and construction uncertainties but the probability of failure remains unknown. In some rare cases (Micic, 1997a, b) the probabilistic approach in the development of design codes has been used by modelling the load and resistance as random variables and selecting appropriate load and resistance factors such that the selected target reliability index is maintained.

Such Limit States methods have been adopted for *design* in both the USA (AASHTO, 1994) and the UK (BS 5400, 1992). In the USA these are classed as Level I or *Load and Resistance Factor (LFRD)* design methods and the relationship between loads and effects is written as:

$$\varphi R_n \geq \sum_i \gamma_i Q_{n,i} \tag{12.5}$$

'Failure' of a bridge could be the result of a single element failure, or it could be due to a number of elements failing simultaneously. Also there is the complication that there are hidden reserves of strength which might mean that the bridge is safe even when a number of elements fail. Other uncertainties such as material strengths, load estimates, lack of relevant statistical data relating to existing bridges, over simplification of the mathematical model of the bridge, etc. All of these issues can be addressed through a *system reliability* approach using probabilistic concepts.

12.4.3 Reliability method

In the previous section the limit states method took into account of uncertainties by way of an FOS, and the loads and the effects of the loads were assumed to be *deterministic* quantities. The reliability method is another way of taking into account of the same uncertainties in a *probabilistic* way.

If R and S are assumed to be independent variables which can be expressed as *known probability distributions* similar to that of Figure 12.2, then the *limit state of failure* has been reached when $R < S$. The *probability of failure (P_f) is equal to the probability that $S > R$* given by:

$$P_f = P[R - S] = P[S > R] = \sum P[S = x] \cdot [R < x] \tag{12.6}$$

where R and S are statistically independent and $[R - S]$ is the safety margin. The degree of variability of either R or S can be measured by a non-dimensional parameter called the Coefficient of Variation:

$$\mathrm{COV} = \frac{\sigma_X}{\mu_X} \tag{12.7}$$

where σ_X and μ_X are the standard deviation and the mean, respectively. The smaller the value of COV the narrower the distribution.

Equation (12.6) describes the probability of failure (P_f) as equal to the summation (over the range of all possible values of S) of the probability that $S = x$ and the probability that $R < x$. This is expressed as an integration from $-\infty$ to $+\infty$,

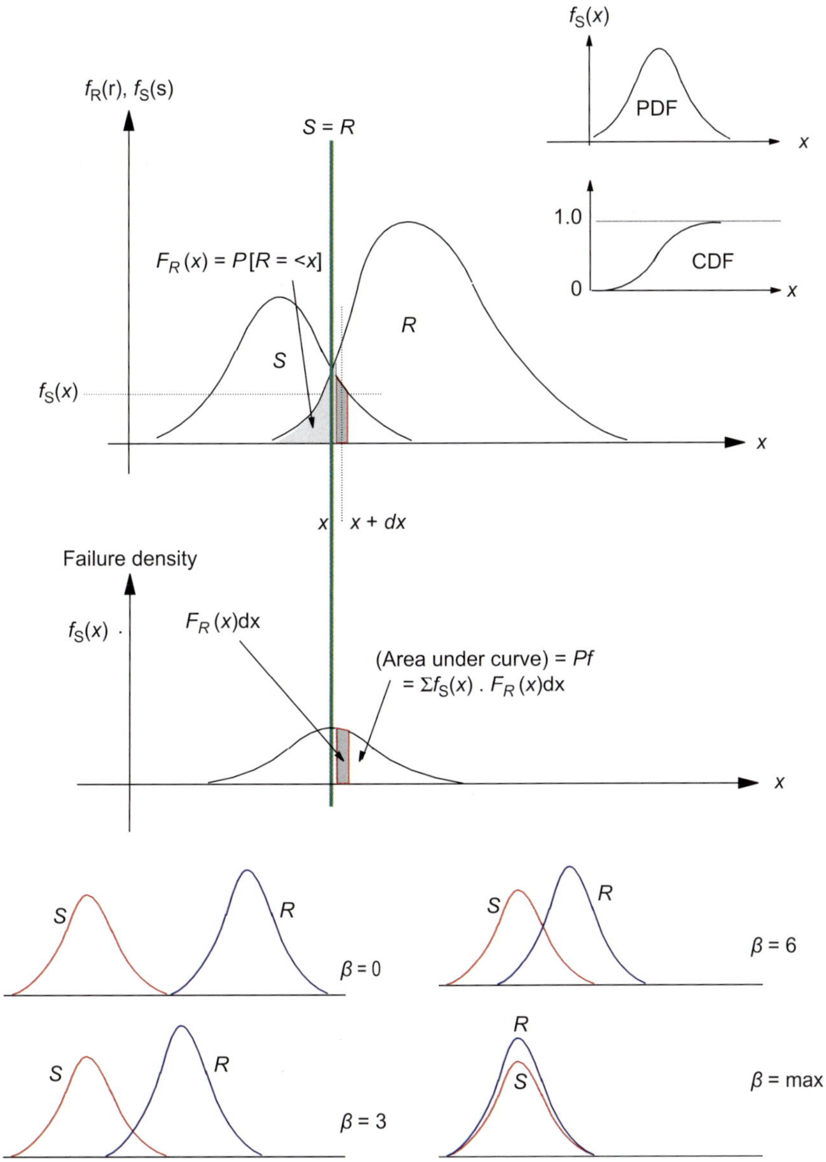

FIGURE 12.2

Probability distribution of R *and* S.

$$P_f = \int_{-\infty}^{\infty} f_S(x) \cdot F_R(x) dx \tag{12.8}$$

where the probability density function (PDF) $f_S(x) = P[x < S < x + dx]$ which is a measure of the relevant frequency of a variable; and the hatched area to the left of the limit state function $R = S$ in Figure 12.2 is the *cumulative probability distribution function* (CDF) which has a value between 0 and 1 where:

$$F_R(x) = P[R < x] \tag{12.9}$$

The failure zone represented by this equation is shown in Figure 12.3.

The design point $[R^*, S^*]$ is defined as the *most probable failure point on the limit state surface* which is nearest to the origin. The safety of the element or structure can then be defined with respect to a limit state function (such as bending failure) by either:

- the probability of failure P_f or
- the reliability index β.

For the normal distribution of random variables the integrals can be evaluated using tables where: Φ – standard normal probability integral or distribution function (see Appendix 12A).

P_f can be expressed as:

$$P_f = \Phi\left(\frac{-(\mu_R - \mu_S)}{\sqrt{\sigma_R^2 + \sigma_S^2}}\right) \tag{12.10}$$

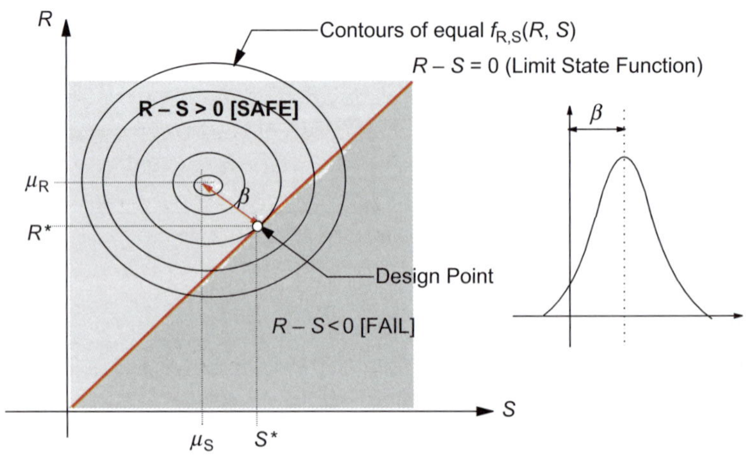

FIGURE 12.3
Failure zone with respect to joint probability distribution of R *and* S.

sometimes expressed as

$$P_f = \Phi(-\beta) \qquad (12.11)$$

or

$$\beta = \Phi^{-1}(1 - P_f)$$

where μ_R and μ_S are the mean values and σ_R and σ_S are the corresponding standard deviations.

It follows then that for a linear limit state function and independent normal variables the *reliability index* β is then defined as:

$$-\beta = \left(\frac{-(\mu_R - \mu_S)}{\sqrt{\sigma_R^2 + \sigma_S^2}} \right) = \frac{\mu_Z}{\sigma_Z} \qquad (12.12)$$

β is the number of standard deviations that the mean is away from the origin and marks the boundary between the *fail* and *safe* regions (see Figure 12.3). Thus β is a way of defining the safety of a structure or structural element. The *higher* the value of β the *higher* the safety and the *lower* the probability of failure P_f. The difference between the R and S curves for different values of β is shown in Figure 12.2. The relationship between P_f and β is often displayed in tabular form for example as in Table 12.3 (Chryssanthopoulos, Micic & Manzocchi, 1997).

The actual design point on the limit state surface is determined by either analytical or numerical methods (Melchers, 1999). The usual methods employed are analytical, either the First Order Reliability Method (FORM) which is widely described in the literature, e.g. Thoft-Christensen and Baker (1982), or the First Order Second Moment (FOSM) method (Madsen and Ditlevsen, 1996). The most popular numerical method (for calculating P_f) is the Iterative Fast Monte Carlo method.

The *reliability index* β (safety factor) is calculated using the FORM method of analysis and is a function of the distance of the design point from the joint *probability design functions* (PDFs) for R and S. Bailey (1996) illustrates this by taking a slice through the limit state surface as shown in Figure 12.4. This value of β is then compared to a *target reliability* $\beta_T = 3.5$

Table 12.3 Relationship Between Reliability Index β and Failure Probability P_f

P_f	10^{-1}	10^{-2}	10^{-3}	10^{-4}	10^{-5}	10^{-6}	10^{-7}	10^{-8}	10^{-9}	10^{-10}
β	1.3	2.3	3.1	3.7	4.2	4.7	5.2	5.6	6.0	6.4

FIGURE 12.4
Three-dimensional representation of the design point (after Bailey, 1996).

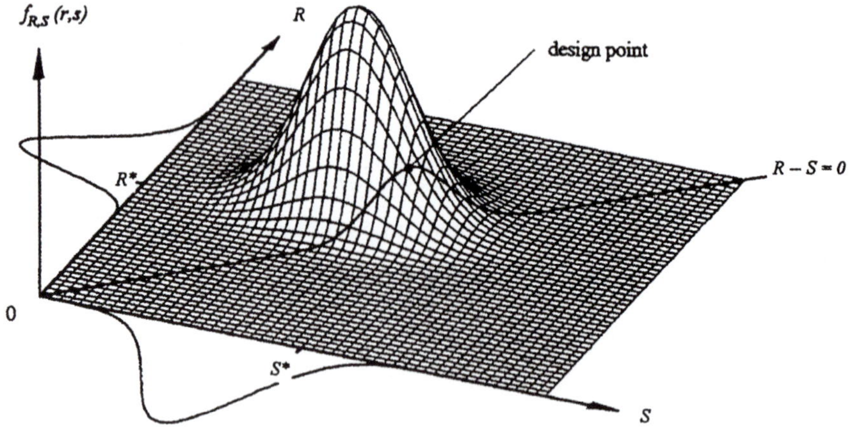

(still the subject of a great deal of discussion, see section 3.2.2). If $\beta > 3.5$ then the bridge can be deemed as 'safe' and it has 'passed' its assessment. This method is designated Level II in the USA (AASHTO, 1994).

Vrouvenvelder and Waarts (1993) relate the target reliability index to a given period of time for which the probability applies. Typical values quoted by them are shown in Table 12.4. Thoft-Christensen and Baker (1982) suggest another target reliability index evaluation as shown in Table 12.5.

According to Bailey (1996) the influence of the load effects can be taken into account by an additional coefficient α_s then in simple terms if $R_a \geq S_a$ (one load and one resistance) the reliability is expressed by:

$$P(R_a > S) = \Phi(-\alpha_S \, \beta_T) \tag{12.13}$$

where Φ is standard normal probability integral, α_s is influence coefficient for load, and β_T is the target reliability index. In Codes of Practice any combination of characteristic value and partial safety factor can be used which leads to S_a, and α_s is found from a complete reliability analysis. In the Swiss Code SIA 160 (1989) $\alpha_{\underline{s}}$ is given as 0.7. The actual value of the

Table 12.4 Typical Value of Target Reliability (after Vrouvenvelder and Waartis, 1993)

Limit State	Reference Period (Years)	Accepted Probability of Failure	Target Reliability Index β_T
ULS	100	10^{-4}	3.6
FLS	100	10^{-4}	3.6
SLS	1	0.5	0

Table 12.5 Target Reliability Index

(Reference Period: One Year)		Type of Failure: Ductile with Reserves	Ductile Without Reserves	Brittle
Safety class	Low	3.1	3.7	4.2
	Normal	3.7	4.2	4.7
	High	4.2	4.7	5.2

reliability index β can be established as mentioned before (section 3.3.2) by an analysis of each of a number of *existing* bridges which are functioning well. The resulting family of reliability indices can then be used to determine the target reliability index β_T for the same type of bridge. This process is called *calibration* for complete bridges or individual elements within a bridge.

12.4.4 Traffic actions

One of the most important loading variables on a bridge is the *live load traffic*. This comprises primarily the static effects which have to be modified to take into account of the dynamic factors (see Table 12.6). The *static effects* assume that the vehicles are travelling at a constant speed along a perfectly flat surface, and the *dynamic effects* are those which come from the action of the vehicle on the bridge and the bridge on the vehicle. To carry out a reliability analysis it is necessary to *simulate* the actual traffic using a particular bridge. The aim of the probabilistic traffic simulation is to obtain the static effects of traffic actions on a given bridge using the local site characteristics. The vehicles are generated randomly from a probabilistic database taking into account of different traffic types and flows which themselves are based

Table 12.6 Static and Dynamic Effects of Vehicles

Static Effects	Dynamic Effects
Traffic composition	Speed of vehicle
Distance between vehicles	Vehicle suspension
Axle group weights	Vehicle weight
Vehicle geometry	Number of vehicles on bridge
Weight variation with lane	Road roughness
	Natural frequency of bridge

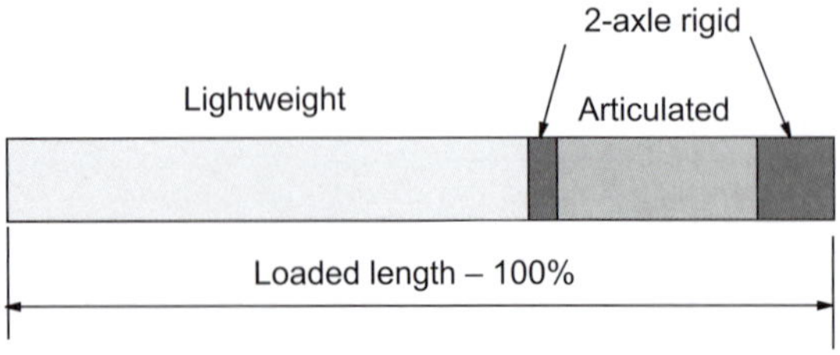

FIGURE 12.5 *Traffic composition.*

on measured probabilistic distributions. To help with modelling or simulating the traffic it is usual to carry out a vehicle survey for the collection of statistical data as in Table 12.6. The data are collected and plotted as histograms and used as the traffic model.

12.4.4.1 Traffic composition

The different categories of vehicle can be defined by the *number of axles, the total weight of each vehicle and the distribution of weight to each axle*. The categories are basically cars, vans, buses and trucks, the last of these often having the most significant effect. The information for rigid lorries, trailer combinations and articulated lorries is available from the *Construction and Use – Road Vehicles (Authorized Weight) Regulations* 1998. The composition on a bridge can then be expressed as a percentage of each different category, e.g. in a given lane a typical load distribution can be described as that shown in Figure 12.5.

12.4.4.2 Distance between vehicles

Traffic flow varies and therefore so will the distance between vehicles. Generally speaking the slower the pace of the traffic the smaller the distance between vehicles, and the faster the pace of the traffic the greater the spacing. This, of course, can be greatly modified in terms of driving habits. In the UK for example there is a tendency for 'stop-go' driving in congested traffic conditions which cause 'bunching' to occur. Thus the flow is not steady but is more akin to *packets of traffic* moving along the highway in the form of practically stationary bunches separated by a stream of vehicles travelling at variable speeds. This could be modelled using a *beta probability distribution* as shown in Figure 12.6a.

12.4.4.3 Axle group weights

Axle group weights are considered as a single point load based on the sum of the axles in the group and experience has shown that the corresponding distribution is best modelled by a bimodal beta probability distribution as shown in Figure 12.6b. This was based on a histogram representing measurements of actual traffic in Switzerland (Bailey, 1996).

12.4.4.4 Vehicle geometry

According to Bailey (1996) this is obtained by measuring the distance between axle groups and the front and rear vehicle overhangs and is best modelled using a beta distribution.

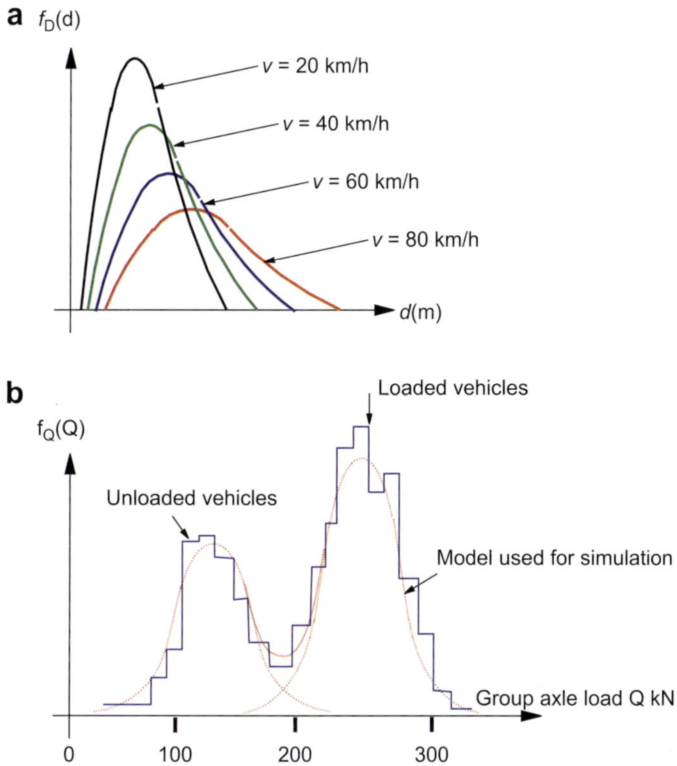

FIGURE 12.6
(a) Distance between vehicles in congested traffic and (b) Distribution of axle group weight.

12.4.4.5 Output

The output is presented in the form of *extreme value* curves for both the mean and the standard deviation of a particular stress resultant (moment, shear etc.). The data are for a specific bridge and can be grouped depending upon the type of flow. Bailey (1996) considers three conditions namely *at rest, congested and free-moving*. Three typical curves are shown in Figure 12.7. This shows that for each of the three traffic conditions the mean maximum effect increases with the number of vehicles. Also, it can clearly be seen that, as expected, the effects of a stationary (at rest) line of vehicles is more critical than a free flowing line. In fact when $N = 10^8$ the 'at rest' value is twice the 'free flowing' value. Which value to choose is a moot point and is not addressed in the literature!

12.4.5 Methodology

The information given in this short chapter may seem daunting and so a typical methodology is suggested here in Figure 12.8 for a given structural element, and then followed by some simple examples.

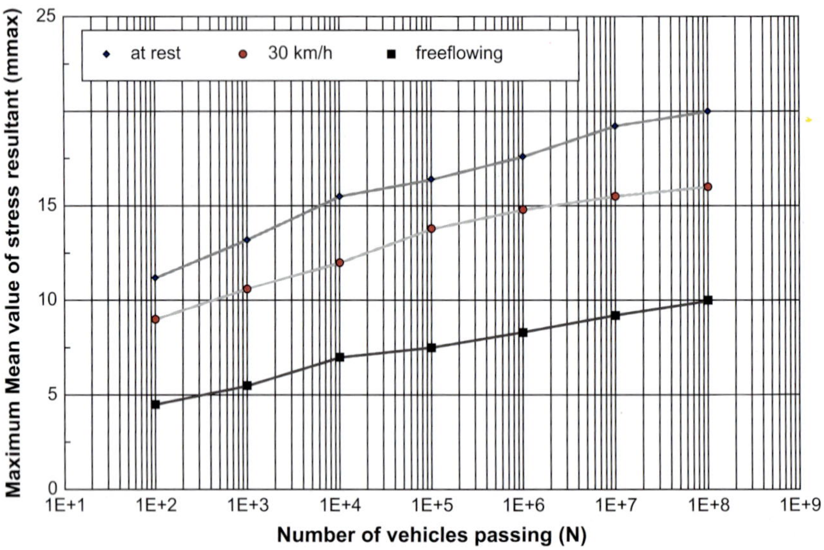

FIGURE 12.7
Mean maximum effect for different traffic conditions.

12.4.6 Examples

The following examples are similar to those presented by Rosowsky (1997) and Ghosn (1996) and serve to illustrate the reliability method in a simple way.

Example 1

The moment capacity (M) of the simply supported beam with a span (L) of 10 m is assumed to be normally distributed with a mean (μ_R) of 25 kN m and a coefficient of variation (COV) of 0.2. Failure occurs if the maximum (design or assessed) moment (M_D) exceeds the moment capacity (M_U). (It is assumed that geometric [span, width, member sizes etc.] and property values (E, I, A etc.) are exactly known.)

(i) What is the probability of failure if: (a) Only the point load ($P = 4$ kN) is applied at midspan? (b) Only the UDL ($w = 1$ kN/m) is applied? (c) Both the point load and the UDL are applied simultaneously?

(ii) What is the probability of safety P_s for a UDL of 0.50 kN/m?

Answer to (i)(a)

$$M_D = PL/4 = 4 \times 10/4 = 10 \text{ kN m}$$
$$\mu_R \times \text{COV} = 25 \times 0.2 = 5.0$$

$$P_f = P[M_U < M_D] = F_M 9(10) = \Phi\left(\frac{10-25}{5}\right) = \Phi(-3.0)$$

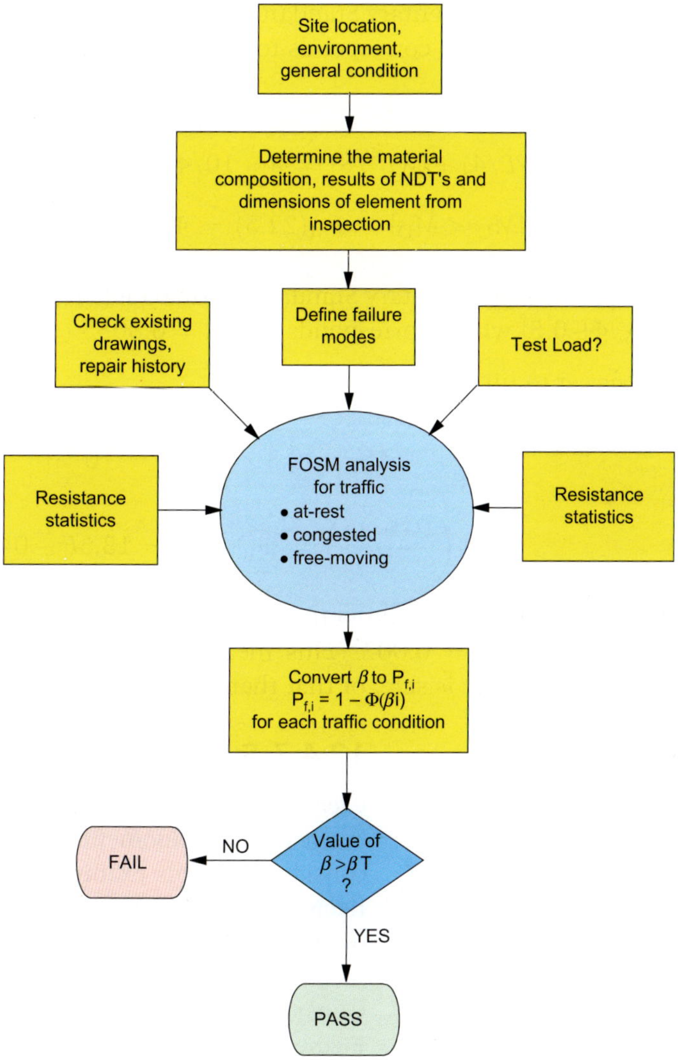

FIGURE 12.8

Typical reliability process.

From the Complementary Standard Normal table reproduced in Appendix 12A, $\Phi(-3.0)$ which corresponds to a $P_f = 0.00135$ (13.5E−04).

Answer to (i)(b)

$$M_D = wL^2/8 = 1 \times 10^2/8 = 12.5 \text{ kN m}$$

$$P_f = P[M_U < M_D] = F_M(12.5) = \Phi\left(\frac{12.5 - 25}{5}\right) = \Phi(-2.5)$$

From the Complementary Standard Normal table reproduced in Appendix 12A, Φ(–2.5) which corresponds to a $P_f = 0.00621$ (62.0E–04).

Answer to (i)(c)

$$M_D = (PL/4) + (wL^2/8) = 4 \times 10/4 + 1 \times 10^2/8 = 22.5 \text{ kN m}$$

$$P_f = P[M_U < M_D] = F_M(22.5) = \Phi\left(\frac{22.5 - 25}{5}\right) = \Phi(-0.5)$$

From the Complementary Standard Normal table reproduced in Appendix 12A, Φ(–0.5) which corresponds to a $P_f = 0.3085$ (3085E–04).

Answer to (ii)

$$M_D = 0.8 \times 10^2/8 = 10$$

$$\Phi\left(\frac{10 - 25}{5}\right) = \Phi(-3.0) = 13.5E - 04$$

The probability of a safe performance $P_s = 1 - 0.00135 = 0.998$ and $P_f = 1 - 0.998 = 0.002$. This means that there is 99.8% probability that the structure is safe, or that there is 0.2% probability that it will fail.

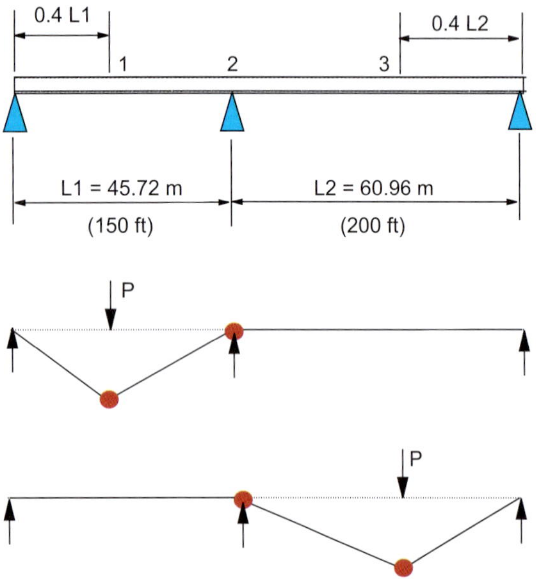

FIGURE 12.9 *Continuous bridge example.*

12.4.7 System reliability

Clearly, since only one type of load and a single beam with one moment failure mechanism was considered in this example, the solution is relatively simple. With a continuous beam having two or more spans, and several possible failure mechanisms, it is much more complicated. Consider for example the two-span bridge in Figure 12.9 (after Ghosn, 1996).

The bridge consists of a concrete slab supported on two steel beams, and it is assumed that any loads are transferred equally to each of the beams. The geometry of the bridge and the section properties are assumed to be exactly known. Ghosn (1996) carried out a *member* reliability analysis for bending at the three critical sections indicated under the action of dead and live loads. His results (utilising a FORM analysis) are shown in Table 12.7. These show a wide disparity, but if

Table 12.7 Results of Member Reliability Analysis

	Section 1	Section 2	Section 3
Member reliability index β	1.9	3.56	2.49
Probability of member failure P_f	2.87%	0.02%	0.64%

the system as a whole is considered along with the two *independent* failure mechanisms, then $\beta_{\text{MODE 1}} = 3.43$ and $\beta_{\text{MODE 2}} = 3.47$. If, however, the two mechanisms are *dependent* then $\beta_{\text{SYSTEM}} = 3.26$ ($P_f = 0.56\%$).

Ghosn (1996) has pioneered a numerical simulation technique which can calculate the reliability of a bridge system as a whole which he calls the '*response function method*'. It uses the results of structural analysis programme able to model the deterministic behaviour of the bridge. A response function (or response surface) is then obtained from a functional relationship between the load that causes system failure (collapse) and the moment capacities and the dead load moments at each of the critical sections. The response surface provides a relationship between the *capacity* of the bridge and the *random variables* considered. A more detailed description is given by Ghosn and Moses (1994) and Moses (1982).

12.4.8 Deemed to have failed

The word 'failed' has been used a lot in this chapter, but what exactly does it mean in terms of reliability analysis? Any bridge may experience an *element* failure or a *system* (total) failure at either the *serviceability* or the *ultimate* limit state. Fortunately, total collapse is a very rare event, but when it does occur it is catastrophic and often results in loss of life. Total system failure at the serviceability limit state is, however, a more frequent event. A classic example of this is the bridge carrying the A31 over the A3 at Guildford. When built in 1930 it had a number steel arch ribs supporting a concrete deck. The steel ribs were provided with a concrete encasement, presumably to protect the steel or maybe to enhance the appearance of the bridge. In 1996 the concrete to the ribs began to deteriorate noticeably and over the following two years continued to get worse. There came a point where lumps of concrete were beginning to drop onto the A3 below and so to avert an accident a polyethylene net was placed around each rib to contain the spalling concrete. Eventually a platform was hung from the deck and refurbishment works carried out by removing the concrete casing completely thus exposing the steel ribs which were then cleaned and given an appropriate paint treatment to prevent corrosion. The bridge had 'failed' in the sense that it was posing a hazard to the general public although it was in no danger of collapse.

Structural failure of individual elements can probably be tolerated for a while if there is enough built-in redundancy to enable a new load path to develop. They should, of course, be replaced as soon as economically and practically possible. Suspect elements should be identified during inspections and a risk check carried out on each one. It is up to the checker (an experienced professional bridge engineer) as to what target reliability index to aim for which will depend on the importance relative to the type, size, location, traffic intensity and general condition of the bridge. This approach makes sense because all bridges are synergistic (the strength of the system is greater than the sum of the strengths of the separate elements).

12.5 TEST LOADS

Test loads in the form of supplementary load tests remove a lot of uncertainty from assessment calculations by indicating a minimum level at which the bridge can operate to satisfy a given limit state. The actual methods used for this and the application to bridges were discussed in Chapter 5 (section 5.2.1).

From a reliability point of view a proof load has the effect of truncating the climbing arm of the probability density (PDF) function for resistance, which has the effect of *increasing* the reliability index.

For example, it is theoretically possible to calculate the critical deflection of a bridge deck under live load and to determine that deflection just as the failure load is reached. If a supplementary load is applied incrementally and it consistently deflects at levels less than the theoretical values up to the load corresponding to normal traffic conditions then it can be concluded that the bridge is *stiffer* or has *better load distribution properties* than expected. If no damage is observed, the test could continue with loads up to a value equivalent to the assessment loads with slightly lower factors of safety than normally used in design to take it close but not up to the ultimate limit state. If the structure is ductile then the load deflection plots can be extrapolated up to failure and the bridge can reasonably be assumed to be safe.

12.6 THE FUTURE

At a recent conference on Structural Reliability in Bridge Engineering (National Science Foundation, 1996) the importance of structural reliability methods was underlined by the staggering fact that of the total stock of about

600,000 bridges in the USA, 40% are deficient in some way and it is estimated that US$75 billion is required just to bring them up to standard. This is on top of the existing budget of £2.7 billion per year spent on repair, replacement and management!

It was recognised that there was a strong incentive to manage the bridge stock as well as possible and to optimise the limited funds available. Seven areas of *research* were identified as:

- Global issues.
- Feedback of bridge performance.
- Models and analytical reliability techniques.
- Inspection techniques.
- Costs.
- Engineering design/evaluation.
- Communication.

Within these general headings 35 topics for action were identified.

It would be satisfying to be able to apply reliability techniques in all of these areas but limited funds mean that this is not possible. In any case trying to convince government that the money is really necessary is nigh on impossible.

Perhaps the most important area is that of *maintenance strategy* and an example of this is work undertaken recently on some viaducts in the UK (Cropper, Jones & Roberts, 1998) where acute chloride induced corrosion was threatening the integrity of the reinforced concrete crossbeams on the piers. Originally a 10-year programme of repair was allowed commencing in 1990, but after sometime it was realised that this was not achievable without a large cash injection. Two objectives were identified to tackle the problem, namely a *rational justification for more time and a means to prioritise repair.*

This would require some idea of the *predicted condition state* of each crossbeam from which *intervention dates* could be defined based on time related deterioration. The corollary is that the work could be prioritised, and also ensure that an 'acceptable risk' was maintained for each of the viaducts. A deterministic approach was adopted to identify the intervention dates (when the capacity ratio = 1.0), and a reliability analysis was the vehicle by which the probability of failure of critical sections was calculated. Using limited data from 'previous repair trials' carried out in 1987 a model was constructed and validated with data from subsequent repair contracts.

The exercise was a success and the use of a deterministic approach in tandem with a reliability approach it was 'possible to define priorities for repair with greater confidence than previously, and to demonstrate that an increase in the length of the repair programme is acceptable' (Cropper et al., 1998).

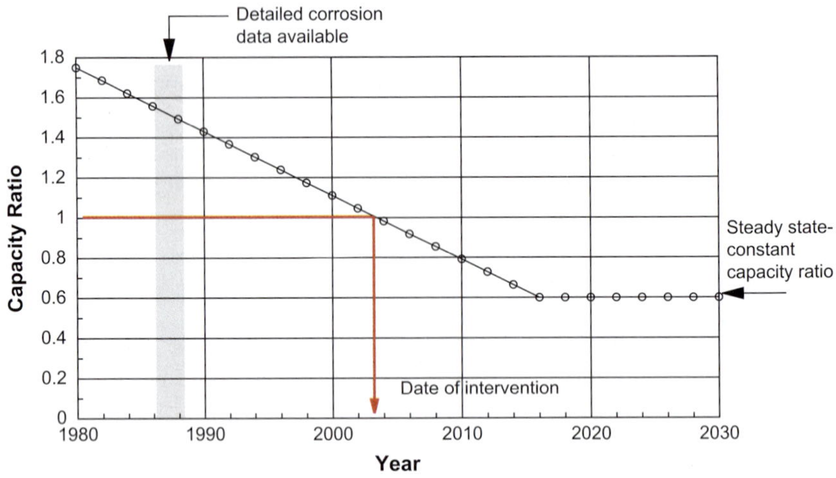

FIGURE 12.10
Deterioration model.

A plot of one of the deterioration models developed is shown in Figure 12.10 for the loss of shear strength with time and illustrates the difficulties that can be encountered in this approach. From carefully reading the paper, it would appear that data on the deterioration were available from previous repair trials as indicated above. Unfortunately the amount and exact nature of the data were not specified, neither was the exact location to which the data referred stated. Assuming that readings were taken each month for a year (1987) of the loss in cross section of the reinforcement and delamination (how these were measured was not stated) then a small window of points could have been plotted and then extrapolated forwards (and backwards?) in time. Furthermore, a sudden change occurs in the year 2016 when the capacity remains constant (presumably because the steel section has fully deteriorated?) but no reason is given. No doubt a lot of engineering judgement was exercised to rationalise the model and the engineers were satisfied with the result. In general a number of questions should be asked:

- Do I have enough data on which to build a model that accurately reflects the past, and accurately predicts the future?
- Is it correct to assume a linear relationship between the capacity and time?
- Can any discontinuity really be forecast?
- Is it possible to forecast a stable situation in the future where, for example, the capacity ratio remains constant?
- Is the degree of confidence placed in the reliability results justified?

FIGURE 12.11

Michael's Bungy Jump from off the Victoria Falls Bridge in Zimbabwe (with Zambezi River below) took place on 18th July 1998. A drop of nearly 350 feet.

- How realistic are the intervention dates?
- How confident am I that the structure will not fail locally?

These questions and many more need to be addressed lest we lull ourselves into a false sense of security because of our trust in the 'black box' approach. Reliability theory is now very sophisticated, but there are limitations. Elms and Turkstra (1992) list no less than 23 practical problems that directly affect its applicability, and they declare that it is not an exhaustive list!

12.7 IN CONCLUSION

Reliability methods are deemed to be the way ahead for a supposedly more rational and realistic approach to the determination of bridge safety. However, it is as well to sound a warning note:

> *'Some of the techniques for calculating the quantities in probabilistic risk assessments are elegant and sophisticated, and they give an impression of authority and precision. However, there are fundamental problems with probabilistic risk assessment, and the results are not necessarily realistic, dependable or even meaningful (Reid, 1992).'*

Reid (1992) then goes on to list the reasons why he holds this view but nevertheless concedes that the whole process can be useful 'as a catalyst for making explicit assessments of risks (accounting for non-quantifiable factors) within a larger review process'. Value judgements (which cannot be quantified) have to be made consciously by individuals, and it is important to remember that reliability methods are there to 'support not supplant engineering judgement'.

Kaschner et al. (2000) also sound a timely warning: 'Although extremely useful, probabilistic theory is sometimes limited. Some parameters are difficult to measure and probability theory calculations quickly become impractical. These considerations are critical in determining what can be expected from probabilistic theory. Probabilities suffer from the fact that they are only estimates of frequencies (sometimes not observable) based upon an evolving set of partial data. Conventional distributions are often used for convenience and, of course, the outcome of a probabilistic approach depends strongly on the assumptions which are made about the uncertainties associated with the variables. If these assumptions are not founded on adequate and appropriate data, estimates of safety will be misleading. Indeed, probabilistic methods are often misused when the variables are not carefully modelled. It is therefore essential that the quality of data and validity of assumptions are borne in mind when using a probabilistic approach to make decisions about the apparent safety of a structure.

REFERENCES

AASHTO. (1994). LFRD *Bridge Design Specifications*. American Association of State Highway and Transportation Officials, Washington, DC.

Bailey, S. F. (1996). Basic principles and load models for the structural safety evaluation of existing road bridges, Thesis No. 1467 (1996), presented to the Department of Civil Engineering, Ecole Polytechnique Federal de Lausanne.

BS 5400. (1992). Steel, concrete and composite bridges.

Chryssanthopoulos, M. K., Micic, T. V., & Manzocchi, G. M. E. (1997). Reliability evaluation of short span bridges, *Joint ICE & Highways Agency Symposium on Safety of Bridges Symposium*, (Ed.) P. G. Das.

Cropper, D., Jones, A., & Roberts, M. (1998). *A risk based maintenance strategy for the Midlands Links viaducts, Management of Highway Structures*. London: ICE Symposium.

Elms, D. G. (1992). In D. Blockley (Ed.), *Risk assessment, Engineering Safety* (pp. 28–46). McGraw-Hill.

Elms, D. G., & Turkstra, C. J. (1992). A critique of reliability theory. In D. Blockley (Ed.), *Engineering Safety* (pp. 427–443). McGraw-Hill.

Ghosn, M. (1996). *Safety evaluation of highway bridge structural systems,* Recent Advances in Bridge Engineering. Barcleona: CIMNE.

Ghosn, M., & Moses, F. (1994). *Redundancy in Highway Bridge Superstructures, Final Report, NCHRP 12–36*. Washington, DC: TRB.

Ghosn, M., Moses, F., & Khedekar, N. (1994). Response functions and system reliability of Bridges, IUTAM Symposium, Probabilistic Structural Mechanics. In P. D. Spanos, & Y. T. Wu (Eds.), *Advances in Structural Reliability Methods*. Heidleberg: Springer-Verlag.

Kaschner, R., Haardt, P., Cremona, C., Cullinton, D., & Daly, A. F. (2000). *Bridge Management in Europe (BRIME): Structural Assessment*. Bridge Management 4, Proc. of Fourth Conference on Bridge Management, Guildford, UK. Thomas Telford Pub. 2000.

Madsen, H. O., & Ditlevsen, O. (1996). *Structural Reliability Methods*. John Wiley.

Melchers, R. (1999). *Structural Reliability – Analysis and prediction* (2nd Edn). Chichester: John Wiley.

Menzies, J. B. (1997). Bridge failures, hazards and societal risk. In P. G. Das (Ed.), *Symposium on the Safety of Bridges* (pp. 36–41). London: ICE.

Micic, T. (1997a). *Structural Reliability methods for Probabilistic Bridge Assessment: Part 1*. London: Note for the Highways Agency.

Micic, T. (1997b). *Structural Reliability methods for Probabilistic Bridge Assessment: Part 2*. London: Note for the Highways Agency.

Moses, F. (1982). System reliability development in structural engineering. *Structural Safety, 1*(1), 3–13.

National Science Foundation. (1996). *Workshop on Structural Reliability in Bridge Engineering*. Boulder: Colorado.

Reid, S. G. (1992). Acceptable risk. In D. Blockley (Ed.), *Engineering Safety* (pp. 138–166). McGraw-Hill.

Rosowsky, D. V. (1997). Structural reliability, *Civil Engineering Handbook*, (Ed.-in-Chief) W. F. Chen, pp. 1773–1805, CRC Press.

Shetty, N. K., Chubb, M. S., & Halden, D. (1997). An overall risk-based assessment procedure for substandard bridges. In P. G. Das (Ed.), *Symposium on the Safety of Bridges* (pp. 225–235). London: ICE.

SIA 160. (1989). Actions on structures. *Societe Suisse des Ingenieers et des Architects*.

Thoft-Christensen, P., & Baker, M. J. (1982). *Structural Reliability Theory and its Applications*. Springer-Verlag. Berlin.

Vrouvenvelder, A. C. W. M., & Waarts, P. H. (1993). Traffic loads on bridges. *Structural Engineering International, 3*(93), 169–177.

FURTHER READING

Blockley, D. (Ed.). (1992). *Engineering Safety*. London: McGraw-Hill.

Kates, R. W. (1978). *Risk Assessment of Environmental Hazards, Report No. 8, Scientific Committee on Problems of the Environment*. Chichester: Wiley.

Melchers, R. (1999). *Structural Reliability – Analysis and Prediction* (2nd Edn). Chichester: John Wiley & Sons.

APPENDIX 12A

Complementary Standard Normal table (after Melchers, 1999)

Table 12A1 $N(0, 1)$ Distribution Defined as $\Phi(-\beta) = 1 - \Phi(\beta)$

β	$\Phi(-\beta)$	β	$\Phi(-\beta)$	β	$\Phi(-\beta)$
0.00	0.5000	0.40	0.3446	0.80	0.2119
0.01	0.4960	0.41	0.3409	0.81	0.2090
0.02	0.4920	0.42	0.3372	0.82	0.2061
0.03	0.4880	0.43	0.3336	0.83	0.2033
0.04	0.4841	0.44	0.3300	0.84	0.2005
0.05	0.4801	0.45	0.3264	0.85	0.1977
0.06	0.4761	0.46	0.3228	0.86	0.1949
0.07	0.4721	0.47	0.3192	0.87	0.1922
0.08	0.4681	0.48	0.3156	0.88	0.1894
0.09	0.4642	0.49	0.3121	0.89	0.1867
0.10	0.4602	0.50	0.3085	0.90	0.1841
0.11	0.4562	0.51	0.3050	0.91	0.1814
0.12	0.4522	0.52	0.3015	0.92	0.1788
0.13	0.4483	0.53	0.2981	0.93	0.1762
0.14	0.4443	0.54	0.2946	0.94	0.1736
0.15	0.4404	0.55	0.2912	0.95	0.1711
0.16	0.4364	0.56	0.2877	0.96	0.1685

Table 12A1 $N(0, 1)$ Distribution Defined as $\Phi(-\beta) = 1 - \Phi(\beta)$ (*continued*)

β	$\Phi(-\beta)$	β	$\Phi(-\beta)$	β	$\Phi(-\beta)$
0.17	0.4325	0.57	0.2843	0.97	0.1660
0.18	0.4286	0.58	0.2810	0.98	0.1635
0.19	0.4247	0.59	0.2776	0.99	0.1611
0.20	0.4207	0.60	0.2743	1.00	0.1587
0.21	0.4168	0.61	0.2709	1.01	0.1563
0.22	0.4129	0.62	0.2676	1.02	0.1539
0.23	0.4091	0.63	0.2644	1.03	0.1515
0.24	0.4052	0.64	0.2611	1.04	0.1492
0.25	0.4013	0.65	0.2579	1.05	0.1469
0.26	0.3974	0.66	0.2546	1.06	0.1446
0.27	0.3936	0.67	0.2514	1.07	0.1423
0.28	0.3897	0.68	0.2483	1.08	0.1401
0.29	0.3859	0.69	0.2451	1.09	0.1379
0.30	0.3821	0.70	0.2420	1.10	0.1357
0.31	0.3783	0.71	0.2389	1.11	0.1335
0.32	0.3745	0.72	0.2358	1.12	0.1314
0.33	0.3707	0.73	0.2327	1.13	0.1292
0.34	0.3669	0.74	0.2297	1.14	0.1271
0.35	0.3632	0.75	0.2266	1.15	0.1251
0.36	0.3594	0.76	0.2236	1.16	0.1230
0.37	0.3557	0.77	0.2207	1.17	0.1210
0.38	0.3520	0.78	0.2177	1.18	0.1190
0.39	0.3483	0.79	0.2148	1.19	0.1170
1.20	0.1151	1.70	0.4457E−01	2.20	0.1390E−01
1.21	0.1131	1.71	0.4363E−01	2.21	0.1355E−01
1.22	0.1112	1.72	0.4272E−01	2.22	0.1321E−01
1.23	0.1094	1.73	0.4182E−01	2.23	0.1287E−01
1.24	0.1075	1.74	0.4093E−01	2.24	0.1255E−01
1.25	0.1057	1.75	0.4006E−01	2.25	0.1222E−01
1.26	0.1038	1.76	0.3921E−01	2.26	0.1191E−01
1.27	0.1020	1.77	0.3836E−01	2.27	0.1160E−01
1.28	0.1003	1.78	0.3754E−01	2.28	0.1130E−01
1.29	0.9853E−01	1.79	0.3673E−01	2.29	0.1101E−01
1.30	0.9680E−01	1.80	0.3593E−01	2.30	0.1072E−01
1.31	0.9510E−01	1.81	0.3515E−01	2.31	0.1044E−01
1.32	0.9342E−01	1.82	0.3438E−01	2.32	0.1017E−01
1.33	0.9176E−01	1.83	0.3363E−01	2.33	0.9903E−02
1.34	0.9013E−01	1.84	0.3289E−01	2.34	0.9642E−02
1.35	0.8851E−01	1.85	0.3216E−01	2.35	0.9387E−02

(*continued*)

Table 12A1 $N(0, 1)$ Distribution Defined as $\Phi(-\beta) = 1 - \Phi(\beta)$ (*continued*)

β	$\Phi(-\beta)$	β	$\Phi(-\beta)$	β	$\Phi(-\beta)$
1.36	0.8692E−01	1.86	0.3144E−01	2.36	0.9138E−02
1.37	0.8535E−01	1.87	0.3074E−01	2.37	0.8894E−02
1.38	0.8380E−01	1.88	0.3005E−01	2.38	0.8657E−02
1.39	0.8227E−01	1.89	0.2938E−01	2.39	0.8424E−02
1.40	0.8076E−01	1.90	0.2872E−01	2.40	0.8198E−02
1.41	0.7927E−01	1.91	0.2807E−01	2.41	0.7976E−02
1.42	0.7781E−01	1.92	0.2743E−01	2.42	0.7760E−02
1.43	0.7636E−01	1.93	0.2680E−01	2.43	0.7550E−02
1.44	0.7494E−01	1.94	0.2619E−01	2.44	0.7344E−02
1.45	0.7353E−01	1.95	0.2559E−01	2.45	0.7143E−02
1.46	0.7215E−01	1.96	0.2500E−01	2.46	0.6947E−02
1.47	0.7078E−01	1.97	0.2442E−01	2.47	0.6756E−02
1.48	0.6944E−01	1.98	0.2385E−01	2.48	0.6569E−02
1.49	0.6811E−01	1.99	0.2330E−01	2.49	0.6387E−02
1.50	0.6681E−01	2.00	0.2275E−01	2.50	0.6210E−02
1.51	0.6552E−01	2.01	0.2222E−01	2.51	0.6037E−02
1.52	0.6426E−01	2.02	0.2169E−01	2.52	0.5868E−02
1.53	0.6301E−01	2.03	0.2118E−01	2.53	0.5703E−02
1.54	0.6178E−01	2.04	0.2068E−01	2.54	0.5543E−02
1.55	0.6057E−01	2.05	0.2018E−01	2.55	0.5386E−02
1.56	0.5938E−01	2.06	0.1970E−01	2.56	0.5234E−02
1.57	0.5821E−01	2.07	0.1923E−01	2.57	0.5085E−02
1.58	0.5706E−01	2.08	0.1876E−01	2.58	0.4940E−02
1.59	0.5592E−01	2.09	0.1831E−01	2.59	0.4799E−02
1.60	0.5480E−01	2.10	0.1786E−01	2.60	0.4661E−02
1.61	0.5370E−01	2.11	0.1743E−01	2.61	0.4527E−02
1.62	0.5262E−01	2.12	0.1700E−01	2.62	0.4397E−02
1.63	0.5155E−01	2.13	0.1659E−01	2.63	0.4269E−02
1.64	0.5050E−01	2.14	0.1618E−01	2.64	0.4145E−02
1.65	0.4947E−01	2.15	0.1578E−01	2.65	0.4025E−02
1.66	0.4846E−01	2.16	0.1539E−01	2.66	0.3907E−02
1.67	0.4746E−01	2.17	0.1500E−01	2.67	0.3793E−02
1.68	0.4648E−01	2.18	0.1463E−01	2.68	0.3681E−02
1.69	0.4552E−01	2.19	0.1426E−01	2.69	0.3573E−02
2.70	0.3467E−02	3.20	0.6871E−03	3.70	0.1077E−03
2.71	0.3364E−02	3.21	0.6636E−03	3.71	0.1036E−03
2.72	0.3264E−02	3.22	0.6409E−03	3.72	0.9956E−04
2.73	0.3167E−02	3.23	0.6189E−03	3.73	0.9569E−04
2.74	0.3072E−02	3.24	0.5976E−03	3.74	0.9196E−04

Table 12A1 $N(0, 1)$ Distribution Defined as $\Phi(-\beta) = 1 - \Phi(\beta)$ (*continued*)

β	$\Phi(-\beta)$	β	$\Phi(-\beta)$	β	$\Phi(-\beta)$
2.75	0.2980E−02	3.25	0.5770E−03	3.75	0.8837E−04
2.76	0.2890E−02	3.26	0.5570E−03	3.76	0.8491E−04
2.77	0.2803E−02	3.27	0.5377E−03	3.77	0.8157E−04
2.78	0.2718E−02	3.28	0.5190E−03	3.78	0.7836E−04
2.79	0.2635E−02	3.29	0.5009E−03	3.79	0.7527E−04
2.80	0.2555E−02	3.30	0.4834E−03	3.80	0.7230E−04
2.81	0.2477E−02	3.31	0.4664E−03	3.81	0.6943E−04
2.82	0.2401E−02	3.32	0.4500E−03	3.82	0.6667E−04
2.83	0.2327E−02	3.33	0.4342E−03	3.83	0.6402E−04
2.84	0.2256E−02	3.34	0.4189E−03	3.84	0.6147E−04
2.85	0.2186E−02	3.35	0.4040E−03	3.85	0.5901E−04
2.86	0.2118E−02	3.36	0.3897E−03	3.86	0.5664E−04
2.87	0.2052E−02	3.37	0.3758E−03	3.87	0.5437E−04
2.88	0.1988E−02	3.38	0.3624E−03	3.88	0.5218E−04
2.89	0.1926E−02	3.39	0.3494E−03	3.89	0.5007E−04
2.90	0.1866E−02	3.40	0.3369E−03	3.90	0.4804E−04
2.91	0.1807E−02	3.41	0.3248E−03	3.91	0.4610E−04
2.92	0.1750E−02	3.42	0.3131E−03	3.92	0.4422E−04
2.93	0.1695E−02	3.43	0.3017E−03	3.93	0.4242E−04
2.94	0.1641E−02	3.44	0.2908E−03	3.94	0.4069E−04
2.95	0.1589E−02	3.45	0.2802E−03	3.95	0.3902E−04
2.96	0.1538E−02	3.46	0.2700E−03	3.96	0.3742E−04
2.97	0.1489E−02	3.47	0.2602E−03	3.97	0.3588E−04
2.98	0.1441E−02	3.48	0.2507E−03	3.98	0.3441E−04
2.99	0.1395E−02	3.49	0.2415E−03	3.99	0.3298E−04
3.00	0.1350E−02	3.50	0.2326E−03	4.00	0.3162E−04
3.01	0.1306E−02	3.51	0.2240E−03	4.05	0.2557E−04
3.02	0.1264E−02	3.52	0.2157E−03	4.10	0.2062E−04
3.03	0.1223E−02	3.53	0.2077E−03	4.15	0.1659E−04
3.04	0.1183E−02	3.54	0.2000E−03	4.20	0.1332E−04
3.05	0.1144E−02	3.55	0.1926E−03	4.25	0.1067E−04
3.06	0.1107E−02	3.56	0.1854E−03	4.30	0.8524E−05
3.07	0.1070E−02	3.57	0.1784E−03	4.35	0.6794E−05
3.08	0.1035E−02	3.58	0.1717E−03	4.40	0.5402E−05
3.09	0.1001E−02	3.59	0.1653E−03	4.45	0.4285E−05
3.10	0.9676E−03	3.60	0.1591E−03	4.50	0.3391E−05
3.11	0.9354E−03	3.61	0.1531E−03	4.55	0.2677E−05

(*continued*)

Table 12A1 $N(0, 1)$ Distribution Defined as $\Phi(-\beta) = 1 - \Phi(\beta)$ *(continued)*

β	$\Phi(-\beta)$	β	$\Phi(-\beta)$	β	$\Phi(-\beta)$
3.12	0.9042E−03	3.62	0.1473E−03	4.60	0.2108E−05
3.13	0.8740E−03	3.63	0.1417E−03	4.65	0.1656E−05
3.14	0.8447E−03	3.64	0.1363E−03	4.70	0.1298E−05
3.15	0.8163E−03	3.65	0.1311E−03	4.75	0.1015E−05
3.16	0.7888E−03	3.66	0.1261E−03	4.80	0.7914E−06
3.17	0.7622E−03	3.67	0.1212E−03	4.85	0.6158E−06
3.18	0.7363E−03	3.68	0.1166E−03	4.90	0.4780E−06
3.19	0.7113E−03	3.69	0.1121E−03	4.95	0.3701E−06
5.00	0.2859E−06	6.00	0.9716E−09	8.00	0.6056E−15
5.05	0.2203E−06	6.10	0.5220E−09	8.10	0.2673E−15
5.10	0.1694E−06	6.20	0.2778E−09	8.20	0.1169E−15
5.15	0.1299E−06	6.30	0.1463E−09	8.30	0.5058E−16
5.20	0.9935E−07	6.40	0.7636E−10	8.40	0.2167E−16
5.25	0.7582E−07	6.50	0.3945E−10	8.50	0.9197E−17
5.30	0.5772E−07	6.60	0.2018E−10	8.60	0.3864E−17
5.35	0.4384E−07	6.70	0.1023E−10	8.70	0.1608E−17
5.40	0.3321E−07	6.80	0.5130E−11	8.80	0.6623E−18
5.45	0.2510E−07	6.90	0.2549E−11	8.90	0.2701E−18
5.50	0.1892E−07	7.00	0.1254E−11	9.00	0.1091E−18
5.55	0.1423E−07	7.10	0.6107E−12	9.10	0.4363E−19
5.60	0.1067E−07	7.20	0.2946E−12	9.20	0.1728E−19
5.65	0.7985E−08	7.30	0.1407E−12	9.30	0.6773E−20
5.70	0.5959E−08	7.40	0.6654E−13	9.40	0.2629E−20
5.75	0.4436E−08	7.50	0.3116E−13	9.50	0.1011E−20
5.80	0.3293E−08	7.60	0.1445E−13	9.60	0.3847E−21
5.85	0.2438E−08	7.70	0.6636E−14	9.70	0.1450E−21
5.90	0.1800E−08	7.80	0.3017E−14	9.80	0.5408E−22
5.95	0.1325E−08	7.90	0.1359E−14	9.90	0.1998E−22

CHAPTER 13

Deterioration Modelling

CONTENTS

13.1 INTRODUCTION

The deterioration of a bridge is a complex phenomenon which occurs during the working life of the bridge. The degradation process may be slow and unseen, but nevertheless it continues inexorably onwards, and if not checked, the materials of the bridge will erode to such an extent that failure ensues (witness the Ynys-y-Gwas Bridge, section 2.3.1.2).

If a rotten apple is placed into a bowl of fresh apples, it will not be long before all of the apples are affected. Unfortunately the reverse is not true. A fresh apple placed into a bowl of rotten apples will not instill its freshness into the others – it too will start to rot. Barring intervention, the end result in both cases is a *bowl of rotten apples*.

13.1.1 Performance profiles

It is probably impossible to be sure that all of the factors that dictate the final cost of a bridge will be foreseen and consequently the final cost will not be known with a high degree of accuracy. Awareness of the fact that bridges cannot last forever, and that problems with bridge deterioration are now better understood, means at least that reasoned attempts can be made to assess the maintenance costs and hence the final cost of a bridge. Great emphasis is rightly placed on ensuring that a bridge is strong enough to carry the loads imposed upon it, but this should not be to the detriment of the *serviceability* of the bridge. It may never reach its ultimate limit state, but it surely will become unserviceable unless proper precautions are taken and every item of the bridge is considered.

To this end the idea of *performance profiles* has been mooted as a possible way ahead. This recognises the fact that some elements of the bridge are *replaceable* whilst others are *permanent*. Both types can be classified with a performance profile which is either *strong* or *weak*. This will undoubtedly affect the conceptual stage of the bridge design process and result in close attention to detail and the elimination as far as possible of all elements that are classified as *weak*. Alternatively, *weak* elements can remain but be made better, for example bearings can be made stronger and more durable and secured in a more robust manner.

Jones (1995) has suggested that *weak* items are expansion joints, metallic bearings, parapet anchorages and reinforced concrete in the presence of salts (deck drainage and half joints could also be added), whilst *strong* items are pre-tensioned concrete beams, polymeric reinforcement, reinforced soil and masonry. A sensible way ahead may be to 'design out' where possible the weak items and build only continuous or integral bridges with materials having good performance profiles.

13.1.2 Deterioration models

Germane to the performance of an item is the manner and rate at which it deteriorates. And this is related to its substance (what it is made of) and what environment it is in.

A *global* model of what is happening is virtually impossible. There are just too many variables to consider, especially in the case of reinforced or pre-stressed concrete. With our current state of knowledge, however, what we can do is to examine the deterioration mechanisms of structural elements when subject to a hostile environment, with a view to finding a suitable *local* model which apes on paper what is happening on the bridge. If this is achievable it could enable *predictions* to be made about the condition of a bridge in the future, and, if necessary preventative action can be taken at the design stage to ensure that deterioration at best does not happen at all, or if it is allowed, then it must be to a certain specified degree.

A possible model for an imperfect *bridge* in a hostile environment is shown in Figure 13.1 where the falling curve from A to B represents, in a general manner, the deterioration of a material, a bridge element, a single bridge or indeed a stock of bridges. If we knew exactly the deteriorating mechanism, then every point on this curve would represent the condition at every point in time. A perfect bridge (i.e. 100% performance) would be represented as a horizontal line.

For a bridge which has been in service for a number of years, and for which deteriorating data have been collected on a regular basis and stored in the BMS database, then the deterioration curve can be plotted accurately up to the present. Comparing the theoretical condition at this point C_1 (i.e. NOW) with the actual measured condition B1 will indicate the difference between

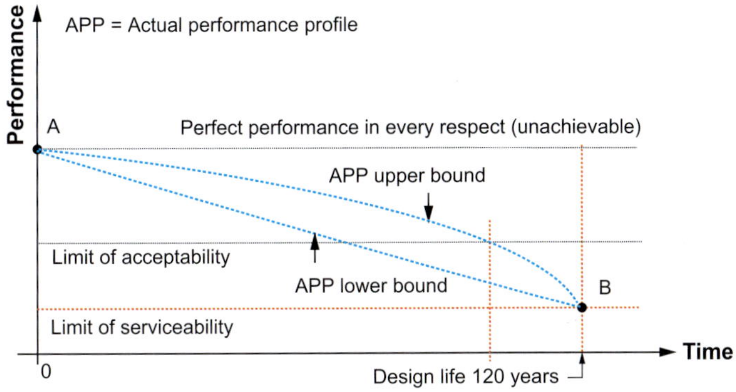

FIGURE 13.1
Deterioration principle.

the theoretical condition and the actual condition, Δe. In order to arrest (or at least lessen) the deterioration, an *intervention* is necessary which is the moment when a decision is taken to change the bridge in some way because it is not meeting the minimum requirements of serviceability and safety (Das & Mimic, 1999).

In the example in Figure 13.2 let us suppose that the load carrying (or strength) of a major RC beam has deteriorated due to corrosion of the steel reinforcement from C_1 to C_2. After considering the options (load restriction notice, do nothing but continue to monitor or strengthen) it is decided to strengthen the bridge back up to its original design value. Thereafter the condition is monitored and if deterioration again reaches an unacceptable level then a repair/strengthening programme is implemented and carried out to bring the bridge back up to standard (that is as close to its original condition as possible). Deterioration would then continue as before and at a certain level (or after a certain time), a second repair/strengthening might be

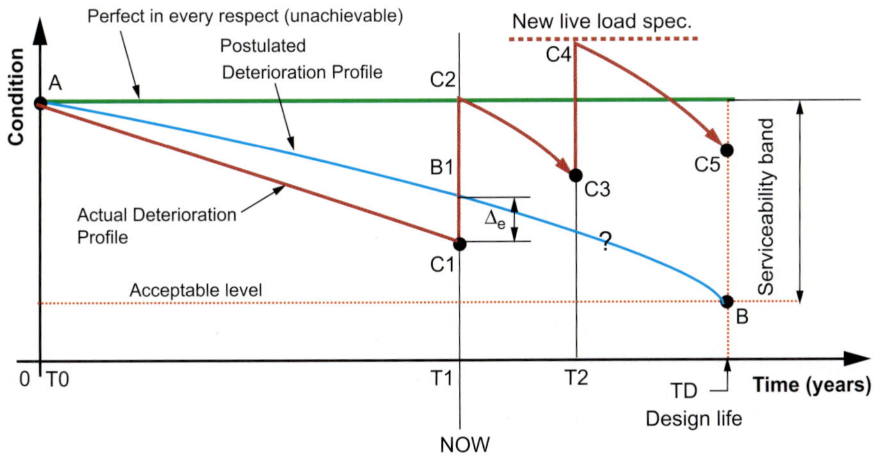

FIGURE 13.2
Moment of intervention.

implemented, but this time it may be necessary to actually improve (that is strengthen) the bridge, for example to carry an increased live load specification or replace a broken tendon.

From the foregoing it is clear that the overall theoretical performance curve *A–B* has little or no value at all, either at the initial design stage or thereafter simply because the bridge has changed due to previous works. It may be useful conceptually but has no practical value whatsoever. The points C1–C6 on the saw-toothed curve indicate the on-going condition before and after intervention. The serviceability band in Figure 13.2 is a measure of the quality of the bridge: a narrow serviceability band implies that the bridge is performing well (with possibly a higher initial cost), and a wider one implies that the bridge is performing badly (with possibly a lower initial cost). The former is difficult to justify unless predicted maintenance costs are high or the discount rate is low, probably of the order of 5% or less. Figure 13.1A in the Appendix shows the repair history of a bridge which has performed well, in that the deviation from the original design carrying capacity W_D is narrow. The initial strength was W_o, and so the bridge began life with some extra capacity. Deterioration commenced on day 1, and after two successive years had practically been arrested. Then in year 5 a legislated increase in live load meant that the bridge had to be strengthened up to W_5. From there, deterioration has again set in, and further decisions about its future are needed. Overall the bridge can be said to be performing well.

In conclusion, the narrower the bandwidth between the strongest and the weakest levels, the higher is the performance.

13.2 LOCAL MODELS

13.2.1 Reinforced concrete

Notwithstanding the huge number of masonry arch bridges in service, the most common building material for bridges and structures is, undoubtedly, *reinforced concrete*. It is used to form structural elements such as beams, columns, slabs (both ground and floor), beam and slab, composite beam and slab, foundations and bridge parapets etc. These can be put together in a recognised geometrical manner to form a bridge. The (basic) material constituents are cement, sand, coarse and fine aggregate, water and steel reinforcing bars.

Reinforced concrete is attacked primarily by *carbonation and chloride* contamination (see sections 4.2.6 and 10.3.8). According to Browne (1986) the attack is in two stages:

- *Stage 1:* when chloride contamination or carbonation has reached the reinforcement.

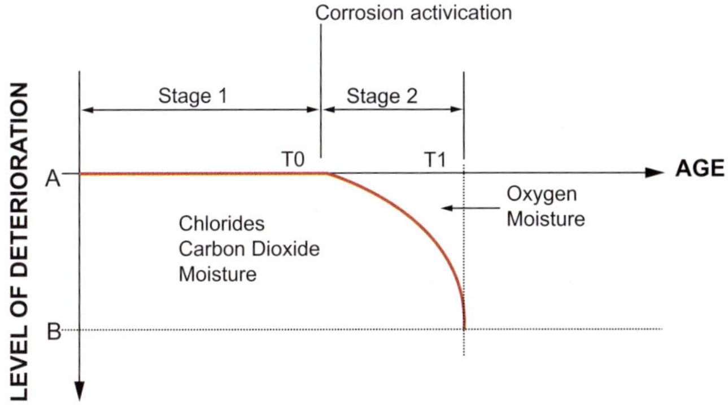

FIGURE 13.3
Deterioration stages (after Browne, 1986).

- *Stage 2:* when corrosion due to oxidisation of the steel has expanded the corrosion products from the steel sufficiently to cause *cracking, spalling* and *delamination* (the corrosion products occupy about five times that of uncorroded steel and tend to cause cracking, spalling and delamination).

These stages are shown in Figure 13.3. For corrosion damage to occur, carbonation or chlorides + water + oxygen + low resistivity concrete would have to be present in sufficient quantity and depth.

According to Browne (1986) the penetration rate for carbonation in the UK can be determined from the simple diffusion law:

$$x = k\sqrt{t} \tag{13.1}$$

Where x is the distance penetrated after time t. (The diffusion constant k is obtained either from experimental work or from a calibration procedure involving a regression analysis of past projects.) According to (Watkins et al., 1993), Eq. (13.1) is based on empirical data only, and the more general form of the equation to take into account of the fractal growth of the calcium carbonate within the pores would be:

$$x = kt^i \tag{13.2}$$

Where i is a fractional parameter between 0.5 and 1.0. Both k and i are obtained from a calibration procedure based on the actual site results.

13.2.2 Research study

The following abbreviated case study, although not for a set of bridges, indicates the procedures to be taken to determine the carbonation depth of reinforced concrete.

In 1993 (Watkins et al., 1993) undertook the survey of 828 multi-storey housing blocks in Hong Kong, varying in age from 5 to 30 years with a view to developing a durability model for the carbonation of reinforced concrete structures. More than 30,000 cores were taken from the buildings for testing to discover:

- depth of carbonation
- saturated surface dry density
- compressive strength
- aggregate/cement ratio

A regression analysis is performed on a scatter diagram of carbonation depth with age for three ranges of compressive strengths, and the results are shown in Table 13.1.

The regression lines for each of the three band widths for strength are shown in Figure 13.4 which clearly show that carbonation depth increases with age and decreases with concrete strength. It was found that the influence of the aggregate/cement ratios was negligible.

As an example of its use, for a 30-MPa strength concrete, and 20 years of age, then from Eq. (13.2):

$$\text{Depth of carbonation} = 4.28 \times 20^{0.592}$$

$$x = 25.2 \text{ mm}$$

From Figure 13.4, $x = 26$ mm

Figure 13.4 represents the distillation of 828 public housing *sites*. If the strength of the concrete being investigated is known, then the depth of the carbonation front for a particular grade of concrete can be found. If this is compared with the depth (cover) to the outermost reinforcing bars, then the likelihood of corrosion initiation can be established. In practice, if the data had *not* been available, then the information required would be obtained from the usual non-destructive test results from a single (or may be a few) core samples.

Table 13.1 Regression of Carbonation Depth with Age

	Equation $x = kt^i$	
Strength of Concrete (MPa)	k	i
15–24.99	6.43	0.570
25–34.99	4.28	0.592
35–45.99	3.07	0.614

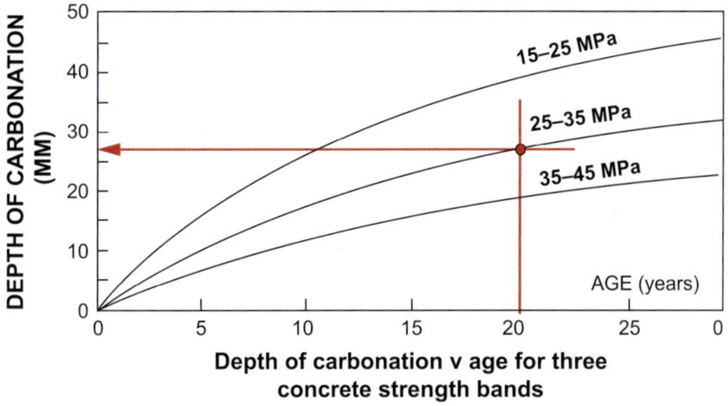

FIGURE 13.4

Variation of penetration depth (x) with time for concrete strength bands.

The massive amount of data that the authors had at their disposal could have been statistically manipulated to examine results for a *single* concrete tower block on each site, or *groups* of tower blocks on each site, or each of the separate sites even. Instead, and no doubt for very good reasons, the research was limited to all of the sites in Hong Kong. The *means* and *standard deviations* of each of the subsets alluded to above, if calculated, would differ from that for the whole *population*, and would be more accurate, which in turn would influence decisions affecting the costs of present and future maintenance requirements.

13.2.3 Research study

More recently Roberts and Atkins (1997) proposed a deterioration process for the reinforced concrete structures in the Midlands Links which consists of five steps as shown in Figure 13.5.

The model they postulate should be able to determine:

- the time at which corrosion starts
- the rate at which corrosion proceeds
- the degree of local corrosion required to cause the cover concrete to delaminate.

This information is then used to predict the loss of section and the modified properties can be determined for a conventional structural analysis.

13.3 CALIBRATION

The most difficult part of the whole process is the *calibration* of the proposed model (Roberts et al. (1998)). There has either to be extant data already

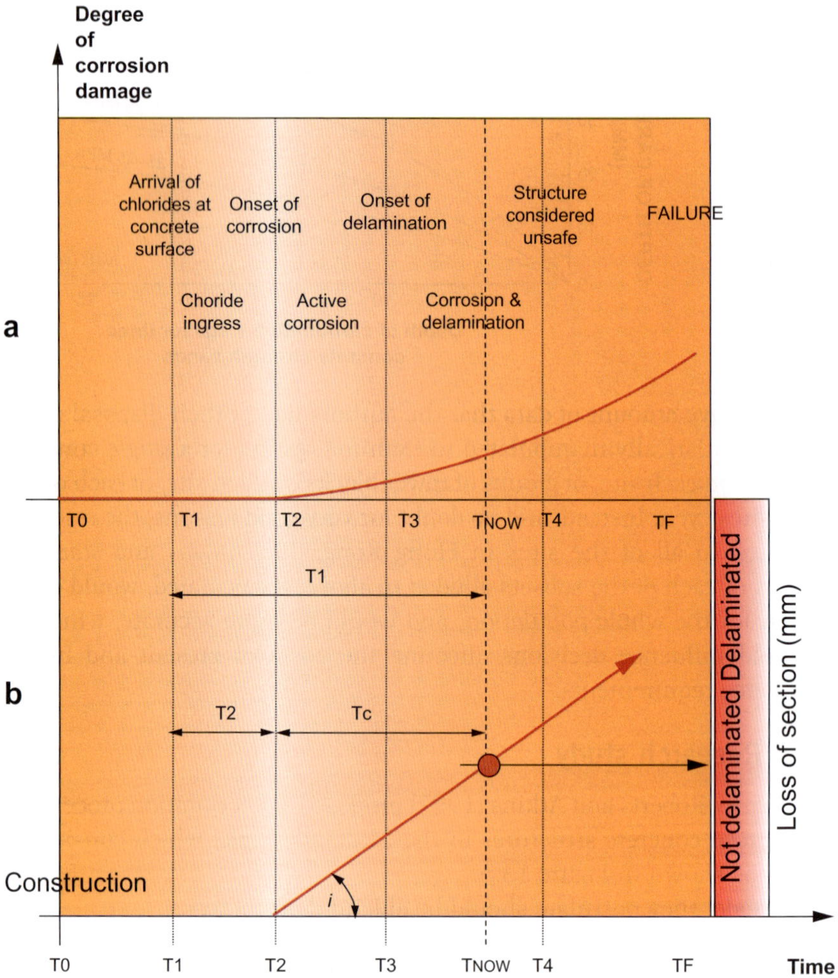

FIGURE 13.5 *Deterioration process for external chloride contamination (top half); and calibration of the deterioration model (bottom half).*

obtained from the bridge being tested or from a similar bridge nearby. Roberts and Atkins (1997) suggest a five-step approach as follows:

Step 1 Obtain chloride profiles (similar to Figure 13.4) to determine the rate of chloride contamination and thus time T1 – the arrival of chlorides at the concrete surface. Such information would only be available if, during past inspections, concrete cores had been taken at strategic locations and tested using a pH indicator such as a ppe for surface carbonation and the curved area sprayed with ppe to determine the depth of carbonation.

Step 2 Carry out a cover survey over the area being inspected to obtain an average cover value, and diffusion constants can be used to determine T2 – the onset of corrosion.

Step 3 Carry out a half-cell potential survey (see section 4.2.1) to locate areas of active corrosion and expose the reinforcement at selected sites to determine the area of steel lost A_{loss} due to corrosion. The rate of corrosion (assumed constant in mm per year) can be calculated from:

$$i = \frac{A_{\text{loss}}}{(T_{\text{now}} - T_2)} \tag{9.6}$$

Step 4 The relationship between corrosion rate and half-cell potential measurements is established and this can be used to determine the likely corrosion rate (i) at any point in the structure from half-cell potentials. Thus the likelihood of steel corrosion can be calculated for any point in time.

Step 5 The areas of corrosion are compared with areas of delamination to determine the value of A_{loss} that causes delamination.

This is a very sophisticated and well thought out methodology and '*for the first time, the risk of continuing deterioration can be quantified,* and this is enabling a rational repair strategy to be developed' (Roberts & Atkins, 1997).

This site-specific model has been further developed into more general corrosion models (one set theoretical and the other empirical) for use on other sites (Roberts, Atkins, Hogg & Middleton, 2000). Although there appears to be wide disparity between the theoretical and empirical models it is acknowledged that the 'theoretical corrosion models have not been validated against real data'. Also in some of the figures presented the authors stress that the diffusion coefficients, surface concentrations (of chlorides) and empirical relations are based on the Midlands Links structures. Nevertheless the paper represents a *modus-operandi* for any one wishing to produce relationships between the time of initiation of corrosion-*vs*-cover, and the area loss ratio of the reinforcement-*vs*-time. This could enable predictions to be made about the future condition of the structure which, hopefully, will help the decision making process on which maintenance strategies to employ and comparison of various whole life costings. In addition there are some very useful definitions in tabular form for establishing the basic parameters such as the corrosion environment, the degree of corrosion, the risk of corrosion and the estimation of the time to arrival of chlorides.

The repair/strengthening work improves the bridge to the level of the original performance profile. Again, if a faster deterioration rate sets in, then a second intervention is required, and so on. In some cases if the bridge is substantially strengthened then it might be possible to accept a condition level higher than the original, thus prolonging the life of the bridge.

This information is then used to predict the loss of section and the modified properties can be determined for a conventional structural analysis.

13.4 STEEL

Modelling the deterioration of a steel bridge involves a three-part process:

- Firstly, the *breakdown of the paint system*, which has no immediate effect on the serviceability limit state.
- Secondly the *corrosion of the steel members.* If they are secondary members (for example diagonal ties or struts in a truss bridge), then if there is a measure of built-in redundancy in the design, an alternative load path is available there should be no cause for alarm. If they are *principal members* such as the tension or compression boom of a truss bridge, then the *strength* of the bridge is threatened.
- Thirdly, *corrosion of riveted, bolted or welded connections*. The likelihood of failure would depend on whether the fastening/gusset is corroding as a whole, or proportionately.

Although this may be true for the more dominant mechanisms, in practice there are a multitude of mechanisms all taking place simultaneously, each one a subset of a larger population. According to van der Toorn and Reij (1990) 'When there are more ageing elements in a structure, the maintenance strategy for the structure as a whole will normally differ from the sum of the individual maintenance strategies'. This is the synergy principle referred to in Chapter 1.

13.4.1 Cables

13.4.1.1 Protective systems

For the last 100 years or so the protective systems for cable-stayed and suspension bridges have remained virtually unchanged and consist of a *four component corrosion system* (see Figure 13.6). The cables themselves consist of small diameter (5 mm) high strength wires which are galvanised and coated in oil and then pasted with a coat of red lead or epoxy resin. The cable is then wrapped with soft, galvanised and annealed steel wires. The wire wrap is then given a final coat of paint.

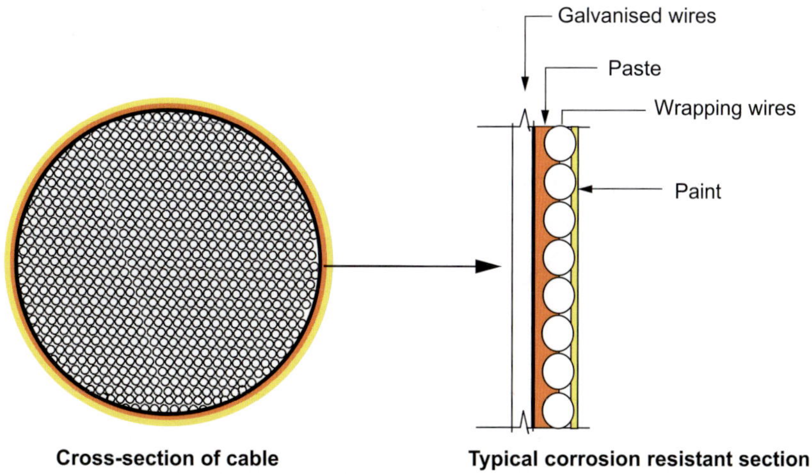

FIGURE 13.6
Protective system for cables (after Nakamura, 2005).

Such a sturdy protection system, however, can still break down and allow water to penetrate into the interstices between the wires. Nakamura and Suzumura (2005) found that in some Japanese bridges, not yet 10 years old, some of the galvanised wires showed signs of corrosion.

13.4.1.2 Deteriorating systems

To provide a deteriorating model, the cause of the breakdown needs to be established, and so a research programme was put in place to study the various *deteriorating mechanisms*.

Water was poured into a model cable (2000 mm long × 650 mm diameter), and was exposed to a natural environment. Temperature and RH readings were taken throughout the day and the results are shown in Figure 13.7.

Tests were carried out to obtain the corrosion performance of the steel wires under the different environments and to compare the corrosion of the steel wires in the field. They were found to be in complete agreement. It was also discovered that the steel wires do not corrode when the relative humidity is below 60%.

Further studies are planned for the future to determine the relationship between the fatigue strength and on-going corrosion of the galvanised steel wires.

13.4.2 Corrosion prevention by dehumidification

To prevent any corrosion at all, a new method of corrosion prevention has been developed by the authors which involves an initial injection of air into the cable over the whole span to remove the water and to keep the

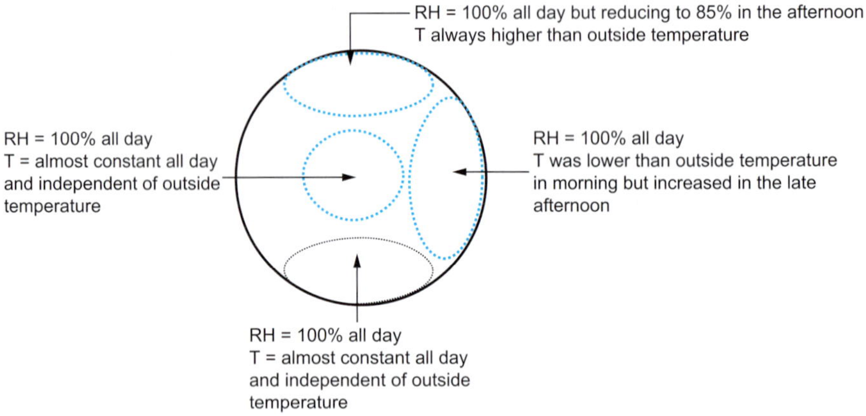

FIGURE 13.7
Estimated environments inside the cable.

environment dry. It has also been modified to remove sodium chloride from the outside air. The device has been installed on the Akashi Kaikyo Bridge in Japan and appears to be working well. The authors stress that regular maintenance and monitoring are important to keep the dehumidification system working well. In effect this amounts to an additional *corrosion resistant component.*

13.4.3 Enhanced wrapping system

It will be appreciated that one weakness of the wrapping layer is that the spirally wound wire is not in total contact with the paste layer (see Figure 13.6). Only a small area of contact exists to prevent entry of water. To rectify this a new 'S-shaped wire wrap' has recently been developed and tested and has been used successfully on a number of suspension bridges in Japan. It is incorporated into a commercially produced product called *'Cableguard'*. Kazinski (2001) describes the wrap thus:

> *'the wrap is based on a cross-linking chlorosulfonated polyethylene polymer which is manufactured into a three-ply (polymer – fabric reinforcement – polymer) laminated construction. A wide variety of fade-resistant colours are available without the application of a painted finish. The specified thickness of the wrap is 1.14 mm (±0.08 mm) v such that after applying the wrap to a bridge cable with a 50% overlap, the total thickness is approximately 2.28 mm'.*

The wrap is applied by means of a 'Skewmaster' wrapping device and can be used on both new construction projects or for the rehabilitation of existing bridges (see Figure 13.8). When applied to existing bridges, it completely encapsulates the cable thereby obviating the need to dispose of hazardous materials.

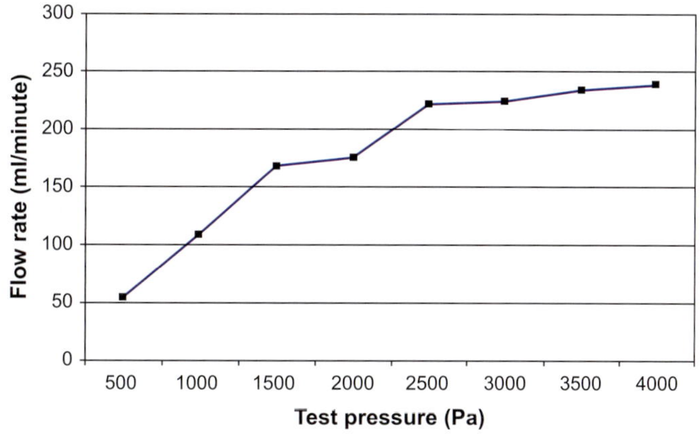

FIGURE 13.8
Cableguard^{TM} elastomeric wrap pressurisation test results. (courtesy of Concrete Engineering).

13.5 TIMBER

In steel and concrete bridges the dominating deteriorating process is *corrosion*. For timber it is *fungal decay and wood-boring insects* which dominate.

13.5.1 Insects

The main aim of any timber treatment is to break the egg-laying cycle of the wood-boring beetle, namely: *'egg – damage – beetle'*. The species of beetle doing the damage depends greatly on:

- The *type* and condition of the timber playing host to the beetle.
- The *density* of the wood – heartwood is less susceptible than sapwood.
- The *degree* of existing decay: in decayed wood, the cycle development may be faster and repeated more often. Some types of woodborer will only attack timber which has started to decay.

Resistance to attack is greatly enhanced by the application of a *preservative* or an *insecticide* or a *mixture* of both, which are capable of *preventing* infestation and *eradicat*ing existing infestation. There are many commercial products on the market and so it is advisable to enlist the services of a specialist so that the most effective product is chosen. According to ProBor (2007) *'modern preservatives are based on a soluble borate and exhibit superb penetration into damp timber'*.

Application is by *brush, spray or pressure injection*, and is best carried out in the Spring when the insects come to the surface to mate. Thereafter treatment should continue annually until attack is terminated.

13.5.2 Fungal decay

Timber structures can suffer severe structural damage at the hands of *wood rotting fungi*. The spawn (of which the fungus plant is composed) lives in the wood and feeds on it by releasing enzymes which enable digestion of the wood. The fungus absorbs the digested material to produce more spawn and continues until there is no more digestible material in the wood. It then dies.

The spawn needs a minimum moisture content of 20% and a low relative humidity in which to germinate and thus initiate decay. Generally, if there is no water present then there will be no decay.

Brown Rot is the result of the fungi feeding on the cellulose wall. It is characterised by wood that is browner than normal and which cracks across the grain. The wood becomes 'spongy', lighter in weight, and is severely weakened. In areas where brown rot has ceased and become dry, it is often erroneously referred to as 'dry rot'.

White Rot feeds on the wood directly, causing the wood to lose its natural colour and to appear whiter than normal. The wood gradually loses its strength and, as in brown rot, it becomes 'spongy' to the touch.

Timber bridges are, for the most part, in a totally exposed environment and subject to *mist, fog, rain and ground water.* The vulnerable areas are at *connections, bearings and expansion joints* which tend to work lose after a period of time, due to vibration from traffic, leaving a gap (be it ever so small), between the timber and steel or concrete. Wind-blown fungal spores can then find their way into the gaps and decay begins.

Treatment of infected areas using preservatives and insecticides is similar to beetle infestation, but in addition, the exposed wood surfaces should be coated with *a fungicide and a water repellant finish*.

13.6 DETERIORATION MODELS

It is not possible to produce a performance curve or a deterioration model as described previously, and so what follows is not so much a model as a 'common-sense procedure'.

Once beetle and/or fungal attack has been identified, then it is necessary to record the location and extent of the attack by using one or all of the tests described in section 2.3.1.5. A judgement then has to be made on whether to intervene or 'do-nothing', if intervention is considered necessary, then a suitable repair is undertaken. From then on a continuous monitoring programme is put in place until the next decision is required. If, of course the damage is so severe that the strength of the bridge is put in jeopardy, then a strengthening programme is put in place. Because timber is not a ductile material the risk of a 'do-nothing' strategy could well result in sudden and catastrophic failure, and should not be considered.

13.7 CONCLUSIONS

Do we need deterioration models? The 'yes' camp argue that since it looks to the future, then it is helpful in predicting what actions *may* need to be taken and thus determine with a certain degree of confidence what costs might be incurred and the level of cost-benefit that might be achieved.

The 'no' camp on the other hand say that the amount of historical information available to provide such models just does not exist and therefore such models cannot be constructed.

They concede, however, that for a new-build, since there would be a large clutch of reliable raw data, then in certain cases it would be possible to begin, at least, such a model.

At the end of the day, probably the surest way of keeping abreast of the *rot* taking place is to *monitor*. This, in conjunction with visual inspections and site measurements, is the most logical alternative to deterioration models. It does not involve guesswork or looking into a crystal ball – it relies on hard data. The information gained can be plotted and studied to ascertain a trend, which will allow the engineer to decide on the moment of intervention.

REFERENCES

Browne, R. D. (1986). Practical considerations in producing durable concrete, Seminar on the Improvement of Concrete Durability. *ICE* 97–116.

Das, P. C., & Micic, T. V. (1999). *Maintaining Highway Structures for Safety, Economy and Sustainability, Ninth Conference on Structural Faults and Repair.* Edinburgh: Engineering Technics Press.

Jones, C. J. F. P. (1995). Performance Profiles and the concept of Zone Defence, One Day Seminar on Whole Life Costing – Concrete Bridges. *Concrete Bridge Development Group* 62–76.

Kazinski, M. (2001). Corrosion Protection of Cables. *Concrete Engineering, Vol. 5*(No. 3), 38–40, Autumn 2001.

Nakamura, S., & Suzumura K. (2005). Corrosion Assessment of Bridge Cables, Bridge Management Five, Fifth International Conference on Bridge Management, University of Surrey, Guildford, UK, pp. 28–36. London: Thomas Telford.

Roberts, M., & Atkins, C. (1997). Deterioration Modelling. *Concrete Engineering International, Vol. 1*(No. 1), 38–40.

Roberts, M. B., Atkins, C., Hogg, V., & Middleton, C. (2000). *A proposed empirical corrosion model for reinforced concrete. Proc. Instn Civ. Engrs Structs & Bldgs, Vol. 140*(No. 1), 1–11.

Van de Toorn, A., & Reij, A. W. F. (1990). A Systematic Approach to Future Maintenance, 1st International Conference on Bridge Management, University of Surrey, Guildford, UK, pp. 215–222. London: Elsevier.

FURTHER READING

Das, P. C. (1997). Whole life performance-based assessment of highway structures: proposed procedure. In P. C. Das (Ed.), *Symposium papers on The Safety of Bridges*. London: Thomas Telford.

De Brito, J., & Branco, F. A. (1996). Whole Life Costing in road bridges applied to service life prediction, 3rd International Conference on Bridge Management, Inspection, Maintenance, Assessment and Repair. University of Surrey, Guildford, UK, pp. 603–612. E and F N Spon.

Departmental Advice Note BA28. (1992). *Evaluation of Maintenance Costs in Comparing Alternative Designs for Highway Structures*. London: SO.

Departmental Standard BD36. (1992). *Evaluation of Maintenance Costs in Comparing Alternative Designs for Highway Structures*. London: SO.

Departmental Standard BA81. (2000). *Whole life performance based assessment of Highway Structures and Structural Components, Design Manual for Roads and Bridges*. London: HMSO.

Holland, D. A., & Dawe, P. H. (1990). Bridge Rehabilitation: Department of Transport's Fifteen-Year Strategy, 1st International Conference on Bridge Management, University of Surrey, Guildford, UK, pp. 145–153. London: Elsevier.

Jones, D. R. B. (1995). The Appraisal Of Traffic Delay Costs, One Day Seminar on Whole Life Costing – Concrete Bridges. *Concrete Bridge Development Group* 27–36.

Jones, M. (1997). At last, an end to the never-ending job, NCE, 11/25 December, p. 4.

Quadro 3. (1996). DMRB Vol. 14, Economic Appraisal of Road Maintenance (plus amendments May 1997 & Nov. 1997), Department of Transport (Now Highways Agency), UK.

Wood, J. M. G., & Johnson, R. A. (1990). Reliability Analysis Applied to Deteriorating Bridge Structures, 1st International Conference on Bridge Management, Inspection, Maintenance, Assessment and Repair. University of Surrey, Guildford, UK, pp. 501–513. London: Elsevier.

APPENDIX 13A

Theory of diffusion

Diffusion refers to the random molecular motion by which matter is transported. In reinforced concrete it refers to the drift of the diffusing substance (carbon dioxide molecules) from a place of higher concentration (air/concrete boundary) to a place of lower concentration within the concrete.

It is illustrated by Ficks Law of Diffusion written in its original form thus:

$$\frac{Q}{t} = \frac{DA(C_1 - C_2)}{h}$$

where Q/t = quantity of substance diffusing at any one time; D is the diffusing coefficient; a constant depending on the nature of the substance diffusing; on the material it is in, and on the temperature; A = area of the surface; $(C_1 - C_2)$ the concentration difference across the material; t = the thickness of the substance across which diffusion is occurring.

Nowadays it is written as:

$$j = D[c_1 - c_2]/1$$

Where 1 is the length of the system (diffusion length); and c_1 and c_2 are the concentrations at each end. This equation is used to determine D from laboratory tests.

CHAPTER 14

Heavy Goods Vehicles (HGVs)

CONTENTS

14.1 INTRODUCTION

The largest vehicles within the caravan of *normal* traffic which cross our bridges are the heavy goods vehicles (HGVs). They are one of the prime movers of a countries raw materials and merchandise, and travel along routes set aside for primary distribution, either alone or in an intermodal configuration (Lowe, 2005). The three most common are shown in Figure 14.1.

The impact they have as they enter the bridge environment is enormous. They pose the greatest single threat to the integrity of our bridges. Not only is their gross weight increasing, but they are able to move faster, and in greater numbers than ever before. No part of any bridge is exempt from the daily dynamic shock waves which reverberate through the superstructure to the foundations, as the wheel axles pass from the roadway to the bridge itself.

They cause damage to the running surface, expansion joints, bearings, concrete, kerbs, steel plates, welds, bolts, hangers, anchors etc. The most vulnerable, directly exposed visible item is the *expansion joint* where the wheels are in direct contact with the joint itself (see Figures 2.12 and 2.13). The less visible damage occurs both beneath and above the road surface in areas of high compression and tension, bending, torsion and shear, causing crushing and cracking in concrete, buckling and twisting of steel plates, weld and bolt failure etc.!

For a given fixed maximum weight, the damage referred to accrues by virtue of the movement of the vehicle. If it is crawling (as in a traffic jam), then the axial loads are to all intents and purposes static. If it is travelling along at a normal speed of say 100 km/h (60 mph) supported on a stiff vehicle suspension, and the road roughness is significant, then the vehicle is subject

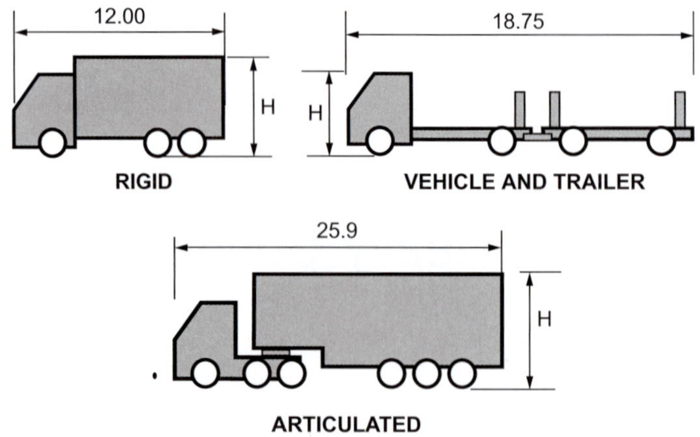

FIGURE 14.1
Most common truck/lorry types.

to a random bouncing and rocking movement which magnifies the axle loads (Ryall, Parke & Harding, 2002).

14.2 DYNAMIC AMPLIFICATION

14.2.1 Divine theory

Maximum magnification of load will be at the precise moment when the HGV enters the bridge domain, as though a huge hammer comes crashing down on the end of the bridge equivalent to a dynamic amplification factor (DAF) of 2.0. Thereafter, it increases to a maximum value and then settles down to a relatively steady-state value. To determine the value of the DAF with any degree of confidence is a complex one, involving, as it does, a number of variables such as road roughness, bridge type and traffic mass and speed... A very interesting theoretical study was carried out by (O'Brien, Gonzalez, Li & Brady, 2005) under the research umbrella of the DIVINE Programme, OECD (1997) with the aim of determining the effect of *speed* on the dynamic amplification of first one, and then two loads crossing a simply supported bridge at the point of resonance. The bridge vibrations are set off when the first load enters the bridge domain, and are further exacerbated with the entry of the second load. The results of the study were very satisfying and have prompted further research with the hope of obtaining more accurate values of the DAF.

14.2.2 Real bridge vibrations

This effect is also borne out by Heinemann Lutzenberger and Baumgartner (2000) who 'measured bridge responses due to an instrumented truck and

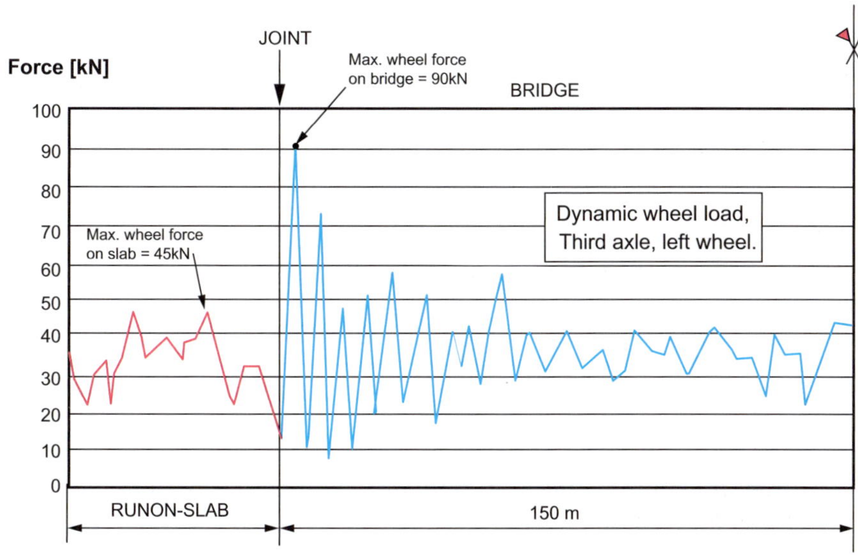

FIGURE 14.2 *Plot of dynamic wheel load (after Lutzenberger & Baumgartner, 2000).*

free traffic' and found that dynamic axle loads at the beginning of the bridge were very localised and resulted in a very high jump compared with the remainder of the bridge.

This sudden increase in the wheel load as it crosses the 'jump' at the bridge joint from the land to the bridge deck can be clearly seen in Figure 14.2. This lower [static-plus-dynamic] force of 45 kN occurs on the land side, and is exactly double at 90 kN on the bridge side.

14.2.3 Drop weight experiment

At this point it is instructive to consider the classical Drop Weight problem of a falling object with a weight W kN about to impact with a structure below it (where the *drop* is akin to a *jump*).

Using the nomenclature of Figure 14.3 and applying Newton's second law of dynamics, the equations of motion are:

$$F = ma = Wa/g$$
$$= W(\sin ce\ g = a) \quad \text{when } [y \leq h] \qquad (14.1)$$

$$F = k(y - h) + W \quad \text{when } [y > h] \qquad (14.2)$$

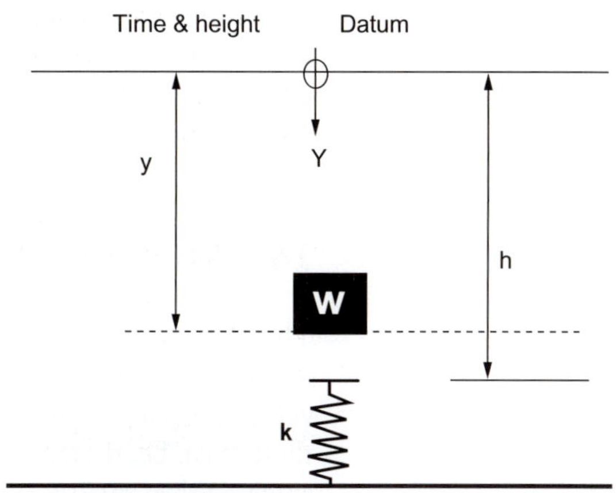

FIGURE 14.3 *Drop weight experiment.*

These equations can be solved to give the maximum deflection (δ) and the maximum force in the spring (F_s), *viz*:

$$d = W/k + (W/k)[1 + (2kh/W)]^{1/2} \tag{14.3}$$

$$F_s = kd = W + W[1 + (2hk/W)]^{1/2} \tag{14.4}$$

Assume that the spring is perfectly rigid, then k = infinity.

Substituting for k, d w and h, then from Eq. (14.4):

$$F_s = W + W[1 + 0] = 2W \tag{14.5}$$

This goes a long way to explaining why shear effects near supports are so critical. For instance, in the same instant of time, the sudden arrival of a 10-kN point load at a simple support is magnified by the order of 2.0 whilst the maximum effect at mid-span is practically zero. For the complete to Eqs (14.3) and (14.4) see reference Marghitu Dan (2001).

14.3 DAMAGE CONTROL

All lorries operating under the auspices of the Vehicle Operator and Services Agency (VOSA) in the UK are plated with the allowed laden weight. Any deviation from this is classed as an absolute offence; that is, one where there is no appeal. In view of the damage amplification effect, vehicle speed and gross weight, they need to be controlled. Government legislation is the means by which control is exercised on both effects.

Trade Standards Officers acting on behalf of the government are ruthless regarding over-loading and are ever ready to commence committal proceedings at the drop of a hat. Punishment is by way of fines which can be quite punitive, and may result in the loss of a job for a driver, and a tarnished reputation for the operator.

14.4 WEIGHT WATCHERS

The first ever government intervention on lorry loads in the UK was the introduction of the Heavy Motor Car Order in 1904 which limited the *gross vehicle weight* (gvw) of a two-axle rigid lorry to 12 tonnes. Since 1904 there have been eight increases (Ryall et al., 2002) the latest being in force since 1 February 2006, and allowing five axle (3 + 2) articulated lorries a gvw of 44 tonnes, and a weight limit of 11.5 tonnes on the drive axle.

In line with EC Directive 96/53/EC information about all the current weight limits for goods vehicles and trailers is enshrined in The Road Vehicles (Construction and Use) Regulations 1986 (C & Use) Regulations 1986 as amended, and The Road Vehicles (Authorised Weight) Regulations 1998. Full details of the restrictions can be found in The Transport Managers & Operators Handbook (Lowe, 2005).

14.5 VEHICLE DIMENSIONS

Allowable dimensions for the length, height and width of HGVs are enshrined in Government legislation which also provide the means by which control is exercised and enforced through the offices of Government Inspectors.

The maximum allowable overall length for each type of HGV (see Figure 14.1) is as follows:

> Generally 12 m for a two-axle rigid vehicle, which may increase to 14.5 m if it is articulated and meets certain turning circle requirements; 18.75 m for a vehicle and Trailer Drawbar Combinations (Road Train), and 25.9 m for an articulated truck subject to minor dimensional restrictions.

For safety reasons and to avoid 'bridge bashing' the height of all HGVs is limited to 3.66 m. There is no statutory limitation at all on the width of the HGV except where for design purposes, the overall width may be taken as 3.5 m (BD37/01).

It is the responsibility of the truck/lorry operator to ensure that his vehicles can negotiate successfully the obstacles likely to be encountered on any specific journey.

Full details of the restrictions can be found in The Transport Managers & Operators Handbook (Lowe, 2005).

14.6 LAW ENFORCEMENT

The greatest aid in helping Government Inspectors to enforce the law on weight is by means of the *weighbridge*.

Superficially it consists of a hole in the ground, rather like a shallow swimming pool, approximately 1 m deep (see Figure 14.4) with the bottom and sides lined with reinforced concrete (RC).

Around the perimeter are strategically placed RC piers to support the ends of a grid of steel beams. Between the underside of the beams and the tops of the piers are load cells. A top platform (or running plate) comprising steel Chequer plate or Durbar plate is secured to the top flanges of the beams.

FIGURE 14.4
Typical weighbridge 'Pit'.

- If the weighbeam is designed to measure loads 'at-rest' (see Figure 14.1), then the HGV leaves the highway, is guided into place, parked on the support platform and each of the load cells is interrogated automatically. If the sum total minus the dead weight of the vehicle is not greater than the allowable live load then the driver is given a certificate of compliance and allowed to continue his journey.
- If the weigh station is designed as a 'weigh-in-motion' facility, then the HGV is driven between guide lines marked on the highway, weighed-in-motion, electronically screened, and signalled to keep going.

Communications between the weighbeam and the vehicle are affected by means of a '*Transponder*' which is an electronic device which receives and

FIGURE 14.5

Weighbeam 'at rest' measurement.

transmits radio and radar signals. It is fast and efficient, and to increase the number of transponder owners some state authorities are offering them free to drivers who want to miss the more traditional 'static' weighbridge.

Road Vehicles (Construction and Use) Regulations 1986 (C & Use) Regulations 1986 as amended, and The Road Vehicles (Authorised Weight) Regulations 1998. Full details of the restrictions can be found in The Transport Managers & Operators Handbook (Lowe, 2005).

14.7 OTHER ISSUES

In the USA there is a tremendous amount of energy in the trucking industry. In fact it is almost a way of life. In addition to the headline subjects of 'size and weight', the activity list is long and practicable. The following topics which might be considered in a typical month are taken from an issue of Oregan State's (2005) own 'Motor Carrier News'.

- **Bend weigh station adds Green Light weigh-in-motion**
  This is a piece about the benefits to be gained by motor carriers if they have a transponder on board, not least because of the journey time saved on account of the weigh-in-motion.

- **New phone system**
  This is about the installation of the Motor Carrier Division's new telephone facility to handle Service Center callers in a more efficient manner. Calls can also be made and logged using 'Trucking Online' *via* the Internet which has more than 6000 users.

  Carriers who operate in other states and Canada and the USA and are members of money saving agreements can save considerable sums of money if they sign up to 'Trucking Online'.

- **CDL Hazmat endorsement rules**
  Hazardous materials must be handled with care and security is of the utmost importance to ensure the safety of the public and the integrity of the carrier. In Oregan, carriers can now obtain an endorsement to their Commercial Driver Licence. Drivers must first be fingerprinted and have a clean background check obtained by specialist professional government and quasi-governmental organisations.

- **Weight-restricted bridges on major routes in Oregan**
  In this article information about restrictions is presented in the form of a map showing the bridge names and locations; a list of highway numbers and the nature of the restriction, for example (SR = Special Restriction – All trucks over 80,000 lbs must stay in the right lane).

14.8 ABNORMAL INDIVISIBLE LOADS (AILs)

These types of loads are those for which special provision has to be made mainly because of their size. They exceed one or all of the codified dimensions of length, width and height, and will probably be of considerable weight.

Their movement has to be very carefully managed as it is almost certain that part of their route will deviate from the existing road infrastructure. Every obstacle along the route has to be identified, logged and recorded in a database. Some typical obstacles are:

- The proximity of overhead electricity transmission cables.
- Bridges crossed in which case, it is necessary to ensure that the strength of each existing bridge is sufficient to resist the loads imposed by the AIL. If it is not, then the bridge must be strengthened. In the UK the bridges are classed by the number of HB units that they can carry, usually 45, 37.5, 30 or 25.
- Utilities such as water, gas, electricity and telephone.
- Building structures.
- Conservation areas.
- Off-road ground conditions.
- Rivers, lakes, canals etc. (In some cases it may be necessary to design a bridge for a 'one-off' event to carry a particular industrial or manufacturing unit!).

There are many other such hazards to contend with, and it is paramount that the inspection is carried out thoroughly, so that a clear uninterrupted route can be marked out and priced with precision. In the UK an AIL notification scheme is in operation whereby hauliers can provide suitable notification of all aspects of the route to the Highways Agency who then carry out checks of their own.

14.9 CONCLUSIONS

This short section highlights some of the bridge management topics which focus more on the vehicles themselves (in terms of the very complex interaction of vehicle/suspension/road surface) and not just on the effect they have on the bridge structure. The next generation of research using the next generation of computers will almost certainly be concentrating into the many contributing factors such as:

- Gross weight of vehicle
- Type of vehicle
- Speed of vehicle

- Type of suspension
- Pavement roughness on the bridge
- Pavement roughness off the bridge
- Stiffness of bridge bearings
- Make-up of vehicles

Gathering data from actual structures along with the results from the modelling of actual structures subject to a consistent traffic flow will result in a greater understanding of vehicle/structure interaction and thereby increase our confidence in deciding our management strategy.

REFERENCES

DIVINE Programme, OECD. (1997). *Dynamic Interaction of heavy vehicles with roads and bridges*. DIVINE Concluding Conference, Ottawa, Canada.

Heinemann Lutzenberger, S., & Baumgartner, W. (2000). *Evaluation of measured bridge responses due to an instrumented truck and free traffic*. London: Thomas Telford. Proc. Bridge Management Four, pp. 72–87.

Lowe, D. (2005). *Intermodal Freight Transport*. Elsevier Butterworth.

Marghitu, Dan., B. (Ed.). (2001). *Mechanical Engineers Handbook*. Academic Press.

O'Brien, E. J., Gonzalez Li, Y. Y., & Brady, S. P. (2005). *The influence of speed on the dynamic amplification of two loads crossing a simply supported bridge*. London: Thomas Telford. Proc. Bridge Management Five, pp. 452–458.

Oregan State. (2005). Motor Carrier News. *June Issue, 20*(70).

Ryall, M. J., Parke., & Harding, J. E. (Eds.). (2002). *Manual of Bridge Engineering*. London: Thomas Telford.

Index

E

F

G

H

I

J

K

L

M

N

O

P

T